Structured Stochastic Matrices of M/G/1 Type and Their Applications

PROBABILITY: PURE AND APPLIED

A Series of Textbooks and Reference Books

Editor

MARCEL F. NEUTS

University of Arizona
Tucson, Arizona

1. Introductory Probability Theory, *by Janos Galambos*
2. Point Processes and Their Statistical Inference, *by Alan F. Karr*
3. Advanced Probability Theory, *by Janos Galambos*
4. Statistical Reliability Theory, *by I. B. Gertsbakh*
5. Structured Stochastic Matrices of M/G/1 Type and Their Applications, *by Marcel F. Neuts*

Other Volumes in Preparation

Structured Stochastic Matrices of M/G/1 Type and Their Applications

MARCEL F. NEUTS

Department of Systems and
Industrial Engineering
University of Arizona
Tucson, Arizona

MARCEL DEKKER, INC. **New York and Basel**

ISBN 0-8247-8283-6

This book is printed on acid-free paper.

MARCEL DEKKER, INC.
270 Madison Avenue, New York, New York 10016

Current printing (last digit):
10 9 8 7 6 5 4 3 2 1

PRINTED IN THE UNITED STATES OF AMERICA

Be admonished: of making many books there is no end; but their study wearies the flesh.

Koheleth, Ch. 12, v. 12.

Preface

Specific stochastic models commonly lead to the examination of specially structured Markov chains or Markov renewal processes. Conversely, cases of such processes for which progress beyond the most general results is possible, all originate in specific classes of applied stochastic models, such as random walks, queues, branching processes, dam and inventory models, and which share structural properties.

In my earlier book *Matrix-Geometric Solutions in Stochastic Models* (The Johns Hopkins University Press, 1981), I examined a class of Markov chains, said to be *of GI / M / 1 type,* after the simplest structured Markov chain of that type. The present book deals with Markov chains and Markov renewal processes, which I shall call *of M / G / 1 type.* The classical $M/G/1$ queueing model is the simplest paradigm for this study, but the general structure to be examined has importance far beyond that elementary example. That structure is shared by a large number of rather complex models in the theories of queues, communications systems, dams and inventories.

As the analysis of the stationary probabilities of the $M/G/1$ queue is less transparent than that of the standard $GI/M/1$ queue, so also, the treatment of Markov chains of $M/G/1$ type in this book is mathematically more demanding than that of the chains of $GI/M/1$ type, discussed earlier. For Markov chains of $M/G/1$ type, the main theoretical results cannot be put in as concise and explicit a form as the matrix-geometric structure of the stationary probability vector for the models of $GI/M/1$ type. The theoretical development in this book is therefore longer, more painstaking and requires greater attention to mathematical detail. In both cases, the nucleus of the mathematical presentation is a nonlinear matrix or matrix-integral equation, studied by function analytic or matrix theoretic methods. In this book, however, the solution of the steady-state equations for the boundary states adds a level of complexity, whose counterpart is elementary for models of $GI/M/1$ type. The examination of important derived distributions, such as those for the waiting times in queueing models, is in general also more involved than corresponding considerations in the earlier book.

Although this book may be viewed as a companion volume to the work published in 1981, it is largely independent of it. Except for occasional references to it in problems and for its abbreviated presentation of the properties of *phase type distributions,* this book may be read without a prior study of matrix-geometric solutions. Given the relative complexity of the development in this text, we must caution the novice reader against doing so. A number of theorems, to be presented here, have simpler corresponding versions and proofs for the $GI / M / 1$ type. The reader who has perused the earlier book will find the present volume more accessible than one who has not yet done so.

This book is written from the same methodological viewpoint and with the same algorithmic concern as its predecessor. Our reasons for giving greater weight to algorithmically useful over formal methods are stated at length in the Preface to the book on matrix-geometric solutions. There should no longer be a need to reiterate them and we therefore refrain from doing so.

On the other hand, the Preface to a book on a novel and alternative methodology is the privileged place for its author to explain the inclusion of or emphasis on some material and the omission or lighter emphasis of some other topics. We shall therefore, at this point, say a few words on the desirability of computer packages for the algorithms we have developed, on the algorithmic problems of time dependent behavior, on comparisons with different approaches and on the need for computational experience as the essential key to an enhanced understanding of the behavior of probabilistic models.

On the first subject, let me say that as a self-taught student of computer programming, I have experienced the early stage of frustration, which is not unlike that of the student of a new language, the next stage of inefficient early competence, the long middle phase where a good computer program is seen as a minor work of art, and the mature stage where it represents an essential but not too appealing part of the solution of an applied mathematical problem. A well-written and trustworthy library routine is a major time saving device and the merit of those who provide these for a wide spectrum of mathematical and statistical uses is beyond praise.

It is indeed possible to write library routines for the solution of a variety of problems in stochastic models and, in particular, for useful models such as the $PH / PH / 1$ and $M / SM / 1$ queues. The need for such routines and the perception of their utility among actual users of stochastic models is growing. That need may yet be filled by the work of some readers of this volume, but over the years when our methodology was developed, we had neither the time nor the personnel to carry out the onerous task of creating and testing packages for general use. In our experience, each application also has special features that add to its difficulty or yield major simplifications. There does not appear to be a unified way of encoding most of these into a computer package. The situation for stochastic models is similar to that in the well-established area of the numerical solution of differential equations, where a variety of methods and codes exist, each best suited to equations with one feature or another.

It needs also be stressed, that the material in this book is not a recipe for the computational solution of a set of probability problems, but rather a unified method for the mathematical analysis of a class of stochastic models. For any given model in that set, the method will clarify the numerical issues to be resolved and this can in a purposeful manner lead to a computer code or package. Depending on the application, the redaction of a good computer code may require a highly varying amount of additional effort and may need to draw on other areas of numerical and computational expertise. It is neither the purpose of this book, nor of our ongoing research in this area, to address these developmental issues in detail.

It has been noted that most of our algorithmic results deal with the steady-state analysis of queues and other models. Questions are often raised about the desirability of algorithmic procedures for the transient behavior of these same or related models. Our brief discussion is an attempt at clarification of these questions. It should first be noted that, at least of subcritical processes, our analysis also yields information on the transient mode. Since our processes are semi-regenerative, properties of the busy cycles are informative about both their transient and steady-state behavior regimes.

Our approach also yields transform solutions for the time dependent distributions of the models. These formal solutions may, in fact, be inverted in terms of matrix-convolution series, which for a limited range of the time parameter t could be used for the evaluation of time dependent solutions. In a similar manner, it is possible to solve the Chapman-Kolmogorov equations (or similar appropriate equations) for the transient behavior, either in a number crunching manner, or by the use of efficient methods, such as the Fast Fourier or Laguerre transforms. For some specific models, where the issues of interest are clear, this has already been done by other workers. In general, however, the utility of detailed and massive information on the transient behavior is not clear, because that behavior is so dependent on the initial conditions that are specified. There is great need for theoretical measures of transience, such as the relaxation time or the parameters of exponential or geometric ergodicity. Where such measures are well-defined, it would be useful to have procedures for their numerical computation, but these are not directly germane to the subject of this book. We hope that other scholars will take up the development and dissemination of such procedures.

As its predecessor, this book therefore deals primarily with the steady-state behavior of positive recurrent models for which the mathematical issues are clear. Which questions on the transient mode are of sufficient interest to warrant their detailed algorithmic investigation is, in my opinion, not yet sufficiently apparent for a definitive treatment.

In Section 1.6 of my 1981 book, I have given a brief discussion of numerical difficulties which are apparently inherent in the classical analysis of a variety of stochastic models by methods of complex analysis. Since then, many research workers have confirmed this by informing me of negative experiences in attempts at numerical implementation of these methods; a few others have sent me examples where no significant difficulties were encountered. Without exception, the latter were rather simple models where matrices of very low order could be used or where the range of parameters was not extensive. For models to which the material in this book is germane, these same difficulties are present in the approach by complex analysis. It is clearly desirable to have alternative numerical methods available, but the task of implementation and the burden of proof rests on the proponents of each method. Dissemination of information on the limitations and failures of each method is as valuable as the reporting of successes.

The most valued return of a detailed numerical study of a stochastic model lies in the insight gained from the interpretation of its numerical results. It took a long development before the notion of random variable broadened the earlier idea of a physical "constant." In our days, a ready understanding of the dynamic randomness inherent in a stochastic process is still uncommon. This is evident in the importance attached to explicit expressions for such quantities as the stationary mean queue length, waiting time, dam content, inventory level and others. For the models treated here, we also obtain expressions - no longer simple ones - for these measures. These are, in fact, valuable mileposts in the computation of the stationary distributions. When taken in isolation, means and even second moments of the stationary distributions may be uninformative and occasionally misleading. Our emphasis on evaluations of distributions rather than a few moments and on the computation and comparison of the stationary distributions of several embedded point processes rather than just one overall steady-state distribution, comes from the wish to understand the stochastic model in its dynamic aspects. We believe that the heavy emphasis in the literature on measures of the average performance and on very general conservation laws has detracted from the difficult, important problems of quantifying path function behavior.

For the models of $M/G/1$ type, treated in this book, the algorithmic analysis requires a rather long sequence of steps, but in return each of these conveys information important to the qualitative interpretation of the numerical results. Embedded in the algorithm, we find, for example, the mean durations of the fundamental and busy periods of queueing models. These give indications of the dependence of the queue on the service times of different types of customers. We may, after a common body of computations, obtain the steady-state distributions at various embedded point processes. These are useful in assessing the effect of some control actions, of server breakdowns and of exceptionally long or short services. From a thorough interpretation of one or more detailed numerical examples, we are usually able to describe qualitatively how a stochastic model will behave and to give reasonable quantitative estimates of the excursions of queue lengths and waiting times. V. Ramaswami, one of my highly numerate former students, once paraphrased an old saying as *"A good numerical example is worth ten thousand words."*

In the book dealing with matrix-geometric solutions, I discussed numerical results for a few examples of highly variable queues. The practice of adding a non-trivial numerical example and its interpretation to journal articles on stochastic models is now widely accepted. At least, authors are no longer routinely asked to delete such material from the published versions of their work. Nevertheless, the generation of sound numerical results and their interpretation remains a time consuming endeavor, whose discussion requires a fair amount of space. Given the length of the theoretical discussion in this book, we have reluctantly decided not to include numerical examples. Years of research in and teaching of algorithmic probability have also taught us that numeracy is gained from practice and less from discussions of examples. The key to a full appreciation of the contents of this book is the application of its methodology to a stochastic model in which the reader is genuinely interested. Since the publication of the earlier monograph in 1981, a number of applied stochastic models have been investigated by use of the matrix-geometric results. If the present material can find a similar number of applications, our effort in writing this book will be amply and truly rewarded.

The organization of the book is as follows. Each chapter treats a body of material, without reference in text to related discussions or to other applications. Following each chapter, there is an extensive set of notes, references and problems. The notes deal with applications, further theoretical extensions and references to other work. The problems are not intended to be used as "textbook" exercises. Many are descriptions of stochastic models which we have encountered in engineering applications or have extracted from articles on such applications. Some of these problems could have been expanded into journal articles had we been motivated to do so. Most of the problems are best approached after a thorough study of the entire book. Indeed, some of them are well suited for Masters theses or class projects which may possibly yield novel research results. We have indeed successfully used similar problems in the early graduate education of our doctoral students.

While we hope that no proofs are lacking in rigor, we have avoided all unnecessary abstractness. In the interest of a smooth presentation, we have often used elliptical terminology, such as the "Markov chain P" or "the Markov renewal process $Q(\cdot)$", where it is clear which sequence of random variables is intended. The repetitive and abstract definition of all random variables involved would have added to the length but not to the substance of this book.

The extensive bibliography requires some comment. The literature on stochastic models, and on queues in particular, is huge and widespread. A seminal model, such as the $M/G/1$ queue, has given rise to many extensions, variants and methodological approaches. The existing theory on the $GI/G/1$ queue is to a degree also applicable to models in this book and there are common themes in articles on dams, traffic models, inventories and such. It has been our intention to compile all references that are actually or potentially related to the treatment in this book. Where that relation is clear, it is mentioned in the notes at the end of the appropriate chapter. In all cases, the reader should be able to extract from the list of references many sources for further study and research documentation. The bibliography was gathered leisurely over a period of several years. Most authors responded to our request for references by bringing additional material to our attention. We are grateful to all who have done so and, in particular, to Professors L. Schrage and Do Le Minh, whose input went considerably beyond their own publications.

The genesis of the material in this book lies in one of my earliest papers, which dealt with the $M/SM/1$ queue and was published in 1966. Most of the matrix analytic methodology, its novel contribution, was developed during the early 1970's after I had discovered for myself the challenges of algorithmic mathematics. Many persons and institutions have lent me encouragement and support in a research effort which spans such a long time period.

My associations as a faculty member or visiting scholar with Purdue University, Cornell University, the University of Delaware, Technion and the University of Stuttgart and currently the University of Arizona have offered me the advantages of a life of scholarship and research. Financial support for myself and for several talented students has been provided, in varying measures, by the National Science Foundation, the Air Force Office of Scientific Research, the Alexander von Humboldt Stiftung (Federal Republic of Germany) and the Lady Davis Foundation (Israel). The support of institutions provides the framework, but to scholarly enquiry the continual stimulation in the meeting of minds with colleagues and doctoral students is equally essential. I acknowledge with profound appreciation the informed interest and collaboration of friends such as Guy Latouche, V. Ramaswami, David Lucantoni, Kathleen Meier-Hellstern, S. Chakravarthy, S. Kumar, Levent Gün and of my current students H. Sitaraman and Y. Chandramouli. Ms. Nia Clark, secretary in the Department of Systems and Industrial Engineering at the University of Arizona, has shared with me the task of entering the text into a word processing file. May she also find words of thanks here which cannot possibly match the size of the task.

A special word of thanks goes to my wife Olga and to our four children. They have offered me the support of a stable and happy family environment and of understanding patience with a mathematical spouse and father, who like all of us, so often seems to dwell in "a galaxy far, far away."

Marcel F. Neuts

Contents

Structured Stochastic Matrices of M/G/1 Type and Their Applications

1

The M/G/1 Queue and Some of Its Variants

1.1. INTRODUCTION

The single server queue with a homogeneous Poisson arrival process and independent, identically distributed service times is known as the $M/G/1$ queue. It is one of the most extensively studied stochastic models; a large variety of elegant mathematical methods have been brought to bear upon its analysis. The principal properties of the $M/G/1$ queue are discussed in all books on the theory of queues and in most texts on stochastic models.

The material, developed in this book, arose largely out of generalizations of a particular way of viewing the elementary $M/G/1$ queue. It is therefore useful to devote this first chapter to a systematic treatment of the $M/G/1$ queue and some of its variants. This treatment differs in details from the presentations commonly found in other sources. If some of our derivations are more lengthy than required by more elegant alternatives, they have the advantage of allowing systematic generalizations to more complex models. It will further be convenient to have the most important formulas and derivations for the $M/G/1$ queue at hand. Several of the calculations for more complex models are matrix analogues of the simpler derivations for the $M/G/1$ queue. In discussing variants of the $M/G/1$ queue, we shall distinguish between the basic structural properties, crucial to the general considerations in the next chapters, and particular properties that are relatively less important.

The *arrival process* in the $M/G/1$ queue is a homogeneous Poisson process of rate λ. Customers are served singly and the successive *service times* are independent, identically distributed nonnegative random variables with the common probability distribution $H(\cdot)$ of positive mean α. In most cases, the mean α will be finite. Unless otherwise noted, the order in which customers are served, *the queue discipline,* is immaterial to our discussion, but we do assume that no service is interrupted in mid-course and that a new service starts as soon as one terminates, provided that customers are waiting for service.

For notational convenience, we shall choose the time origin to correspond to the end of a service. Let T_n, $n \geqslant 0$, denote the times of successive service completions with $T_0 = 0$, and I_n, the number of customers *in the system* immediately after the n-th service completion. If we define τ_n by $\tau_n = T_n - T_{n-1}$, for $n \geqslant 1$, then it is readily seen that the sequence $\{(I_n, \tau_n), n \geqslant 0\}$ is a *Markov renewal sequence* on the state space $\{i \geqslant 0\} \times \{0,\infty\}$. This follows from the fact that

$$I_{n+1} = (I_n - 1)^+ + \nu_{n+1}, \quad \text{for } n \geqslant 0,$$

where ν_{n+1} is the number of arrivals during the $(n+1)$-st service time. Under the assumption of Poisson arrivals, the random variables ν_n, $n \geqslant 1$, are independent and identically distributed. The random variables $\{\tau_1, \tau_2, \ldots, \tau_k\}$ are also *conditionally independent,* given $\{I_0, \ldots, I_{k-1}\}$ and this for every $k \geqslant 2$.

The transition probability matrix $Q(\cdot)$ with elements

$$Q_{ii'}(x) = P\{I_n = i', \tau_n \leqslant x \mid I_{n-1} = i\},$$

for $i \geqslant 0$, $i' \geqslant 0$, $x \geqslant 0$, is given by

$$Q_{0i'}(x) = \int_0^x \lambda e^{-\lambda u} Q_{1i'}(x-u)\,du, \quad \text{for } i' \geqslant 0, \tag{1.1.1}$$

$$Q_{ii'}(x) = \int_0^x e^{-\lambda u} \frac{(\lambda u)^{i'-i+1}}{(i'-i+1)!} dH(u), \quad \text{for } i \geqslant 1, \quad i' \geqslant i-1,$$

$$Q_{ii'}(x) = 0, \quad \text{for } i \geqslant 1, \quad i' < i-1.$$

In this book, we shall use the term *probability mass-function* for any non-decreasing function, taking values between 0 and 1, but which does not necessarily tend to 0 at $-\infty$ or to 1 at $+\infty$. The elements of a semi-Markov matrix are probability mass-functions. If we define the probability mass-functions $A_\nu(\cdot)$ by

$$A_\nu(x) = \int_0^x e^{-\lambda u} \frac{(\lambda u)^\nu}{\nu!} dH(u), \quad \text{for } \nu \geqslant 0, x \geqslant 0, \tag{1.1.2}$$

and set

$$B_\nu(x) = \int_0^x \lambda e^{-\lambda u} A_\nu(x-u)\,du = \int_0^x [1 - e^{-\lambda(x-u)}]\,dA_\nu(u),$$

for $\nu \geqslant 0$, $x \geqslant 0$, then we see that the matrix $Q(\cdot)$ has the important structural form

$$Q(x) = \begin{vmatrix} B_0(x) & B_1(x) & B_2(x) & B_3(x) & B_4(x) & \cdots \\ A_0(x) & A_1(x) & A_2(x) & A_3(x) & A_4(x) & \cdots \\ 0 & A_0(x) & A_1(x) & A_2(x) & A_3(x) & \cdots \\ 0 & 0 & A_0(x) & A_1(x) & A_2(x) & \cdots \\ \cdot & \cdot & \cdot & \cdot & \cdot & \\ \cdot & \cdot & \cdot & \cdot & \cdot & \end{vmatrix} \qquad (1.1.3)$$

The structural form (1.1.3) is preserved in many variants of the $M/G/1$ queue, whose analysis differs only in details, but not in essential ideas from that of the simpler model. For purposes of illustration and for later use, we now describe a number of such variants.

Variant 1: The $M/G/1$ Queue with Group Arrivals. This is the case where groups of customers arrive according to a Poisson process of rate λ and each group consists of k customers with probability p_k, $k \geqslant 1$. The successive group sizes are assumed to be independent with the common discrete density $\{p_k\}$. Let $\phi_\nu(t)$, $\nu \geqslant 0$, $\tau \geqslant 0$, be the probability that ν customers arrive in $(0,t]$, then we readily see that

$$\phi_0(t) = e^{-\lambda t},$$

$$\phi_\nu(t) = \sum_{r=1}^{\nu} e^{-\lambda t} \frac{(\lambda t)^r}{r!} p_\nu^{(r)}, \quad \text{for } \nu \geqslant 1,$$

where $\{p_\nu^{(r)}\}$ is the r-fold convolution of the density $\{p_\nu\}$. We note that for $\nu \geqslant 1$,

$$\phi_\nu(t) = \int_0^t \lambda e^{-\lambda\tau} \sum_{k=1}^{\nu} p_k \phi_{\nu-k}(t-\tau)\,d\tau.$$

With I_n and τ_n defined as before, the transition probability matrix $Q(\cdot)$ of the embedded Markov renewal process has the form (1.1.3) with

$$A_\nu(x) = \int_0^x \phi_\nu(u)dH(u),$$

$$B_\nu(x) = \sum_{r=1}^{\nu+1} p_r \int_0^x [1 - e^{-\lambda(x-u)}]dA_{\nu-r+1}(u),$$

for $\nu \geqslant 0$, $x \geqslant 0$.

Variant 2: Special Service after an Idle Period. In some applications, the first customer to join the queue when the server is idle has a different service time distribution $H_0(\cdot)$ of mean α_0. This may come about because a set-up time is required for such services. The only difference between the matrix $Q(\cdot)$ for this case and that for the ordinary $M/G/1$ queue is that the mass-functions $B_\nu(\cdot)$ are now given by

$$B_\nu(x) = \int_0^x [1 - e^{-\lambda(x-u)}]\, e^{-\lambda u} \frac{(\lambda u)^\nu}{\nu!}\, dH_0(u), \quad \text{for } \nu \geqslant 0.$$

Variant 3: The $M/G/1$ Queue with the N-Policy. In a mode of operation, known as the N-policy, an idle server does not start serving customers unless there are $N \geqslant 1$ customers waiting for service. Customers are still served one at a time. The case $N = 1$ corresponds to the ordinary $M/G/1$ queue.

The $Q(\cdot)$-matrix for this case again differs from that of the $M/G/1$ queue in the elements of the first row, which are given by

$$B_\nu(x) = 0, \qquad \text{for } 0 \leqslant \nu \leqslant N-2,$$
$$= E_N(\lambda,\cdot) * A_{\nu-N+1}(x), \qquad \text{for } \nu \geqslant N-1,$$

where $E_N(\lambda,\cdot)$ is the *Erlang distribution* of order N and $*$ denotes convolution.

The analyses of Variants 2 and 3 are only trivially different from that of the $M/G/1$ queue because they differ only in their behavior at the state 0. We shall call the state 0 a *boundary state* and the states

$i \geqslant 1$, *non-boundary states*. In the general analysis, to be presented in the subsequent chapters, we shall separate the discussion of the non-boundary states from that of the boundary states and so study several variants of a given model in a unified and efficacious manner. The specific analytic form of the mass-functions $A_\nu(\cdot)$ is also of little consequence to the general discussion. This is illustrated by the next variant.

Variant 4: The $M/G/1$ Queue with Limited Admission. One procedure to control large build-ups in an $M/G/1$ queue with occasional long service times is to limit the number of *new* customers allowed to join the queue during each service to be at most N. The matrix $Q(\cdot)$ for this case is of the form (1.1.3) with

$$A_\nu(x) = \int_0^x e^{-\lambda u} \frac{(\lambda u)^\nu}{\nu!} \, dH(u), \quad \text{for } 0 \leqslant \nu < N,$$

$$A_N(x) = \sum_{\nu=N}^{\infty} \int_0^x e^{-\lambda u} \frac{(\lambda u)^\nu}{\nu!} \, dH(u),$$

$$A_\nu(x) = 0, \qquad \text{for } \nu > N,$$

$$B_\nu(x) = \int_0^x [1 - e^{-\lambda(x-u)}] dA_\nu(u), \quad \text{for } \nu \geqslant 0.$$

It is clear from these and other variants of the $M/G/1$ queue, that Markov renewal processes with a transition probability matrix $Q(\cdot)$ of the form (1.1.3) deserve to be studied in general. We shall do so in the next section, but first we impose assumptions to guarantee the irreducibility of $Q(\infty)$ and also to eliminate some trivial cases.

We require that the sums

$$A(x) = \sum_{\nu=0}^{\infty} A_\nu(x), \qquad B(x) = \sum_{\nu=0}^{\infty} B_\nu(x),$$

are bona fide probability distributions on $[0,\infty)$, that are not degenerate at zero.

Writing A_ν and B_ν for $A_\nu(\infty)$ and $B_\nu(\infty)$ respectively, we shall assume that $B_0 < 1$, $A_0 > 0$, and $A_0 + A_1 < 1$. This guarantees that the embedded Markov chain with transition probability matrix $P = Q(\infty)$, is *irreducible*. The case where $A_0 > 0$, and $A_0 + A_1 = 1$, is trivial. It arises, for example, in Variant 4 when $N = 1$. Unless otherwise noted, we shall exclude it from consideration.

1.2. THE FUNDAMENTAL PERIOD

Let $T(i+r, i)$ be the first passage time from the state $i + r$, $r \geqslant 1$, to the state $i \geqslant 0$, in the Markov renewal process $Q(\cdot)$. In order to reach the state i from state $i + r$, a Markov renewal process with a transition probability matrix $Q(\cdot)$ of the form (1.1.3) must visit all intermediate states $i+r-1, \ldots, i+1$ at least once. We may therefore write

$$T(i+r, i) = \sum_{h=1}^{r} T(i+h, i+h-1).$$

Similarly, let $V(i+r, i)$ be the number of transitions involved in the first passage from state $i + r$ to state i. Equivalently $V(i+r, i)$ is the first passage time from $i + r$ to i in the embedded Markov chain $P = Q(\infty)$. By the same argument, we have that

$$V(i+r, i) = \sum_{h=1}^{r} V(i+h, i+h-1).$$

By the Markov property of the Markov renewal process at transition epochs, we obtain that the pairs of random variables $[T(i+h, i+h-1), V(i+h, i+h-1)]$, $1 \leqslant h \leqslant r$, are independent. Furthermore, during a first passage from $i+h$ to $i+h-1$, the process visits only states $i+h+\nu$, $\nu \geqslant 0$. It therefore follows from the spatial homogeneity of the process that the pairs $[T(i+h, i+h-1), V(i+h, i+h-1)]$, $1 \leqslant h \leqslant r$, are also identically distributed and that their joint distribution does not depend on i. A first passage time $T(i, i-1)$ will be called a *fundamental period*. The random variable $V(i, i-1)$ is the *number of transitions during a fundamental period*. As the joint distribution of $T(i, i-1)$ and $V(i, i-1)$ is the same for all $i \geqslant 1$, we may, for convenience of notation, suppress the state indices.

The probabilities

$$G_k(x) = P\{T \leqslant x,\ V = k\}, \tag{1.2.1}$$

defined for $x \geqslant 0$ and $k \geqslant 1$, play a crucial role in the theory of Markov renewal processes of $M/G/1$ type.

Let the probability mass-functions $G_k^{(r)}(\cdot)$, $k \geqslant r$, $r \geqslant 1$, be defined by the convolution formula

$$G_k^{(r)}(x) = \sum G_{k_1} * G_{k_2} * \cdots * G_{k_r}(x),$$

where the summation is over the set of indices $(k_1, \ldots, k_r)$ satisfying $k_1 \geqslant 1, \ldots, k_r \geqslant 1$, $k_1 + \cdots + k_r = k$. Then clearly

$$G_k^{(r)}(x) = P\{T(i+r,i) \leqslant x,\ V(i+r,i) = k\}.$$

For convenience, we set $G_k^{(r)}(x) = 0$, for $0 \leqslant k < r$, $r \geqslant 0$, and $G_0^{(0)}(x) = U(x)$, where $U(\cdot)$ is the probability distribution degenerate at 0.

Theorem 1.2.1: The probabilities $G_k(x)$ satisfy the equations

$$G_1(x) = A_0(x),$$

$$G_k(x) = \sum_{\nu=1}^{\infty} A_\nu(\cdot) * G_{k-1}^{(\nu)}(x), \quad \text{for } k \geqslant 2. \tag{1.2.2}$$

The joint transform $\tilde{G}(z,s) = E[\exp(-sT)z^V]$, defined for $|z| \leqslant 1$, Re $s \geqslant 0$, satisfies the equation

$$\tilde{G}(z,s) = z \sum_{\nu=0}^{\infty} \tilde{A}_\nu(s)\, \tilde{G}^\nu(z,s), \tag{1.2.3}$$

where

$$\tilde{A}_\nu(s) = \int_0^\infty e^{-sx}\, dA_\nu(x), \quad \text{for } \nu \geqslant 0.$$

Proof: The first equation in (1.2.2) is obvious. For $k \geqslant 2$, the first transition from state i must be to one of the states $i+\nu-1$, $\nu \geqslant 1$, and in the remaining $k-1$ transitions, the process must reach state $i-1$ from $i+\nu-1$. By conditioning on the time and the destination of the first transition and applying the law of total probability, we obtain the stated formula. Formula (1.2.3) follows by routine computation of the transform. •

Remark *a*. In terms of the generating function

$$\tilde{A}(z,s) = \sum_{\nu=0}^{\infty} \tilde{A}_{\nu}(s) z^{\nu}, \tag{1.2.4}$$

equation (1.2.3) may be written as

$$\tilde{G}(z,s) = z\, \tilde{A}[\tilde{G}(z,s),s]. \tag{1.2.5}$$

In formula (1.2.3), $\tilde{G}^{\nu}(z,s)$ is clearly the ν-th power $[\tilde{G}(z,s)]^{\nu}$ of $\tilde{G}(z,s)$. This is not to be confused with the functional iterates of $\tilde{G}(z,s)$.

Remark *b*. For the $M/G/1$ queue and for Variants 2 and 3, we have

$$\tilde{A}_{\nu}(s) = \int_0^{\infty} e^{-(\lambda+s)u} \frac{(\lambda u)^{\nu}}{\nu!}\, dH(u), \text{ for } \nu \geqslant 0,$$

so that (1.2.5) may be written as

$$\tilde{G}(z,s) = z\, h[s+\lambda-\lambda\tilde{G}(z,s)],$$

where $h(s)$ is the Laplace-Stieltjes transform of $H(\cdot)$. This is the classical *Takács equation* for the joint transform of the duration of and the number of customers served during the busy period in the $M/G/1$ queue. For the $M/G/1$ queue, the busy period is also a fundamental period, but this is not generally the case for Variants 2 and 3. In general, it is necessary to distinguish between the busy and the fundamental periods.

In Chapter 2, where a matrix analogue of equation (1.2.3) is considered, we shall show that for $0 \leqslant z \leqslant 1$, $s \geqslant 0$, the solution of probabilistic interest is the ***minimal nonnegative solution*** of that equation. In order to construct that solution, we consider the sequence of functions, defined for $0 \leqslant z \leqslant 1$, $s \geqslant 0$, by

$$\tilde{G}_0(z,s) = 0,$$

$$\tilde{G}_{n+1}(z,s) = z \sum_{\nu=0}^{\infty} \tilde{A}_\nu(s)\tilde{G}_n^{\nu}(z,s), \quad \text{for } n \geqslant 0.$$

It is readily seen that each of the functions $\tilde{G}_n(z,s)$ is a transform of the form

$$\tilde{G}_n(z,s) = \sum_{k=1}^{\infty} \int_0^{\infty} e^{-sx}\, dG_k(n;x)\, z^k,$$

where the $G_k(n;x)$ are probability mass-functions. Their probabilistic significance is discussed in Chapter 2. For each pair (z,s) with $0 \leqslant z \leqslant 1$, $s \geqslant 0$, the sequence $\tilde{G}_n(z,s)$ is nondecreasing and bounded above by one. Its point-wise limit $\tilde{G}(z,s)$ is the minimal nonnegative solution to equation (1.2.3). We shall show by a probabilistic argument that it is the desired transform of the mass-functions $G_k(x)$.

The quantity $\eta = \sum_{k=1}^{\infty} G_k(\infty)$, is the probability that the fundamental period ends in finite time or equivalently that the state i is eventually reached from state $i+1$. In order for the Markov renewal process $Q(\cdot)$ to be ***recurrent*** it is clearly ***necessary*** that $\eta = 1$.

Theorem 1.2.2: The quantity η is equal to one if and only if $\rho = \sum_{k=1}^{\infty} k\, A_k \leqslant 1$. It satisfies $0 < \eta < 1$ otherwise.

Proof: By Abel's theorem, we have that $\eta = \tilde{G}(1-,0+)$. For $s > 0$ and $0 < z \leqslant 1$, the equation

$$\varsigma = z \sum_{\nu=0}^{\infty} \tilde{A}_\nu(s)\varsigma^{\nu},$$

has a unique solution $\varsigma(z,s)$ in (0,1). This is obvious by consideration of the graph of the right hand side for $0 \leqslant \varsigma \leqslant 1$. The right hand side is continuous in z, s and ς. It is ***increasing*** in z for every $s \geqslant 0$ and ς

with $0 \leqslant \varsigma \leqslant 1$ and *decreasing* in s for every z and ς satisfying $0 < z \leqslant 1$ and $0 \leqslant \varsigma \leqslant 1$.

As $s \to 0+$, and $z \to 1-$, the root $\varsigma(z,s)$ converges (increasingly in z, decreasingly in s) to the smallest solution η in (0,1] of the equation

$$\varsigma = \sum_{\nu=0}^{\infty} A_\nu \varsigma^\nu,$$

By considering the graph of the right hand side of that equation, it follows that $\eta = 1$, if and only if $\sum_{k=1}^{\infty} kA_k = \rho \leqslant 1$. When the contrary inequality holds, we have $0 < A_0 < \eta < 1$. The essential property of the graph is that $\sum_{k=0}^{\infty} A_k \varsigma^k$ is a strictly convex function of ς in [0,1] which attains the value 1 at $\varsigma = 1$. •

By differentiation in the equations

$$\tilde{G}(z,0) = z \sum_{\nu=0}^{\infty} A_\nu \tilde{G}^\nu(z,0), \tag{1.2.6}$$

$$\tilde{G}(1,s) = \sum_{\nu=0}^{\infty} \tilde{A}_\nu(s)\tilde{G}^\nu(1,s), \tag{1.2.7}$$

and letting $z \to 1-$, and $s \to 0+$, respectively, we obtain that

$$\tilde{\mu}_1 = E(V) = (1-\rho)^{-1},$$

provided that $\rho < 1$. Clearly $\tilde{\mu}_1 = \infty$, when $\rho = 1$. Also

$$\tilde{\mu}_1^* = E(T) = (1-\rho)^{-1} \int_0^\infty u\,dA(u),$$

provided that $\rho < 1$ and $\alpha = \int_0^\infty u\,dA(u) < \infty$. In the contrary case, $\tilde{\mu}_1^* = \infty$.

Remark: The equations (1.2.6) and (1.2.7) are respectively the transform versions of the nonlinear *difference equation*

$$G_1 = A_0,$$
$$G_k = \sum_{\nu=1}^{k-1} A_\nu G_{k-1}^{(\nu)}, \quad \text{for } k \geqslant 2, \tag{1.2.8}$$

where $G_k^{(\nu)} = G_k^{(\nu)}(\infty)$, for $k \geqslant 1$, $\nu \geqslant 1$, and of the nonlinear *integral equation*

$$G(x) = \sum_{\nu=0}^{\infty} \int_0^x dA_\nu(u) G^{(\nu)}(x-u), \quad \text{for } x \geqslant 0, \tag{1.2.9}$$

where $G(x) = \sum_{k=1}^{\infty} G_k(x)$. The sequence $\{G_k\}$ is the (possibly defective) probability density of V and $G(\cdot)$ is the (possibly defective) probability distribution of the fundamental period T.

By further differentiations in (1.2.6) and (1.2.7) we may evaluate higher factorial moments of $\{G_k\}$ and higher noncentral moments of $G(\cdot)$. The second moments are explicitly given by

$$\tilde{\mu}_2 = \sum_{k=2}^{\infty} k(k-1)G_k = \frac{2\rho}{(1-\rho)^2} + \frac{\rho_2}{(1-\rho)^3},$$

and

$$\tilde{\mu}_2^* = \int_0^\infty x^2 dG(x) = \frac{\alpha_2}{(1-\rho)} - \frac{2\alpha}{(1-\rho)^2} \sum_{\nu=1}^{\infty} \nu \tilde{A}'_\nu(0) + \frac{\alpha\rho_2}{(1-\rho)^3},$$

where $\rho_2 = \sum_{k=2}^{\infty} k(k-1)A_k$, and $\alpha_2 = \int_0^\infty u^2 dA(u)$. $\tilde{\mu}_2$ is finite provided that $\sum_{k=1}^{\infty} k^2 A_k < \infty$; $\tilde{\mu}_2^*$ is finite provided that $\sum_{k=1}^{\infty} k^2 A_k < \infty$ and $\alpha_2 < \infty$.

By use of Fáa di Bruno's formula applied to (1.2.5) we may express the n-th order moments recursively in terms of moments and cross-moments of lower order of T and V. We shall omit the details and the rather involved analytic expressions that result, but we note that this readily shows that $EV^n < \infty$, provided that $\sum_{k=1}^{\infty} k^n A_k < \infty$, and $ET^n < \infty$, provided that $\sum_{k=1}^{\infty} k^n A_k < \infty$ *and* $\int_0^{\infty} x^n \, dA(x)$ converges. Recursive computation of the high order moments of T and V is numerically delicate and needs to be programmed with great care.

1.3. THE RECURRENCE TIME OF THE STATE 0

Let T_0 be the recurrence time of the boundary state 0 and V_0 the number of transitions between returns to the state 0. The probabilities $K_k(x)$ are defined by

$$K_k(x) = P\{T_0 \leqslant x, V_0 = k\}, \quad \text{for } k \geqslant 1, x \geqslant 0.$$

By applying the law of total probability with conditioning on the time and the destination of the first transition, we readily obtain that the transform

$$\tilde{K}(z,s) = E[\exp(-sT_0) z^{V_0}] = \sum_{k=1}^{\infty} \int_0^{\infty} e^{-sx} dK_k(x) z^k,$$

is given by

$$\tilde{K}(z,s) = z \sum_{\nu=0}^{\infty} \tilde{B}_\nu(s) \tilde{G}^\nu(z,s), \tag{1.3.1}$$

for $|z| \leqslant 1$, $Re\ s \geqslant 0$, where $\tilde{B}_\nu(s)$ is the Laplace-Stieltjes transform of $B_\nu(\cdot)$ for $\nu \geqslant 0$. Letting $s \to 0+$, $z \to 1-$, we see that $\tilde{K}(1-,0+) = 1$, if and only of $\eta = \tilde{G}(1-,0+) = 1$.

The following theorem gives the recurrence criteria for the Markov renewal process $Q(\cdot)$ and its embedded Markov chain P.

Theorem 1.3.1: The Markov renewal process $Q(\cdot)$ is *positive recurrent* if and only if

$$\rho < 1, \quad \alpha = \int_0^{\infty} x dA(x) < \infty, \quad \beta = \int_0^{\infty} x dB(x) < \infty, \quad \sum_{\nu=1}^{\infty} \nu B_\nu < \infty.$$

It is *null recurrent* if and only if $\rho = 1$, *or* $\rho < 1$, and at least one of the quantities α, β, or $\sum_{\nu=1}^{\infty} \nu B_\nu$ is infinite. It is *transient* if and only if $\rho > 1$. The embedded Markov chain with transition probability matrix $P = Q(\infty)$, is *positive recurrent* if and only if

$$\rho < 1, \quad \text{and} \quad \sum_{\nu=1}^{\infty} \nu B_\nu < \infty.$$

It is *null-recurrent* if and only if $\rho = 1$, *or* $\rho < 1$, and $\sum_{\nu=1}^{\infty} \nu B_\nu$ is infinite. It is *transient* if and only if $\rho > 1$.

Proof: Since the Markov renewal process $Q(\cdot)$ is irreducible, all its states have the same recurrence properties. It is therefore sufficient to classify the state 0.

If $\rho > 1$, the state 0 is transient as the return time of that state is infinite with probability $1 - \tilde{K}(1-, 0+) > 0$. The embedded chain and therefore also the Markov renewal process are transient.

If $\rho \leqslant 1$, we examine the mean recurrence times $\tilde{\mu}_{00}^*$ and $\tilde{\mu}_{00}$ of state 0, respectively in the Markov renewal process and in its embedded Markov chain. From formula (1.3.1), we obtain

$$\left[\frac{\partial \tilde{K}(z,s)}{\partial s}\right]_{z=1, s=0} = -\beta - \frac{\alpha}{1-\rho} \sum_{\nu=1}^{\infty} \nu B_\nu = -\tilde{\mu}_{00}^*, \tag{1.3.2}$$

provided that $\rho < 1$ and α, β, and $\sum_{\nu=1}^{\infty} \nu B_\nu$ are finite. In the contrary case, the mean recurrence time of the state 0 is infinite.

Similarly

$$\left[\frac{\partial \tilde{K}(z,s)}{\partial z}\right]_{z=1, s=0} = 1 + \frac{1}{1-\rho} \sum_{\nu=1}^{\infty} \nu B_\nu = \tilde{\mu}_{00}, \tag{1.3.3}$$

provided that $\rho < 1$ and $\sum_{\nu=1}^{\infty} \nu B_\nu$ is finite. In the contrary case, the embedded Markov chain is null-recurrent. •

Remark *a*. For the ordinary $M/G/1$ queue and for Variant 3, $\rho = \lambda\alpha$, and the process $Q(\cdot)$ and the Markov chain P are transient, positive or null recurrent depending on whether $\rho > 1$, $\rho < 1$, or $\rho = 1$. From Variant 2, we learn that even with $\rho < 1$, the queue may be null recurrent when the mean of $H_0(\cdot)$ is infinite.

Remark *b*. For Variant 4, let $\phi_N(\cdot)$ be defined by

$$\phi_N(u) = \sum_{\nu=1}^{N-1} \nu e^{-\lambda u} \frac{(\lambda u)^\nu}{\nu!} + N\left[1 - \sum_{\nu=0}^{N-1} e^{-\lambda u} \frac{(\lambda u)^\nu}{\nu!}\right],$$

for $u \geqslant 0$, then

$$\rho = \int_0^\infty \phi_N(u)\,dH(u) = \sum_{\nu=1}^\infty \nu B_\nu.$$

Since $\phi_N(u)$ is strictly increasing in N for every $u > 0$, there exists a largest (possibly infinite) value $N^*(\lambda)$ for which $\rho < 1$. $N^*(\lambda)$ is infinite if and only if $\lambda\alpha \leqslant 1$.

For $N = 1$, we obtain a trivial case since $A_\nu = 0$, for $\nu \geqslant 2$. We verify that the corresponding embedded Markov chain is always positive recurrent. We also have $\rho = 1 - h(\lambda) < 1$, where $h(\cdot)$ is the Laplace-Stieltjes transform of $H(\cdot)$. This further shows that $N^*(\lambda) \geqslant 1$. We notice that when the mean α is *infinite*, the Markov renewal process is never positive recurrent. This has far-reaching implications for the behavior of the corresponding queue.

By enumerating all possibilities for the first transition, we may similarly study the first passage time from state 1 to itself. If we denote the joint transform corresponding to $\tilde{K}(z,s)$ by $\tilde{K}_1(z,s)$, we obtain

$$\tilde{K}_1(z,s) = z\tilde{A}_0(s)[1 - z\tilde{B}_0(s)]^{-1} \sum_{\nu=1}^\infty z\tilde{B}_\nu(s)\tilde{G}^{\nu-1}(z,s)$$
$$+ \sum_{\nu=1}^\infty z\tilde{A}_\nu(s)\tilde{G}^{\nu-1}(z,s).$$

The first term corresponds to the case where the process visits the state

0 between returns to state 1; the second, to the case where it does not.

By routine differentiations, we compute the mean recurrence times of the state 1 in the Markov renewal process $Q(\cdot)$ and in the Markov chain P. We obtain respectively

$$\tilde{\mu}_{11}^{*} = \frac{A_0}{1 - B_0} [\beta + \frac{\alpha}{1 - \rho} \sum_{\nu=0}^{\infty} \nu B_\nu] = \frac{A_0}{1 - B_0} \tilde{\mu}_{00}^{*}, \tag{1.3.4}$$

and

$$\tilde{\mu}_{11} = \frac{A_0}{1 - B_0} [1 + \frac{1}{1 - \rho} \sum_{\nu=0}^{\infty} \nu B_\nu] = \frac{A_0}{1 - B_0} \tilde{\mu}_{00}, \tag{1.3.5}$$

The recurrence times of the states $i \geqslant 2$ may, in principle, be studied by a similar enumeration, but this technique becomes increasingly more involved and is impractical. In the next two sections, we shall obtain expressions for the mean recurrence times of these states by much simpler methods.

1.4. THE STATIONARY PROBABILITY VECTOR OF P

When the embedded Markov chain P is positive recurrent, there is a unique probability vector $\mathbf{x}$, which satisfies

$$\mathbf{x} P = \mathbf{x}, \quad \mathbf{x} \mathbf{e} = 1, \tag{1.4.1}$$

and $\mathbf{x}$ is strictly positive. When the Markov chain P is aperiodic, the components x_i, $i \geqslant 0$, of $\mathbf{x}$ are given by $x_i = \tilde{\mu}_{ii}^{-1}$, where $\tilde{\mu}_{ii}$ is the mean recurrence time of the state i in the Markov chain P. In exceptional applications where the Markov chain P is periodic, x_i is given by $x_i = d\, \tilde{\mu}_{ii}^{-1}$, where $d > 1$ is the period. In that case, minor modifications are needed to the statements of the results in this section. We shall leave these to the initiative of the reader and assume henceforth that P is aperiodic.

Since $\tilde{\mu}_{00}$ was obtained in the formula (1.3.3) we immediately obtain that

$$x_0 = [1 + \frac{1}{1 - \rho} \sum_{\nu=1}^{\infty} \nu B_\nu]^{-1}. \tag{1.4.2}$$

The expanded form

$$x_i = x_0 B_i + \sum_{\nu=1}^{i+1} x_\nu A_{i-\nu+1}, \quad \text{for } i \geqslant 0, \tag{1.4.3}$$

of the equation $\mathbf{x} P = \mathbf{x}$, shows that the other components x_i, $i \geqslant 1$, may be recursively computed. We note in passing that

$$x_1 = A_0^{-1}(1 - B_0)x_0,$$

which is consistent with formula (1.3.5).

The recursion formula

$$x_{i+1} = A_0^{-1}\left[x_i - x_0 B_i - \sum_{\nu=1}^{i} x_\nu A_{i-\nu+1} \right], \quad \text{for } i \geqslant 0,$$

may, however, suffer from loss of significance in computer implementation of practical interest. The reasons for this will be readily apparent to the numerate reader. The quantity inside the brackets is the difference of the positive quantity x_i, which becomes small for large i, and a positive sum of comparable magnitude.

It is possible to transform the relation (1.4.3) into a mathematically equivalent form which is highly stable and ideally suited for numerical computation. To that end, we introduce the notation

$$X(z) = \sum_{\nu=0}^{\infty} x_\nu z^\nu, \quad A(z) = \sum_{\nu=0}^{\infty} A_\nu z^\nu, \quad B(z) = \sum_{\nu=0}^{\infty} B_\nu z^\nu,$$

$$\hat{A}_\nu = 1 - \sum_{r=0}^{\nu} A_r, \quad \hat{B}_\nu = 1 - \sum_{r=0}^{\nu} B_r, \quad \text{for } \nu \geqslant 0,$$

and note that

$$\sum_{\nu=0}^{\infty} \hat{A}_\nu z^\nu = \frac{1 - A(z)}{1 - z}, \quad \sum_{\nu=0}^{\infty} \hat{B}_\nu z^\nu = \frac{1 - B(z)}{1 - z}, \quad \text{for } 0 \leqslant z < 1.$$

The equations (1.4.3) readily lead to

$$[z - A(z)]\, X(z) = x_0\, [zB(z) - A(z)], \tag{1.4.4}$$

which is equivalent to

$$X(z) = \frac{1 - A(z)}{1 - z}\, X(z) + x_0 + x_0 \left\{ z\, \frac{1 - B(z)}{1 - z} - \frac{1 - A(z)}{1 - z} \right\}.$$

Upon series expansion and noting that $\hat{A}_0 = 1 - A_0$, we obtain the recursion formula

$$x_i = A_0^{-1} [\, x_0 \hat{B}_{i-1} + \sum_{\nu=1}^{i-1} x_\nu \hat{A}_{i-\nu}\,], \quad \text{for } i \geqslant 1. \tag{1.4.5}$$

Formula (1.4.5) is highly recommended for the accurate and efficient computation of the probabilities x_i.

The moments of the stationary probability density $\{x_i\}$ may be computed by successive differentiations in formula (1.4.4). The formal results are given in the following theorem.

Theorem 1.4.1: Provided that the $(N+1)$–st moments of the probability densities $\{A_k\}$ and $\{B_k\}$ exist, the N–th moment of the probability density $\{x_i\}$ exists and the n-th factorial moments $X^{(n)}(1-)$ are recursively given by

$$X^{(0)}(1-) = 1, \tag{1.4.6}$$

$$X'(1-) = \frac{1}{2(1 - \rho)} \left\{ A''(1-) + x_0[2B'(1-) + B''(1-) - A''(1-)] \right\},$$

$$X^{(n)}(1-) = \frac{1}{(n+1)(1-\rho)} \left\{ \sum_{\nu=2}^{n+1} \binom{n+1}{\nu} A^{(\nu)}(1-) X^{(n-\nu+1)}(1-) \right.$$

$$\left. + x_0 \, [B^{(n+1)}(1-) + (n+1) B^{(n)}(1-) - A^{(n+1)}(1-)] \right\},$$

for $2 \leqslant n \leqslant N$, where $A^{(\nu)}(1-)$ and $B^{(\nu)}(1-)$ are the ν-th factorial moments of $\{A_k\}$ and $\{B_k\}$ respectively.

Proof: By differentiating n times in formula (1.4.4), we obtain

$$[z - A(z)]\, X^{(n)}(z) + n\,[1 - A'(z)]\, X^{(n-1)}(z) - \sum_{\nu=2}^{n} \binom{n}{\nu} A^{(\nu)}(z) X^{(n-\nu)}(z)$$

$$= x_0[\, zB^{(n)}(z) + nB^{(n-1)}(z) - A^{(n)}(z)\,].$$

Moving all terms but the first to the right hand side and letting $z \to 1-$, we obtain an indeterminate form, as (by induction) the new right hand side vanishes. Rewriting the equation as

$$\frac{z - A(z)}{z - 1}\, X^{(n)}(z) = \frac{1}{z-1}\Bigg\{-nX^{(n-1)}(z) + \sum_{\nu=1}^{n} \binom{n}{\nu} A^{(\nu)}(z) X^{(n-\nu)}(z)$$

$$+ x_0\,[zB^{(n)}(z) + nB^{(n-1)}(z) - A^{(n)}(z)]\Bigg\},$$

and letting $z \to 1-$, we obtain the stated recursion formulas by Abel's theorem, l'Hospital's rule and some simplifications. •

Remark: The recursion formula may be obtained heuristically and with less calculation by differentiating $n + 1$ times in (1.4.4) and setting $z = 1$. The formal limit argument is needed for $n = N$, as we have no guarantee that $[z - A(z)]\, X^{(N+1)}(z)$ tends to zero as $z \to 1-$. This remark should be borne in mind when the much more involved derivation for the matrix analogue of the formulas (1.4.6) is discussed.

The probability distribution $R_D(\cdot)$, defined by

$$R_D(x) = \sum_{i=0}^{\infty} x_i\, G^{(i)}(x), \quad \text{for } x \geqslant 0, \tag{1.4.7}$$

where $G^{(i)}(\cdot)$ is the i-fold convolution of $G(\cdot)$, is the stationary distribution of the time to reach the state 0 after a transition epoch. Similarly in the $M/G/1$ queue and its variants, discussed earlier, the probability distribution $W_D(\cdot)$, defined by

$$W_D(x) = \sum_{i=0}^{\infty} x_i H^{(i)}(x), \quad \text{for } x \geqslant 0, \tag{1.4.8}$$

is the stationary distribution of *the server backlog or virtual waiting time immediately after a service completion.* These probability distributions are, in general, not easily computable because of the required numerical convolutions.

1.5. THE MARKOV RENEWAL MATRIX OF $Q(\cdot)$

This section contains material preliminary to the discussion of the stationary distributions of the queue length at an arbitrary time and of the virtual waiting time. These probability distributions are conveniently studied by relating the state of the queue at time t to its embedded Markov renewal process. This relationship is most transparently expressed in terms of the Markov renewal matrix corresponding to $Q(\cdot)$.

For ease of notation and without loss of generality we shall assume that time $t = 0$ corresponds to a transition epoch in the Markov renewal process $Q(\cdot)$ and that the state at time $t = 0+$ is $i_0 \geqslant 0$. For $t \geqslant 0$, let $M_{i_0 i}(t)$ denote the conditional expected number of visits to the state i in the interval $[0,t]$. The matrix $M(t) = \{M_{i_0 i}(t), i_0 \geqslant 0, i \geqslant 0\}$ is called the *Markov renewal matrix* of the Markov renewal process $Q(\cdot)$.

If we denote by $U(\cdot)$, the infinite diagonal matrix with the degenerate probability distribution as its diagonal elements, then by conditioning on the time and the destination of the first transition, we see that the matrix $M(\cdot)$ satisfies the matrix renewal equation

$$M(t) = U(t) + Q(\cdot) * M(t), \quad \text{for } t \geqslant 0, \tag{1.5.1}$$

It is shown in the general theory of Markov renewal processes that the unique solution to (1.5.1) in the set of matrices bounded in any interval $[0,t)$ is given by the Neumann series

$$M(t) = \sum_{\nu=0}^{\infty} Q^{(\nu)}(t), \tag{1.5.2}$$

where $Q^{(\nu)}(\cdot)$, $\nu \geqslant 1$, is the ν-fold matrix convolution product of the matrix $Q(\cdot)$ with itself and $Q^{(0)}(\cdot) = U(\cdot)$.

The increment $dM_{i_0 i}(t)$ of $M_{i_0 i}(t)$ may be interpreted as the elementary conditional probability that the Markov renewal process $Q(\cdot)$ enters the state i in the interval $(t, t+dt)$, given that the departure at time 0 leaves i_0 customers behind. The matrix $M(t)$ plays a crucial role in the derivation of analytic expressions for the *time dependent distributions* of the waiting time in queues related to the $M/G/1$ model. These probability distributions depend not only on the embedded Markov renewal process, but also on the nature of the arrival process between departure epochs. As we shall see, the derivations of these probability distributions are different for each variant, but there are nevertheless a number of common general steps which we shall now discuss.

It is easily seen that the matrix $M(t)$, as given by (1.5.2), also satisfies the integral equation

$$M(t) = U(t) + M(\cdot) * Q(t), \quad \text{for } t \geqslant 0,$$

In expanded form, that equation is equivalent to the system

$$M_{i_0 i}(t) = \delta_{i_0 i}\, U(t) + M_{i_0 0}(\cdot) * B_i(t) + \sum_{\nu=1}^{i+1} M_{i_0 \nu}(\cdot) * A_{i-\nu+1}(t), \quad (1.5.3)$$

where $i_0 \geqslant 0$, $i \geqslant 0$, $t \geqslant 0$. $\delta_{i_0 i}$ is the Kronecker delta and $U(\cdot)$ stands here for the degenerate distribution.

If we introduce the transforms

$$m_{i_0 i}(\xi) = \int_0^\infty e^{-\xi t}\, dM_{i_0 i}(t), \qquad m_{i_0}(z,\xi) = \sum_{i=0}^{\infty} m_{i_0 i}(\xi) z^i,$$

$$\tilde{A}(z,\xi) = \sum_{i=0}^{\infty} \tilde{A}_i(\xi) z^i, \qquad \tilde{B}(z,\xi) = \sum_{i=0}^{\infty} \tilde{B}_i(\xi) z^i,$$

for $|z| < 1$, Re $\xi \geqslant 0$, or $|z| \leqslant 1$, Re $\xi > 0$, then (1.5.3) leads successively to

$$m_{i_0 i}(\xi) = \delta_{i_0 i} + m_{i_0 0}(\xi)\tilde{B}_i(\xi) + \sum_{\nu=1}^{i+1} m_{i_0 \nu}(\xi)\tilde{A}_{i-\nu+1}(\xi),$$

for $i_0 \geqslant 0$, $i \geqslant 0$, and

$$[z - \tilde{A}(z,\xi)]\, m_{i_0}(z,\xi) = z^{i_0+1} + m_{i_00}(\xi)\, [z\tilde{B}(z,\xi) - \tilde{A}(z,\xi)]. \tag{1.5.4}$$

The transform $m_{i_00}(\xi)$ may be expressed in terms of quantities we have already discussed. The successive visits to the state 0 form a renewal process and $m_{i_00}(\xi)$ is the Laplace-Stieltjes transform of the corresponding renewal function $M_{i_00}(t)$. The Laplace-Stieltjes transform of the initial interval is given by $\tilde{G}^{i_0}(1,\xi)$, where $\tilde{G}(z,\xi)$ is the transform discussed in Theorem 1.2.1. The Laplace-Stieltjes transform of the time between returns to the state 0 is given by $\tilde{K}(1,\xi)$, where $\tilde{K}(z,\xi)$ is given in formula (1.3.1). We readily obtain that

$$m_{i_00}(\xi) = \tilde{G}^{i_0}(1,\xi)\, [1 - \tilde{K}(1,\xi)]^{-1}, \quad \text{for Re } \xi > 0. \tag{1.5.5}$$

It follows that

$$m_{i_0}(z,\xi) = \{z^{i_0+1} + \tilde{G}^{i_0}(1,\xi)[1 - \tilde{K}(1,\xi)]^{-1} \tag{1.5.6}$$

$$\cdot\, [z\tilde{B}(z,\xi) - \tilde{A}(z,\xi)]\, \}[z - \tilde{A}(z,\xi)]^{-1},$$

for $|z| < 1$, Re $\xi \geqslant 0$, or $|z| \leqslant 1$, Re $\xi > 0$.

Equation (1.5.5) shows that the renewal function $M_{i_00}(t)$ may *in principle* be computed by solving the equations

$$M_{i_00}(t) = \int_0^t G^{(i_0)}(t-u)\,dM_{00}(u),$$

$$M_{00}(t) = U(t) + \int_0^t K(t-u)\,dM_{00}(u), \tag{1.5.7}$$

$$K(t) = \sum_{\nu=0}^{\infty} \int_0^t B_\nu(t-u)\,dG^{(\nu)}(u),$$

where $G(\cdot)$ is the solution to the nonlinear integral equation (1.2.9). By use of the equations (1.5.3), the renewal functions $M_{i_0 i}(t)$ may, again in principle, be computed in terms of $M_{i_0 0}(t)$. The task of carrying out these computations numerically is formidable. It is the principal reason for the fact that numerical computations for the time dependent distributions, even of a simple model as the $M/G/1$ queue, have only been discussed for very special cases. This should be borne in mind when assessing the computational utility of many formulas and procedures in priority queues and in control of queues. These frequently involve auxiliary quantities which require the time dependent probability distributions for the $M/G/1$ queue and related models.

The preceding discussion is particularly useful in *the derivation of the limit distributions* as $t \to \infty$ of the time dependent distributions in the $M/G/1$ queue and its variants. In order to avoid uninteresting cases, we shall henceforth assume that the probability distribution $K(\cdot)$ of the recurrence time of the state 0 is *non-lattice*. This is clearly the case for the $M/G/1$ queue and its variants since the distribution of the idle time is absolutely continuous. Since the Markov renewal process $Q(\cdot)$ is irreducible, it follows that the recurrence time distributions of all states $i \geqslant 0$ are non-lattice.

By a classical result on positive recurrent Markov renewal processes, we may relate the *mean recurrence time* $\tilde{\mu}_{ii}^*$ of the state i to the stationary probabilities x_i, discussed in Section 1.4.

The inner product E of the stationary probability vector $\mathbf{x}$ and the vector of row sum means $\int_0^\infty x dQ(x)\, \mathbf{e}$ of the matrix $Q(\cdot)$ is called the *fundamental mean* of the Markov renewal process. By the particular form (1.1.3) of the matrix $Q(\cdot)$ and by formula (1.4.2), we see that E is given by

$$E = x_0 \beta + (1 - x_0)\alpha = \frac{\beta(1-\rho) + \alpha \sum_{\nu=1}^{\infty} \nu B_\nu}{1 - \rho + \sum_{\nu=1}^{\infty} \nu B_\nu}. \tag{1.5.8}$$

The mean recurrence times $\tilde{\mu}_{ii}^*$ are then given by

$$\tilde{\mu}_{ii}^* = \frac{E}{x_i}, \quad \text{for } i \geqslant 0. \tag{1.5.9}$$

As an application, let us consider the state $I(t)$ visited by the Markov renewal process at the last transition prior to time t. The conditional probabilities

$$V_i(t) = P\{I(t) = i \mid I(0) = i_0\},$$

for $i \geqslant 0$, $t \geqslant 0$, are expressed in terms of the Markov renewal matrix $M(\cdot)$ by

$$\begin{aligned} V_i(t) &= \int_0^t dM_{i_0 i}(u)[1 - B(t-u)], \quad \text{for } i = 0, \\ &= \int_0^t dM_{i_0 i}(u)[1 - A(t-u)], \quad \text{for } i \geqslant 1, \end{aligned}$$

where $A(\cdot) = \sum_{\nu=0}^{\infty} A_\nu(\cdot)$, and $B(\cdot) = \sum_{\nu=0}^{\infty} B_\nu(\cdot)$.

By the key renewal theorem, the limits $\lim_{t \to \infty} V_i(t) = V_i$, exist and are given by

$$\begin{aligned} V_i &= (\tilde{\mu}^*_{ii})^{-1}\beta = E^{-1}x_0\beta, \quad \text{for } i = 0, \\ &= (\tilde{\mu}^*_{ii})^{-1}\alpha = E^{-1}x_i\alpha, \quad \text{for } i \geqslant 1. \end{aligned}$$

In queueing models, $\{V_i\}$ is the steady-state probability density of the queue length at the last service completion prior to time t. For the ordinary $M/G/1$ queue, we have

$$\beta = \lambda^{-1} + \alpha, \qquad E = \lambda^{-1},$$

so that

$$V_0 = \lambda x_0(\lambda^{-1} + \alpha) = x_0(1 + \rho) = 1 - \rho^2,$$

$$V_i = \lambda x_i \alpha = x_i \rho, \qquad \text{for } i \geqslant 1.$$

Remarks: The quantity E^{-1} may be interpreted as the rate at which transitions occur in the stationary version of the Markov renewal process $Q(\cdot)$. In the $M/G/1$ queue and its variants, E^{-1} is therefore *the rate of departures of served customers* in the steady-state regime of the queue.

For the $M/G/1$ queue with group arrivals (Variant 1), we have $\rho = \lambda\alpha\bar{p}$, where $\bar{p}$ denotes the mean group size. In addition, we have $\sum_{\nu=1}^{\infty} \nu B_\nu = \rho + \bar{p} - 1$, and therefore $x_0 = (1-\rho)(\bar{p})^{-1}$, and $E = (\lambda\bar{p})^{-1}$.

For Variant 2, $\rho = \lambda\alpha$, $\sum \nu B_\nu = \lambda\alpha_0$, and therefore

$$x_0 = (1-\lambda\alpha)(1-\lambda\alpha+\lambda\alpha_0)^{-1}, \quad \text{and} \quad E = \lambda^{-1}.$$

For Variant 3, $\rho = \lambda\alpha$, $\sum \nu B_\nu = \lambda\alpha + N - 1$, and therefore $x_0 = (1-\lambda\alpha)\, N^{-1}$, and $E = \lambda^{-1}$.

The queue in Variant 4 is stable if and only if $\alpha < \infty$, and

$$\rho = \sum_{i=1}^{N-1} i \int_0^\infty e^{-\lambda u} \frac{(\lambda u)^i}{i!} dH(u) + N\left[1 - \sum_{i=0}^{N-1} \int_0^\infty e^{-\lambda u} \frac{(\lambda u)^i}{i!} dH(u)\right] < 1.$$

Since $\sum \nu B_\nu = \rho$, we obtain that $x_0 = 1 - \rho$, and therefore $E = \lambda^{-1}(1 + \lambda\alpha - \rho)$.

1.6. THE QUEUE LENGTH AT TIME t

In studying the queue length at an arbitrary time t in the $M/G/1$ queue and its variants, it is no longer enough to consider only the embedded Markov renewal process $Q(\cdot)$. We need also to take the behavior of the queue length process *between* the transitions in the process $Q(\cdot)$ into account. During the intervals between transitions, the behavior of this queue is very simple as the arrival process is Poisson. As we shall see later, the same basic argument, used in this section, may be applied to much more complex cases. In this section, we discuss the probability density of the queue length at time t for Variant 3 (the N-policy). As a particular case, we shall by setting $N = 1$, obtain the classical results for the $M/G/1$ queue. The other variants may be handled in a similar manner and are left to the initiative of the reader.

As before, we assume that time $t = 0$ corresponds to a departure epoch and that there are i_0 customers in the system at that time. By $P_i(t)$, we denote the (conditional) probability that there are i customers in the system at time t. The probabilities $P_i(t)$, $i \geq 0$, are related to the renewal functions $M_{i_0 i}(\cdot)$ by the formulas

$$P_0(t) = \int_0^t dM_{i_0 0}(u) e^{-\lambda(t-u)},$$

$$P_i(t) = \int_0^t dM_{i_0 0}(u) e^{-\lambda(t-u)} \frac{[\lambda(t-u)]^i}{i!} \tag{1.6.1}$$

$$+ \sum_{k=1}^{i} \int_0^t dM_{i_0 k}(u) e^{-\lambda(t-u)} \frac{[\lambda(t-u)]^{i-k}}{(i-k)!} [1 - H(t-u)],$$

$$\text{for } 1 \leq i \leq N-1,$$

$$P_i(t) = \int_0^t dM_{i_0 0}(u) \int_0^{t-u} e^{-\lambda v} \frac{(\lambda v)^{N-1}}{(N-1)!} \lambda \, dv$$

$$\cdot e^{-\lambda(t-u-v)} \frac{[\lambda(t-u-v)]^{i-N}}{(i-N)!} [1 - H(t-u-v)]$$

$$+ \sum_{k=1}^{i} \int_0^t dM_{i_0 k}(u) e^{-\lambda(t-u)} \frac{[\lambda(t-u)]^{i-k}}{(i-k)!} [1 - H(t-u)],$$

$$\text{for } i \geq N.$$

The arguments leading to each of these formulas are similar. For the case $i \geq N$, there are two alternatives. If at the last transition prior to t (it occurs at some time u, $0 \leq u < t$), the queue is empty, then at some time $u+v$, $0 \leq v \leq t-u$, the N-th arrival must have occurred and a service has started at that time. That service is still in course at time t, so that its duration must exceed $t-u-v$. During the interval $(u+v, t)$, $i - N$ customers arrive so that the queue length at time t is i.

If at the last transition prior to t, there are k customers present, then it is clearly necessary that $1 \leq k \leq i$. The service, starting at the time u of that transition, is still in course at time t and during the interval (u, t), $i - k$ arrivals occur to bring the queue length to i at time t. The stated formula is now clearly an application of the law of total probability with each alternative contributing a term.

The probabilities, given in the formulas (1.6.1), and their limits as $t \to \infty$, may be further studied by deriving analytic expressions for appropriate transforms. The required calculations are classical, but as they are among the simplest examples of much more involved manipulations needed later, we discuss the main steps of the derivations in full detail.

Some preliminary remarks are in order. The transforms $\tilde{A}(z,\xi)$ and $\tilde{B}(z,\xi)$, defined in Section 1.5, are explicitly given by

$$\tilde{A}(z,\xi) = h(\xi+\lambda-\lambda z), \qquad \tilde{B}(z,\xi) = z^{N-1}\left(\frac{\lambda}{\xi+\lambda}\right)^{N} h(\xi+\lambda-\lambda z), \quad (1.6.2)$$

where $h(\cdot)$ is the Laplace-Stieltjes transform of the service time distribution $H(\cdot)$. The joint transform $\tilde{G}(z,\xi)$ is now the familiar transform of the *number of customers* served during a busy period and of the *duration* of the busy period in the $M/G/1$ queue. It satisfies Takács' equation

$$\tilde{G}(z,\xi) = z\, h[\xi+\lambda-\lambda\tilde{G}(z,\xi)]. \quad (1.6.3)$$

It is clear from formula (1.3.1) that

$$\tilde{K}_N(z,\xi) = \left(\frac{\lambda}{\xi+\lambda}\right)^{N} \tilde{G}^N(z,\xi). \quad (1.6.4)$$

The subscript N is added to emphasize the dependence on the parameter of the N-policy. Upon substitution into the formulas (1.5.5) and (1.5.6), we obtain

$$m_{i_0}(z,\xi) = [z - h(\xi+\lambda-\lambda z)]^{-1} \quad (1.6.5)$$

$$\cdot \left\{ z^{i_0+1} + m_{i_0}(\xi) h(\xi+\lambda-\lambda z)\, [z^N\left(\frac{\lambda}{\xi+\lambda}\right)^{N} - 1] \right\},$$

where

$$m_{i_0 0}(\xi) = \tilde{G}^{i_0}(1,\xi)\left\{1 - \left(\frac{\lambda}{\xi+\lambda}\right)^N \tilde{G}^N(1,\xi)\right\}^{-1}. \tag{1.6.6}$$

Substitution into the formulas (1.4.2) and (1.4.4) yields

$$x_0 = (1 - \rho)/ N, \tag{1.6.7}$$

and

$$X_N(z) = N^{-1}(1 - \rho)\frac{(z^N - 1)h(\lambda - \lambda z)}{z - h(\lambda - \lambda z)} \tag{1.6.8}$$

$$= \frac{(z^N - 1)}{N(z-1)}\,\frac{(1-\rho)(z-1)h(\lambda - \lambda z)}{z - h(\lambda-\lambda z)} = \frac{(z^N-1)}{N(z-1)}X_1(z).$$

We note that the stationary queue length density after departures in the $M/G/1$ queue with the N-policy is simply the convolution of the corresponding density for the ordinary $M/G/1$ queue with a *uniform* density on the integers $0, \cdots, N-1$.

We now return to the formulas (1.6.1) and define the Laplace transforms

$$P_i^*(\xi) = \int_0^\infty P_i(t)e^{-\xi t}\,dt, \quad \text{for } i \geqslant 0, \quad \text{Re } \xi > 0,$$

of the probabilities $P_i(t)$. The formulas (1.6.1) lead to

$$P_0^*(\xi) = \frac{1}{\xi + \lambda}m_{i_0 0}(\xi),$$

$$P_i^*(\xi) = \frac{1}{\lambda}\left(\frac{\lambda}{\xi+\lambda}\right)^{i+1} m_{i_0 0}(\xi) \tag{1.6.9}$$

$$+ \sum_{k=1}^{i} m_{i_0 k}(\xi)\int_0^\infty e^{-(\xi+\lambda)u}\frac{(\lambda u)^{i-k}}{(i-k)!}[1 - H(u)]\,du,$$

$$\text{for } 1 \leqslant i \leqslant N-1,$$

$$P_i^*(\xi) = \left(\frac{\lambda}{\xi+\lambda}\right)^N m_{i_0 0}(\xi)\int_0^\infty e^{-(\xi+\lambda)u}\frac{(\lambda u)^{i-N}}{(i-N)!}[1-H(u)]du$$

$$+\sum_{k=1}^{i} m_{i_0 k}(\xi)\int_0^\infty e^{-(\xi+\lambda)u}\frac{(\lambda u)^{i-k}}{(i-k)!}[1-H(u)]du,$$

$$\text{for } i \geqslant N.$$

Next we evaluate the generating function

$$P^*(z,\xi) = \sum_{i=0}^{\infty} P_i^*(\xi)\, z^i,$$

and by noting that

$$\sum_{i=1}^{\infty} z^i \int_0^\infty e^{-(\xi+\lambda)t}\frac{(\lambda t)^i}{i!}[1-H(t)]dt$$

$$= \int_0^\infty e^{-(\xi+\lambda-\lambda z)t}[1-H(t)]dt = \frac{1-h(\xi+\lambda-\lambda z)}{\xi+\lambda-\lambda z},$$

we obtain

$$P^*(z,\xi) = \frac{1}{\lambda} m_{i_0 0}(\xi)\sum_{i=1}^{N}\left(\frac{\lambda}{\xi+\lambda}\right)^i z^{i-1}$$

$$+ m_{i_0 0}(\xi)\, z^N\left(\frac{\lambda}{\xi+\lambda}\right)^N \frac{1-h(\xi+\lambda-\lambda z)}{\xi+\lambda-\lambda z}$$

$$+ [m_{i_0}(z,\xi) - m_{i_0 0}(\xi)]\frac{1-h(\xi+\lambda-\lambda z)}{\xi+\lambda-\lambda z}.$$

Upon substitution by the expressions in (1.6.5) and (1.6.6) and simplifying, we obtain

$$P^*(z,\xi) = \frac{z^{i_0+1}}{z-h(\xi+\lambda-\lambda z)}\,\frac{1-h(\xi+\lambda-\lambda z)}{\xi+\lambda-\lambda z} \qquad (1.6.10)$$

$$+ \tilde{G}^{i_0}(1,\xi) \left[1 - \left(\frac{\lambda}{\xi+\lambda}\right)^N \tilde{G}^N(1,\xi)\right]^{-1} \left\{\frac{1}{\lambda}\sum_{i=1}^{N}\left(\frac{\lambda}{\xi+\lambda}\right)^i z^{i-1}\right.$$

$$\left. + \frac{1 - h(\xi+\lambda-\lambda z)}{\xi+\lambda-\lambda z} \frac{z}{z - h(\xi+\lambda-\lambda z)} \left[z^N\left(\frac{\lambda}{\xi+\lambda}\right)^N - 1\right]\right\}.$$

For $N = 1$, the expression in (1.6.10) reduces to the well-known transform formula for the $M / G / 1$ queue. The practical utility of formulas such as (1.6.10) is limited; they serve primarily as a vehicle for the study of the limiting probabilities as $t \to \infty$, but as we shall see in the proof of Theorem 1.6.1, most of the preceding calculations can still be obviated when the stationary probabilities are of sole interest.

In the course of the proof of Theorem 1.6.1, we shall show that the limits $\lim_{t \to \infty} P_i(t) = P_i$, exist for $i \geqslant 0$. By appealing to Abel's theorem for Laplace transforms and the dominated convergence theorem, it then follows that

$$P^*(z) = \sum_{i=0}^{\infty} P_i z^i = \lim_{\xi \to 0+} \xi P^*(z,\xi).$$

The evaluation of this limit is elementary once we notice that for $\rho < 1$,

$$\lim_{\xi \to 0+} \xi \left[1 - \left(\frac{\lambda}{\xi+\lambda}\right)^N \tilde{G}^N(1,\xi)\right]^{-1} = \left[\frac{N}{\lambda} + \frac{N\alpha}{1-\rho}\right]^{-1} = N^{-1}\lambda(1 - \rho),$$

by the key renewal theorem. For $\rho \geqslant 1$, the corresponding limit and therefore $P^*(z)$ is zero. This leads to the following theorem, of which also a more concise proof will be presented.

Theorem 1.6.1: In the stable $M / G / 1$ queue with the N-policy, the stationary probability density $\{P_i\}$ of the queue length at an arbitrary time is the same as the stationary probability density $\{x_i\}$ of the queue length following service completions. Its probability generating function $P^*(z) = X_N(z)$, is given by formula (1.6.8).

Proof: We recall that in Variant 3, the fundamental mean E of the Markov renewal process $Q(\cdot)$ is given by $E = \lambda^{-1}$. In the formulas (1.6.6), each $P_i(t)$ is a finite sum of terms to which the key renewal theorem may be applied. The limits P_i, as $t \to \infty$, therefore exist and satisfy the equations

$$P_0 = \frac{1}{\lambda}\frac{x_0}{E} = x_0,$$

$$P_i = \lambda x_0 \int_0^\infty e^{-\lambda u} \frac{(\lambda u)^i}{i!} du + \lambda \sum_{k=1}^{i} x_k \int_0^\infty e^{-\lambda u} \frac{(\lambda u)^{i-k}}{(i-k)!} [1 - H(u)] du, \quad \text{for } 1 \leqslant i \leqslant N-1,$$

$$P_i = \lambda x_0 \int_0^\infty e^{-\lambda u} \frac{(\lambda u)^{i-N}}{(i-N)!} [1 - H(u)] du + \lambda \sum_{k=1}^{i} x_k \int_0^\infty e^{-\lambda u} \frac{(\lambda u)^{i-k}}{(i-k)!} [1 - H(u)] du, \quad \text{for } i \geqslant N.$$

Upon evaluating the generating function $P^*(z)$ and using formula (1.6.8), we obtain in a straightforward manner that $P^*(z) = X_N(z)$. •

Remark: In the expressions for the quantities $P_i(t)$ with $0 \leqslant i \leqslant N-1$, the first terms correspond to the cases where there are i customers in the system *and the server is idle*. It is easily seen that each of these terms converges to the same limit x_0, so that in the stationary version of the queue, the fraction of time the server is idle is given by $Nx_0 = 1 - \rho$.

1.7. THE VIRTUAL WAITING TIME

The virtual waiting time $W(t)$ is the length of time a customer arriving at time t would have to wait before entering service. In studying the virtual waiting time, we again need to take the behavior of the queue *between* the transitions in the process $Q(\cdot)$ into account. It is also necessary to specify the queue discipline. In order to have an interesting model, which includes the standard $M/G/1$ queue as a particular case, we shall study the virtual waiting time process for the $M/G/1$ queue with N-policy and with services in the order of arrival.

There are several alternative derivations of the (conditional) distribution of $W(t)$ and of its limiting distribution as $t \to \infty$. In this book, we shall consider the random variable $W(t)$ as the first passage time in a stochastic process that is intimately related to the embedded Markov renewal process $Q(\cdot)$. This will enable us to obtain an expression for the limiting distribution of $W(t)$ again by applying the key renewal theorem. Our approach may involve more belabored calculations than some, but it has the advantage of being highly systematic and will later be extended to considerably more complex models.

As before, we choose the origin of the time axis to coincide with a departure epoch. At time $t = 0+$, there are $i_0 \geq 0$, customers in the system. We shall write the probability

$$P\{W(t) \leq x \mid i_0\}, \tag{1.7.1}$$

for $t \geq 0$, $x \geq 0$, as the sum of three terms. A more concise expression for an appropriate transform will then be obtained after some calculations.

The event $\{W(t) \leq x\}$ may be partitioned into the following three events:

Event A: At time t, the server is idle and there are i, $0 \leq i \leq N-1$, customers in the system. The virtual customer will then wait until there are N customers, himself included, in the system and for i customers to be completed thereafter.

By the law of total probability, this first alternative contributes the term

$$\sum_{i=0}^{N-2} \int_{0_{(u)}}^{t} \int_{0_{(v)}}^{x} dM_{i_0 0}(u)\, e^{-\lambda(t-u)} \frac{[\lambda(t-u)]^i}{i!} e^{-\lambda v} \frac{(\lambda v)^{N-i-2}}{(N-i-2)!} \lambda dv\; H^{(i)}(x-v)$$

$$+ \int_{0_{(u)}}^{t} dM_{i_0 0}(u)\, e^{-\lambda(t-u)} \frac{[\lambda(t-u)]^{N-1}}{(N-1)!} H^{(N-1)}(x), \tag{1.7.2}$$

to the conditional probability in (1.7.1).

Event B: At time t, the server is busy and the last state visited by the embedded Markov renewal process is 0. In this case, the service in course at time t is the first of a busy period of the queue. The number of customers in the system is therefore equal to N plus the number (i) of arrivals during the first service of that busy period, but prior to time t. If the total number of customers at t is $N + i$, the virtual customer will wait out the residual service time of the customer in process and the total service time of the $N + i - 1$ additional customers, who precede him in the queue.

This alternative contributes the term

$$\sum_{i=0}^{\infty} \int_{0_{(u)}}^{t} \int_{0_{(v)}}^{t-u} \int_{0_{(\tau)}}^{x} dM_{i_0 0}(u)\, e^{-\lambda v} \frac{(\lambda v)^{N-1}}{(N-1)!} \lambda dv \tag{1.7.3}$$

$$\cdot\, e^{-\lambda(t-u-v)} \frac{[\lambda(t-u-v)]^i}{i!}\, dH(t+\tau-u-v)\, H^{(N+i-1)}(x-\tau),$$

to the conditional probability in (1.7.1).

Event C: The server is busy at time t and the last state visited by the embedded Markov renewal process is some state $k \geqslant 1$. This means that at the beginning of the service in course at time t, there are k customers in the system. An additional number, say $i-k$, may arrive prior to time t. With i customers in the system at t, the virtual waiting time consists of the residual service time of the customer in process plus $i-1$ additional service times.

The contribution of this alternative to the conditional probability in (1.7.1) is given by

$$\sum_{i=1}^{\infty} \sum_{k=1}^{i} \int_{0_{(u)}}^{t} \int_{0_{(\tau)}}^{x} dM_{i_0 k}(u) e^{-\lambda(t-u)} \frac{[\lambda(t-u)]^{(i-k)}}{(i-k)!} \tag{1.7.4}$$

$$\cdot\, dH(t+\tau-u) H^{(i-1)}(x-\tau).$$

As the events A, B and C are mutually exclusive, the conditional probability $P\{W(t) \leqslant x \mid i_0\}$ is given by the sum of the expressions in the formulas (1.7.2-4).

By evaluating the transform

$$W^*(\xi, s) = \int_0^\infty e^{-\xi t}\, dt \int_0^\infty e^{-sx}\, d_x P\{W(t) \leqslant x \mid i_0\},$$

we may rewrite the preceding expressions in analytically more concise forms. We shall evaluate the transforms for each of the three terms separately and simplify their sum thereafter.

The term contributed by (1.7.2) is evaluated by simple interchanges of integrations and is given by

$$(\xi + \lambda)^{-1} m_{i_0 0}(\xi) \sum_{i=0}^{N-1} \left(\frac{\lambda}{\xi + \lambda}\right)^i \left(\frac{\lambda}{s + \lambda}\right)^{N-i-1} h^i(s). \tag{1.7.5}$$

After three elementary interchanges of integration, the transform of the expression in (1.7.3) becomes

$$\sum_{i=0}^{\infty} m_{i_0 0}(\xi) \left(\frac{\lambda}{\xi + \lambda}\right)^N \int_0^\infty \int_0^\infty e^{-(\xi + \lambda)u - s\tau} \frac{(\lambda u)^i}{i!} du\ dH(u + \tau) h^{N+i-1}(s),$$

and that expression is further simplified as follows:

$$m_{i_0 0}(\xi) \left(\frac{\lambda}{\xi + \lambda}\right)^N h^{N-1}(s) \int_0^\infty \int_0^\infty e^{-(\xi + \lambda)u - s\tau + \lambda h(s) u}\, du\ dH(u + \tau)$$

$$= m_{i_0 0}(\xi) \left(\frac{\lambda}{\xi + \lambda}\right)^N h^{N-1}(s) \int_0^\infty e^{-sw}\, dH(w) \int_0^w e^{-[\xi + \lambda - \lambda h(s) - s]u}\, du \tag{1.7.6}$$

$$= m_{i_0 0}(\xi) \left(\frac{\lambda}{\xi + \lambda}\right)^N h^{N-1}(s) \frac{h(s) - h[\xi + \lambda - \lambda h(s)]}{\xi + \lambda - \lambda h(s) - s}.$$

The transform of the expression in (1.7.4) is evaluated by similar manipulations. After two interchanges of integrals, the main steps are as follows:

$$\sum_{i=1}^{\infty} \sum_{k=1}^{i} m_{i_0k}(\xi) h^{i-1}(s) \int_0^{\infty}\int_0^{\infty} e^{-(\xi+\lambda)u - s\tau} \frac{(\lambda u)^{i-k}}{(i-k)!} du \; dH(u+\tau)$$

$$= \sum_{k=1}^{\infty} m_{i_0k}(\xi) h^{k-1}(s) \int_0^{\infty}\int_0^{\infty} e^{-(\xi+\lambda)u - s\tau + \lambda h(s) u} du \; dH(u+\tau)$$

$$= \frac{[m_{i_0}[h(s),\xi] - m_{i_00}(\xi)]}{h(s)} \frac{h(s) - h[\xi+\lambda-\lambda h(s)]}{\xi+\lambda-\lambda h(s) - s}.$$

The quantity $m_{i_0}[h(s),\xi]$ is obtained by replacing z by $h(s)$ in formula (1.5.6). Upon substitution and simplification, the preceding expression is given by

$$\frac{h^{i_0}(s) + m_{i_00}(\xi)\left\{\left(\frac{\lambda}{\xi+\lambda}\right)^N h^{N-1}(s) h[\xi+\lambda-\lambda h(s)] - 1\right\}}{\xi+\lambda-\lambda h(s) - s}. \tag{1.7.7}$$

The expressions in (1.7.5 - 7) are now added and further simplified to yield the final expression for the transform $W^*(\xi,s)$.

$$W^*(\xi,s) = \frac{h^{i_0}(s)}{\xi+\lambda-\lambda h(s)-s} \tag{1.7.8}$$

$$+ m_{i_00}(\xi)\left\{\frac{\left(\frac{\lambda}{\xi+\lambda}\right)^N h^N(s) - 1}{\xi+\lambda-\lambda h(s) - s} + \frac{1}{\xi+\lambda}\sum_{i=0}^{N-1}\left(\frac{\lambda}{\xi+\lambda}\right)^i\right.$$

$$\left. \cdot \left(\frac{\lambda}{s+\lambda}\right)^{N-i-1} h^i(s)\right\}.$$

We recall that $m_{i_00}(\xi)$ is given by formula (1.6.6).

As in the case treated in Section 1.6, the key renewal theorem and the bounded convergence theorem guarantee that the limit as $t \to \infty$ of each of the expressions (1.7.2-4) exists and is finite. With the formula (1.7.8) available, we may calculate the Laplace-Stieltjes transform $W^*(s)$ of the limit

$$\tilde{W}(x) = \lim_{t \to \infty} P\{W(t) \leqslant x \mid i_0\},$$

by evaluating the limit $\lim_{\xi \to 0+} \xi W^*(\xi, s)$. By noting, as in Section 1.6, that

$$\begin{aligned} \lim_{\xi \to 0+} \xi m_{i_0 0}(\xi) &= N^{-1}\lambda(1-\rho), \qquad \text{for } \rho < 1, \\ &= 0, \qquad \text{for } \rho \geqslant 1. \end{aligned}$$

we obtain the following result:

Theorem 1.7.1: When $\rho \geqslant 1$, $\tilde{W}(x) = 0$, for all $x \geqslant 0$. In the stable queue, the transform $W^*(s)$ is given by

$$W^*(s) = \frac{\lambda(1-\rho)}{N}\left\{\frac{1-h^N(s)}{s-\lambda+\lambda h(s)} + \frac{1}{\lambda}\sum_{i=0}^{N-1}\left(\frac{\lambda}{s+\lambda}\right)^{N-i-1} h^i(s)\right\}. \tag{1.7.9}$$

Remarks: Formula (1.7.9) may be rewritten in the following form, which leads to a convenient inversion procedure.

$$W^*(s) = (1-\rho)\frac{1}{N}\sum_{i=0}^{N-1}\left(\frac{\lambda}{s+\lambda}\right)^{N-i-1} h^i(s) \tag{1.7.10}$$

$$+ \rho\,\frac{1-h^N(s)}{N\alpha s}\left\{(1-\rho)\left[1-\rho\,\frac{1-h(s)}{\alpha s}\right]^{-1}\right\}.$$

We first observe that for $N = 1$,

$$W^*(s) = W_1^*(s) = (1 - \rho)\left[1 - \rho\,\frac{1 - h(s)}{\alpha s}\right]^{-1}. \qquad (1.7.11)$$

This is the classical *Pollaczek-Khinchin formula* for the transform of the virtual waiting time in the ordinary $M/G/1$ queue. It implies that

$$W_1^*(s) = \sum_{\nu=1}^{\infty} (1 - \rho)\,\rho^{\nu}\left[1 - \rho\,\frac{1 - h(s)}{\alpha s}\right]^{\nu},$$

so that

$$W_1^*(s) = \sum_{\nu=1}^{\infty} (1 - \rho)\,\rho^{\nu}\,H_1^{*(\nu)}(x), \quad \text{for } x \geqslant 0, \qquad (1.7.12)$$

where

$$H_1^*(x) = \int_0^x \alpha^{-1}[1 - H(u)]du, \quad \text{for } x \geqslant 0,$$

The probability distribution $H_1^*(\cdot)$ is sometimes called the *random modification* of $H(\cdot)$.

By routine calculations, we may now verify that (1.7.11) is equivalent to the integral equation

$$\tilde{W}_1(x) = 1 - \rho + \lambda\int_0^x [1 - H(x-u)]\,\tilde{W}_1(u)du, \quad \text{for } x \geqslant 0. \qquad (1.7.13)$$

This is a classical *Volterra integral equation of the second kind.* Its numerical solution may be performed by any one of several available library routines.

From formula (1.7.10) it is apparent that $W_N^*(s) = W^*(s)$, is a mixture with mixing parameters $1 - \rho$ and ρ of two probability distributions $K_N(\cdot)$ and $L_N(\cdot)$. The distribution $L_N(\cdot)$ is given by

$$L_N(x) = \int_0^x (N\alpha)^{-1}\,[1 - H^{(N)}(x-u)]\,\tilde{W}(u)du, \quad \text{for } x \geqslant 0,$$

and $K_N(\cdot)$ is the mixture with the Laplace-Stieltjes transform

$$\frac{1}{N}\sum_{i=0}^{N-1}\left(\frac{\lambda}{s+\lambda}\right)^{N-i-1} h^i(s).$$

For purposes of numerical computation, it is convenient to evaluate the probability distributions $K_N(\cdot)$ recursively by

$$K_1(x) = U(x),$$

$$K_{N+1}(x) = \frac{1}{N+1} E_N(\lambda;x) + \frac{N}{N+1}\int_0^x K_N(x-u)\,dH(u),$$

for $N \geqslant 1$, where $U(\cdot)$ and $E_N(\lambda;\cdot)$ are respectively the degenerate and the Erlang distributions (of order N).

The mean $\overline{W}_N$ of the waiting time distribution is readily computed by differentiation in (1.7.10). One obtains the simple formulas:

$$\overline{W}_1 = \frac{\lambda\alpha_2}{2(1-\rho)},$$

$$\overline{W}_N = \overline{W}_1 + \frac{N-1}{2\lambda}, \quad \text{for } N \geqslant 1.$$

The analytic manipulations leading to the transform $W^*(\xi,s)$ are usually more involved than in the present case and the resulting formula is often of limited practical interest. In cases where the limiting distribution $\tilde{W}(\cdot)$ is of primary interest, it is more expeditious to evaluate the limits as $t \to \infty$ of the expressions given in the formulas (1.7.2-4). The interchange of the infinite summations and the passage to the limit may routinely be justified by appealing to the dominated convergence theorem. The existence of the limits of the integrals in the individual terms is guaranteed by the key renewal theorem. The finiteness of the mean service time α is crucial to both steps. In Variant 4, neither the queue length at time t, nor the virtual waiting time at time t admit proper limiting distributions when α is infinite, yet the embedded Markov chain P has a non-trivial stationary probability vector when $\rho < 1$.

There is a different construction which leads directly to the analytic form of the stationary (or limiting) distribution of the virtual waiting time. It is applicable if and only if the embedded Markov renewal process is positive recurrent. There is then a unique and standard choice of the initial conditions of the Markov renewal process for which it becomes a stationary process, called the ***stationary version*** of the Markov renewal process. We shall not discuss this construction in any detail, as it leads to the same analytic results as the passage to the limit as $t \to \infty$. If we construct the virtual waiting time process with respect to the stationary version of the embedded Markov renewal process, we find that the process $\{W(t),\, t \geqslant 0\}$ is itself stationary. Substantially the same probabilistic arguments that led to the expressions (1.7.2-4), now lead to analytic expressions which agree with those obtained by the passage to the limit as $t \to \infty$. In that case, the distribution of $W(t)$ does not depend on t and is clearly identical to the limiting distribution.

The key to the derivation of the stationary distribution is that for the stationary version of the Markov renewal process, the expected number $M_k(t)$ of visits to the state k in $[0,t)$ is given by

$$M_k(t) = \frac{t}{\tilde{\mu}^*_{kk}} = E^{-1} x_k t, \quad \text{for } k \geqslant 0, \quad t \geqslant 0.$$

We stress that the initial state and the forward recurrence time at time 0 now need to be chosen in a specific manner in accordance with the construction of the stationary version. In queues with other than Poisson arrivals, it is necessary to choose also the initial conditions of the arrival process appropriately to obtain the stationary version of the entire queueing process. As the emphasis of this book is on analytic and algorithmic results, we shall refer the interested reader to sources where such matters are treated in great generality.

For the sake of completeness, we shall list the expressions obtained by letting $t \to \infty$ in (1.7.2-4). Some obvious simplifications have already been made. By evaluating the Laplace-Stieltjes transform of the sum of these three expressions and by using formula (1.6.8), one obtains the expression for $W^*(s)$, given in Theorem 1.7.1.

The expressions for the limits in (1.7.2-4) are respectively as follows:

$$x_0 \sum_{i=0}^{N-2} \int_0^x e^{-\lambda v} \frac{(\lambda v)^{N-i-2}}{(N-i-2)!} \lambda\, H^{(i)}(x-v)\, dv + x_0 H^{(N-1)}(x),$$

$$x_0 \sum_{i=0}^{\infty} \int_0^{\infty} \int_0^{\infty} e^{-\lambda v} \frac{(\lambda v)^i}{i!} \lambda H^{(N+i-1)}(x-\tau)\, dH(\tau+v)\, dv,$$

$$\sum_{i=0}^{\infty} \sum_{k=0}^{i} x_k \int_0^{\infty} \int_0^{x} e^{-\lambda v} \frac{(\lambda v)^{i-k}}{(i-k)!} \lambda H^{(i-1)}(x-\tau)\, dH(\tau+v)\, dv.$$

The *server backlog* $\hat{W}(t)$ is closely related to the virtual waiting time $W(t)$. It is defined as the length of time required to complete all the work present in the system at time t. The behavior of the server backlog process $\{\hat{W}(t), t \geqslant 0\}$ depends also on the queue discipline and will be briefly discussed here for the first-come, first-served mode of operation. With $N \geqslant 2$, the $M/G/1$ queue with the N-policy is possibly the simplest variant of the $M/G/1$ queue for which the random variables $W(t)$ and $\hat{W}(t)$ are different. We study the server backlog by imagining that the arrival of customers is cut off at time t and by considering how long it would take to clear all remaining work when this occurs.

In all known cases, the arguments that lead to the time dependent and limiting distributions of the server backlog are very similar to those for the virtual waiting time. To illustrate this, we note that the discussion of the server backlog differs from the one earlier in this section, only in the treatment of the event A, which corresponds to the situation where the server is idle at time t. If all further arrivals are cut off at time t, the server will start, at that time, to serve those customers who are present.

The term corresponding to the event A in formula (1.7.2) is given by

$$\sum_{i=0}^{N-1} \int_0^{t} dM_{i_0 0}(u) e^{-\lambda(t-u)} \frac{[\lambda(t-u)]^i}{i!} H^{(i)}(x).$$

The other two terms are identical to those for the virtual waiting time.

With the remaining calculations being entirely similar, we shall only record the expression for the Laplace-Stieltjes transform $\hat{W}^*(s)$ of the limiting distribution of $\hat{W}(t)$, which again exists if and only if the queue is stable. Written in the transparent form of (1.7.10), $\hat{W}^*(s)$ is given by

$$\hat{W}^*(s) = \frac{(1-\rho)}{N} \sum_{i=0}^{N-1} h^i(s) \tag{1.7.14}$$
$$+ \rho \frac{1-h^N(s)}{N\alpha s} \left\{ (1-\rho) \left[1 - \rho \frac{1-h(s)}{\alpha s} \right]^{-1} \right\}.$$

It is clear that for $N \geqslant 2$, the corresponding probability distribution differs from $\tilde{W}(x)$. For $N = 1$, and under the first-come, first-served discipline, the virtual waiting time and the server backlog measure the same quantity.

1.8. THE RELATION TO THE GALTON-WATSON PROCESS

The Markov renewal process $Q(\cdot)$ has itself an embedded Markov renewal process which has a close relationship to the classical Galton-Watson process. Let $J_0 = I_0$, be the state of the Markov renewal process at time $t = 0+$, and let $\tau'_0 = 0$. The pairs (J_n, τ'_n), $n \geqslant 1$, are recursively defined as follows. If $J_n > 0$, τ'_{n+1} is the additional time required to perform J_n transitions in the Markov renewal process $Q(\cdot)$. J_{n+1} is then given by the state of the Markov renewal process $Q(\cdot)$ at the end of these J_n transitions. If $J_n = 0$, τ'_{n+1} is the additional time required to perform one transition in the process $Q(\cdot)$ and J_{n+1} is the state of the process $Q(\cdot)$ after that transition. It is routinely verified that $\{J_n, \tau'_n), n \geqslant 0\}$ is a Markov renewal sequence.

By $R(x)$, we shall denote the transition probability matrix with elements

$$R_{ii'}(x) = P\{J_n = i', \tau'_n \leqslant x \mid J_{n-1} = i\} \quad \text{for } i \geqslant 0, i' \geqslant 0, x \geqslant 0.$$

The elements of the matrix $R(\cdot)$ may be expressed in terms of the sequences of mass-functions $\{A_\nu(\cdot)\}$ and $\{B_\nu(\cdot)\}$. We define the sequences $\{A_\nu^{(r)}(\cdot)\}$ of mass-functions for $r \geqslant 0$, by the convolution formula

$$A_\nu^{(0)}(x) = \delta_{0\nu} U(x),$$

$$A_{\nu}^{(r+1)}(x) = \sum_{k=0}^{\nu} A_k(\cdot) * A_{\nu-k}^{(r)}(x), \quad \text{for } \nu \geq 0.$$

The matrix $R(x)$ is then given by

$$R(x) = \begin{vmatrix} B_0(x) & B_1(x) & B_2(x) & B_3(x) & \cdots \\ A_0^{(1)}(x) & A_1^{(1)}(x) & A_2^{(1)}(x) & A_3^{(1)}(x) & \cdots \\ A_0^{(2)}(x) & A_1^{(2)}(x) & A_2^{(2)}(x) & A_3^{(2)}(x) & \cdots \\ A_0^{(3)}(x) & A_1^{(3)}(x) & A_2^{(3)}(x) & A_3^{(3)}(x) & \cdots \\ A_0^{(4)}(x) & A_1^{(4)}(x) & A_2^{(4)}(x) & A_3^{(4)}(x) & \cdots \\ \cdot & \cdot & \cdot & \cdot & \end{vmatrix} \qquad (1.8.1)$$

The Markov renewal process $R(\cdot)$ arises naturally in the $M/G/1$ queue. If there are $i_0 \geq 1$ customers in the queue at time $t = 0$, we consider the queue length J_1 at the time τ'_1 when all i_0 customers complete service (under the first-come, first service discipline). J_1 is then simply the number of new customers who arrive during the interval $(0, \tau'_1)$. If $J_1 > 0$, we consider the epoch $\tau'_1 + \tau'_2$ when these J_1 customers complete service and J_2 is the queue length at that time. The pairs (J_n, τ'_n) are recursively defined in this manner. If $J_n = 0$, we consider the queue at the end of the first service following the idle period and define J_{n+1} as the queue length at that time.

By analogy with the Galton-Watson process, we refer to the customers served during the successive intervals $\tau'_1, \tau'_2, \cdots$ as the first, second, $\cdots$ *generations* of customers. We shall also use this graphic terminology in our discussion of the Markov renewal process $R(\cdot)$.

By repeating verbatim the corresponding argument for the Galton-Watson process, we may show that the joint transform $E[z^V \exp(-sT)]$ of the first passage time T from state 1 to state 0 and of the total number of transitions in the process $Q(\cdot)$ (services) during that first passage time, is given by the function $\tilde{G}(z,s)$, discussed in Section 1.2. The corresponding transform for the first passage from state i to state 0 is given by $\tilde{G}^i(z,s)$, for $i \geq 1$, and the corresponding transform for the duration and the number of services of the recurrence time of the state 0 is given by $\tilde{K}(z,s)$, the function discussed in Section 1.3.

It follows that the recurrence criteria of the Markov renewal process $R(\cdot)$ and of its embedded Markov chain with transition probability matrix $R = R(\infty)$ are exactly the same as for the Markov renewal process $Q(\cdot)$ and for its embedded Markov chain.

Theorem 1.8.1: When $\rho < 1$, the generating function $U(z) = \sum_{i=0}^{\infty} u_i z^i$, of the stationary probabilities $\{u_i\}$ of the stochastic matrix R, is given by

$$U(z) = u_0 \{ 1 + \sum_{\nu=0}^{\infty} \{B [A_{(\nu)}(z)] - B [A_{(\nu)}(0)]\} \}, \tag{1.8.2}$$

where the generating functions $A_{(\nu)}(z)$, $\nu \geqslant 0$, are recursively defined by

$$A_{(0)}(z) = z, \qquad A_{(1)}(z) = A(z) = \sum_{i=0}^{\infty} A_i z^i,$$

$$A_{(\nu)}(z) = A [A_{(\nu-1)}(z)] = A_{(\nu-1)} [A(z)], \quad \text{for } \nu \geqslant 2,$$

and $B(z) = \sum_{i=0}^{\infty} B_i z^i$.

The quantity u_0 is given by

$$u_0 = \{ 1 + \sum_{\nu=0}^{\infty} [1 - B [A_{(\nu)}(0)]] \}^{-1}. \tag{1.8.3}$$

When $0 < A_0 + A_1 < 1$, the inequality

$$u_0 > x_0, \tag{1.8.4}$$

holds, where x_0 is given by formula (1.4.2).

Proof: Upon taking the generating function, using the equations

$$u_i = u_0 B_i + \sum_{\nu=1}^{\infty} u_\nu A_i^{(\nu)}, \quad \text{for } i \geqslant 0,$$

we obtain

$$U(z) = U[A(z)] - u_0[1 - B(z)], \tag{1.8.5}$$

for $0 \leqslant z \leqslant 1$. Successively substituting z by $A_{(\nu)}(z)$, $\nu \geqslant 1$, and adding the resulting equations, we see that

$$U(z) = U[A(z)] - u_0 \sum_{\nu=0}^{n-1} [1 - B[A_{(\nu)}(z)]], \quad \text{for } n \geqslant 1.$$

If $n \to \infty$, we readily see by considering the graph of $A(z)$ that for every z satisfying $0 \leqslant z < 1$, the sequence $A_{(n)}(z)$ converges to the smallest solution η of the equation $z = A(z)$ in [0,1].

If $\eta < 1$, equation (1.8.5) yields that $u_0 = 0$. If $\eta = 1$, or equivalently $\rho \leqslant 1$, we obtain, as $n \to \infty$, that

$$U(z) = 1 - u_0 \sum_{\nu=0}^{\infty} [1 - B[A_{(\nu)}(z)]]. \tag{1.8.6}$$

The quantity

$$\sum_{\nu=0}^{\infty} [1 - B[A_{(\nu)}(0)]],$$

is the expected number of generations until extinction in a Galton-Watson process with $\{A_\nu\}$ as the density of the number of offspring per individual and $\{B_\nu\}$ as the density of the initial population size. It is well-known that this quantity is finite if $\rho < 1$ and infinite if $\rho = 1$.

Setting $z = 0$, in (1.8.6) we see that u_0 is positive if and only if $\rho < 1$ and is as given by formula (1.8.3).

For $\rho < 1$ and under the commonly satisfied assumption $0 < A_0 + A_1 < 1$, the functions $B[A_{(\nu)}(z)]$, $\nu \geqslant 1$, are strictly convex on [0, 1] and their graphs therefore lie entirely above the line tangent at the point (1, 1). This readily yields the inequality

$$B\ [A_{(\nu)}(0)] > 1 - \rho^{\nu} \sum_{k=1}^{\infty} k\ B_k, \quad \text{for } \nu \geqslant 1.$$

The (weak) inequality for $\nu = 0$, is obvious.

By adding the inequalities (appropriately rewritten) we obtain

$$\sum_{\nu=0}^{\infty} [1 - B\ [A_{(\nu)}(0)]\ ] < (1 - \rho)^{-1} \sum_{k=1}^{\infty} k\ B_k,$$

which is tantamount to $u_0 > x_0$. Upon substituting u_0 in (1.8.6) and forming the difference of the two convergent series, we obtain the expression for $U(z)$, given in (1.8.2). The preceding argument further shows, that the series for the mean number of generations, while not summable in closed form, nevertheless converges at least as fast as a geometric series of ratio ρ. This is an important property for numerical computation. •

By routine differentiation in (1.8.2), we see that, when $\rho < 1$, the mean $U'(1-)$ of the density $\{u_i\}$ is given by

$$U'(1-) = u_0(1 - \rho)^{-1} \sum_{k=1}^{\infty} k\ B_k, \tag{1.8.7}$$

It is interesting to note that the finiteness of the mean $U'(1-)$ does not involve conditions on the second moments of the densities $\{A_k\}$ and $\{B_k\}$. This should be compared with the requirements for the existence of the mean $X'(1-)$ of $\{x_i\}$ as given in formula (1.4.5).

The fundamental mean E^* of the Markov renewal process $R(\cdot)$ is given by

$$E^* = u_0\beta + \sum_{i=1}^{\infty} iu_i\,\alpha = u_0(1 - \rho)^{-1}\,[\beta(1 - \rho) + \sum_{k=1}^{\infty} k\ B_k]. \tag{1.8.8}$$

Upon comparison with formula (1.5.8) and recalling the expression for x_0 in (1.4.2), we obtain that

$$\frac{E^*}{u_0} = \frac{E}{x_0}.$$

That equality was of course to be anticipated. Both sides of the equality are equal to the mean recurrence time of the state 0 in the Markov renewal process $Q(\cdot)$ or $R(\cdot)$. From the definition of the process $R(\cdot)$, it is clear that both (mean) recurrence times are the same.

The function $B[A_{(\nu)}(z)]$, $\nu \geqslant 0$, is the probability generating function of the number in the ν-th generation in a Galton-Watson process with initial population density $\{B_i\}$ and where the size of the number of offspring per individual has density $\{A_i\}$. If we write

$$B[A_{(\nu)}(z)] = \sum_{r=0}^{\infty} \psi_r(\nu) z^r, \quad \text{for } \nu \geqslant 0,$$

then we obtain from formula (1.8.2) that

$$u_i = u_0 \sum_{\nu=0}^{\infty} \psi_i(\nu), \quad \text{for } i \geqslant 1. \tag{1.8.9}$$

The term u_0 may easily be computed by use of (1.8.3), since the series

$$\sum_{\nu=0}^{\infty} [1 - B[A_{(\nu)}(0)]],$$

converges faster than a geometric series of ratio $\rho < 1$. The computation of the u_i, $i \geqslant 1$, is more involved. The sequences $\{\psi_r(\nu), r \geqslant 0\}$ are recursively computed by use of the formulas

$$\psi_r(0) = B_r, \qquad \text{for } r \geqslant 0, \tag{1.8.10}$$

$$\psi_r(\nu) = \sum_{k=0}^{\infty} \psi_k(\nu-1) A_r^{(k)}, \quad \text{for } r \geqslant 0, \ \nu \geqslant 1,$$

and by adding term by term over ν. We also accumulate the sum

$$S_N = \sum_{\nu=0}^{N} \sum_{i=1}^{\infty} \psi_i(\nu),$$

which converges to $u_0^{-1} - 1 = \sum_{\nu=0}^{\infty} [1 - B[A_{(\nu)}(0)]]$, as $N \to \infty$. We stop at the index N for which the computed S_N is sufficiently close to the limit value, which was evaluated earlier.

The implementation of the recursive computation in (1.8.10) involves a large number of arithmetic operations and needs to be programmed with care. There are several alternate ways of coding the algorithm. These typically involve a trade-off of storage against execution time. The design of computer algorithms for the evaluation of the probability density $\{u_i\}$ is a fine exercise for a novice programmer.

NOTES, REFERENCES AND COMMENTS ON CHAPTER 1

Section 1.1

The classical $M/G/1$ queueing model has a long history and an extensive literature. It has played a seminal role in the development of many modern ideas of stochastic modeling and continues to be of interest to a wide variety of applications. Its significance as a source of theorems, generalizations and applications is comparable to that of the much older "Gambler's Ruin Problem" and the subject of random walks that developed from it.

The $M/G/1$ queueing model is at the origin of wide ranging theoretical developments in the study of embedded Markov chains and Markov renewal processes, of combinatorial methods in probability [T-019], of Markov renewal decision theory [H-051], of branching processes [N-019], of inventory theory and of algorithmic methodology for stochastic models. It still serves as the primary test and as an illustrative elementary example for many theorems and methods of greater generality.

The generality of the service time distribution makes it an appealing model for many applications with a single service mechanism. The assumption of Poisson arrivals is often less tenuous than that of exponential service times; its usefulness in applications is therefore greater than that of the elementary $GI/M/1$ model. With a variety of tractable service disciplines, it is widely used in engineering studies of message buffers in communications and computer modeling, of waiting lines in manufacturing and of queues at traffic lights or access ramps to mention only a few. With special features, such as service interruptions, it frequently serves as an approximate model to components of queueing networks, which are otherwise so complex as to defy analysis.

The $M/G/1$ queue is a standard subject in books on stochastic processes [B-050,C-026,C-027,K-004,M-026,P-040,P-045], probability [H-003,H-082,H-083,R-047], operations research [H-051,R-046] and queueing theory [B-032,B-044,C-037,C-060,C-066,G-029,G-047,K-072,R-032,S-010,T-006,T-047]. As it lies at the core of so many useful ideas and methods, there are a number of alternative presentations of the theory of the $M/G/1$ queue. While the results in Chapter 1 are classical, we know of no book that treats the $M/G/1$ queue by systematically relating its features to the embedded Markov renewal process and by exploiting the structure of that process as fully as is done here. As that approach is essential to the remainder of this book, we have devoted the first chapter to our preferred presentation of the theory of the $M/G/1$ queue. For the derivation of some results, other approaches are occasionally more incisive or elegant, but for our purpose, the present exposition is most effective. The informed reader or diligent student of any subject is clearly familiar with many ways of viewing a problem area and an eclectic knowledge is essential to the research worker. As a discussion of specialized methods and problems, this book does not treat all alternative approaches to the classical $M/G/1$ queue. It shares that limitation with all other works extant.

Results for Variant 1 are almost trivial extensions of those for the ordinary $M/G/1$ queue. For modifications of the Poisson input process with other than simple group arrivals, the theory is no longer elementary and usually requires the general methods of this book. Such models are treated in Chapter 5.

Variant 2, where the first service of a busy period includes a "warm-up" time, has been discussed by many authors. It is one of few variations of the $GI/G/1$ queue to remain analytically somewhat tractable. Only very simple boundary behavior can be handled by the invariance arguments needed in the usual analysis of the $GI/G/1$ queue and of the related random walk. See Bhat [B-040], Yeo [Y-005], and for generalizations, Rossberg and Siegel [R-054-R-057] and Siegel [S-059-S-061].

Variant 3, the $M/G/1$ queue with the $N-$policy, is but one of many variants in which the queue length is to exceed a threshold before service is initiated. See Shanthikumar [S-042], Teghem Jr. [T-048]. A related variant is the $T-$policy. In it, the server remains idle for a period of possibly random length T and initiates service if customers are present at the end of that period. Several idle periods may succeed each other, as long as no customers request service. Under the names of "server vacations" or "enforced idle time", many variations of the $T-$policy have been described. See e.g. Balachandran and Tijms [B-016], Heyman [H-050], Levy and Yechiali [L-025], Lucantoni, Meier-Hellstern and Neuts [L-053], Schellhaas [S-014], Scholl and Kleinrock [S-019], and in particular the survey article by Doshi [D-055]. Vacation models are used in communications engineering as approximate descriptions of a processor which performs other tasks whenever a primary queue becomes empty.

Problem 1.1.1: The derivation of the precise analytic expressions for the mass-functions $A_k(\cdot)$ and $B_k(\cdot)$ often requires some work. When they are obtained as transforms, there is usually some challenge in translating these to procedures useful in computations. As an example, consider the $M/G/1$ queue with multiple vacations in which vacations and service times have respectively the probability distributions $V(\cdot)$ and $H(\cdot)$. An application of the law of total probability will give you an expression for $B_k(x)$ which involves several summations and integrals of convolution type, so that it is natural to take appropriate transforms. Show that the generating function of the Laplace-Stieltjes transforms of the functions $A_k(\cdot)$ and $B_k(\cdot)$ are given respectively by $h(s-\lambda+\lambda z)$ and

$$\frac{v(s+\lambda-\lambda z)-v(s+\lambda)}{z[1-v(s+\lambda)]}\,h(s-\lambda+\lambda z),$$

where $v(\cdot)$ and $h(\cdot)$ are the Laplace-Stieltjes transforms of V and H. Discuss how the elements A_k and B_k of the transition probability matrix P of the embedded Markov chain may be computed efficiently. In particular, when the probability distributions V and H are *discrete* or distributions *of phase type*, (see Chapter 5), these quantities may be evaluated *without numerical integrations*. Write the necessary code to do so and build as many internal accuracy checks into your program as possible. •

The $M/G/1$ queue with limited admission (Variant 4) is analytically more complex than the other variants. The limitation described in the text is known as the $N-$rule. One obtains a related variant by allowing admissions only during the first T units of time of each service ($T-$rule). Both are discussed in detail in Neuts [N-058]. With $N = 1$, the $N-$rule implies that, at any time, not more than one customer can be waiting. The model is then equivalent to the $M/G/1/1$ queue which is clearly always stable.

Section 1.2

The classical notion of the *busy period* of a single server queue always requires consideration of the boundary states and is therefore more involved than that of the *fundamental period.* For the classical $M/G/1$ queue, these notions agree, but for all more complex models, the distinction is essential. As we have illustrated, consideration of the fundamental period permits a unified treatment of models which differ only in their boundary behavior.

A thorough treatment of the busy period for the ordinary $M/G/1$ queue is given in Takács [T-006], a book which provided most of our earliest methodological insights. For the ordinary $M/G/1$ queue, it is possible to obtain explicit inversions for several transforms related to the busy period. These are based on Lagrange's expansion or may be obtained directly by various elegant combinatorial arguments, using the ballot theorem. These explicit formulas were obtained and extensively discussed by Takács [T-002,T-006,T-007,T-010,T-019,T-034]. They do not have tractable generalizations to more complex models and occasionally present difficulties in numerical implementation. Combinatorial approaches to the busy period of $M/G/1$ have interesting applications in statistics and counting problems. We refer to Takács' book [T-019] for many further developments and as a source of largely unexplored algorithmic problems. The reader who is acquainted with these explicit formulas may enjoy the following problem whose simple answer requires a clever derivation.

Problem 1.2.1: Consider a homogeneous Poisson process of rate λ and let $N(t)$ denote the number of events in $[0,t)$ with $N(0) = 0$. Determine the probability

$$P\{N(k) = k, \quad \text{for some } k > 0\}. \qquad \bullet$$

The nonlinear Volterra integral equation for the distribution of the busy period of the $M/G/1$ queue and its matrix generalizations deserve greater attention from the viewpoint of numerical analysis. For the scalar case, solution by discretization with or without use of the Fast Fourier Transform algorithm or by approximations of the service time distribution, works quite well over a wide range of parameter values. For discussions of these various approaches, we refer to Ackroyd and Kanyangarara [A-006], Heimann and Neuts [H-037] and Neuts [N-030].

Takács' transform equation for the duration of the busy period (in real time or in number of services) and a number of its variants are well-suited for the computation of *moments* by means of Fáa di Bruno's formula for the n-th derivative of the composition of two functions. For higher moments, the implementation of this formula poses challenging algorithmic problems. See Klimko and Neuts [K-071].

For high traffic intensity, the behavior of the density of the number served during a busy period or (if it exists) of the duration of the busy period of the stable $M/G/1$ queue is often astonishing.

Even for a traffic intensity of 0.95, few busy periods last for more than ten services, but those that do may extend over several hundreds. This phenomenon is qualitatively similar to that noted earlier for coin tossing and discussed in detail in Feller [F-010].

The independence and common distribution of the first passage times $T(i+h, i+h-1)$ and $V(i+h, i+h-1)$, $h = 1, \ldots, r$, is essential to the discussion of the fundamental and busy periods. The matrix case, discussed in the next chapters, involves a corresponding *conditional independence* property. Variants, even of the elementary $M/M/1$ queue, in which these properties are lost, are usually complicated or intractable. For an example occurring in communications engineering, in which Poisson arrivals to a queue are allowed to enter the system only at equally spaced time points (a clock-pulse operated gate), we refer to Jung [J-031,J-032].

Further references dealing with aspects of the busy period in the $M/G/1$ queue are Boxma [B-076,B-080], Cohen [C-032,C-038], Craven and Shanbhag [C-082], Daley [D-010], Daley and Jacobs [D-011], De Meyer and Teugels [D-033], Enns [E-013,E-016], Erlander [E-017], Haight [H-002], Heyman [H-049], Morimura [M-060], Neuts [N-010,N-030], Ott [O-016], Rosenlund [R-039,R-041,R-043,R-045], Serfozo [S-032], Shaw [S-055], Takács [T-002,T-006,T-007,T-010,T-019,T-034], and Teugels and Neuts [T-052].

The following problems deal with variants of the $M/G/1$ queue in which is server is subject to failure. Related models are treated in Cao [C-003], Cao and Cheng [C-002] and Keilson [K-013].

Problem 1.2.2: Consider the variant of the $M/G/1$ queue in which the server may fail, but only when busy. The time-to-failure, measured only in busy time, has an exponential distribution with mean d^{-1}. When a failure occurs, the customer in service is lost and the server enters a repair period with distribution $C(\cdot)$ of finite mean. At the end of the repair period, a new customer is taken into service if one is present. Show that this variant is equivalent to an $M/G/1$ queue with a modified service time distribution. Determine the equilibrium condition and the expected values of the duration of the busy period and of the total repair time during a busy period. •

Problem 1.2.3: Modify the model in Problem 1.2.2 so that the server may also fail when idle. Discuss the distribution of the busy period, which is measured from the time the idle server begins repair or service until the first time thereafter that he is both idle and operational. •

Problem 1.2.4: Consider the variant of the model in Problem 1.2.2 where a service, interrupted by failures of the server, is *resumed* whenever the server is repaired. Successive repair times are independent and identically distributed and completion of a repair returns the server to a "good-as-new" condition. Treat the total time between the beginning and the end of a customer's service as an augmented service time. Discuss the same quantities as in Problem 1.2.2. •

The "augmented service time" for the model in Problem 1.2.4 plays an important role in the study of $M/G/1$ queues with preemptive-resume priorities and is extensively used in the book by Jaiswal [J-012].

Problem 1.2.5: For the standard $M/G/1$ queue, let $\Psi(x;y,k)$ be the conditional probability that a busy period, which starts with an *amount of work* x at time $t = 0$, terminates no later than time y, $y \geqslant x$, and involves the service of k, $k \geqslant 0$, additional customers. Show by a direct probabilistic argument that

$$\sum_{k=0}^{\infty} \int_{x}^{\infty} e^{-sy} \, d\Psi(x;y,k) \, z^k = \exp\{-[s + \lambda - \lambda \tilde{G}(z,s)] \, x\},$$

and use this formula to derive Takács' equation for $\tilde{G}(z,s)$. •

Problem 1.2.6: Derive the joint transform of the residual busy period and the virtual waiting time at time t for the $M/G/1$ queue [N-017]. •

Section 1.3

In this section and throughout the remainder of this book, we often use classical properties of Markov renewal processes. We recommend the survey paper by Çinlar [C-025] as a readily accessible and comprehensive source. We shall also refer to the matrix formulas for moments in Hunter [H-077,H-079]. Other standard, informative references on Markov renewal theory are Pyke [P-053,P-054] and Nollan [N-078].

The next few problems serve to illustrate the widely varying algorithmic difficulties in solving design problems of superficially similar nature.

Problem 1.3.1: The N-policy may be used to avoid the start-and-stop processing of work associated with short busy periods. How would you determine the value of N to assure that the expected number of services during a busy period exceeds a given value (much) larger that $1/(1-\rho)$? Similarly for the mean duration (in real time) of the busy period. Next, suppose we wish to choose the smallest value of N such that, with a given probability θ, at least M services are dispensed per busy period. What algorithmic steps are needed to determine N? This question clearly requires computation of the probabilities G_k. The corresponding question in which we wish to guarantee that, once started, the server will remain busy with probability θ for a length at least T, is still more demanding; it requires the numerical solution of the integral equation (1.2.9)! •

Problem 1.3.2: The nonlinear difference equation (1.2.8) may be solved by successive substitutions, preferably starting with an initial solution $\{G_k(0)\}$, such as $\{A_k\}$, which is stochastically smaller than $\{G_k\}$. Evaluate the successive approximations by adapting Horner's algorithm to compute the necessary convolution polynomials. Trim off a small probabilty at each iteration to avoid carrying unnecessarily long sequences. Show that the successive iterates converge monotonically to the desired sequence. •

Problem 1.3.3: For the $M/G/1$ queue with $\rho = 1$, the sequence $\{G_k\}$ is a proper probability density with infinite mean. Because of the high random variability of the number of services in a busy period, the higher percentiles of the density $\{G_k\}$ are large. Write a subroutine to evaluate the smallest integer K for which $G_1 + \cdots + G_K \geqslant 0.99$. Next, study the effect of the $N-$rule of Variant 4, by determining the largest value of the parameter N for which the 99-th percentile of the corresponding density $\{G_k\}$ is reduced to $0.8\,K$. Also print the values of the traffic intensity of the $M/G/1$ queue with the $N-$rule and the fraction of customers who are not admitted. Implement your algorithm for various service time distributions of phase type (see Section 5.1) and for various discrete service time distributions, as for these the set-up computations are easy. By carefully planning ahead, one can obtain a very fast and efficient computer code. The corresponding problem for the duration of the busy period in real time is much more laborious. •

Section 1.4

The problem of loss of significance in the recursive computation of the stationary queue length density is often underestimated as it is not apparent in many sample calculations for the $M/G/1$ queue performed on computers with a large word length. It is, however, easily demonstrated on minicomputers or, by appropriate choices of the entries of the matrix Q, also on larger computers. A number of algorithms proposed in the literature, apparently without extensive implementation, suffer from loss of significance for substantially the same reason. The easy cure, which leads to the useful formula (1.4.4), was suggested to me by Paul J. Burke. Most recently, V. Ramaswami [R-016] has obtained a suitable analogue of Burke's recursion formula, which holds out great promise for improved efficiency in the numerical computations for the matrix cases.

Burke's suggestion is further useful in the following manner. If we consider the stationary density of the number of customers, who are waiting immediately after the beginning of a service, we see that it satisfies the discrete version of the Pollaczek-Khinchin equation for the waiting time distribution (prove this !). The terms of that density may be evaluated by a stable recurrence. For several variants of the $M/G/1$ queue, other queue length densities may then be computed by additional elementary operations, such as shifts and convolutions. For detailed examples, see Ali and Neuts [A-014], and Neuts [N-058,N-062]. Burke's suggestion is seen to be a particular example of this technique.

The particular structure of the embedded Markov chain of the $M/G/1$ queue also arises in continuous parameter Markov processes of applied interest. The translation of results for the discrete time to the continuous parameter case is elementary and will not be treated in detail in this book. It is discussed for block-partitioned Markov chains of $GI/M/1$ type in Section 1.7 of Neuts [N-042]. The following problem provides both an example and a nice, yet elementary algorithmic project.

Problem 1.4.1: Assume that in the $M/M/c$ queue, customers arrive in groups. The sizes of the successive groups are independent, identically distributed. Study this model as a continuous parameter Markov process. Notice the similarity in structure between its infinitesimal generator and the transition probability matrix of $M/G/1$. Determine the necessary and sufficient condition for the queue to be stable. Examine the steady-state equations and construct an efficient algorithm for their numerical computation.

For the stable $M/M/c$ queue with group arrivals, examine the queue length and waiting time densities in steady-state at various embedded point processes. Specifically, find the stationary queue length density at departures, before arrivals and after arrivals. Find the stationary distributions of the virtual waiting time, of the total amount of work in the system and of the r-th customer in a group arriving together. Show how the analytic results lead to simple algorithms for the numerical computation of these quantities. Write and test a program to carry out the required computations. (Although the methods required to solve this problem are classical, we know of no reference where the results, to be derived here, may be found in full detail and with an algorithmic emphasis. The problem is suitable for an undergraduate research project.) •

In studying variants of the $M/G/1$ queue, the staircase pattern of the elements (upper Hessenberg form) of the matrix in (1.1.3) and the spatial homogeneity of the transition probabilities, reflected in the fact that the same elements occur in the rows from the second onward, are both essential to the analysis presented here. If the upper Hessenberg form is preserved, but the successive rows are different, we have a Markov renewal process typical of the $M/G/1$ queue with state dependent service times. The literature on these is extensive, but there are few analytically tractable cases.

If the service time distribution depends on the queue length i only for $i < N+1$, we obtain a variant of the $M/G/1$ queue whose analysis is straightforward. It was discussed in Suzuki [S-079] and in Harris [H-011,H-013,H-014,H-015]. For algorithmic analyses, see Schellhaas [S-016,S-017]. The $M/G/1$ queue where the first customer of a busy period has a different service time distribution, is a simple case of state dependence. Mine, Ohno and Koizumi [M-042] discuss the more complicated case where the service time distributions of the successive customers during each busy period may depend on the rank number of that service in the busy period. If only the first m customers of each busy period can have different service time distributions, one obtains an interesting model which may be analyzed by the methods of this book. It is the subject of Problem 2.3.8 and was treated in greater generality in Ali and Neuts [A-016].

A case of state dependence, which is highly tractable, is treated in Aleksandrov [A-011], Falin [F-005], Keilson, Cozzolino and Young [K-027] and Neuts and Ramalhoto [N-056]. Customers arrive into a pool according to a homogeneous Poisson process of rate λ and issue a sequence of calls for access to a single server. The sequences of calls of unserved customers are independent Poisson streams of a rate σ. When the server becomes free, the first call issued thereafter determines which customer is taken into service. Service times of successive customers are independent, identically distributed and have a finite mean. In [A-011], [F-005] and [K-027], a customer who arrives when the pool is empty and the server free, enters service immediately; in [N-056], that customer has to issue requests for service as any other customer. The treatment in Falin [F-005] is particularly thorough and detailed.

Problem 1.4.2: Consider the Markov renewal sequences, embedded at the departure epochs, for the models we have just described and set up their transition probability matrices $Q(\cdot)$. Before reading the references, carry out the analysis of the stationary queue length distributions as far as possible. •

In text books, the results for the $M/G/1$ queue and its elementary variants are often illustrated by providing more detailed formulas for such particular service time distributions as the Erlang or hyperexponential distributions. See, for example, Gross and Harris [G-047] or Kleinrock [K-070]. These distributions are, however, particular cases of probability distributions of phase type (PH-distributions) and for these, the stationary queue length and waiting time distributions are given by explicit matrix formulas in Neuts [N-042,N-047]. The major simplifications in the general equations for queues of $M/G/1$ type, when the service time distribution is of phase type, are discussed in Chapter 5.

Section 1.5

The Markov renewal matrix of $Q(\cdot)$ provides the key to the study of the time dependent behavior of the $M/G/1$ queue. Time dependent distributions of queue length and waiting times may be related by rather simple convolution formulas to the elements of the matrix $M(t)$, which plays a role similar to that of the Green's function of analysis, Keilson [K-010,K-021,K-023]. The computation of $M(t)$, and particularly of its counterpart for the matrix case, is a monumental task. In addition, the performance of any algorithm for the transient behavior of the queue depends strongly on the particular initial conditions. Studies dealing with numerical procedures for the transient behavior of non-Markovian queues have addressed a limited range of problems and parameter values. It would be interesting to clarify precisely which aspects of the transient behavior are of practical utility, so that further investigations could proceed in a more purposeful manner. One of the most promising tools to that end appears to be the matrix Laguerre transform of Sumita [S-078].

The idea of systematically relating the features of a probability model to those of a mathematically simpler embedded process finds many uses in applied probability. It is the basis for many derivations in this book and has been discussed in general in the articles of

Jankiewicz [J-016-J-019] and Schäl [S-008,S-009].

Section 1.6

The relationships between the stationary queue length (and waiting time) distributions at various embedded point processes and at arbitrary time have been the subject of extensive investigations. A thorough and general discussion of this subject may be found in Franken, König, Arndt and Schmidt [F-033]. Such relationships are algorithmically useful whenever it is possible to compute one stationary distribution to which all the others are (tractably) related. For the models in this book, this is the stationary density of the embedded Markov chain of the Markov renewal process of $M/G/1$ type. In deriving related distributions, we shall not strive for the level of generality at which such relations are established in [F-033]. Here, in all cases it suffices to invoke the key renewal theorem after relating the distributions of interest to the matrix renewal function of the embedded Markov renewal process. This argument which recurs throughout this book is not repeated in detail every time.

Problem 1.6.1: The stationary probability density of the queue length at an arbitrary time t is obtained in Section 1.6. For the standard $M/G/1$ queue, use a similar argument based on the key renewal theorem to obtain the stationary probabilities that at time t, there are $i \geqslant 1$ customers in the system and the *remaining service time* of the customer is service is at most x. See Takács [T-006]. Find the Laplace-Stieltjes transform of the joint stationary conditional distribution of the elapsed and remaining service times of the customer in service at time t, given that the queue is not empty. Show that both marginal distributions are the same. Do there exist service time distributions for which the stationary elapsed and remaining service times are independent? •

The next problem combines server vacations with a limited state dependence.

Problem 1.6.2: Consider the $M/G/1$ queue in which the server rests for a period of random duration after each busy period lasting for N or more services. Rest periods are independent, identically distributed with finite mean duration. Study the Markov renewal process embedded at departure epochs and obtain the stationary density of its embedded Markov chain. Express stationary queue length densities at other embedded epochs to it in an analytically transparent manner. Consider only the case of single vacations, i.e. if at the end of a vacation, the queue is still empty, the server awaits the arrival of customers to initiate service.

Hint: Imagine a switch which is turned on as soon as, following a departure, it is clear that the current busy period will be followed by a vacation. The switch is turned off at the beginning of vacations. The empty state (following departures) is to be split into states 0 and $0'$ depending on whether the switch was on or off. For the queue lengths $1, \ldots, N-2$ (following departures) we need to keep track of the number of services already completed during the current busy period if the switch is still off. This is done by considering states (i,j) with $i = 1, \ldots, N-2$ and $j = 1, \ldots, N-i-1$. We also have states i' with $i' > 0$ to indicate that the switch is on and i customers are in the queue. Note that the structure of the embedded Markov chain differs from the elementary case only in the more complicated boundary states and that its transition probability matrix is expressed in terms of two discrete densities only. •

Problem 1.6.3: For the $M/G/1$ queue, derive the stationary density of the number of customers present at the *start of a service*. Derive the joint Laplace-Stieltjes transform of the durations of two consecutive time intervals between starts of services. Show that, for the $M/M/1$ queue, these intervals do not have exponential distributions and are *dependent*. •

The point of Problem 1.6.3 is to draw attention to the care needed in making statements about point processes associated, even with the simplest queues. A classical result (Burke's theorem) states that the departure process of the stationary $M/M/1$ queue is a Poisson process. In considering starts of services, we make only a minor modification in that the ends of busy periods are replaced by the starts of the next busy periods. Yet, this entirely destroys the simple nature of the output process and even changes the marginal distribution of single intervals. Detailed discussions of this matter are found in Falin [F-006] and Hunter [H-086].

Section 1.7

By noting that the virtual waiting time $W(t)$ is a Markov process with parameter t, the conditional distribution of $W(t)$ may be shown to satisfy an integro-differential equation. This was observed by Takács and is discussed in [T-004,T-007]. A number of extensions of Takács' integro-differential equation to other queueing and dam models have been proposed in the literature. An extension of that equation also plays a basic role in the theory of insurance risk. See, for example, Keilson and Kooharian [K-008], Keilson [K-017-K-019], Prabhu [P-039-P-041,P-045] and Takács [T-019]. The rigorous discussion of these equations is technically quite difficult. In many cases in queueing theory, the distributions of interest may be obtained directly by considering suitable first passage times, as was done in the text for the $M/G/1$ queue with N-policy. This avoids the difficulties of integro-differential equations and yields the stationary waiting time distributions by a direct renewal argument.

The Pollaczek-Khinchin integral equation for the stationary waiting time distribution in the $M/G/1$ queue is of fundamental importance. For most known variants of the $M/G/1$ queue the waiting time distribution satisfies a similar linear Volterra integral equation, with a slightly different kernel, but most commonly only with a different inhomogeneous term. This unifying feature, which is not readily evident from transform solutions, is discussed in detail in Neuts [N-062]. Cases with only a different inhomogeneous term lead to waiting time distributions which admit a factorization into the convolution product of the stationary waiting time distribution for $M/G/1$ and one or more probability distributions with useful interpretations. Such factorizations are sometimes algorithmically useful and provide additional theoretical insight. See Ali and Neuts [A-014], Doshi [D-055], Fuhrmann [F-036], Fuhrmann and Cooper [F-037], Lucantoni, Meier-Hellstern and Neuts [L-053], Neuts [N-062], Neuts and Ramalhoto [N-056], and Ott [O-019]. For the $M/G/1$ queue with N-policy, the stationary waiting time distribution is a simple mixture of identifiable terms. This result was obtained by a different method by Shanthikumar [S-042].

The literature on priority queues offers many transforms for stationary waiting time distributions. Upon inspection, many of these are seen to be equivalent to linear Volterra integral equations with a kernel involving the distribution of the busy period in a related $M/G/1$ queue. The reason for this is discussed in detail in Jaiswal [J-012]. In brief, the time added to the service time of a customer of a given priority class may often be viewed as the busy period in an $M/G/1$ queue "serving" only the customers of higher priority. As pointed out in Neuts [N-062], the integral equation is much more transparent than the equivalent transform expressions. It shows that the (largely unexplored) numerical solution of priority queues needs to concentrate on the computation of the kernel, and therefore on the distribution of the $M/G/1$ busy period. The solution of the linear Volterra equation itself is algorithmically elementary.

When a stochastic model has an embedded positive recurrent Markov renewal process, the least computational derivation of many distributions of interest, such as the stationary distribution of the queue length and the virtual waiting time in the $M/G/1$ queue, proceeds by the construction of the stationary version of the Markov renewal process as described, for example, in Pyke [P-053,P-054] and Çinlar [C-025]. As the Markov renewal matrix $M(t)$ is now very simple, the derivations shown in Sections 1.6 and 1.7 are then straightforward. For the $M/G/1$ queue, the initial conditions of the stationary version correspond to the joint probabilities of the queue length and the time until the next departure, viewed at an arbitrary time.

The proof of Theorem 1.7.1 shows that the joint stationary distribution of the queue length and the residual service time at an arrival epoch plays a crucial role. That distribution is discussed in detail in Boxma [B-083]. It is the subject of the following problem.

Problem 1.7.1: From the proof of Theorem 1.7.1, extract the expressions needed to derive the joint transform $S(z,s)$ of the joint stationary distribution of the queue length and the residual service time seen at a (real or virtual) arrival epoch. From $S(z,s)$, obtain by a direct argument the joint transform $W(z,s)$ of the joint stationary distribution of the queue length and waiting time. Verify that the Pollaczek-Khinchin formula is obtained by setting $z = 1$. As noted in Neuts [N-062], the simple relation between $S(z,s)$ and $W(z,s)$ does not carry over to the matrix case. •

Problem 1.7.2: The N-policy trades off the advantages of longer busy periods against possibly longer waiting times. Illustrate this by plotting the ratio of the mean waiting time at arrivals and the mean duration of the busy period for selected values of N and for $M/G/1$ queues with simple service time distributions of your choice. You may assume known that the stationary waiting time distributions of real and virtual customers are the same. (This follows from a general result for queues with Poisson arrivals, which is known by the acronym *PASTA* : " ***Poisson arrivals see time averages***

Section 1.8

The relationship between the $M/G/1$ queue and the Galton-Watson process was pointed out by Good in the discussion following Kendall [K-044]. It was taken as the point of departure for an alternative analysis of the $M/G/1$ queue in Neuts [N-019] and also led him to consider the Markov renewal branching process [N-025], which corresponds to the sequence of generations during a fundamental period for the matrix case.

In the literature on communications models, service by generations is known as *gating.* It is often useful in applications where the items to be served are entered into a buffer so that the server may reassign priorities to a rather small group of customers, who all arrive at approximately the same time. Examples of such priority rules are discussed in Nair and Neuts [N-001,N-004] by transform methods, which limit the usefulness of the results obtained. A numerical examination of the merits of these priority rules would be useful and interesting. In the analysis here, we have assumed that there is no time required to enter the next generation into the buffer to which the server has access. In practical situations, the transfer time may add significantly to the virtual and actual waiting times. Cases, where some form of transfer time is incorporated into the model, are treated in Takács [T-021], Ali and Neuts [A-014].

Further references dealing with models related to this section are Breny [B-088], Cooper and Murray [C-062], Cooper [C-063], Nair [N-002], Neuts [N-030], Shanthikumar [S-041].

Problem 1.8.1: Discuss the stationary density of the numbers of customers in successive generations for the stable $M/G/1$ queue operating under the $N-$policy. •

Problem 1.8.2: Consider a stable $M/G/1$ queue with service by generations and operating under the $N-$policy. In order to bring the next generation inside the gate for service, the server expends a transfer time. The successive transfer times are of independent, identically distributed durations with the common distribution $F(\cdot)$ of finite mean. Derive the Laplace-Stieltjes transform for the stationary distribution of the waiting time of a customer arriving at time t. Compare the mean waiting time to that of the $M/G/1$ queue with $N-$policy given in the text. •

Problem 1.8.3: Consider a stable $M/G/1$ queue with service by generations in which the server has an unlimited supply of secondary customers on hand. These have the same service time distribution as the ordinary customers. Whenever a generation is of size less than N, $N > 1$, as many secondary customers as needed are added to make up a generation of size N. These customers are served after the ordinary customers, but before the gate is opened again to let in the next generation. Prove that this additional input to the queue does not affect its stability condition. For the stable queue, obtain an expression for the probability that an arbitrary service involves an ordinary customer.

This will also give the fraction of secondary customers receiving service. Study the stationary waiting time distribution of an ordinary customer and the invariant probability vectors of the embedded Markov chains. Clarify as much as possible the effect of the secondary input on these descriptors of the queue and discuss several criteria for the choice of the parameter N. •

This interesting model has not been previously discussed in the literature. It offers a rich choice of algorithmic problems to be explored with only a modest amount of computational effort.

2

Markov Chains of M/G/1 Type

2.1. SOME ILLUSTRATIVE EXAMPLES

The structural form (1.1.3) of the transition probability matrix $Q(x)$ is common to the embedded Markov renewal processes of many useful stochastic models, provided we allow the objects $A_\nu(\cdot)$ and $B_\nu(\cdot)$ to be *matrices* of mass-functions. In this chapter, we shall discuss such Markov renewal processes in general, with special emphasis on the fundamental period. In Chapter 3, we shall develop further theoretical results shared by all such Markov renewal processes. This formal development is lengthy and occasionally arduous. It is therefore useful to have a small number of specific examples available beforehand. These should serve to show that the Markov renewal processes under study arise naturally in widely disparate applications and that most of the steps in our analysis have specific probabilistic interpretations of independent interest. In cases where the $A_\nu(\cdot)$ and $B_\nu(\cdot)$ are matrices, there are also certain mathematical difficulties without a counterpart in the scalar case. Some of our examples illustrate these difficulties and therefore serve to motivate their study. At this stage, we will not present the examples in their most general setting. Rather to the contrary, we prefer to choose simpler versions of the stochastic models of which our examples are representative. These should show that the mathematical efforts required in this and the next chapter can rarely be avoided. The issues, treated in these chapters, are germane and essential even to the simpler examples.

Example A : The $M/\,SM/\,1$ Queue. In this natural generalization of the $M/\,G/\,1$ queue, the service times of the successive customers form a Markov renewal process with a finite number m of states. The abbreviation SM stands for semi-Markovian (services). As is well-known, semi-Markov and Markov renewal processes are intimately related and substantially equivalent ways of viewing the process obtained from a Markov chain by allowing the sojourn times in the successive states to have general probability distributions which depend (only) on the current state and the state to be visited next.

The transition probability matrix $H(\cdot)$ of the Markov renewal process, which describes the types and the service times of the successive customers, is an $m \times m$ matrix of probability mass-functions on $[0, \infty)$. By assumption, its row sums

$$H_j(x) = \sum_{j'=1}^{m} H_{jj'}(x),$$

are nondegenerate, proper probability distributions of finite means α_j, $1 \leqslant j \leqslant m$, and the matrix $H = H(\infty)$, is an *irreducible* stochastic matrix.

In the sequel we shall discuss several concrete examples of queues with semi-Markovian services. The following is but one simple example to motivate the study of the $M/SM/1$ queue. Consider a variant of the $M/G/1$ queue in which after completing each m-th service, the server is required to break for a maintenance period. The durations of the successive maintenance periods are independent, identically distributed and also independent of the service times. Their common distribution function is denoted by $K(\cdot)$.

It is convenient to view the maintenance periods as augmenting the service times of every m-th customer and this readily leads to an $M/SM/1$ queue for which the matrix $H(\cdot)$ is given by

$$H(\cdot) = \begin{vmatrix} 0 & H(\cdot) & 0 & \cdots & 0 & 0 \\ 0 & 0 & H(\cdot) & \cdots & 0 & 0 \\ & \cdots & & & \cdots & \\ 0 & 0 & 0 & \cdots & 0 & H(\cdot) \\ H{*}K(\cdot) & 0 & 0 & \cdots & 0 & 0 \end{vmatrix},$$

where the scalar function $H(\cdot)$ is the ordinary service time distribution.

If we consider the $M/SM/1$ queue immediately after the successive service completions and form the trivariate sequence $\{I_n, J_n, X_n\}$, where I_n denotes the queue length, $J_n \in \{1, \ldots, m\}$ the state of the Markov renewal process $H(\cdot)$ and X_n the time between the n-th and the $(n+1)$-st departures, we obtain, as for the $M/G/1$ queue, the embedded Markov renewal process of the $M/SM/1$ queue. Its transition probability matrix $Q(x)$, $x \geqslant 0$, is given by

$$Q(x) = \begin{vmatrix} B_0(x) & B_1(x) & B_2(x) & B_3(x) & B_4(x) & \cdots \\ A_0(x) & A_1(x) & A_2(x) & A_3(x) & A_4(x) & \cdots \\ 0 & A_0(x) & A_1(x) & A_2(x) & A_3(x) & \cdots \\ 0 & 0 & A_0(x) & A_1(x) & A_2(x) & \cdots \\ \cdot & \cdot & \cdot & \cdot & \cdot & \\ \cdot & \cdot & \cdot & \cdot & \cdot & \end{vmatrix}, \tag{2.1.1}$$

where the matrices $A_\nu(\cdot)$ and $B_\nu(\cdot)$, $\nu \geqslant 0$, are given by

$$A_\nu(x) = \int_0^x e^{-\lambda u} \frac{(\lambda u)^\nu}{\nu!} dH(u), \quad \text{for } \nu \geqslant 0,\ x \geqslant 0, \tag{2.1.2}$$

$$B_\nu(x) = \int_0^x \lambda e^{-\lambda u} A_\nu(x-u) du = \int_0^x [1 - e^{-\lambda(x-u)}]\, dA_\nu(u),$$

In this book, we shall frequently write Lebesgue-Stieltjes integrals with respect to matrices of mass-functions. The significance of such matrices will always be clear from the context. In the preceding expression for $A_\nu(x)$, for example, it is obvious that the matrix $A_\nu(x)$ has its elements given by

$$[A_\nu(x)]_{jj'} = \int_0^x e^{-\lambda u} \frac{(\lambda u)^\nu}{\nu!} dH_{jj'}(u), \quad \text{for } 1 \leqslant j, j' \leqslant m.$$

The appealing structure of the matrix $Q(\cdot)$ is brought out by listing the states (i,j), $i \geqslant 0$, $1 \leqslant j \leqslant m$, in the *lexicographic order.* While the ordering of the states of a Markov renewal process is clearly immaterial to its general properties, it is important for all models treated in this book to define and order the states in such a manner as to bring out the block structure of the matrix $Q(\cdot)$ in the most convenient way. For some of the more involved models, this requires both care and experience.

The irreducibility of the matrix H is inherited by all the nonnegative matrices $A_\nu = A_\nu(\infty)$, and $B_\nu = B_\nu(\infty)$, for $\nu \geqslant 0$. In Example C, we shall discuss a case where the blocks lack this useful property.

Example B: Bailey's Bulk Queue. This classical queueing model deals with the waiting line at a bus terminal which is visited by a bus at the epochs of a renewal process of underlying distribution $H(\cdot)$. At each arrival of the bus, up to $m \geqslant 1$ customers are removed from the queue, which is fed by a Poisson arrival process of rate λ. We denote by ξ_n, $n \geqslant 0$, the number of customers left behind by the n-th departure of the bus. We let the time $t = 0$, correspond to a departure epoch and define X_n, $n \geqslant 1$, to be the time between the $(n-1)$-st and the n-th departures. We set $X_0 = 0$, and define $\tau(X_n)$, $n \geqslant 1$, to be the number of customers who arrive at the terminal during the time span X_n.

The successive interdeparture times X_n, $n \geqslant 1$, are independent and

$$\xi_n = \max\{0, \xi_{n-1} + \tau(X_n) - m\}, \quad \text{for } n \geqslant 1.$$

It is therefore easily seen that the sequence $\{(\xi_n, X_n), n \geqslant 0\}$ is a Markov renewal sequence. We display its transition probability matrix $Q(\cdot)$ for $m = 3$, from which the particular structure for a general value of $m \geqslant 1$ is readily apparent.

$$Q(x) = \left| \begin{array}{cccccc} a_0(x) & a_1(x) & a_2(x) & a_3(x) & a_4(x) & \cdots \\ a_0(x) & a_1(x) & a_2(x) & a_3(x) & a_4(x) & \cdots \\ a_0(x) & a_1(x) & a_2(x) & a_3(x) & a_4(x) & \cdots \\ a_0(x) & a_1(x) & a_2(x) & a_3(x) & a_4(x) & \cdots \\ 0 & a_0(x) & a_1(x) & a_2(x) & a_3(x) & \cdots \\ 0 & 0 & a_0(x) & a_1(x) & a_2(x) & \cdots \\ 0 & 0 & 0 & a_0(x) & a_1(x) & \cdots \\ \cdot & \cdot & \cdot & \cdot & \cdot & \end{array} \right|, \qquad (2.1.3)$$

where

$$a_\nu(x) = \int_0^x e^{-\lambda u} \frac{(\lambda u)^\nu}{\nu!} \, dH(u), \quad \text{for } \nu \geqslant 0, x \geqslant 0,$$

The transition probability matrix $P = Q(\infty)$ of the embedded Markov chain is very similar to that of the $M/G/1$ queue, which indeed corresponds to the case $m = 1$. It is therefore natural that in the original discussion of this model by N. T. J. Bailey and in the treatment of many closely related models in the subsequent literature, the stationary distributions have been studied by the methods of generating functions and of complex analysis. This leads, via a classical analyticity argument presented in many texts, to the consideration of the roots in the unit disk $|z| \leqslant 1$, of the equation

$$z^m = h(\lambda - \lambda z), \tag{2.1.4}$$

where $h(s)$ is the Laplace-Stieltjes transform of the probability distribution $H(\cdot)$.

The present model is amenable to analysis by the methods of this book, which avoid complex analysis. To see this, we partition the matrix $Q(x)$ into $m \times m$ blocks to obtain

$$Q(x) = \begin{vmatrix} B_0(x) & B_1(x) & B_2(x) & B_3(x) & B_4(x) & \cdots \\ A_0(x) & A_1(x) & A_2(x) & A_3(x) & A_4(x) & \cdots \\ 0 & A_0(x) & A_1(x) & A_2(x) & A_3(x) & \cdots \\ 0 & 0 & A_0(x) & A_1(x) & A_2(x) & \cdots \\ \cdot & \cdot & \cdot & \cdot & \cdot & \\ \cdot & \cdot & \cdot & \cdot & \cdot & \end{vmatrix}.$$

The matrices $B_\nu(\cdot)$ have identical rows with elements

$$a_{m\nu}(x),\ a_{m\nu+1}(x),\ \ldots,\ a_{m\nu+m-1}(x).$$

The matrices $A_\nu(\cdot)$ for $\nu \geqslant 1$, are given by

$$A_\nu(x) = \begin{vmatrix} a_{m\nu}(x) & a_{m\nu+1}(x) & \cdots & a_{m\nu+m-1}(x) \\ a_{m\nu-1}(x) & a_{m\nu}(x) & \cdots & a_{m\nu+m-2}(x) \\ \cdot & \cdot & & \cdot \\ \cdot & \cdot & & \cdot \\ a_{m\nu-m-1}(x) & a_{m\nu-m-2}(x) & \cdots & a_{m\nu}(x) \end{vmatrix},$$

and $A_0(x)$ is the upper triangular matrix

$$A_0(x) = \begin{vmatrix} a_0(x) & a_1(x) & \cdots & a_{m-1}(x) \\ 0 & a_0(x) & \cdots & a_{m-2}(x) \\ \cdot & \cdot & & \cdot \\ \cdot & \cdot & & \cdot \\ 0 & 0 & \cdots & a_0(x) \end{vmatrix}.$$

The following observations are in order:

a. Particular features of the matrices $A_\nu(\cdot)$, which account for simplifications over the general case in the usual analytic treatment, should also lead to major simplifications in the methods and in the ensuing algorithms, proposed here. We shall later show that this is indeed the case.

b. There are many seemingly different stochastic models, which lead to an embedded Markov renewal process with a transition probability matrix $Q(\cdot)$ differing from that of Bailey's bulk service queue only in the definition of the elements of the first m rows. One such variant is known as the *M/G/1 queue with a general bulk service rule,* which is treated in detail in Section 4.2. According to that rule, the server may only start a service if there are at least $N \geq 1$ customers in the queue. If there are fewer, he waits for the queue length to reach N and then serves a group of size N. If after the n-th departure, the queue length ξ_n is at least N, a group of size $\min(\xi_n, m)$ with $m \geq N$ is served. The duration of a service is allowed to depend on the group size, but for any $k \geq 2$, the durations of the first k services are mutually conditionally independent, given the sizes of the groups served. The distribution of the service time of a group is size j, $N \leq j \leq m$, is denoted by $H_j(\cdot)$, but for $j = m$, the subscript is deleted.

All rows of the matrix $Q(\cdot)$ for the Markov renewal process embedded at departures are the same as for Bailey's queue, except for the first m which are given by

$$\begin{vmatrix} b_{00}(x) & b_{01}(x) & b_{02}(x) & \cdots \\ b_{10}(x) & b_{11}(x) & b_{12}(x) & \cdots \\ \cdot & \cdot & \cdot & \\ \cdot & \cdot & \cdot & \\ b_{m-1,0}(x) & b_{m-1,1}(x) & b_{m-1,2}(x) & \cdots \end{vmatrix},$$

with the mass-functions $b_{ij}(\cdot)$, $0 \leq i \leq m-1$, $j \geq 0$, given by

$$b_{ij}(x) = \int_0^x e^{-\lambda(x-u)} \frac{[\lambda(x-u)]^{N-i-1}}{(N-i-1)!} \lambda du \int_0^{x-u} e^{-\lambda v} \frac{(\lambda v)^j}{j!} dH_N(v),$$

for $0 \leqslant i < N$, and by

$$b_{ij}(x) = \int_0^x e^{-\lambda v} \frac{(\lambda v)^j}{j!} dH_i(v), \quad \text{for } N \leqslant i \leqslant m-1.$$

The boundary behavior of this model is much more complex than that of Bailey's queue, but there remains the fact that the parameter N appears only in the elements of the first m rows. In this book, the treatment of the boundary behavior is largely separated from that of other states. The effect of the parameter N on various probability distributions related to this model may therefore be studied in a particularly efficacious manner.

Example C**: A Dam with Markovian Input.** Several models, proposed for the study of the content of a dam, have embedded Markov chains similar in structure to those arising in queueing theory. The model to be described here is known as the *Odoom-Lloyd-Ali Khan-Gani dam.* It arose from the need to model a dam with seasonally varying inflows of water.

The dam is considered at equally spaced times and the difference between these successive times is chosen as the unit of time. The amount of water removed per unit of time is assumed to be constant and is chosen as the unit in which both the amounts of the inflows and the content of the dam are expressed. It is further assumed that the inflow X_n during the time interval $(n-1, n)$ is added *in toto* to the dam just prior to time n and that one unit is drained from the dam at time n, if such is possible. The inflows $\{X_n, n \geqslant 0\}$ form a Markov chain with states $0, 1, \ldots, m$ and transition probability matrix $B = \{b_{jj'}\}$, $0 \leqslant j, j' \leqslant m$. The stochastic matrix B is *irreducible.*

The content of the dam after all possible transactions at time n have been completed is denoted by Y_n, $n \geqslant 0$. By virtue of the relation

$$Y_{n+1} = \max(0, Y_n + X_{n+1} - 1),$$

the sequences $\{(Y_n, X_n), n \geqslant 0\}$ and $\{(Y_n, X_{n+1}), n \geqslant 0\}$ are both Markov chains with states (i, j), $i \geqslant 0$, $0 \leqslant j \leqslant m$. Their transition probability matrices P_1 and P_2 are displayed for $m = 2$. The structure for a general value of m is readily apparent. The matrix P_1 is given by

	00	01	02	10	11	12	20	21	22	30	31	32	...
00	b_{00}	b_{01}				b_{02}							
01	b_{10}	b_{11}				b_{12}							...
02	b_{20}	b_{21}				b_{22}							
10	b_{00}				b_{01}				b_{02}				...
11	b_{10}				b_{11}				b_{12}				
12	b_{20}				b_{21}				b_{22}				...
20				b_{00}				b_{01}				b_{02}	
21				b_{10}				b_{11}				b_{12}	...
22				b_{20}				b_{21}				b_{22}	
30							b_{00}				b_{01}		...
31							b_{10}				b_{11}		
32							b_{20}				b_{21}		...
		...			...			...			...		

, (2.1.5)

and the matrix P_2 by

	00	01	02	10	11	12	20	21	22	30	31	32	...
00	b_{00}	b_{01}	b_{02}										
01	b_{10}	b_{11}	b_{12}										...
02				b_{20}	b_{21}	b_{22}							
10	b_{00}	b_{01}	b_{02}										...
11				b_{10}	b_{11}	b_{12}							
12							b_{20}	b_{21}	b_{22}				...
20				b_{00}	b_{01}	b_{02}							
21							b_{10}	b_{11}	b_{12}				...
22										b_{20}	b_{21}	b_{22}	
30							b_{00}	b_{01}	b_{02}				...
31										b_{10}	b_{11}	b_{12}	
32													...
		...			...			...			...		

, (2.1.6)

Both matrices have the common structure

$$\begin{vmatrix} A_0+A_1 & A_2 & A_3 & A_4 & A_5 & \cdots \\ A_0 & A_1 & A_2 & A_3 & A_4 & \cdots \\ 0 & A_0 & A_1 & A_2 & A_3 & \cdots \\ 0 & 0 & A_0 & A_1 & A_2 & \cdots \\ \cdot & \cdot & \cdot & \cdot & \cdot & \end{vmatrix},$$

with $A_\nu = 0$, for $\nu > m$. For $0 \leqslant \nu \leqslant m$, the matrix A_ν in P_1 shares the column of index ν with the matrix B and its other columns are zero. In P_2, the same prevails for the rows.

The matrix P_1 was used in the treatment by Ali Khan and Gani, while the discussion of the matrix P_2 was presented by Odoom and Lloyd. Every property of P_1 is readily translated into a corresponding property of the matrix P_2, so that both approaches to the dam model are essentially equivalent. As we shall show, in our approach, the matrix P_1 has a property, not shared by P_2, which leads to major simplifications in numerical computations. Particularly for large values of m, numerical computations for P_1 may be carried out more efficiently than for P_2. This is a first example, with others to follow later, of a model for which equivalent mathematical descriptions are quite different in the effort required by a given algorithmic procedure.

The other point of this example is that in both cases the matrices $A_0, \ldots, A_m$ are clearly reducible. The sum $B = A_0 + \cdots + A_m$ is still an irreducible stochastic matrix. As we shall see, the particular reducible structure of A_0 in the matrix P_1 leads to computational simplification, but this requires that we discuss the issues of *reducibility* in general. That general discussion is quite involved and is given in Sections 3.4 and 3.5. Finally, we note that in the matrix P_1 the elements in the columns labeled $(0,j)$, $2 \leqslant j \leqslant m$, are zero. This shows that the corresponding states $(0,j)$ are *ephemeral,* that is, except as an initial state, the state $(0,j)$ is never visited by the Markov chain. In order to make the Markov chain P_1 irreducible, we would have to delete the rows and columns corresponding to those pairs $(0,j)$. However, in order to preserve the convenient matrix notation, it is preferable not to do this. In all matrix calculations, the probabilities of visits to the ephemeral states will automatically become zero, so that no difficulties arise by proceeding formally as though the chain were irreducible. In the (rare) other cases, where ephemeral states arise, the same comment applies.

Example D: A Case with a Complex Boundary. In many practical applications of Markov renewal processes of $M/G/1$ type, the physical behavior of the model in the boundary states is very different from that away from the boundary states, where the spatial homogeneity of the process prevails. This is commonly reflected in the different dimensions and definition of the matrices $B_\nu(\cdot)$, $\nu \geqslant 0$. To illustrate this, we discuss a model that arose in the study of overload control for a message buffer. This model, chosen from among many, will be described as a variant of the $M/G/1$ queue.

Consider an $M/G/1$ queue with service time distribution $H(\cdot)$. The arrival rate of the Poisson input process is controlled as follows. There is an overload control switch (OC), which can be either OFF or ON. The possible control actions are taken at the completions of services and they are the following:

a. If the queue length (after a departure) is at most $K \geqslant 1$ and $OC = OFF$, then the control switch remains OFF and the arrival rate is λ^*.

b. If the queue length exceeds K and $OC = OFF$, the overload control switch is set to ON. The arrival rate λ^* prevails for an exponentially distributed time with mean $1/\theta$ and then changes to a (usually lower) value λ.

c. If the queue length exceeds L, with $0 \leqslant L \leqslant K$, and $OC = ON$, the overload control switch and the arrival rate remain as they are.

d. If the queue length is at most L and the condition $OC = ON$ exists, the overload control switch is set to OFF and the arrival rate is returned to the value λ^*.

The interest in this model was to examine suitable values for the thresholds K and L in the presence of the switch OC, whose effect on the arrival rate is not felt for a random length of time. The durations of the switch-overs to the arrival rate λ are assumed to have an exponential distribution of parameter θ.

The Markov renewal process embedded at departure epochs has the states

$$\begin{aligned} &i, && 0 \leqslant i \leqslant K, \\ &(i,0), && i \geqslant L+1, \\ &(i,1), && i \geqslant L+1. \end{aligned}$$

The state i with $0 \leqslant i \leqslant K$ signifies that there are i customers in the queue with $OC = OFF$. The states $(i,0)$ and $(i,1)$ signify that there are

	0	1	2	3	4	5
0	$b_0'(x)$	$b_1'(x)$	$b_2'(x)$	$b_3'(x)$	$b_4'(x)$	$b_5'(x)$
1	$a_0'(x)$	$a_1'(x)$	$a_2'(x)$	$a_3'(x)$	$a_4'(x)$	$a_5'(x)$
2	0	$a_0'(x)$	$a_1'(x)$	$a_2'(x)$	$a_3'(x)$	$a_4'(x)$
3	0	0	$a_0'(x)$	$a_1'(x)$	$a_2'(x)$	$a_3'(x)$
4	0	0	0	$a_0'(x)$	$a_1'(x)$	$a_2'(x)$
5	0	0	0	0	$a_0'(x)$	$a_1'(x)$
3,0	0	0	$c^0_{00}+c^0_{01}$	0	0	0
3,1	0	0	c^0_{11}	0	0	0
4,0	0	0	0	0	0	0
4,1	0	0	0	0	0	0
5,0	0	0	0	0	0	0
5,1	0	0	0	0	0	0
6,0	0	0	0	0	0	0
6,1	0	0	0	0	0	0
			. . .			

(2.1.7)

	3,0	3,1	4,0	4,1	5,0	5,1	6,0	6,1	. . .
0	0	0	0	0	0	0	$b_6'(x)$	0	. . .
1	0	0	0	0	0	0	$a_6'(x)$	0	
2	0	0	0	0	0	0	$a_5'(x)$	0	. . .
3	0	0	0	0	0	0	$a_4'(x)$	0	
4	0	0	0	0	0	0	$a_3'(x)$	0	. . .
5	0	0	0	0	0	0	$a_2'(x)$	0	
3,0	c^1_{00}	c^1_{01}	c^2_{00}	c^2_{01}	c^3_{00}	c^3_{01}	c^4_{00}	c^4_{01}	. . .
3,1	0	c^1_{11}	0	c^2_{11}	0	c^3_{11}	0	c^4_{11}	
4,0	c^0_{00}	c^0_{01}	c^1_{00}	c^1_{01}	c^2_{00}	c^2_{01}	c^3_{00}	c^3_{01}	. . .
4,1	0	c^0_{11}	0	c^1_{11}	0	c^2_{11}	0	c^3_{11}	
5,0	0	0	c^0_{00}	c^0_{01}	c^1_{00}	c^1_{01}	c^2_{00}	c^2_{01}	. . .
5,1	0	0	0	c^0_{11}	0	c^1_{11}	0	c^2_{11}	
6,0	0	0	0	0	c^0_{00}	c^0_{01}	c^1_{00}	c^1_{01}	. . .
6,1	0	0	0	0	0	c^0_{11}	0	c^1_{11}	
		. . .					. . .		

i customers in the system and that the OC switch is respectively in its initial exponential phase where the arrival rate is still λ^* or has already reached the operational phase where the arrival rate is λ.

The transition probability matrix $Q(\cdot)$ is displayed for $L = 2$ and $K = 5$. Its structure is again representative of that of the case for general values of L and K with $0 \leqslant L \leqslant K$. Because of the size of the matrix, we shall display the columns with state labels 0, 1, $\cdots$, 5 and the columns corresponding to the states $(i,0)$ and $(i,1)$, for $i \geqslant 3$, in separate arrays. The first array corresponds to the first column of blocks in the transition probability matrix $Q(\cdot)$, partitioned into the standard structural form of $M/G/1$ type.

The relevant elements, which are all probability mass-functions, are explicitly given by

$$b_\nu'(x) = \int_0^x e^{-\lambda^*(x-u)} \lambda^* a_\nu'(u)\, du,$$

$$a_\nu'(x) = \int_0^x e^{-\lambda^* u} \frac{(\lambda^* u)^\nu}{\nu!}\, dH(u), \quad \text{for } \nu \geqslant 0.$$

$$c_{00}^\nu(x) = \int_0^x e^{-(\theta+\lambda^*)u} \frac{(\lambda^* u)^\nu}{\nu!}\, dH(u), \tag{2.1.8}$$

$$c_{01}^\nu(x) = \sum_{r=0}^{\nu} \int_0^x dH(v) \int_0^v e^{-(\theta+\lambda^*)u} \frac{(\lambda^* u)^r}{r!}\, \theta\, e^{-\lambda(v-u)} \frac{[\lambda(v-u)]^{\nu-r}}{(\nu-r)!}\, du$$

$$= \theta \int_0^x dH(v) \int_0^v e^{-(\theta+\lambda^*)u-\lambda(v-u)} \frac{[\lambda^* u + \lambda(v-u)]^\nu}{\nu!}\, du,$$

$$c_{11}^\nu(x) = \int_0^x e^{-\lambda u} \frac{(\lambda u)^\nu}{\nu!}\, dH(u), \quad \text{for } \nu \geqslant 0.$$

The character ν in $c_{00}^\nu(x)$, $c_{01}^\nu(x)$ and $c_{11}^\nu(x)$ is clearly a superscript and not an exponent. A careful examination of the stated expressions will show that they indeed account for the arrivals and, where appropriate, the change in the phase of the switch, which may occur during a transition of the Markov renewal process.

In order to see that $Q(x)$ is of the form of the transition probability matrices, studied in this book, we partition that matrix as

$$Q(x) = \begin{vmatrix} B_0(x) & B_1(x) & B_2(x) & B_3(x) & B_4(x) & \cdots \\ C_0(x) & A_1(x) & A_2(x) & A_3(x) & A_4(x) & \cdots \\ 0 & A_0(x) & A_1(x) & A_2(x) & A_3(x) & \cdots \\ 0 & 0 & A_0(x) & A_1(x) & A_2(x) & \cdots \\ \cdot & \cdot & \cdot & \cdot & \cdot & \end{vmatrix},$$

where $B_0(x)$ is the square matrix of order $K+1$ in the upper left hand corner of $Q(x)$. The matrices $B_\nu(x), \nu \geqslant 1$, are of dimensions $(K+1) \times 2$ and $B_\nu(x) = 0$, for $1 \leqslant \nu \leqslant K-L$. For $\nu > K-L$, the second column of $B_\nu(x)$ is zero. The elements of the first column of $B_\nu(x)$ are given in order by

$$b_{\nu+L}{}'(x),\ a_{\nu+L}{}'(x),\ a_{\nu+L-1}{}'(x),\ \ldots,\ a_{\nu+L-K+1}{}'(x).$$

The matrices $A_\nu(x)$ are 2×2 upper triangular matrices, given by

$$A_\nu(x) = \begin{vmatrix} c_{00}^{\nu}(x) & c_{01}^{\nu}(x) \\ 0 & c_{11}^{\nu}(x) \end{vmatrix}, \qquad \text{for } \nu \geqslant 0.$$

The matrix $C_0(x)$ is of dimensions $2 \times (K+1)$ and only its column of index L differs from zero. That column is given by the vector $A_0(x)\mathbf{e}$.

This model exhibits a number of interesting features, which are reflected in the special structure of the matrix $Q(\cdot)$. The states $0, 1, \ldots, K$ appear clearly as the boundary states. Their number is now unrelated to the order $m = 2$ of the matrices $A_\nu(\cdot)$. The matrices $A_\nu(\cdot), \nu \geqslant 0$, and also their sum $\sum_{\nu=0}^{\infty} A_\nu(\cdot)$, are all reducible matrices. The matrix $C_0(\cdot)$ has a highly special structure. At the appropriate place in the analysis, this leads to substantial analytic and algorithmic simplifications.

Levels: As in the case of Bailey's queue, the indexing of the blocks differs from the labeling of the states. The latter is related to the queue length. It is convenient to introduce nomenclature that refers to the indexing of the blocks. In cases where that indexing does not agree with a variable of direct interest, such as the queue length, it is an easy matter to translate from one terminology to the other. The *sets of states* which correspond to the blocks in the partitioned form of $Q(\cdot)$ are called *levels*. For the model in Example D, level 0 corresponds to the set of states $\{0, \ldots, K\}$. Level i, with $i \geqslant 1$, corresponds to the pair of states $(i+L, 0)$ and $(i+L, 1)$.

Canonical Form: The examples $A - D$, chosen from among many, show that there is merit in a general study of Markov renewal processes with a transition probability matrix $Q(\cdot)$, which may be partitioned into blocks, so as to bring out their common structure. The common form of all such transition probability matrices is the canonical form of Markov renewal processes (or Markov chains) of $M/G/1$ type. It is given by

$$Q(x) = \left| \begin{array}{cccccc} B_0(x) & B_1(x) & B_2(x) & B_3(x) & B_4(x) & \cdots \\ C_0(x) & A_1(x) & A_2(x) & A_3(x) & A_4(x) & \cdots \\ 0 & A_0(x) & A_1(x) & A_2(x) & A_3(x) & \cdots \\ 0 & 0 & A_0(x) & A_1(x) & A_2(x) & \cdots \\ \cdot & \cdot & \cdot & \cdot & \cdot & \end{array} \right|, \tag{2.1.9}$$

Such a Markov renewal process is said to be *of $M/G/1$ type*. We also refer to the matrix $Q(\cdot)$ and to the stochastic matrix $Q(\infty)$, respectively as a *semi-Markov matrix of $M/G/1$ type* and as a *stochastic matrix of $M/G/1$ type*.

The matrices $A_\nu(\cdot)$, $\nu \geqslant 0$, are square matrices of order m. The matrix $B_0(\cdot)$ is a square matrix of order m_1. The matrices $B_\nu(\cdot)$, $\nu \geqslant 1$, and $C_0(\cdot)$ are respectively of dimensions $m_1 \times m$ and $m \times m_1$. The sum $\sum_{\nu=0}^{\infty} A_\nu(x)$ is a stochastic semi-Markov matrix $A(x)$. The fact that $Q(\infty)$ is stochastic is equivalent to the conditions

$$B_0(\infty)\mathbf{e} + \sum_{\nu=1}^{\infty} B_\nu(\infty)\mathbf{e} = \mathbf{e},$$

$$C_0(\infty)\mathbf{e} + \sum_{\nu=1}^{\infty} A_\nu(\infty)\mathbf{e} = \mathbf{e},$$

$$A(\infty)\mathbf{e} = \mathbf{e}.$$

Henceforth, we shall denote $A_\nu(\infty)$ by A_ν and similarly for all the other matrices.

In all the examples (except for the ephemeral states in Example C), the Markov renewal process $Q(\cdot)$ is *irreducible.* This is typical of all the properly formulated, specific models to which the theory presented in this book is applicable. As shown by Example C, there are cases where all the matrices A_ν, $\nu \geqslant 0$, are reducible yet their sum A is irreducible. As shown by Example D, there are cases where also the matrix A is reducible.

In studying Markov renewal process of $M/G/1$ type, we proceed along the same lines as in our discussion of the $M/G/1$ queue in Chapter 1. Next, we identify the analogue of the fundamental period and devote the remainder of this chapter to this concept which is crucial to all further developments.

2.2. THE FUNDAMENTAL PERIOD

The state space of a Markov renewal process of $M/G/1$ type consists of the m_1 states $(0,1), \ldots, (0,m_1)$ in level 0 and of the lattice points (i,j), $i \geqslant 1$, $1 \leqslant j \leqslant m$, in a semi-infinite strip. Level $\mathbf{i}$, $i \geqslant 1$, consists of the m states (i,j), $1 \leqslant j \leqslant m$. The structure of the matrix $Q(\cdot)$ in (2.1.9) implies that the process is *left skip-free for levels,* that is, any path leading from a state in level $\mathbf{i}'$ to a state in level $\mathbf{i}$, with $i' > i \geqslant 0$, must visit every intermediate level at least once. Away from the boundary level 0, the transition mechanism is also *spatially homogeneous,* that is, the transition probabilities from level $\mathbf{i}$ are the same for all $i \geqslant 2$.

The following are elementary, but essential observations. In order that the process $Q(\cdot)$ be *irreducible,* it is necessary that from any state (i,j), $1 \leqslant j \leqslant m$, it is possible with positive probability to reach any lower level and in particular level $\mathbf{i} - 1$, $i \geqslant 1$. In order that the process $Q(\cdot)$ be *recurrent,* it is necessary that from any state (i,j), $1 \leqslant j \leqslant m$, in level $\mathbf{i}$, the level $\mathbf{i} - 1$ be reached eventually with probability one.

Let $T(i+r,j;i,j')$ be the first passage time from the state $(i+r,j)$, $r \geqslant 1$, $1 \leqslant j \leqslant m$, to the state (i,j'), $i \geqslant 1$, $1 \leqslant j' \leqslant m$, with the additional requirement that (i,j') is the first state in level $\mathbf{i}$

to be visited. Similarly, $V(i+r, j; i, j')$ is the number of transitions involved in going from the state $(i+r, j)$ to the first hitting state (i, j') in the level i.

Since the process $Q(\cdot)$ is left skip-free for levels, every path from $(i+r, j)$ to (i, j') must visit the intermediate levels $i+r-1, \ldots, i+1$, at least once. Let the random variables $\xi_{i+r-1}, \ldots, \xi_{i+1}$ with values in $\{1, \ldots, m\}$ be the indices of the hitting states in the levels $i+r-1, \ldots, i+1$, at the first passages to successively lower levels. In other words, starting in the state $(i+r, j)$, the level $i+r-1$ is first visited at the state $(i+r-1, \xi_{i+r-1})$. From that state, the first visit to $i+r-2$ occurs at the state $(i+r-2, \xi_{i+r-2})$, and so on.

It is now clear that we may write

$$T(i+r, j; i, j') = \sum_{h=1}^{r} T(i+h, \xi_{i+h}; i+h-1, \xi_{i+h-1}), \tag{2.2.1}$$

$$V(i+r, j; i, j') = \sum_{h=1}^{r} V(i+h, \xi_{i+h}; i+h-1, \xi_{i+h-1}),$$

with $\xi_{i+r} = j$, and $\xi_i = j'$.

Moreover, by the Markov property of the Markov renewal process $Q(\cdot)$ at transition epochs, it is evident that, for $1 \leqslant h \leqslant r$, the pairs of random variables

$$[\, T(i+h, \xi_{i+h}; i+h-1, \xi_{i+h-1}), V(i+h, \xi_{i+h}; i+h-1, \xi_{i+h-1}) \,],$$

are mutually conditionally independent, given the random variables $\xi_{i+r}, \xi_{i+r-1}, \ldots, \xi_i$. This holds for every $i \geqslant 1$ and $r \geqslant 1$.

The joint distribution of the terms in the sum in (2.2.1) is therefore completely determined by the conditional probabilities

$$P\{\, T(i+h, \xi_{i+h}; i+h-1, \xi_{i+h-1}) \leqslant x, \tag{2.2.2}$$

$$V(i+h, \xi_{i+h}; i+h-1, \xi_{i+h-1}) = k \mid \xi_{i+h} = j, \xi_{i+h-1} = j' \}$$

for $x \geqslant 0$, $k \geqslant 1$, $1 \leqslant j, j' \leqslant m$.

Finally, the spatial homogeneity of the transition probabilities away from the boundary implies that for $i+h \geqslant 2$, the probabilities in (2.2.2) do not depend on the value of $i+h$.

We now introduce the probabilities

$$G_{jj'}^{(r)}(k;x) = P\{T(i+r,j;i,j') \leqslant x,\ V(i+r,j;i,j') = k\}, \quad (2.2.3)$$

for $r \geqslant 1$, $x \geqslant 0$, $k \geqslant 0$, $1 \leqslant j, j' \leqslant m$. These play a crucial role in the study of Markov renewal processes of $M/G/1$ type. It is further convenient to introduce the matrices $G^{(r)}(k;x) = \{G_{jj'}^{(r)}(k;x)\}$ and to define $G_{jj'}^{(0)}(k;x)$ by

$$G_{jj'}^{(0)}(k;x) = \delta_{jj'} \cdot \delta_{0k}\, U(x), \quad (2.2.4)$$

where $U(\cdot)$ is the probability distribution degenerate at 0 and δ is the Kronecker delta.

Lemma 2.2.1: For $r \geqslant 2$, $k \geqslant r$, the matrix $G^{(r)}(k;x)$ is given by the matrix-convolution formula

$$G^{(r)}(k;x) = \sum G^{(1)}(k_1;\cdot) * G^{(1)}(k_2;\cdot) * \cdots * G^{(1)}(k_r;x), \quad (2.2.5)$$

where the summation extends over all r-tuples $(k_1, \ldots, k_r)$ satisfying $k_1 \geqslant 1, \cdots k_r \geqslant 1$, $k_1 + \cdots + k_r = k$.

Proof: This is immediate from the formulas (2.2.1) and the conditional independence of the pairs

$$[\,T(i+h,\xi_{i+h};i+h-1,\xi_{i+h-1}), V(i+h,\xi_{i+h};i+h-1,\xi_{i+h-1})\,],\quad 1 \leqslant h \leqslant r.$$

It suffices to condition on the times of the successive first passages to the levels $i+r-1, \ldots, i+1$ and on the states hit in each level. Formula (2.2.5) then follows directly by the law of total probability. •

We now introduce the joint transform matrix

$$\tilde{G}(z,s) = \sum_{k=1}^{\infty} z^k \int_0^{\infty} e^{-sx}\, dG^{(1)}(k;x), \quad (2.2.6)$$

which is defined for $|z| \leqslant 1$, and Re $s \geqslant 0$. We shall also drop the superscript (1) in $G^{(1)}(k;x)$ and we note that by virtue of Lemma 2.2.1, the sequence of matrices $\{G^{(r)}(k;x), k \geqslant r\}$ is constructed in a routine manner from the sequence $\{G(k;x), k \geqslant 1\}$. It is simply the r-th matrix-convolution power of the sequence $\{G(k;x), k \geqslant 1\}$.

Upon evaluation of the joint transform matrix

$$\tilde{G}^{(r)}(z,s) = \sum_{k=r}^{\infty} z^k \int_0^{\infty} e^{-sx}\, dG^{(r)}(k;x),$$

formula (2.2.5) readily yields the basic relation

$$\tilde{G}^{(r)}(z,s) = [\, \tilde{G}(z,s)\,]^r. \tag{2.2.7}$$

We see that, by the consistent definition in (2.2.4), this formula holds for all $r \geqslant 0$.

Theorem 2.2.1: The matrices of mass-functions $G(k;x)$ satisfy the integro-difference equations

$$G(1;x) = A_0(x), \quad G(k;x) = \sum_{\nu=1}^{\infty} A_\nu(x) * G^{(\nu)}(k-1;x), \quad \text{for } k \geqslant 2. \tag{2.2.8}$$

The transform matrix $\tilde{G}(z,s)$ satisfies the equation

$$\tilde{G}(z,s) = z \sum_{\nu=0}^{\infty} \tilde{A}_\nu(s) \tilde{G}^{\nu}(z,s), \tag{2.2.9}$$

where

$$\tilde{A}_\nu(s) = \int_0^{\infty} e^{-sx}\, dA_\nu(x), \quad \text{for } \nu \geqslant 0.$$

Proof: As in Theorem 1.2.1, the equation for $G(k;x)$, $k \geqslant 2$, follows by the law of total probability, conditioning on the duration of the first transition and the state visited by it. Formula (2.2.9) is obtained by routine evaluation of transforms and the use of formula (2.2.7). •

The matrix $G(x)$, defined by

$$G(x) = \sum_{k=1}^{\infty} G(k;x), \tag{2.2.10}$$

is a matrix of probability mass-functions. Since clearly

$$G(x)\mathbf{e} \leqslant \mathbf{e},$$

for all $x \geqslant 0$, $G(x)$ is a possibly substochastic semi-Markov matrix. We call $G(x)$, the *matrix-distribution* of the *fundamental period* T.

By adding the equations in (2.2.8) or by setting $z = 1$ in (2.2.9) and inverting the transforms, we see that $G(\cdot)$ satisfies the nonlinear Volterra matrix-integral equation

$$G(x) = \sum_{\nu=1}^{\infty} \int_0^x dA_\nu(u) G^{(\nu)}(x-u), \quad \text{for } x \geqslant 0. \tag{2.2.11}$$

This is the matrix analogue of Takács' equation for the busy (and fundamental) period of the ordinary $M/G/1$ queue.

The sequence of matrices $\{G(k), k \geqslant 1\}$, with $G(k) = G(k, \infty)$, for $k \geqslant 1$, may be considered as the density matrix corresponding to a semi-Markov matrix of lattice type. That is a semi-Markov matrix in which all entries are lattice mass-functions on a common lattice, which is here the positive integers. We call the sequence $\{G(k)\}$, the *matrix-density* of the *number of transitions* V *during the fundamental period* T.

The elements $G_{jj'}$, $1 \leqslant j, j' \leqslant m$, of the matrix $G = G(\infty) = \sum_{k=1}^{\infty} G(k)$, are the conditional probabilities that the Markov renewal process $Q(\cdot)$ will eventually hit the set $\mathbf{i}$ in the state (i, j'), given that it starts in the state $(i+1, j)$, $i \geqslant 1$.

From the interpretation of the elements of G and from the elementary observations made at the beginning of this section, the following statements are obvious.

Corollary 2.2.1: If the Markov renewal process $Q(\cdot)$ is irreducible, the matrix G does not have zero rows. If the Markov renewal process $Q(\cdot)$ is recurrent, the matrix G is stochastic.

The following is a related technical result which will be needed later.

Corollary 2.2.2: If the Markov renewal process $Q(\cdot)$ is irreducible, the matrix G has an eigenvalue η of maximum modulus which is *positive.*

Proof: From the Perron-Frobenius theory of nonnegative matrices, either G has a positive eigenvalue of maximum modulus or all eigenvalues of G are zero. This holds for any nonnegative matrix G. It therefore suffices to show that G cannot have all its eigenvalues equal to zero. If it did, its canonical equation would reduce to $\lambda^m = 0$, and the Cayley-Hamilton theorem then implies that $G^m = 0$. In an irreducible process $Q(\cdot)$, this is impossible, since $G^m = 0$, implies that no path starting in state $(i+m, j)$, $1 \leqslant j \leqslant m$, can reach the level i with positive probability. •

The eigenvalue of maximum modulus of G will henceforth be called the *spectral radius* of G. It will be denoted by η or by $sp(G)$, depending on the context.

We next discuss the system of equations (2.2.8) and the equivalent transform equation (2.2.9) in detail. We shall state properties, whenever appropriate, both in terms of matrices of probability mass-functions and in terms of the corresponding matrix transforms. The first terminology brings out the probabilistic interpretation most clearly; the second is occasionally more convenient for analytic manipulations.

We begin by noting that the equations (2.2.8) are *recursive.* By substitution for $G(1,x)$, we immediately obtain that

$$G(2,x) = \sum_{\nu=1}^{\infty} A_\nu(\cdot) * A_0^{(\nu)}(x),$$

and by continued substitutions, we may in principle express any matrix $G(k;x)$ in terms of the coefficient matrices $A_\nu(\cdot)$, $\nu \geqslant 0$. The resulting expressions rapidly become too complicated for use and the non-commutativity of the matrix product precludes any hope of simplification. This is in contrast with the ordinary $M/G/1$ queue, where the particular form of the mass-functions $A_\nu(\cdot)$ leads, via elegant combinatorial arguments due to Takács, to explicit formulas for the (scalar) functions $G(k;x)$.

A transparent discussion of the system (2.2.8) requires a brief excursion into nonlinear functional analysis. The result which is essential to the sequel is stated in Theorem 2.2.2. At first reading, some readers may wish to skip the details of the discussion that follows.

Let S be the set of sequences $\{S(k;\cdot), k \geqslant 1\}$ of $m \times m$ matrices $S(k;\cdot)$, whose elements are functions of bounded variation on $[0,\infty)$. It is readily verified that S is a vector space. Let C be the subset of S consisting of the sequences of matrices, whose elements are *mass-functions* on $[0,\infty)$. Finally, C_0 is the subset of C for which

$$\sum_{k=1}^{\infty} S(k;\infty)\mathbf{e} \leqslant \mathbf{e}.$$

It is evident that the set C is a *cone,* which spans the vector space S. The cone C will be called the cone of nonnegative elements of S and C_0 is the set of *substochastic* elements in C.

The equations

$$S_\sigma(1;x) = A_0(x), \tag{2.2.12}$$

$$S_\sigma(k;x) = \sum_{\nu=1}^{\infty} A_\nu(\cdot) * S^{(\nu)}(k-1;x), \quad \text{for } k \geqslant 2,$$

define a mapping σ of S into itself. Theorem 2.2.1 may be paraphrased to say that the mapping σ has a fixed point $\{G(k;\cdot)\}$ in the set C_0.

The set S may be endowed with a *partial order relation* $<\cdot$, by saying that

$$\{S(k;\cdot)\} <\cdot \{T(k;\cdot)\},$$

if and only if the sequence $\{T(k;\cdot) - S(k;\cdot)\}$ belongs to C.

The *minimal nonnegative solution* to the equations (2.2.8) is constructed as follows. The sequence $\{T_0(k;\cdot)\}$ with $T_0(k;x) = 0$, for $k \geqslant 1$, $x \geqslant 0$, is trivially in C_0. Applying the mapping σ to that sequence, we obtain the sequence $\{T_0(k;\cdot)\}$, with $T_1(k;\cdot) = A_0(\cdot)$, $T_1(k;\cdot) = 0$, $k \geqslant 2$. Clearly $\{T_1(k;\cdot)\}$ is also in C_0.

We now define the sequences $\{T_\nu(k;\cdot)\}$, $\nu \geqslant 1$, by

$$\{T_{\nu+1}(k;\cdot)\} = \sigma\,\{T_\nu(k;\cdot)\}, \quad \text{for } \nu \geqslant 0,$$

and establish the following of its properties:

Lemma 2.2.2: *a.* The sequence $\{T_\nu(k;\cdot)\}$ belongs to C_0 for all $\nu \geqslant 0$.

b. The successive iterates are non-decreasing, that is

$$\{T_\nu(k;\cdot)\} <\cdot \{T_{\nu+1}(k;\cdot)\}, \quad \text{for } \nu \geqslant 0.$$

c. The sequence $\{G(k;\cdot)\}$ is an upper bound, that is

$$\{T_\nu(k;\cdot)\} <\cdot \{G(k;\cdot)\}, \quad \text{for } \nu \geqslant 0.$$

d. The iterates converge to a sequence $\{T_\nu(k;\cdot)\}$ in C_0, which satisfies the equations (2.2.8).

e. The sequence $\{T(k;\cdot)\}$ is the *minimal* nonnegative solution of (2.2.8), that is, for any solution $\{S(k;\cdot)\}$ in C, we have

$$\{T(k;\cdot)\} <\cdot \{S(k;\cdot)\}.$$

Proof: If $\sum_{k=1}^{\infty} T_\nu(k;\infty)\mathbf{e} \leqslant \mathbf{e}$, then trivially

$$\sum_{k=1}^{\infty} T_\nu^{(r)}(k;\infty)\mathbf{e} = \sum_{k=1}^{\infty} T_\nu^{r}(k;\infty)\mathbf{e} \leqslant \mathbf{e}, \quad \text{for } r \geqslant 0,$$

since the matrices $T_\nu(k;\infty)$ are substochastic. It follows that

$$\sum_{k=1}^{\infty} T_{\nu+1}(k;\infty)\mathbf{e} = A_0(\infty) + \sum_{k=2}^{\infty}\sum_{r=1}^{\infty} A_r(\infty)\, T_\nu^{r}(k-1;\infty)\mathbf{e}$$

$$\leqslant \sum_{r=0}^{\infty} A_r(\infty)\mathbf{e} \leqslant \mathbf{e},$$

This proves the statement *a.* since the induction is clear.

To establish *b.*, we prove a more general inequality. If $\{R(k;\cdot)\} <\cdot \{S(k;\cdot)\}$ and if $\{R(k;\cdot)\}$ and $\{S(k;\cdot)\}$ are in C and if their images under σ exist, then $\sigma\{R(k;\cdot)\} <\cdot \sigma\{S(k;\cdot)\}$. If $B(\cdot)$ and

$C(\cdot)$ are $m \times m$ matrices of mass-functions on $[0,\infty)$ and if the inequality $B(x) \leqslant C(x)$, holds element-wise for all $x \geqslant 0$, then the r-fold matrix convolutions $B^{(r)}(\cdot)$ and $C^{(r)}(\cdot)$ satisfy $B^{(r)}(x) \leqslant C^{(r)}(x)$ for $r \geqslant 1$, $x \geqslant 0$. This is readily proved by induction, since $B(x) \leqslant C(x)$, implies $B^{(r)}(x) \leqslant C(\cdot) * B^{(r-1)}(x)$, and $B^{(r-1)}(x) \leqslant C^{(r-1)}(x)$, implies $C(\cdot) * B^{(r-1)}(x) \leqslant C^{(r)}(x)$, for $x \geqslant 0$.

By repeated application of the preceding property, it is clear from (2.2.12) that the mapping σ has the stated monotonicity property and Statement *b.* is now readily obtained by induction.

Since trivially $\{T_0(k;\cdot)\} <\cdot \{G(k;\cdot)\}$, it follows by induction that

$$\{T_{\nu+1}(k;\cdot)\} = \sigma\{T_\nu(k;\cdot)\} <\cdot \sigma\{G(k;\cdot)\} = \{G(k;\cdot)\},$$

so that *c.* is valid.

For every $k \geqslant 1$, the matrices $T_\nu(k;\cdot)$ converge element-wise and in the sense of weak convergence of mass-functions to a matrix $T(k;\cdot)$ as $\nu \to \infty$. The sequence $\{T(k;\cdot)\}$ is clearly in C_0. A straightforward application of the dominated convergence theorem shows that the sequence $\{T(k;\cdot)\}$ satisfies the equations (2.2.8). If the sequence $\{S(k;\cdot)\}$ is a fixed point of the mapping σ in C, then it is clear by induction that $\{T_\nu(k;\cdot)\} <\cdot \{S(k;\cdot)\}$, for $\nu \geqslant 0$, and therefore $\{T(k;\cdot)\} <\cdot \{S(k;\cdot)\}$. •

Theorem 2.2.2: The sequence $\{G(k;\cdot)\}$ is the minimal nonnegative solution of the equations (2.2.8). It is their unique solution in the set C.

Proof: Upon inspection of the equations (2.2.8) and (2.2.10), we see that for $\nu \geqslant 1$, and $1 \leqslant k \leqslant \nu$, $T_\nu(k;\cdot) = G(k;\cdot)$. Since the iterates $\{T_\nu(k;\cdot)\}$ converge to $\{T(k;\cdot)\}$, the matrices $T(k;\cdot)$, and $G(k;\cdot)$ are equal for all $k \geqslant 1$. The same argument also shows the uniqueness of the solution in C.

The transform matrices $\tilde{T}_\nu(z,s)$ are defined by

$$\tilde{T}_\nu(z,s) = \sum_{k=1}^{\infty} z^k \int_0^\infty e^{-sx}\, dT_\nu(k;x),$$

and clearly satisfy

$$\tilde{T}_0(z,s) = 0, \quad \tilde{T}_{\nu+1}(z,s) = z \sum_{r=0}^{\infty} \tilde{A}_r(s)\tilde{T}_\nu^r(z,s), \tag{2.2.13}$$

for $\nu \geqslant 0$, $|z| \leqslant 1$, Re $s \geqslant 0$. The fact that the iterates $\{T_\nu(k;\cdot)\}$ are non-decreasing - or a simple direct argument - implies that the sequence of matrices $\{\tilde{T}_\nu(z,s)\}$ is non-decreasing for $0 \leqslant z \leqslant 1$, $s \geqslant 0$. The convergence of the iterates $\{T_\nu(k;\cdot)\}$ to the sequence $\{G(k;\cdot)\}$ implies, by the dominated convergence theorem, that the sequence $\{\tilde{T}_\nu(z,s)\}$ converges to $\tilde{G}_\nu(z,s)$ for all z and s, satisfying $|z| \geqslant 1$, Re $s \geqslant 0$. •

A Probabilistic Interpretation of the Matrices $T_\nu(k;x)$: The matrices $T_\nu(k;x)$ have a probabilistic interpretation, which was used by D. M. Lucantoni to construct a different proof of Theorem 2.2.2. This interpretation, which apparently had not been noted, even for the classical $M/G/1$ queue, offers a way of understanding the recursive scheme in (2.2.13) in terms of monotone sequences of events. It will not be further needed in this book.

The element $[T_\nu(k;x)]_{jj'}$, is the conditional probability that, starting in the state $(i+1,j)$, the Markov renewal process hits the level i for the first time after exactly k transitions and no later than time x, *by proceeding along a path restricted to the set of paths* $\tau_\nu(j,j')$. The sets of paths $\tau_\nu(j,j')$ are recursively defined for $\nu \geqslant 1$, as follows. The set $\tau_1(j,j')$ contains only the path which goes from $(i+1,j)$ to (i,j') in one transition. The set $\tau_{\nu+1}(j,j')$ contains the path in $\tau_1(j,j')$ and all paths which in their first step lead to some state $(i+r,j'')$, $r \geqslant 1$, and from that state to (i,j') by r consecutive paths belonging to the sets $\tau_\nu(j^\circ, j^{\circ\circ})$, where the indices j° and $j^{\circ\circ}$ range over $\{1, \ldots, m\}$. A careful consideration of these matrices of probabilities leads to the equations (2.2.13). Lucantoni's proof then proceeds, by a combinatorial argument, to show that the matrices $T_\nu(k;\cdot)$ converge weakly to the matrix $G(k;\cdot)$, as ν tends to infinity.

Restricting our attention to the sequence $\{G(k)\}$, where $G(k) = G(k;\infty)$, we see that its elements satisfy

$$G(1) = A_0, \quad G(k) = \sum_{\nu=1}^{k-1} A_\nu G^{(\nu)}(k-1), \quad \text{for } k \geqslant 2. \tag{2.2.14}$$

By repeating verbatim the proofs of Lemma 2.2.2 and Theorem 2.2.2, it follows that the sequence $\{G(k)\}$ is the unique solution to these equations in the set of sequences of nonnegative matrices of order m.

By summing the equations (2.2.14), we obtain that the matrix G satisfies

$$G = \sum_{\nu=0}^{\infty} A_\nu G^\nu, \tag{2.2.15}$$

and also that the sums $G_r = \sum_{k=1}^{\infty} G_r(k)$, where the sequence $\{G_r(k), k \geq 1\}$ is the r-th iterate obtained by successive substitutions in (2.2.14), correspond to the matrices obtained by forming the successive substitutions

$$G_0 = 0, \quad G_r = \sum_{\nu=0}^{\infty} A_\nu G_{r-1}^\nu, \quad \text{for } r \geq 1,$$

in (2.2.15). This readily implies that the important matrix G is the minimal solution of (2.2.15) in the set of nonnegative matrices. We know that matrix G is substochastic.

In general, G is not the *unique* nonnegative matrix which satisfies (2.2.15). This is already seen in the scalar case. If the ordinary $M/G/1$ queue is unstable, the equation

$$z = h(\lambda - \lambda z),$$

to which (2.2.15) then reduces, has two roots $z = g < 1$, and $z = 1$, in the interval [0,1]. In the matrix case, the minimal solution G may be strictly substochastic, yet Brouwer's fixed point theorem guarantees the existence of a second solution to (2.2.15) in the set of *stochastic* matrices. It is therefore of utmost importance that the matrix G, which is of interest to the discussions in this book, be identified as the *minimal* nonnegative solution to (2.2.15).

In a similar manner, by summing the equations (2.2.8) on k, we see that the matrix $G(\cdot)$ is the monotone limit of the matrices $G_r(\cdot)$, obtained by forming the successive substitutions

$$G_0(x) = 0, \quad G_r(x) = \sum_{\nu=0}^{\infty} A_\nu(\cdot) * G_{r-1}^{(\nu)}(x), \quad \text{for } r \geq 1, x \geq 0.$$

This implies that the matrix $G(\cdot)$ is the *minimal* solution to the integral equation

$$G(x) = \sum_{\nu=0}^{\infty} A_\nu(\cdot) * G^{(\nu)}(x), \quad \text{for } x \geqslant 0,$$

in the set of $m \times m$ matrices, whose elements are mass-functions on $[0,\infty)$. The matrix $G(\cdot)$ belongs to the subset of matrices $S(\cdot)$ for which $S(\infty)\,\mathbf{e} \leqslant \mathbf{e}$. These matrices are called (substochastic) *semi-Markov matrices*.

2.3. THE MATRIX G

In this section, we make a detailed study of the matrix $G = G(\infty) = \sum_{k=1}^{\infty} G(k)$. We shall henceforth limit our attention to the case where the Markov renewal process $Q(\cdot)$ is *irreducible*. As is well-known, $Q(\cdot)$ is irreducible if and only if the embedded Markov chain with transition probability matrix $Q = Q(\infty)$ is irreducible. From Corollaries 2.2.1 and 2.2.2, we know that the matrix G then cannot have zero rows and that its spectral radius η is positive. We shall now concentrate on the conditions under which G is *stochastic*.

With the sequence of matrices $\{A_\nu, \nu \geqslant 0\}$, we associate the matrix of generating functions

$$A^*(z) = \sum_{\nu=0}^{\infty} A_\nu z^\nu, \quad \text{for } |z| \leqslant 1, \tag{2.3.1}$$

and the stochastic matrix

$$A = \sum_{\nu=0}^{\infty} A_\nu = A^*(1-). \tag{2.3.2}$$

The matrix G satisfies the equation

$$G = \sum_{\nu=0}^{\infty} A_\nu G^\nu, \tag{2.3.3}$$

and is, by virtue of the discussion following Theorem 2.2.2, its minimal nonnegative solution.

By the probabilistic interpretation of the matrix G, we see that a first ***necessary*** condition for ***recurrence*** of the Markov renewal process $Q(\cdot)$ in (2.1.9) is that the ***minimal solution to*** (2.3.3) ***in the set of nonnegative matrices is stochastic.*** It is of considerable interest to derive ***tractable analytic conditions*** which are equivalent to that somewhat elusive condition.

Under the additional assumption that the stochastic matrix A is ***irreducible,*** it is possible to derive a simple explicit condition for G to be stochastic. We now proceed to discuss that case, which covers most applications, in detail. The specific analysis needed to handle the cases where A is reducible will be discussed thereafter. When the matrix A is irreducible, there exists a unique (positive) vector $\boldsymbol{\pi}$, which satisfies

$$\boldsymbol{\pi} A = \boldsymbol{\pi}, \quad \boldsymbol{\pi}\mathbf{e} = 1. \tag{2.3.4}$$

The vector $\boldsymbol{\pi}$ is the ***invariant probability vector*** of the matrix A. The vector $\boldsymbol{\beta}$ is defined by

$$\boldsymbol{\beta} = \sum_{\nu=1}^{\infty} \nu A_{\nu}\mathbf{e}. \tag{2.3.5}$$

The components of $\boldsymbol{\beta}$ are either finite and nonnegative or $+\infty$. The inner product $\boldsymbol{\pi\beta}$ is therefore finite if and only if all components of the vector $\boldsymbol{\beta}$ are finite.

The discussion that follows is aimed at proving that with the matrix A ***irreducible,*** the matrix G is stochastic if and only if the inequality

$$\boldsymbol{\pi\beta} \leqslant 1, \tag{2.3.6}$$

holds. Before we embark upon this rather lengthy discussion, we examine the inequality (2.3.6) for the illustrative examples in Section 2.1.

Example A : For the $M/\ SM/\ 1$ queue, the vector $\boldsymbol{\pi}$ is the invariant probability vector of the matrix $H = H(\infty) = A$. The vector $\boldsymbol{\beta}$ is then given by

$$\boldsymbol{\beta} = \sum_{\nu=0}^{\infty} \nu \int_0^{\infty} e^{-\lambda u} \frac{(\lambda u)^{\nu}}{\nu!}\, dH(u)\mathbf{e} = \lambda \int_0^{\infty} u\ dH(u)\mathbf{e} = \lambda\, \boldsymbol{\alpha},$$

where α is the column vector of the row sum means of $H(\cdot)$. The inequality (2.3.6) then says that $\lambda\pi\alpha$, the mean number of arrivals during an "average" service time, should not exceed one.

Example B: For Bailey's bulk queue and its variants, the stochastic matrix A is a *circulant,* that is a matrix of the form

$$\begin{vmatrix} \tilde{A}_0 & \tilde{A}_1 & \tilde{A}_2 & \cdots & \tilde{A}_{m-1} \\ \tilde{A}_{m-1} & \tilde{A}_0 & \tilde{A}_1 & \cdots & \tilde{A}_{m-2} \\ \tilde{A}_{m-2} & \tilde{A}_{m-1} & \tilde{A}_0 & \cdots & \tilde{A}_{m-3} \\ \cdots & & \cdots & & \cdots \\ \tilde{A}_1 & \tilde{A}_2 & \tilde{A}_3 & \cdots & \tilde{A}_0 \end{vmatrix} .$$

In the notation of Example B of Section 2.1, the quantities $\tilde{A}_r$, $0 \leqslant r \leqslant m-1$, are given by

$$\tilde{A}_r = \sum_{\nu=0}^{\infty} a_{m\nu+r}(\infty).$$

A stochastic circulant is clearly doubly stochastic, so that the vector π is given by

$$\pi_j = \frac{1}{m}, \quad \text{for } 0 \leqslant j \leqslant m-1.$$

The component β_j, $0 \leqslant j \leqslant m-1$, of the vector β is given by

$$\beta_j = \sum_{\nu=1}^{\infty} \nu \sum_{r=0}^{m-1} a_{m\nu+r-j}(\infty).$$

If N is an integer-valued random variable with the probability density $\{ a_r(\infty) \}$, it is easily seen that

$$\beta_j = E\left[\frac{N+j}{m} \right], \quad \text{for } 0 \leqslant j \leqslant m-1.$$

where $[\cdot]$ is the integer part function. It follows that

$$\pi\beta = \frac{1}{m}\sum_{j=0}^{m-1} E[\frac{N+j}{m}] = \frac{1}{m} E \sum_{j=0}^{m-1} [\frac{N+j}{m}] = \frac{1}{m}E(N).$$

For Bailey's bulk queue, the inequality (2.3.6) reduces to the familiar equilibrium condition $\lambda\alpha \leqslant m$, where α is the mean of the probability distribution $H(\cdot)$.

Example C: For the dam model with Markovian inputs, let π be the invariant probability vector of the matrix B. For the matrix P_1, the column vector β has the components

$$\beta_j = \sum_{r=0}^{m} rb_{jr}, \quad \text{for } 0 \leqslant j \leqslant m,$$

so that

$$\pi\beta = \sum_{r=1}^{m} \sum_{j=0}^{m} r\pi_j b_{jr} = \sum_{r=1}^{m} r\pi_r .$$

For the matrix P_2, the column vector β has the components $0, 1, \ldots, m$, so that, as is to be expected

$$\pi\beta = \sum_{r=1}^{m} r\pi_r .$$

In either case, the inequality (2.3.6) states that the average inflow per time unit should not exceed the outflow of one unit of water.

For Example D, the matrix A is reducible, so that the preceding discussion is not germane.

Lemma 2.3.1: When the matrix A is irreducible, the right eigenvector $\mathbf{u}$ of the substochastic matrix G, corresponding to its maximal eigenvalue η, may be chosen to be *positive.* The Perron eigenvalue of the matrix $A^*(\eta)$ is η.

Proof: Since G is a nonnegative matrix with a positive maximal eigenvalue η, the corresponding right eigenvector $\mathbf{u}$ may be chosen to be nonnegative. Since G is substochastic, we have that $0 < \eta \leqslant 1$. Postmultiplying in the equation (2.3.3) by $\mathbf{u}$, we obtain that

$$G\mathbf{u} = A^*(\eta)\,\mathbf{u} = \eta\mathbf{u}.$$

Because A is irreducible and $0 < \eta \leqslant 1$, the matrix $A^*(\eta)$ is irreducible. It has η as one of its eigenvalues and the corresponding (right) eigenvector is nonnegative. It follows from the Perron-Frobenius theory that η is therefore the maximal eigenvalue of $A^*(\eta)$ and also that the vector $\mathbf{u}$ is strictly positive. •

Corollary 2.3.1: When A is irreducible, the maximal eigenvalue η of G is of geometric multiplicity one.

Proof: The statement is equivalent to saying that to the eigenvalue η there cannot correspond two linearly independent eigenvalues of G. If we assume the contrary, then there is a vector $\mathbf{v}$, linearly independent of $\mathbf{u}$, for which $G\mathbf{v} = \eta\,\mathbf{v}$. As in the proof of Lemma 2.3.1, this implies that $A^*(\eta)\,\mathbf{v} = \eta\,\mathbf{v}$. However, it is well-known that the irreducible matrix $A^*(\eta)$ cannot have two linearly independent (right) eigenvectors corresponding to its Perron root. The vectors $\mathbf{u}$ and $\mathbf{v}$ are therefore linearly dependent and the corollary follows by contradiction. •

Lemma 2.3.2: The matrix G is the unique nonnegative matrix which solves the equation (2.3.3) and has the maximal eigenvalue η.

Proof: Let G° be a nonnegative matrix of maximal eigenvalue η which satisfies (2.3.3), then $G^\circ \geqslant G$, since G is the minimal nonnegative solution. If $\mathbf{v}$ is a nonnegative right eigenvector of G°, then also $A^*(\eta)\,\mathbf{v} = \eta\,\mathbf{v}$. Since η is the Perron eigenvalue of the irreducible nonnegative matrix $A^*(\eta)$, it follows that $\mathbf{v}$ is a multiple of the right eigenvector $\mathbf{u}$ of G. Since $\mathbf{u}$ is positive, it follows that the obvious equality $(G^\circ - G)\,\mathbf{u} = 0$, can only hold if $G^\circ = G$. •

Corollary 2.3.2: With A irreducible, if the matrix G is stochastic, it is the unique solution to equation (2.3.3) in the set of substochastic matrices.

Proof: Since the matrix G is the minimal nonnegative solution to (2.3.3), there cannot exist a solution which is properly substochastic. Lemma 2.3.2 implies that there cannot be another nonnegative solution with spectral radius one, and hence no other stochastic solution. •

The matrix $A^*(z)$ is irreducible for $0 < z \leqslant 1$, if A is irreducible. Let $\chi(z)$ be the Perron eigenvalue of $A^*(z)$ for $0 < z \leqslant 1$, and let $\chi(0) = sp\,(A_0)$, by right hand continuity. It then follows from Lemma 2.3.1 and the uniqueness of the Perron eigenvalue of $A^*(\eta)$, that

$$\eta = \chi(\eta). \tag{2.3.7}$$

We shall obtain further information on $\eta = sp(G)$, by examining the roots in [0,1] of the equation

$$z = \chi(z). \tag{2.3.8}$$

Lemma 2.3.3: The left hand derivative $\chi'(1-)$ of $\chi(z)$ at $z = 1$ is given by

$$\chi'(1-) = \boldsymbol{\pi}\boldsymbol{\beta}. \tag{2.3.9}$$

Proof: For $0 < z \leqslant 1$, let $\mathbf{u}(z)$ and $\mathbf{v}(z)$ be respectively left and right eigenvectors of $A^*(z)$, corresponding to its maximal eigenvalue $\chi(z)$. The vectors $\mathbf{u}(z)$ and $\mathbf{v}(z)$ may be chosen so that their components are continuously differentiable functions of z and such that the normalizations

$$\mathbf{u}(z)\mathbf{v}(z) = \mathbf{u}(z)\mathbf{e} = 1, \quad \mathbf{u}(1-) = \boldsymbol{\pi}, \quad \mathbf{v}(1-) = \mathbf{e}, \tag{2.3.10}$$

hold for $0 < z \leqslant 1$.

Differentiation of the terms in the equation

$$A^*(z)\mathbf{v}(z) = \chi(z)\mathbf{v}(z), \tag{2.3.11}$$

leads to

$$[\chi(z)I - A^*(z)]\mathbf{v}'(z) = [A^{*\prime}(z) - \chi'(z)I]\,\mathbf{v}(z). \tag{2.3.12}$$

Premultiplying by $\mathbf{u}(z)$ in (2.3.12), we see that

$$\chi'(z) = \mathbf{u}(z)\,A^{*\prime}(z)\,\mathbf{v}(z), \quad \text{for } 0 < z \leqslant 1. \tag{2.3.13}$$

Now letting z tend to one from below, we obtain (2.3.9). •

Remark: The higher derivatives of $\chi(z)$, $\mathbf{u}(z)$ and $\mathbf{v}(z)$ at $z = 1$ are of interest to other aspects of Markov renewal processes of $M/G/1$ type. They are discussed in the Appendix.

Lemma 2.3.4: The equation $z = \chi(z)$, has at most two roots in [0,1], one of which is $z = 1$. If $\pi\beta \leqslant 1$, $z = 1$ is the only root. If $\pi\beta > 1$, there is a second root $z = \xi$, $0 < \xi < 1$, provided that for some z_0, in (0,1), we have

$$z\chi'(z) < \chi(z), \quad \text{for } 0 < z < z_0. \tag{2.3.14}$$

If $\chi(0) > 0$, condition (2.3.14) is always satisfied.

Proof: By the change of variable $z = e^{-s}$, for $s \geqslant 0$, the equation (2.3.8) may be written as

$$s = \log \chi(e^{-s}), \quad \text{for } s \geqslant 0. \tag{2.3.15}$$

Since the matrix A is stochastic, that equation is satisfied by $s = 0$. An important theorem of J. F. C. Kingman, of which the proof may be found in the Appendix, asserts that the right hand side of (2.3.15) is a *concave* function of s for $s \geqslant 0$. It is clearly also positive and increasing for $s > 0$. The derivative of the right hand side at $s = 0+$ is equal to $\chi'(1-) = \pi\beta$.

If $\pi\beta \leqslant 1$, the graph of the right hand side never crosses above the bisectrix line, so that $s = 0$, is the unique solution of (2.3.15).

If $\pi\beta > 1$, and $\chi(0) > 0$, the graph of the right hand side swings above the bisectrix and then approaches a horizontal asymptote as $s \to +\infty$. There is then a unique second intersection at some value $s_0 > 0$, and the corresponding solution ξ to (2.3.8) in (0, 1) is given by $\xi = e^{-s_0}$.

If $\pi\beta > 1$, and $\chi(0) = 0$, there will be a unique second intersection at some point $s_0 > 0$, provided that the derivative of the right hand side is eventually less than one. This is equivalent to the requirement (2.3.14). •

Lemma 2.3.5: The maximal eigenvalue η of G is the smallest positive root of equation (2.3.8).

Proof: We know that in an irreducible Markov renewal process of $M/G/1$ type, the matrix G has a positive maximal eigenvalue η (Corollary 2.2.2). Let η_0 be the smallest positive solution of (2.3.8) and we denote by $\mathbf{u}^\circ$ a positive right eigenvector of the irreducible matrix $A^*(\eta_0)$, corresponding to its maximal eigenvalue $\chi(\eta_0) = \eta_0$.

Now consider the sequence of matrices $\{G_\nu\}$ defined by

$$G_0 = 0, \quad G_{\nu+1} = \sum_{r=0}^{\infty} A_r G_\nu^r \quad \text{for } \nu \geq 0.$$

We then trivially have that $G_0 \mathbf{u}^\circ < \eta_0 \mathbf{u}^\circ$. Furthermore, if $G_\nu \mathbf{u}^\circ \leq \eta_0 \mathbf{u}^\circ$, then

$$G_{\nu+1} \mathbf{u}^\circ = \sum_{r=0}^{\infty} A_r G_\nu^r \mathbf{u}^\circ \leq A^*(\eta_0) \mathbf{u}^\circ = \eta_0 \mathbf{u}^\circ.$$

The matrices G_ν converge monotonically to the matrix G and therefore

$$G \mathbf{u}^\circ \leq \eta_0 \mathbf{u}^\circ.$$

Since the vector $\mathbf{u}^\circ$ is positive, this implies that $sp(G) = \eta \leq \eta_0$. Since η is also a positive solution to (2.3.8), it follows that $\eta = \eta_0$. •

Remark: The condition (2.3.14) serves to preclude the highly degenerate case where the minimal nonnegative solution G to equation (2.3.3) has all its eigenvalues equal to zero. It is easy to give examples of coefficient matrices A_ν, $\nu \geq 0$, for which this occurs, but by Lemma 2.2.2, the Markov renewal process $Q(\cdot)$ is necessarily reducible in such cases.

The essential points of the preceding lemmas may be summarized as follows:

Theorem 2.3.1: If the matrix A is irreducible, the matrix G is stochastic if and only if the inequality $\pi\beta \leq 1$, holds. That inequality is a necessary condition for the recurrence of the Markov renewal process $Q(\cdot)$.

Structural Properties of the Matrix G: We shall now examine some structural properties of the matrix G. The following elementary observation is useful in many applications. It does not depend on the irreducibility of A.

Lemma 2.3.6: To every zero column of the matrix A_0, there is a corresponding zero column in the matrix G.

Proof: This follows directly from the fact that G is the minimal nonnegative solution of equation (2.3.3). If we consider the sequence of iterates

$$G_0 = 0, \quad G_{\nu+1} = \sum_{r=0}^{\infty} A_r G_\nu^r, \quad \text{for } \nu \geqslant 0,$$

then clearly $G_1 = A_0$, and the matrix

$$G_2 = \sum_{r=0}^{\infty} A_r A_0^r = [\, I + \sum_{r=1}^{\infty} A_r A_0^{r-1} \,] \, A_0,$$

inherits all the zero columns from A_0. It follows by induction that the zero columns of A_0 are preserved in all the matrices G_ν and therefore in G. •

Remark *a.* Lemma 2.3.6 immediately implies that in Example C. of Section 2.1, the matrix G for the Markov chain P_1 has its first column positive and all others zero. Provided the condition $\sum_{r=1}^{m} r \pi_r \leqslant 1$, holds, all elements in the first column of G are equal to one. By contrast, when $b_{0j} > 0$ for $0 \leqslant j \leqslant m$, the iterate G_m for the Markov chain P_2 is strictly positive and therefore also the matrix G. As we have noted, the Markov chains P_1 and P_2 are equivalent descriptions of the same stochastic model, but in the recurrent case, the matrix G for P_1 is obtained without any computation.

Remark *b.* The irreducibility of any iterate G_{ν_0} implies that all iterates G_ν with $\nu \geqslant \nu_0$ are irreducible and therefore also the matrix G. In the $M/\ SM/\ 1$ queue of Example A, the matrix A_0 is irreducible since we have assumed that the stochastic matrix $H(\infty) = A$ is irreducible. The matrix G is therefore irreducible. For that case, we may rewrite the equation (2.3.3) as

$$G = \int_0^{\infty} dH(u) \exp[-\lambda u \, (I - G)],$$

in terms of the matrix exponential function. However, by a classical property of the matrix exponential, the matrix $\exp\,[-\lambda u \, (I - G)]$ is strictly positive for $u > 0$. For the $M/\ SM/\ 1$ queue, the matrix G is therefore *strictly positive.* We may often use this same argument with minor modifications to show that the matrices G of other queueing models are positive. We leave it to the initiative of the reader to show, for example, that the matrix G for Bailey's queue is always positive. This may, in fact, already be seen by examining the matrix G_2.

Remark *c.* A necessary and sufficient condition on the locations of the positive elements of the coefficient matrices A_ν, $\nu \geqslant 0$, which guarantees irreducibility of the matrix G is not yet known. The positive elements of the matrix G may be located by replacing the matrices A_ν by the corresponding incidence matrices $\hat{A}_\nu$, $\nu \geqslant 0$, and by examining the equation

$$\hat{G} = \sum_{\nu=0}^{\infty} \hat{A}_\nu \hat{G}^\nu,$$

in which all operations are performed in *Boolean arithmetic*. That equation may be solved by successive substitutions starting with the matrix $\hat{A}_0$ (in Boolean arithmetic). It is easily seen that *in fewer than* m^2 *iterations,* a pattern of zeros and ones will emerge which is repeated in all future iterations. It is therefore always possible to determine the pattern of zeros and ones in $\hat{G}$ and therefore the "class structure" of the matrix G. In all applications to date, the structure of the matrix G has been clear by inspection or by elementary arguments. In particular, cases where G is irreducible or positive are usually identifiable by applying the elementary Theorem 2.3.2. A combinatorial examination of the complete structure of G in terms of that of the coefficient matrices A_ν remains an open problem of interest.

Theorem 2.3.2: If the matrix $(I-A_1)^{-1}A_0$ is irreducible, so is G. If in addition, for some $\nu_0 \geqslant 1$, the matrix $\sum_{k=1}^{\infty} A_k G_{\nu_0}^{k-1}$ is irreducible, the matrix G is positive.

Proof: If the matrix $I - A_1$ is singular, some subset of each level i, $i \geqslant 1$, is a recurrent class and the Markov chain Q is therefore reducible. Since we have assumed that Q is irreducible, the inverse of $I - A_1$ exists. The matrix $(I - A_1)^{-1}A_0$ is the first iterate in the equivalent form

$$G = (I-A_1)^{-1} \sum_{\nu \neq 1} A_\nu G^\nu,$$

of equation (2.3.3). It follows that $G \geqslant (I-A_1)^{-1} A_0$, and therefore that G is irreducible whenever the matrix $(I-A_1)^{-1} A_0$ is.

If for some $\nu_0 \geqslant 1$, the matrix $\sum_{k=1}^{\infty} A_k G_{\nu_0}^{k-1}$ is irreducible, then since $G \geqslant G_{\nu_0}$, the same is true for the matrix $\sum_{k=1}^{\infty} A_k G^{k-1}$. We write the equation (2.3.3) as

$$G = [I - \sum_{k=1}^{\infty} A_k G^{k-1}]^{-1} A_0. \tag{2.3.16}$$

If G is irreducible, then clearly A_0 cannot have zero columns. We shall show that the inverse in (2.3.16) exists and is positive. This clearly implies that G is positive.

To show that the inverse exists, consider a positive right eigenvector $\mathbf{u}$ of G corresponding to its maximal eigenvalue η, then

$$\sum_{k=1}^{\infty} A_k G^{k-1} \mathbf{u} = \eta^{-1} [A^*(\eta) - A_0] \mathbf{u} \leqslant \mathbf{u},$$

with strict inequality for some components. Since $\mathbf{u}$ is positive, this implies that the irreducible matrix $\sum_{k=1}^{\infty} A_k G^{k-1}$ has spectral radius less than one, so that the inverse in (2.3.16) exists.

The matrix $C = \sum_{k=1}^{\infty} A_k G^{k-1}$, is irreducible if and only if the matrix $I + C + \cdots + C^{m-1}$ is positive, but clearly $(I-C)^{-1} \geqslant I + C + \cdots + C^{m-1}$, so that the inverse is positive. •

Corollary 2.3.3: When the inverse of $I - \sum_{k=1}^{\infty} A_k G^{k-1}$ exists, the matrices A_0 and G have the same rank. This is always the case when the matrix A is irreducible.

Proof: This follows immediately from equation (2.3.16). •

The same arguments as used in the proof of Theorem 2.3.2 lead to information on the positivity of the remaining elements of G when certain columns of A_0 and therefore of G are zero. Suppose that A_0 has some zero columns, then by appropriately relabeling rows and columns, we may cast A_0 and G into the forms

$$A_0 = \begin{vmatrix} A_0(1) & 0 \\ A_0(3) & 0 \end{vmatrix}, \qquad G = \begin{vmatrix} G(1) & 0 \\ G(3) & 0 \end{vmatrix}.$$

Equation (2.3.3) then leads to

$$G(1) = \sum_{k=0}^{\infty} A_k(1) G^k(1) + \sum_{k=1}^{\infty} A_k(2) G(3) G^{k-1}(1), \tag{2.3.17}$$

$$G(3) = \sum_{k=0}^{\infty} A_k(3) G^k(1) + \sum_{k=1}^{\infty} A_k(4) G(3) G^{k-1}(1),$$

where $A_k(2)$ and $A_k(4)$ are the rightmost blocks in

$$A_k = \begin{vmatrix} A_k(1) & A_k(2) \\ A_k(3) & A_k(4) \end{vmatrix}, \qquad \text{for } k \geqslant 1.$$

The first equation in (2.3.17) implies that

$$G(1) \geqslant \sum_{k=0}^{\infty} A_k(1) G^k(1),$$

so that, by the same argument as in the proof of Theorem 2.3.2, the irreducibility of $[I - A_1(1)]^{-1} A_0(1)$ shows that the matrix $G(1)$ is irreducible. Similarly, if $\sum_{k=1}^{\infty} A_k(1) G^{k-1}(1)$ is irreducible, the inequality

$$G(1) \geqslant [I - \sum_{k=1}^{\infty} A_k(1) G^{k-1}(1)]^{-1} A_0(1),$$

implies that $G(1)$ is positive. •

A complete examination of the reducibility structure of the matrix G is belabored, even for the case where the matrix A is irreducible. For most applications, the results we have discussed are adequate to identify the cases where G is irreducible and even positive. When the matrix A_0 has zero columns, we can usually establish the irreducibility of $G(1)$. When $G(1)$ is positive, it usually follows in an elementary fashion that the matrix $G(3)$ is also positive.

Reducibility of the Matrix A: Cases where the matrix A is reducible occur in rare, but important practical applications. While the mathematical theory of such cases is more involved, their particular structure usually leads to major algorithmic simplifications. We begin by appealing to the general decomposition theorem from the theory of Markov chains with a finite number of states. It is well-known that by a common permutation of the row and column indices, the reducible stochastic matrix A may be cast into the form

$$A = \begin{vmatrix} A(1) & 0 & \cdots & 0 & 0 \\ 0 & A(2) & \cdots & 0 & 0 \\ \cdots & & \cdots & & \cdots \\ 0 & 0 & \cdots & A(c) & 0 \\ & & & & \\ T(1) & T(2) & \cdots & T(c) & T(0) \end{vmatrix}, \qquad (2.3.18)$$

where the matrices $A(1), \ldots, A(c)$ are *irreducible* stochastic matrices of orders $m_1, \ldots, m_c$, with $m_1 + \cdots + m_c \leqslant m$. The bottom row, which may be absent, consists of the substochastic matrix $T(0)$ of order m_0 and of matrices $T(1), \ldots, T(c)$ of dimensions $m_0 \times m_j$, $1 \leqslant j \leqslant c$. In the terminology of Markov chains, the matrices $A(1), \ldots, A(c)$ correspond to the irreducible recurrent classes, while the row indices of the bottom row correspond to the transient states. When transient states are present, the fact that the Markov chain cannot stay in the set of transient states forever readily implies the nonsingularity of the matrix $I - T(0)$.

If the matrix A has the block-partitioned form (2.3.18), the same is true for all the matrices A_k, $k \geqslant 0$, that is, each A_k has zero blocks where A has zero blocks. In individual matrices A_k, the diagonal blocks $A_k(j)$, $1 \leqslant j \leqslant c$, need not be irreducible.

Let us refer to the structure displayed in (2.3.18) as the *class structure*. It is now readily verified that all the matrices G_ν, $\nu \geqslant 1$, generated by the recurrence relation

$$G_1 = A_0, \quad G_{\nu+1} = \sum_{r=0}^{\infty} A_r G_\nu^r, \quad \text{for } \nu \geqslant 1,$$

have the same class structure as A. The minimal nonnegative solution G of equation (2.3.3) is therefore a matrix of the form

$$G = \begin{vmatrix} G(1) & 0 & \cdots & 0 & 0 \\ 0 & G(2) & \cdots & 0 & 0 \\ \cdots & & \cdots & & \cdots \\ 0 & 0 & \cdots & G(c) & 0 \\ \hat{G}(1) & \hat{G}(2) & \cdots & \hat{G}(c) & \hat{G}(0) \end{vmatrix}, \qquad (2.3.19)$$

and it is immediate that the matrices $G(1), \ldots, G(c)$ are the minimal nonnegative solutions to the c nonlinear matrix equations

$$G(j) = \sum_{r=0}^{\infty} A_r G^r(j), \quad \text{for } 1 \leqslant j \leqslant c. \qquad (2.3.20)$$

The necessary and sufficient condition for the matrix $G(j)$ to be *stochastic* is now immediately obtained from Theorem 2.3.1. If the vector $\pi(j)$ is the solution of the equations

$$\pi(j)A(j) = \pi(j), \quad \pi(j)\mathbf{e} = 1,$$

and the vector $\beta(j)$ is defined by

$$\beta(j) = \sum_{\nu=1}^{\infty} \nu A_\nu(j)\mathbf{e},$$

then $G(j)$ is stochastic if and only if

$$\pi(j)\beta(j) \leqslant 1. \qquad (2.3.21)$$

It is clear that for the matrix G to be stochastic, the inequality (2.3.21) must hold for all j with $1 \leqslant j \leqslant c$. We shall now show that this is also the *sufficient* condition for G to be a stochastic matrix.

Theorem 2.3.3: When the matrix A is reducible and written in the form (2.3.18), the necessary and sufficient condition for G to be stochastic is that

$$\pi(j)\beta(j) \leqslant 1, \quad \text{for } 1 \leqslant j \leqslant c.$$

Proof: The theorem is obviously valid when the matrix A is block-diagonal, that is when the bottom row of blocks is absent. When there are transient states in A, it suffices to prove that if the matrices $G(j)$, $1 \leqslant j \leqslant c$, are all stochastic, then

$$\sum_{r=0}^{c} \hat{G}(r)\mathbf{e} = \mathbf{e}.$$

This requires some calculation. The k-th power of the nonnegative matrix G in (2.3.19) has the same class structure as G and has the diagonal blocks

$$G^k(1), \ldots, G^k(c), \hat{G}^k(0).$$

If we denote the first c blocks in the bottom row of G^k by

$$\hat{G}(1), \ldots, \hat{G}(c),$$

the matrices $\hat{G}_k(j)$, $1 \leqslant j \leqslant c$, are given recursively in k by

$$\hat{G}_1(j) = \hat{G}(j), \quad \hat{G}_{k+1}(j) = \hat{G}(j)G^k(j) + \hat{G}(0)\hat{G}_k(j), \quad \text{for } k \geqslant 1,$$

or explicitly as the sum

$$\hat{G}_k(j) = \sum_{h=0}^{k-1} \hat{G}^h(0)\hat{G}(j)G^{k-h-1}(j), \quad \text{for } k \geqslant 1, \tag{2.3.22}$$

It follows from (2.3.3) that the matrices $\hat{G}(j)$ satisfy the equations

$$\hat{G}(j) = \sum_{k=0}^{\infty} T_k(j)G^k(j) + \sum_{k=1}^{\infty} T_k(0)\hat{G}_k(j), \quad \text{for } 1 \leqslant j \leqslant c, \tag{2.3.23}$$

and

$$\hat{G}(0) = \sum_{k=0}^{\infty} T_k(0)\hat{G}^k(0). \tag{2.3.24}$$

The matrix $\hat{G}(0)$ is substochastic. Its maximal eigenvalue is less than one. To see this, suppose that $\hat{G}(0)$ has maximal eigenvalue one, then there would exist a non-zero, nonnegative vector $\mathbf{u}$ for which $\hat{G}(0)\mathbf{u} = \mathbf{u}$. By equation (2.3.24), this would lead to $\mathbf{u} = T(0)\mathbf{u}$, which is a contradiction since $I - T(0)$ is nonsingular. The matrix $I - \hat{G}(0)$ is therefore nonsingular.

Since the matrix $G(j)$ is stochastic, we obtain from (2.3.22) that

$$\hat{G}_k(j)\mathbf{e} = \sum_{h=0}^{k-1} \hat{G}^h(0)\hat{G}(j)\mathbf{e} = [I - \hat{G}^k(0)]\,[I - \hat{G}(0)]^{-1}\hat{G}(j)\mathbf{e}.$$

Upon substitution into (2.3.23), this in turn implies that

$$\hat{G}(j)\mathbf{e} = T(j)\mathbf{e} + \sum_{k=1}^{\infty} T_k(0)[I - \hat{G}^k(0)]\,[I - \hat{G}(0)]^{-1}\hat{G}(j)\mathbf{e} \qquad (2.3.25)$$

$$= T(j)\mathbf{e} + [T(0) - \sum_{k=0}^{\infty} T_k(0)\hat{G}^k(0)]\,[I - \hat{G}(0)]^{-1}\hat{G}(j)\mathbf{e}$$

$$= T(j)\mathbf{e} + [T(0) - \hat{G}(0)]\,[I - \hat{G}(0)]^{-1}\hat{G}(j)\mathbf{e}.$$

Set now

$$[I - \hat{G}(0)]^{-1} \sum_{j=1}^{c} \hat{G}(j)\mathbf{e} = \mathbf{v},$$

then upon summation over j, the equations (2.3.25) yield

$$[I - \hat{G}(0)]\,\mathbf{v} = \sum_{j=1}^{c} T(j)\mathbf{e} + [T(0) - \hat{G}(0)]\,\mathbf{v},$$

which immediately yields that $[I - T(0)]\,(\mathbf{v} - \mathbf{e}) = 0$, and therefore $\mathbf{v} = \mathbf{e}$. This is equivalent to $\sum_{j=0}^{c} \hat{G}(j)\mathbf{e} = \mathbf{e}$, which was to be proved. •

Remark *a*. We see that Theorem 2.3.3 reduces the conditions for G to be stochastic to the explicit result obtained for the case where the matrix A is irreducible. The matrix G has the highly structured form (2.3.19), in which the blocks $G(j)$, $1 \leqslant j \leqslant c$, may be computed by solving nonlinear matrix equations of the form (2.3.3) for their minimal nonnegative solutions.

Remark *b*. The matrices $\hat{G}(j)$, $1 \leqslant j \leqslant c$, may be computed by successive substitutions in the equations (2.3.23). We shall have occasion to discuss the solution of matrix equations of a similar form in detail later, but it is of interest to note here that the row sums $\mathbf{v}(j) = \hat{G}(j)\,\mathbf{e}$, $1 \leqslant j \leqslant c$, are explicitly given by

$$\mathbf{v}(j) = [I - \hat{G}(0)]\,[I - T(0)]^{-1}T(j)\mathbf{e}.$$

Once the matrix $\hat{G}(0)$ has been computed by an iterative solution of equation (2.3.24), we evaluate the row sum vectors $\mathbf{v}(j)$. These then provide us with a stopping criterion or an accuracy check in the more belabored computation of the matrices $\hat{G}(j)$.

Example: The Case where A is Lower Triangular: The generality of the preceding discussion may leave the impression that cases where the matrix A is reducible require a very involved analysis. This is not the case in actuality, as the class structure of the reducible matrices A that occur in practice is usually very simple. Insight into the general case is, of course, essential in exploiting that simplicity to the fullest. In a commonly occurring case of reducibility, A is upper or lower triangular as in Example *D.* of Section 2.1. Both cases are handled in the same manner. We treat the case where A is lower triangular for consistency of notation with the general discussion. For convenience in displaying matrices, we choose the order m of A to be four and we also assume that only one of the diagonal elements of A is equal to one. That assumption also is purely for convenience of discussion.

The matrix A, which is given by

$$A = \begin{vmatrix} 1 & 0 & 0 & 0 \\ a_{21} & a_{22} & 0 & 0 \\ a_{31} & a_{32} & a_{33} & 0 \\ a_{41} & a_{42} & a_{43} & a_{44} \end{vmatrix},$$

is partitioned as

$$A = \begin{vmatrix} A(1) & 0 \\ T(1) & T(0) \end{vmatrix},$$

and the matrices A_k, $k \geqslant 1$, given by

$$A_k = \begin{vmatrix} a_{11,k} & 0 & 0 & 0 \\ a_{21,k} & a_{22,k} & 0 & 0 \\ a_{31,k} & a_{32,k} & a_{33,k} & 0 \\ a_{41,k} & a_{42,k} & a_{43,k} & a_{44,k} \end{vmatrix},$$

are similarly partitioned as

$$A_k = \begin{vmatrix} A_k(1) & 0 \\ T_k(1) & T_k(0) \end{vmatrix}.$$

By virtue of Theorem 2.3.3, the matrix G is stochastic if and only if $\sum_{k=1}^{\infty} k\ a_{11,k} \leqslant 1$, and it is then of the form

$$G = \begin{vmatrix} 1 & 0 & 0 & 0 \\ g_{21} & g_{22} & 0 & 0 \\ g_{31} & g_{32} & g_{33} & 0 \\ g_{41} & g_{42} & g_{43} & g_{44} \end{vmatrix},$$

which is partitioned as

$$G = \begin{vmatrix} G(1) & 0 \\ \hat{G}(1) & \hat{G}(0) \end{vmatrix}.$$

The lower triangular form of A is preserved in the matrix G. This leads to some further simplifications. For example, the diagonal elements of $\hat{G}(0)$ are the unique solutions in (0, 1) of the equations

$$z = \sum_{k=0}^{\infty} a_{jj,k} z^k, \quad \text{for } 2 \leqslant j \leqslant m.$$

Once these roots g_{jj} have been determined, it is an easy matter to compute the other non-zero elements of $\hat{G}(0)$. The reader may wish to explore various alternative ways of doing this. Since G is stochastic, the computation of $\hat{G}(1)$, which here reduces to a column vector, is effortless.

2.4. RECURRENCE AND THE BOUNDARY STATES

In Sections 2.2 and 2.3, we have discussed the first passage time distributions from states $(i + r, j)$ to a lower level i, $i \geqslant 1$. We now study the first passage time distributions from states $(1,j)$ to the boundary level 0, as well as the return time distributions of states in the level 0. With the help of these, we shall be able to establish the necessary and sufficient condition for the recurrence of Markov renewal processes of $M/G/1$ type. The additional conditions for *positive* recurrence will be obtained in Chapter 3.

Let $L_{jj'}(k;x)$ be the conditional probability that the Markov renewal process, starting in the state $(1,j)$, $1 \leqslant j \leqslant m$, reaches the set 0 after exactly $k \geqslant 1$ transitions and no later than time $x \geqslant 0$, by hitting the state $(0,j')$, $1 \leqslant j' \leqslant m_1$. The matrices $L(k;x)$ have the elements $L_{jj'}(k;x)$ and are of dimensions $m \times m_1$.

We define the joint transform matrix

$$\tilde{L}(z,s) = \sum_{k=1}^{\infty} z^k \int_0^{\infty} e^{-sx}\, dL(k;x), \tag{2.4.1}$$

for $|z| \leqslant 1$, Re $s \geqslant 0$.

Lemma 2.4.1: The matrix $\tilde{L}(z,s)$ satisfies

$$\tilde{L}(z,s) = z\,\tilde{C}_0(s) + \sum_{\nu=1}^{\infty} z\,\tilde{A}_\nu(s)\tilde{G}^{\nu-1}(z,s)\,\tilde{L}(z,s), \qquad (2.4.2)$$

where

$$\tilde{C}_0(s) = \int_0^\infty e^{-sx}\,dC_0(x),$$

and is explicitly given by

$$\tilde{L}(z,s) = [\,I - \sum_{\nu=1}^{\infty} z\,\tilde{A}_\nu(s)\tilde{G}^{\nu-1}(z,s)]^{-1}\,z\,\tilde{C}_0(s). \qquad (2.4.3)$$

If the matrix G is stochastic, we have the equality

$$\tilde{L}(1-,0+)\mathbf{e} = \mathbf{e}. \qquad (2.4.4)$$

Proof: Formula (2.4.2) is obtained by a routine application of the law of total probability for which we consider all possible states of the Markov renewal process after the first transition. The set 0 may be visited after one transition. This contributes the term $z\tilde{C}_0(s)$ to the transform. The first transition may take the process to some level ν. In order to reach the level 0, the process must necessarily return to the level 1 and thence reach the set 0. This contributes the terms

$$z\,\tilde{A}_\nu(s)\tilde{G}^{\nu-1}(z,s)\tilde{L}(z,s), \quad \text{for } \nu \geqslant 1,$$

to the right hand side of (2.4.2).

In order to justify formula (2.4.3), it suffices to show that the inverse exists. For $|z| \leqslant 1$, and s with Re $s \geqslant 0$, the absolute value of each element of the matrix

$$\sum_{\nu=1}^{\infty} z\,\tilde{A}_\nu(s)\tilde{G}^{\nu-1}(z,s)$$

is at most equal to the corresponding element of the matrix $\sum_{\nu=1}^{\infty} A_\nu G^{\nu-1}$. It is therefore sufficient to prove that the matrix $I - \sum_{\nu=1}^{\infty} A_\nu G^{\nu-1}$ is nonsingular. We have already shown this for the case where the matrix A is irreducible in the course of proving Theorem 2.3.2. When A is reducible, we need to take into account the class structure of the matrices A_ν and G. By using (2.3.18) and (2.3.19) it is readily verified that the matrix $\sum_{\nu=1}^{\infty} A_\nu G^{\nu-1}$ inherits the class structure of G. The proof that each of the diagonal blocks in $I - \sum_{\nu=1}^{\infty} A_\nu G^{\nu-1}$ is nonsingular is then entirely routine.

Set $L = \tilde{L}(1-,0+)$, then (2.4.2) leads to

$$L = C_0 + \sum_{\nu=1}^{\infty} A_\nu G^{\nu-1} L. \tag{2.4.5}$$

Setting $L\mathbf{e} = \mathbf{v}$, and noting that $C_0\mathbf{e} = A_0\mathbf{e}$, we see that

$$\mathbf{v} = A_0\mathbf{e} + \sum_{\nu=1}^{\infty} A_\nu G^{\nu-1} \mathbf{v}.$$

If G is stochastic, this equation is clearly satisfied by $\mathbf{v} = \mathbf{e}$. Since the coefficient matrix $I - \sum_{\nu=1}^{\infty} A_\nu G^{\nu-1}$ is nonsingular, that is also the only solution. •

Remark *a*. It is easily seen that the matrix $\tilde{L}^{[r]}(z,s)$ which is the transform matrix of the first passage times from states (r,j), $1 \leqslant j \leqslant m$, in level **r** with $r \geqslant 1$ to the level 0, is given by

$$\tilde{L}^{[r]}(z,s) = \tilde{G}^{r-1}(z,s)\tilde{L}(z,s). \tag{2.4.6}$$

This first passage time transform is of interest in many applications. We shall discuss it in greater detail in the sequel.

Remark *b*. Formulas (2.4.4) and (2.4.6) readily imply that, if G is stochastic, the set 0 is eventually reached from any state (i,j) with $i \geqslant 1$ with probability one.

Next we consider the first passage time distributions (return time distributions) of the set 0 to itself. We now define the quantities $K_{jj'}(k;x)$, $k \geqslant 1$, $x \geqslant 0$, $1 \leqslant j \leqslant m_1$, $1 \leqslant j' \leqslant m_1$. $K_{jj'}(k;x)$ is the conditional probability that the Markov renewal process, starting in the state $(0,j)$, $1 \leqslant j \leqslant m_1$, returns to the set 0 after exactly $k \geqslant 1$, transitions and no later than time $x \geqslant 0$, by hitting the state $(0,j')$, $1 \leqslant j' \leqslant m_1$.

We consider the square matrices $K(k;x) = \{K_{jj'}(k;x)\}$, $k \geqslant 1$, $x \geqslant 0$, and introduce the $m_1 \times m_1$ transform matrix

$$\tilde{K}(z,s) = \sum_{k=1}^{\infty} z^k \int_0^{\infty} e^{-sx}\, dK(k;x), \tag{2.4.7}$$

for $|z| \leqslant 1$, Re $s \geqslant 0$. •

Lemma 2.4.2: The matrix $\tilde{K}(z,s)$ is given by

$$\tilde{K}(z,s) = z\, \tilde{B}_0(s) + \sum_{\nu=1}^{\infty} z\, \tilde{B}_\nu(s)\tilde{G}^{\nu-1}(z,s)\, \tilde{L}(z,s). \tag{2.4.8}$$

If the matrix G is stochastic, so is the matrix $K = \tilde{K}(1-,0+)$.

Proof: Formula (2.4.8) is proved in the same manner as (2.4.2). Starting in a state in 0, the Markov renewal process may visit the set 0 again after the first transition or it may jump to some level ν with $\nu \geqslant 1$. In the latter case, the transform matrix of the first passage time from ν to 0 is given by (2.4.6).

The matrix K is given by

$$K = B_0 + \sum_{\nu=1}^{\infty} B_\nu G^{\nu-1} L, \tag{2.4.9}$$

and is clearly stochastic when G is stochastic. •

Corollary 2.4.1: If the Markov renewal process $Q(\cdot)$ is irreducible, so is the matrix K.

Proof: Consider the Markov renewal process $Q(\cdot)$ at the epochs of its successive visits to the set 0. The bivariate sequence of the states visited and of the times between successive visits forms an m_1-state Markov renewal process, embedded in the Markov renewal process

$Q(\cdot)$. It is clear that the matrix $K(\cdot)$ with transform $\tilde{K}(1,s)$ is the transition probability matrix of that embedded process.

If the matrix K were reducible, there would exist states $(0,j)$ and $(0,j')$ such that, starting in the state $(0,j)$, the probability of ever visiting the set 0 at the state $(0,j')$ is zero. This is clearly equivalent to saying that in the Markov renewal process $Q(\cdot)$, the state $(0,j')$ cannot be reached from $(0,j)$, so that $Q(\cdot)$ is reducible. This contradiction shows that the matrix K must be irreducible. •

NOTES, REFERENCES AND COMMENTS ON CHAPTER 2

Section 2.1

References to the theory of the $M/SM/1$ queue and to bulk service queues are given at the end of Chapter 4, where these models are discussed in detail. The dam with Markovian inputs is treated in Chapter 6. Example D is discussed, with minor additional modelling features, in Neuts [N-057].

In most specific applications, the structural form (2.1.9) of the transition probability matrix $Q(\cdot)$ is readily apparent. In some others, it is necessary to display the particular matrix or, on occasion, to try alternative formalizations of the state space before that structure emerges in the most convenient form. In the examples and problems that follow, the emphasis is on the formulation of the models as Markov chains of $M/G/1$ type.

Problem 2.1.1: An $M/G/1$ queue has the following queue length dependent service discipline. Waiting items are stored in bins of size m. Upon completing a service, the server selects the next bin for processing and must choose an incomplete bin, if there is one. All customers in the selected bin are served together as a group and the service time distribution of a bin depends on the number of customers in it. The successive service times are conditionally independent, given the successive group sizes. Consider the Markov chain embedded at departures, write its transition probability matrix Q and show that it is of the form (2.1.9). •

The model described in Problem 2.1.1 was discussed for $m = 2$ by transform methods in Neuts [N-015]. It served there as an example related to a problem in the theory of state-dependent queues which was significant at that time. The pervasiveness of the structure (2.1.9) was not realized until later.

Problem 2.1.2: A gambler believes that the following behavior will bring him luck. He starts out with a pile of $C > 0$ chips, which he divides up into piles of m chips. The chips that are left over are his incomplete pile. In successive games, he always bets the number of chips in his incomplete pile, if there is one. If his current capital is a multiple of m, he picks the number of chips for his next bet at random from $1, \ldots, m$. On a bet of i chips, he wins double his ante back with probability $p(i)$ and loses it with probability $1 - p(i)$, $i = 1, \ldots, m$. Show that the gambler's capital after successive games is a Markov chain of $M/G/1$ type with the absorbing state 0. Identify the blocks in the most convenient partition of the transition probability matrix. How is this problem similar to Problem 2.1.1 ? •

Problem 2.1.3: The following are closely related models for a single server queue in which after a certain number m of services have been dispensed, the server is required to take out an extra period for maintenance, rest, vacation, etc. We suggest that the reader set up the transition probability matrix $Q(x)$ for the first of these models and then describe how the others may be obtained from it by minor modifications, mostly of the boundary matrices only. In all cases, we have a single server queue with Poisson arrivals and service time distribution $H(\cdot)$. The durations of the successive maintenance periods are independent of the service times and have the distribution $K(\cdot)$, unless otherwise noted. We suggest treating the maintenance period as an addition to the service time which it follows and to consider the embedded Markov renewal process at the completions of regular or augmented services. Other embedded sequences may also be considered.

a. A maintenance period is required after every m-th service. This is the example of an $M/SM/1$ queue, given in the text.

b. A maintenance period with distribution $F(\cdot)$ starts every time the queue becomes empty. If at the end of that period there are customers waiting, a service starts; if not, the server waits for the first Poisson arrival thereafter. The counter which keeps track of the number of services dispensed since the last maintenance is set to zero and every m-th service is followed by a maintenance period of distribution $K(\cdot)$ until the queue becomes once again empty.

c. The same operating procedure as in *b*. except that after a maintenance period with distribution $F(\cdot)$, the counter is not necessarily set to zero. If the queue is empty after an m-th service, the maintenance period with distribution $F(\cdot)$ is expended and not the regular type with distribution $K(\cdot)$.

d. The same operating procedure as in *b*. except that if no customers have arrived during the maintenance period, the server starts another maintenance period with distribution $F(\cdot)$ and continues to do so until customers are present.

e. The same operating procedure as in *a*. except that during each service a Bernoulli trial with probability p is performed. If a "success" occurs, the maintenance period with distribution $K(\cdot)$ is to be performed immediately after that service. Consider the two cases, where this resets the counter to zero and where it does not. •

It is clear that by combining models for the single server queue with maintenance with a threshold rule such as the N-policy, described in Chapter 1, a very large number of additional variants may be produced. It should be noted that all these are variations upon the theme of the $M/SM/1$ queue which affect only the boundary matrices. The formulation and analysis of these models is left to the initiative of the reader, although their addition to the huge journal literature on queues is not necessarily encouraged.

Problem 2.1.4: Consider the Markov chains for the dam model in Example C for the case where up to M units of water are removed at each time point. If fewer than M units are available, the dam is drained completely. Write the form of the matrices for $M = 2$ and $M = 3$ and infer the form for a general value of M from it. Discuss the blocks in the partitioned form of the matrix, with careful attention to the boundary matrices. •

Problem 2.1.5: Consider a discrete $GI/G/1$ queue in which both the interarrival and service times have lattice distributions on $1, 2, \ldots,$ and assume that the interarrival time distribution has finite support. Obtain the standard Lindley relation between the waiting time of the $(n+1)$-st and the n-th customers and see that it defines a Markov chain on the nonnegative integers. Show that this Markov chain has the same structure as that of Bailey's bulk queue, but with appropriate modifications for the boundary states. Show that the transition probability matrix is of the form (2.1.9). •

For a computational approach to this problem, based on transform methods, see Ponstein [P-026] and the related discussion in Konheim [K-086]. The discrete $GI/G/1$ queue as an approximation to the general version of that model is treated in Kimura [K-056] and Wolf [W-018]. The utility of discrete versions of some classical queues is discussed in Balmer [B-020], Dafermos and Neuts [D-001], Kobayashi and Konheim [K-080] and Takács [T-026]. Markov chains with the structure of Bailey's bulk queue are discussed in Abolnikov [A-001], Abolnikov and Dzalalov [A-002], Abolnikov and Postan [A-003], Bailey [B-011], Moran [M-055,M-057], Neuts [N-032], Powell and Humblet [P-029,P-034], Powell [P-030-P-033]

Problem 2.1.6: Consider the queueing model $M/D/m$, in which a Poisson stream of arrivals is served by m independent, identical servers all with the same constant holding time a. Notice that at time $t + a$, all the customers in service at time t and no others will have left the system. Use this to relate the queue length $X(t+a)$ at time $t + a$ to the queue length $X(t)$ and to the (independent) number of arrivals during $(t, t+a]$. Show that if the stationary queue length density at an arbitrary time point exists, it is given by the stationary density of a Markov chain of $M/G/1$ type. Set up the transition probability matrix of that Markov chain and see that, when partitioned into $m \times m$ blocks, it is of the type treated in this book. Use the special form of its elements to obtain the probability generating function of the stationary queue length density and derive the necessary and sufficient condition for its existence, Crommelin [C-085,C-086]. •

The next six problems deal with a queue in which extra delays may occur because the server needs to be resupplied with parts to be fitted to each customer during a service. The server is equipped with a container of limited capacity and whenever, upon completion of a service, only K parts remain, a call for replenishment is issued. The requested number of new parts arrive after an exponentially distributed delay (of mean $1/c$). Only one part is dispensed to each customer. Service may continue during the exponential delay for parts for as long as there are parts available. If the container becomes depleted before replenishment, no further service may be initiated until after the parts order has arrived. Assume that customers arrive according to a homogeneous Poisson process and that the service times of the successive customers are independent with the common distribution $H(\cdot)$. The analysis of some of these problems requires material from later sections or from Neuts [N-042].

Problem 2.1.7: Let the container have a capacity $m > K$ and assume that at every resupply epoch the container is refilled to capacity. Set up the transition probability matrix of the Markov renewal process embedded at service completions and discuss the necessary and sufficient condition for the matrix G to be stochastic. •

Problem 2.1.8: Similarly discuss the variant where at every resupply epoch $m - K$ parts are added. •

Problem 2.1.9: The container have capacity $m + K$ and each reorder is of size m. Set up the transition probability matrix of the embedded Markov renewal process and derive the explicit condition for the matrix G to be stochastic. •

Problem 2.1.10: For exponential service times, the variants in Problems 2.1.7-9 may be studied as continuous parameter Markov processes of the type discussed in Neuts [N-042]. For each case, determine the equilibrium condition explicitly and discuss the computation of the matrix-geometric solution. •

Problem 2.1.11: The matrix-geometric analysis in Problem 2.1.10 can be extended to the case where the service time distribution

$H(\cdot)$ is of phase type. Set up the infinitesimal generator of the corresponding Markov process for each of the three variants, but note the high dimension of the matrix R for cases where m and the number r of phases, needed to describe $H(\cdot)$ are large. Discuss in detail how the special structure of the matrices may be exploited in the computation of the matrix R. •

Problem 2.1.11 offers a nice example of a problem which can be treated both by the matrix-geometric solution and by the methods of this book. In the first case, we have an analytically simpler method, but the computations involve matrices of typically high dimensions; in the second case, the more involved methods in Chapter 3 lead to an algorithm operating on much smaller matrices.

Problem 2.1.12: Discuss the model of Problem 2.1.7 for the case where the delay in receiving an order for parts has a distribution of phase type. Note that this extension requires a careful choice of the state description and full acquaintance with the material in Chapter 2 of [N-042]. This is a challenging problem in model formulation. •

Problem 2.1.13: In an $M/M/1$ queue with arrival rate λ, the server prefers to work at a rate μ_1 when there are at most K customers in the system and at rate μ_2, when more than K customers are present. However, there is an exponential delay with parameter θ in relaying the information that the threshold K has been crossed (either way) to the server. Upon receiving such information, the service rate is changed instantaneously. Formulate models to study the effect of that delay on the stationary queue length and waiting time distributions. Note that several models can be formulated depending on whether threshold crossings during a delay trigger a change in service rate or not. Write the infinitesimal generators for the Markov chains in block partitioned forms and discuss their equilibrium conditions. •

Problem 2.1.14: Customers of two types A and B arrive at a single server queue according to a homogeneous Poisson process. The customers arrive singly and they are of type A with probability p or of type B with probability $q = 1 - p$, $0 < p < 1$, according to independent Bernoulli trials. The customers queue up in the order of their arrival. At the end of each service, the server may take up to two customers into service provided that they are not both of type A. If the next two customers are both of type A, he will serve one of them. The next service is then of a pair (A,B) or of the single customer of type A. There are four possible types of service: (A), (B), (A,B) and (B,B). A service of type (B) is possible only if there is only one customer waiting and he is of type B. Let each of these four types of service have its own service time distribution and make the customary assumption of conditional independence of the successive service times given the compositions of the groups being served. Define the state space appropriately so that, following service completions, this queue has an embedded Markov renewal process of $M/G/1$ type. Write its transition probability matrix, identify the blocks in its partitioned form and determine the necessary and sufficient condition for the matrix G to be stochastic. •

Problem 2.1.15: Consider a system of two queues with Poisson arrivals of rates λ_1 and λ_2 respectively. The (independent) service times have probability distributions $H_1(\cdot)$ and $H_2(\cdot)$ respectively. The second queue has a waiting room of *finite capacity* K, so that at most $K+1$ customers can ever be present in that portion of the system. Customers (of the second kind) who arrive in excess of that number are lost. The server alternatingly processes a customer from each queue, whenever that is possible, that is, he checks the other queue after each service completion and will serve a customer if one is present. If not, he instantaneously returns to the other queue to initiate a service there. When both queues are empty, he will serve the first customer to arrive in either queue. This model has an embedded Markov renewal process of $M/G/1$ type, but it requires some care to choose the sequence of embedded events and the precise definition of the matrix $Q(\cdot)$ involves quite a few alternatives. We consider the queue at the end of *tasks*. A task consists of a service in Unit 1, followed by a service in Unit 2, if that is possible. It may also be a service in Unit 1, provided Unit 2 is empty at the end of that service; similarly, when there are no customers in Unit 1, a task will consist of a service in Unit 2, if possible. Finally, if both units are empty, the task consists of a service in Unit 2 or by a service in Unit 1, possibly followed by one in Unit 2, depending on whether a customer of type 2 arrives during that service. The state of the embedded Markov renewal process is the pair (i,j), where i is the queue length in Unit 1 and j the number remaining in Unit 2.

First give the precise definition of the matrices $A_k(\cdot)$. You will find that they are of upper Hessenberg form and that only the analytic form of the elements in their first rows is somewhat involved. Next, give the analytic expressions for the matrices $B_k(\cdot)$. These require major care as several cases of transitions need to be considered. •

The analysis of this modest variation on the $M/G/1$ queue is quite difficult. There is a growing literature on the model with two unbounded queues, which has been investigated by the mathematically demanding generalization of the Rouché analysis to functions of two complex variables. See Cohen and Boxma [C-052] and Eisenberg [E-009]. The case $K = 1$, is somewhat more tractable and may be viewed as a minor variation on the $M/SM/1$ queue. In setting up the matrices, we recommend working through a generic case, such as $K = 3$ and to infer the general case from it.

Section 2.2

For the earliest discussion of the matrix equations for the fundamental period, though largely by transform methods, see Neuts [N-025]. The proof that the solution of interest is the minimal nonnegative solution, as discussed in this section, is new. The first general proof is given in Lucantoni [L-051]. It is based on the combinatorial interpretation of the probabilities corresponding to the iterates in (2.2.13). For a purely analytic, but more restrictive treatment, based on a contraction mapping argument, see Purdue [P-047].

Problem 2.2.1: Discuss the probabilistic significance of the iterates in formula (2.2.13) for the classical $M/G/1$ queue. Examine explicit inversions of the transforms for the $M/M/1$ queue. •

Problem 2.2.2: Discuss the probabilistic significance of the matrices $T_\nu(k;x)$ and show that they satisfy the equations (2.2.13). Complete the proof of Theorem 2.2.2 by following the approach in Lucantoni [L-051]. •

The need to stress that, barring reducibility of the Markov chain, the minimal solution G of equation (2.2.13) has at least one positive eigenvalue is easily illustrated by cases of coefficient matrices for which the minimal solution has all eigenvalues equal to zero.

Let the 2×2 matrices A_0 and A_1 be defined by

$$A_0 = \begin{vmatrix} 0 & 0 \\ a & 0 \end{vmatrix}, \qquad A_1 = \begin{vmatrix} b & 0 \\ c & 0 \end{vmatrix},$$

where a, b and c are positive. Let (even) all matrices A_k for $k > 1$ be positive, then by direct verification we see that $G = A_0$, and that both eigenvalues of G are equal to zero.

Section 2.3

For Markov chains with a matrix-geometric solution, there are a few useful examples for which the rate matrix R may be written explicitly without being trivial. See Colard and Latouche [C-055], Neuts [N-042] and Ramaswami and Latouche [R-014]. There do not appear to be such cases for the matrix G. In the recurrent case of the dam model in Section 2.1, the matrix G corresponding to the Markov chain in (2.1.5) has a first column of ones and all its other elements are zero. In this case, the "explicit" form of G is trivially simple. In Section 3.3 of Neuts [N-042], a case where G is nearly explicit is discussed. In the following problem, we exhibit a case with an explicit solution which serves to illustrate an important theoretical point.

Problem 2.3.1: Let the matrix A_0 be given by $A_0 = \mathbf{c}\,\mathbf{a}$, where $\mathbf{a}$ is a probability row vector and $\mathbf{c}$ is a nonnegative column vector and set $X = \mathbf{e}\,\mathbf{a}$. Show that X is a stochastic solution to the equation (2.3.3), but not necessarily its minimal nonnegative solution. When is $G = X$? •

Problem 2.3.1 makes an important point. By Brouwer's Fixed Point Theorem, the basic equation (2.3.3) always has a stochastic solution. The essential point of our argument in this chapter is that the solution of probabilistic interest is the *minimal solution* and the issue is to decide when it is stochastic. We have occasionally found it difficult to persuade some that all the developments here are necessary and cannot be avoided by a sweeping appeal to a great theorem. The example in the problem should make this clear.

Problem 2.3.2: Show that the matrix G for the model described in Problem 2.1.1 is positive. •

Problem 2.3.3: For the cases $m = 3$ and $m = 4$, find explicitly the necessary and sufficient condition for the matrix G in the model of Problem 2.1.2 to be stochastic. Show that when all the $p(i)$ are equal, this reduces to the condition one should anticipate. •

Problem 2.3.4: Consider the Markov chain of Problem 2.1.2 and let $p(1) = \cdots = p(m) = p$, where $0 < p < 1$. Show that the matrix G is stochastic if p does not exceed $1/2$. *Hint:* Treat the cases where m is even and m is odd separately and carefully see how the matrix A is constructed. In doing so, very little calculation is required. •

The very useful property of the Perron eigenvalue of a matrix of Laplace-Stieltjes transforms, used in Lemma 2.3.4, is due to Kingman [K-058]. His proof is reproduced in the Appendix.

The positivity of the exponential of a useful class of matrices, used in showing that the matrix G for the $M/SM/1$ and related queues is positive, is established by several different arguments in Bellman [B-028].

Problem 2.3.5: Construct an example where the matrix A is positive, A_0 is irreducible, but the matrix G is not strictly positive. (The matrix G is clearly irreducible. One can find an example using 2×2 matrices and with only A_0, A_1 and A_2 different from the zero matrix). •

Examples of reducibility in the matrices G or A : Most early cases where the matrices G or A *and* G are reducible could be handled by simple, direct arguments. The papers dealing with general aspects of Markov renewal processes of $M/G/1$ type, such as [N-025] and [N-028], were limited for convenience to the common case where the matrix G is irreducible. With some models in communications engineering leading to more complex cases of reducibility and also to round out the general theoretical development, the issues of reducibility are treated here and further in Sections 3.4 and 3.5. See also Neuts [N-054].

Cases where the matrix A is reducible have an interesting and possibly quite complicated physical behavior. Away from the boundary level 0, the state space is partitioned into two or more semi-infinite strips. One of these corresponds to the "transient" states of the stochastic matrix A ; the others correspond to the irreducible recurrent classes. Once the Markov chain reaches one of the strips which corresponds to a recurrent class, it can only move to another such strip via a visit to the boundary level 0. The boundary set 0 holds the separate semi-infinite strips together in an overall *irreducible* Markov chain.

The following problems deal with specific models for which the matrix A is reducible and not lower triangular. They are typical of the practical situations of buffers with internal thresholds, that lead to reducibility of the matrix A in the models of $M/G/1$ type for a variety of problems in communications engineering.

Problem 2.3.6: Customers arrive at a single server station according to a Poisson process and receive a primary service of exponential duration. With probability p, a customer who has completed the primary service goes into a buffer to await a secondary service. The secondary services also have exponential durations, possibly with a different parameter. With probability $1 - p$, the customer departs the system after the primary service. The buffer, where customers wait for the second stage of service, is of finite capacity N. When it becomes full (at the end of a primary service), the server must go and process K, $K = 1, \ldots, N$, secondary jobs and returns to the primary queue thereafter. In addition, whenever the primary queue becomes empty, the processor starts service in the secondary queue, but must return to the primary queue as soon as a customer arrives there. When both queues are empty, the server awaits an arrival in the primary queue.

This model has a few variants, which require minor modifications, mostly of the boundary matrices only. For each case, set up the rate matrix (infinitesimal generator) of the continuous parameter Markov process which describes this queue and find the condition for the matrix G to be stochastic. Note that with $K < N$, the matrix A is reducible and of dimension $N + K$.

a. As described, customers who arrive to an empty primary queue, have preemptive priority over secondary services. Treat also the case where that priority is non-preemptive.

b. The buffer may become full at the same time that the primary queue becomes empty. The operation of the server may be determined by the "full buffer" or by the "empty queue" rules at these points in time. Treat the two cases, where one or the other rule has precedence.

With exponential service times, the model as decribed up to this point may be solved by the methods treated in Neuts [N-042], although the reducibility of A leads to some special considerations as in Section 1.4 of that book. We suggest that the reader carry out the necessary analysis in detail. The treatment of the matrix G, given here, should be compared with the related discussion in Section 3.3 of [N-042]. For block tridiagonal transition probability matrices, where both the analyses of $GI/M/1$ and of $M/G/1$ type matrices is applicable, the former is usually simpler and more efficient. The model in this problem is no exception. The choice between the two methods is no longer open in the next variant.

c. Let the arrival process to the primary queue be a Poisson process with group arrivals. The infinitesimal generator is now, in general, no longer of $GI / M / 1$ type. Examine the consequences of the reducibility of the matrix A . •

Problem 2.3.7: Give a complete discussion of the structure of the rate matrix R of the matrix-geometric solution of the Markov process in Problem 2.3.6 for the case where the customers arriving to an empty primary queue have preemptive priority over the secondary customers. Determine the necessary and sufficient condition for $sp(R) < 1$, and describe and implement an efficient algorithm for the computation of the matrix R . (This problem requires acquaintance with the material in Chapter 3 of [N-042]; a fairly thorough discussion of the matrix R is given in Ali and Neuts [A-015]). •

Problem 2.3.8: Use the same situation as in Problem 2.3.6. Customers who arrive at an empty primary queue have non-preemptive priority over those being served in the buffer. All service times are independent. Those in the primary queue have the common distribution $H(\cdot)$; those in the secondary buffer have the common distribution $F(\cdot)$ and the arrival process is Poisson. Show that the queue has an embedded Markov renewal process at the epochs of service completions in the primary queue and find its transition probability matrix $Q(\cdot)$. Verify that it is of $M / G / 1$ type and that the matrix A is reducible. Determine the necessary and sufficient condition for the matrix G to be stochastic and describe the structure of G in detail. •

Problem 2.3.9: Consider an $M / G / 1$ queue in which the service times of successive customers are allowed to depend on their rank number in the busy period during which they are served. Specifically, let the service time distribution of the $i-$th customer in a busy period be $H(j;\cdot)$ for $j = 1, \ldots, m$, and $H(\cdot)$ for any customer for which already m or more services have been completed during the current busy period. All service time distributions have finite means. Show that this model has a Markov renewal process of $M / G / 1$ type embedded at the epochs of the successive service completions. Let the states be 0, and $(i,1), \ldots, (i,m)$, for $i > 0$. For $j = 1, \ldots, m-1$, the state (i,j) signifies that the service just completed is the $j-$th in the current busy period. The state (i,m) signifies that with the service just completed, the current busy period consists of more than m completed services. In all cases, i indicates the number of customers in the system.

Verify that the matrices A_k, $k > 0$, all have the structure, illustrated here for $m = 10$, and deduce that the matrix G has the reducible structure that is displayed.

$$A_k = \left|\begin{array}{cccccccccc}
\cdot & * & \cdot & \cdot & \cdot & \cdot & \cdot & \cdot & \cdot & \cdot \\
\cdot & \cdot & * & \cdot & \cdot & \cdot & \cdot & \cdot & \cdot & \cdot \\
\cdot & \cdot & \cdot & * & \cdot & \cdot & \cdot & \cdot & \cdot & \cdot \\
\cdot & \cdot & \cdot & \cdot & * & \cdot & \cdot & \cdot & \cdot & \cdot \\
\cdot & \cdot & \cdot & \cdot & \cdot & * & \cdot & \cdot & \cdot & \cdot \\
\cdot & \cdot & \cdot & \cdot & \cdot & \cdot & * & \cdot & \cdot & \cdot \\
\cdot & \cdot & \cdot & \cdot & \cdot & \cdot & \cdot & * & \cdot & \cdot \\
\cdot & \cdot & \cdot & \cdot & \cdot & \cdot & \cdot & \cdot & * & \cdot \\
\cdot & \cdot & \cdot & \cdot & \cdot & \cdot & \cdot & \cdot & \cdot & * \\
\cdot & \cdot & \cdot & \cdot & \cdot & \cdot & \cdot & \cdot & \cdot & *
\end{array}\right|,$$

$$G = \left|\begin{array}{cccccccccc}
\cdot & * & * & * & * & * & * & * & * & * \\
\cdot & \cdot & * & * & * & * & * & * & * & * \\
\cdot & \cdot & \cdot & * & * & * & * & * & * & * \\
\cdot & \cdot & \cdot & \cdot & * & * & * & * & * & * \\
\cdot & \cdot & \cdot & \cdot & \cdot & * & * & * & * & * \\
\cdot & \cdot & \cdot & \cdot & \cdot & \cdot & * & * & * & * \\
\cdot & \cdot & \cdot & \cdot & \cdot & \cdot & \cdot & * & * & * \\
\cdot & \cdot & \cdot & \cdot & \cdot & \cdot & \cdot & \cdot & * & * \\
\cdot & \cdot & \cdot & \cdot & \cdot & \cdot & \cdot & \cdot & \cdot & * \\
\cdot & \cdot & \cdot & \cdot & \cdot & \cdot & \cdot & \cdot & \cdot & *
\end{array}\right|.$$

Show that the matrix G is stochastic if and only if the traffic intensity of the corresponding ordinary $M/G/1$ queue does not exceed one. Discuss how the non-zero elements of the matrix G may be efficiently be computed, Ali and Neuts [A-016]. •

Section 2.4

Problem 2.4.1: Discuss the boundary equations in detail for the particular case where $m = 1$ and the matrix C_0 is of the form

$$C_0 = |\ 0\,0\,0\ \cdots\ 0\,A_0\ |. \quad \bullet$$

This highly special case nevertheless covers most cases of the state-dependent $M/G/1$ queue where the state dependence affects only a finite number of states, as for example in Harris [H-010,H-011], Suzuki [S-079], or in Problem 1.6.1.

Problem 2.4.2: Discuss the simplifications in the boundary equations for the $M/SM/1$ queue. •

Problem 2.4.3: Define the embedded Markov renewal process at departure epochs for the $M/SM/1$ queue operating under the N-policy, discussed in Chapter 1 for the $M/G/1$ queue. Derive the corresponding boundary equations and discuss any simplifications that may arise as compared to the general case. •

Problem 2.4.4: In the treatment in this book, the analysis of the behavior of the Markov chain in the boundary and non-boundary states is largely separated. The advantage of this approach lies in the unified methodology by which models, differing only in their boundary behavior, can be handled. On the other hand, it is clear that, for example, in queueing models with very light traffic, the embedded Markov renewal process will rarely leave the lower levels. For such cases, excellent approximations can be obtained by truncating the Markov renewal process to a small number of levels. It is advantageous to recognize such cases and to handle them separately by an elementary analysis and by algorithms requiring much less effort. Define various measures of the time spent by the embedded Markov renewal process in the levels 0, or in the levels $0, \ldots, i$, where i is a small positive integer. One set of such measures are the expected times until the process leaves the set of "lower" levels. How can these be easily computed? Another measure is the Perron-Frobenius eigenvalue of the matrix obtained by truncating $Q(\infty)$ at some level i. Discuss the physical significance of that quantity and why it is useful in identifying cases where a truncated version of the model may be adequate for most applications. •

For discussions of first passage time distributions in finite structured stochastic matrices, see the papers by Gaver, Jacobs and Latouche [G-016,L-005] and Wikarski [W-014]. The Perron-Frobenius eigenvalue of a principal submatrix of a stochastic matrix and its associated left eigenvector, normalized to be a probability vector, are of great interest. The normalized left eigenvector is known as *the quasi-stationary distribution* over the corresponding set of states. When that vector, augmented by zeros, is chosen as the initial probability vector of the Markov chain, this can be interpreted as the choice of initial conditions correponding to an epoch at which the chain has already spent a long time in that subset of the state space.

3

Positive Recurrence

3.1. MOMENTS OF THE FUNDAMENTAL PERIOD

In this chapter, we shall obtain the necessary and sufficient conditions for *positive recurrence* of Markov renewal processes of $M/G/1$ type. The invariant probability vector $\mathbf{x}$ of the stochastic matrix $Q(\infty)$ will also be studied in detail. To these ends, we need to discuss the means (and higher moments) of various first passage times and, in particular, of the fundamental period. Cases where the matrix G is reducible require more belabored discussions, which proceed along lines similar to those of the simpler (and common) case where G is irreducible. Henceforth, we study only the case where the Markov renewal process $Q(\cdot)$ is *recurrent* and, until stated differently, we assume that the stochastic matrix G is *irreducible*.

In such cases, the matrix A is also irreducible and by Theorem 2.3.1, the inequality

$$\rho = \pi\beta \leqslant 1, \tag{3.1.1}$$

holds. We start from the basic transform equation

$$\tilde{G}(z,s) = z\sum_{\nu=0}^{\infty}\tilde{A}_{\nu}(s)\tilde{G}^{\nu}(z,s), \tag{3.1.2}$$

established in Theorem 2.2.1, and study the matrices of means defined by

$$\tilde{M}_1 = \left[\frac{\partial\tilde{G}(z,s)}{\partial z}\right]_{z=1,s=0}, \tag{3.1.3}$$

and

$$\tilde{M}_1^* = \left[-\frac{\partial\tilde{G}(z,s)}{\partial s}\right]_{z=1,s=0}, \tag{3.1.4}$$

If the embedded Markov chain $Q(\infty)$ is positive recurrent, the matrix $\tilde{M}_1$ is finite. Similarly if the Markov renewal process $Q(\cdot)$ is positive recurrent, the matrix $\tilde{M}_1^*$ is also finite. These statements follow from general results on taboo probabilities in Markov renewal processes. Even when the matrix G is stochastic, the occurrence of infinite elements in $\tilde{M}_1$ or $\tilde{M}_1^*$ implies that the corresponding process is null-recurrent.

The vectors $\tilde{\mu}_1 = \tilde{M}_1\mathbf{e}$, and $\tilde{\mu}_1^* = \tilde{M}_1^*\mathbf{e}$, are the vectors of row sums of $\tilde{M}_1$ and $\tilde{M}_1^*$. The vector $\mathbf{g}$, which is the unique solution of

$$\mathbf{g} = \mathbf{g}G, \qquad \mathbf{ge} = 1, \tag{3.1.5}$$

is the invariant probability vector of the irreducible stochastic matrix G. A classical property of such matrices guarantees that the matrix $I - G + \mathbf{eg}$ is *nonsingular*.

Upon differentiating in (3.1.2) and taking limits, we see that the matrices $\tilde{M}_1$ and $\tilde{M}_1^*$, if they are finite, satisfy the equations

$$\tilde{M}_1 = G + \sum_{\nu=1}^{\infty} A_\nu \sum_{k=0}^{\nu-1} G^k \tilde{M}_1 G^{\nu-k-1}, \tag{3.1.6}$$

$$\tilde{M}_1^* = -\sum_{\nu=0}^{\infty} A_\nu'(0+)G^\nu + \sum_{\nu=1}^{\infty} A_\nu \sum_{k=0}^{\nu-1} G^k \tilde{M}_1^* G^{\nu-k-1}. \tag{3.1.7}$$

The nonnegative matrix $C^*(1) = -\sum_{\nu=0}^{\infty} A_\nu'(0+)G^\nu$, is finite if and only if the vector

$$\beta^* = -\sum_{\nu=0}^{\infty} A_\nu'(0+)\mathbf{e} = \int_0^{\infty} x\,dA(x)\mathbf{e}, \tag{3.1.8}$$

is finite. The equations (3.1.6) and (3.1.7) readily lead to

$$[\,I - \sum_{\nu=1}^{\infty} A_\nu \sum_{k=0}^{\nu-1} G^k\,]\,\tilde{\mu}_1 = \mathbf{e}, \tag{3.1.9}$$

and

$$[I - \sum_{\nu=1}^{\infty} A_\nu \sum_{k=0}^{\nu-1} G^k]\,\tilde{\boldsymbol{\mu}}_1^* = \boldsymbol{\beta}^*. \tag{3.1.10}$$

The following theorem leads to major simplifications in various expressions in the sequel. It is also basic to the algorithmic procedures to be discussed.

Theorem 3.1.1: When $\rho < 1$, we have

$$[I - \sum_{\nu=1}^{\infty} A_\nu \sum_{k=0}^{\nu-1} G^k]^{-1} = (I - G + \mathbf{eg})\,[I - A + (\mathbf{e} - \boldsymbol{\beta})\mathbf{g}]^{-1}. \tag{3.1.11}$$

The vectors $\tilde{\boldsymbol{\mu}}_1$ and $\tilde{\boldsymbol{\mu}}_1^*$ are then given by

$$\tilde{\boldsymbol{\mu}}_1 = (I - G + \mathbf{eg})\,[I - A + (\mathbf{e} - \boldsymbol{\beta})\mathbf{g}]^{-1}\mathbf{e}, \tag{3.1.12}$$

$$\tilde{\boldsymbol{\mu}}_1^* = (I - G + \mathbf{eg})\,[I - A + (\mathbf{e} - \boldsymbol{\beta})\mathbf{g}]^{-1}\boldsymbol{\beta}^*, \tag{3.1.13}$$

and the equalities

$$\mathbf{g}\tilde{\boldsymbol{\mu}}_1 = (1 - \rho)^{-1}, \qquad \mathbf{g}\tilde{\boldsymbol{\mu}}_1^* = (1 - \rho)^{-1}\,\boldsymbol{\pi}\boldsymbol{\beta}^*, \tag{3.1.14}$$

hold. For $\rho = 1$, the matrix $I - \sum_{\nu=1}^{\infty} A_\nu \sum_{k=0}^{\nu-1} G^k$ is singular. The linear equations (3.1.6) and (3.1.7) have unique solutions provided that $\rho < 1$, and $\boldsymbol{\beta}^*$ is finite (this condition is required for (3.1.7) only).

Proof: It is clear that

$$\begin{aligned}
&[I - \sum_{\nu=1}^{\infty} A_\nu \sum_{k=0}^{\nu-1} G^k]\,(I - G + \mathbf{eg}) \\
&= I - G + \mathbf{eg} - \sum_{\nu=1}^{\infty} A_\nu (I - G^\nu) - \sum_{\nu=1}^{\infty} \nu A_\nu\,\mathbf{eg} \\
&= I - A + (\mathbf{e} - \boldsymbol{\beta})\mathbf{g}.
\end{aligned}$$

When $\rho = 1$, noting that $\boldsymbol{\pi}\boldsymbol{\beta} = \rho = 1$, we have

$$\pi[I - A + (\mathbf{e} - \boldsymbol{\beta})\mathbf{g}] = 0,$$

so that $I - \sum_{\nu=1}^{\infty} A_\nu \sum_{k=0}^{\nu-1} G^k$ is singular.

Let now $\rho < 1$, and suppose that for some column vector $\mathbf{u} \neq 0$, we have

$$\mathbf{u} - A\,\mathbf{u} + (\mathbf{g}\mathbf{u})(\mathbf{e} - \boldsymbol{\beta}) = 0.$$

Premultiplying by π, we obtain $(1 - \rho)\mathbf{g}\mathbf{u} = 0$, so that $\mathbf{g}\mathbf{u} = 0$, and therefore $\mathbf{u} = A\,\mathbf{u}$. Since A is irreducible, this in turn implies that $\mathbf{u} = k\mathbf{e}$, for some constant k. Since $\mathbf{g}\mathbf{u} = 0$, $\mathbf{g}\mathbf{e} = 1$, it follows that $k = 0$, and hence $\mathbf{u} = 0$. It follows from this contradiction that the matrix $I - A + (\mathbf{e} - \boldsymbol{\beta})\,\mathbf{g}$, and therefore also the coefficient matrix in (3.1.9) and (3.1.10), are nonsingular. The formulas (3.1.12) and (3.1.13) are now immediate and (3.1.14) is obtained by routine calculations using the relation

$$(1 - \rho)\,\mathbf{g}[I - A + (\mathbf{e} - \boldsymbol{\beta})\mathbf{g}]^{-1} = \pi. \tag{3.1.15}$$

The equations (3.1.6) and (3.1.7) are both of the form

$$X = C + \sum_{\nu=1}^{\infty} A_\nu \sum_{k=0}^{\nu-1} G^k X G^{\nu-k-1}, \tag{3.1.16}$$

where C is a nonnegative matrix for which the row sum vector $C\mathbf{e} = \mathbf{c}$, is positive. The preceding arguments also yield that for $\rho < 1$, the vector $\mathbf{x}^* = X\mathbf{e}$, is given by

$$\mathbf{x}^* = (I - G + \mathbf{e}\mathbf{g})\,[I - A + (\mathbf{e} - \boldsymbol{\beta})\mathbf{g}]^{-1}\mathbf{c},$$

and it is clear that $\mathbf{x}^* \geqslant \mathbf{c}$.

Let $\mathbf{X}$ be the set of nonnegative matrices X of order m, for which $X\mathbf{e} \leqslant \mathbf{x}^*$. Clearly $C \in \mathbf{X}$. By straightforward comparisons, the sequence of matrices $\{X_r\}$ which is recursively defined by $X_0 = 0$, and

$$X_{r+1} = C + \sum_{\nu=1}^{\infty} A_\nu \sum_{k=0}^{\nu-1} G^k X_r G^{\nu-k-1}, \quad \text{for } r \geqslant 0, \tag{3.1.17}$$

is seen to be element-wise nondecreasing. Moreover, if $X_r \in \mathbf{X}$, formula (3.1.17) implies that

$$X_{r+1}\mathbf{e} = \mathbf{c} + \sum_{\nu=1}^{\infty} A_\nu \sum_{k=0}^{\nu-1} G^k X_r \mathbf{e} \leqslant \mathbf{c} + \sum_{\nu=1}^{\infty} A_\nu \sum_{k=0}^{\nu-1} G^k \mathbf{x}^* = \mathbf{x}^*,$$

so that $X_{r+1} \in \mathbf{X}$. It follows that $X_r \in \mathbf{X}$ for all $r \geqslant 0$. Since the set $\mathbf{X}$ is compact, $\lim_{r\to\infty} X_r = X^*$ exists, belongs to $\mathbf{X}$ and, by continuity, satisfies (3.1.16).

Writing $\sum_{\nu=1}^{\infty} A_\nu \sum_{k=0}^{\nu-1} G^k = B$, we see that (3.1.16) implies that $B\mathbf{x}^* < \mathbf{x}^*$, so that the spectral radius θ of B is at most one. Since $I - B$ is nonsingular, we have $\theta < 1$. Let now Y be any other solution of (3.1.16). We denote by $| X^* - Y |$, the matrix with elements $| X^*_{jj'} - Y_{jj'} |$, $1 \leqslant j, j' \leqslant m$. It is obvious that

$$| X^* - Y | \leqslant \sum_{\nu=1}^{\infty} A_\nu \sum_{k=0}^{\nu-1} G^k | X^* - Y | G^{\nu-k-1}.$$

Setting $\mathbf{v} = | X^* - Y | \mathbf{e}$, we obtain $\mathbf{v} \leqslant B\mathbf{v}$, and therefore $\mathbf{v} \leqslant B^r\mathbf{v}$, for $r \geqslant 0$. Since $\theta < 1$, it follows by letting r tend to infinity that $\mathbf{v} = 0$, and hence that $Y = X^*$. The equation (3.1.16) therefore has the unique solution X^*. •

Remark *a*. Formulas (3.1.12) and (3.1.13) are matrix generalizations of the elementary formulas for $\tilde{\mu}_1$ and $\tilde{\mu}_1^*$, given after the proof of Theorem 1.2.2.

Remark *b*. The formulas (3.1.14) provide useful numerical accuracy checks by relating the vectors $\mathbf{g}$, $\tilde{\boldsymbol{\mu}}_1$ and $\tilde{\boldsymbol{\mu}}_1^*$, which require a fair amount of computation, to easily computable functions of the data.

Remark *c*. The condition $\rho < 1$ is *necessary* for the positive recurrence of the Markov chain $Q(\infty)$. In addition, finiteness of $\boldsymbol{\beta}^*$ is necessary for the positive recurrence of the Markov renewal process $Q(\cdot)$.

Remark *d*. When computing the matrices $\tilde{M}_1$ and $\tilde{M}_1^*$, it is convenient to evaluate the vectors $\tilde{\boldsymbol{\mu}}_1$ and $\tilde{\boldsymbol{\mu}}_1^*$ first from the formulas (3.1.12) and (3.1.13). The equations (3.1.6) and (3.1.7) are in a convenient form for solution by successive substitutions. Any real matrices may be used as starting solutions, but there is some practical advantage in starting with matrices $\tilde{M}_1(0)$ and $\tilde{M}_1^*(0)$ which satisfy $\tilde{M}_1(0)\mathbf{e} = \tilde{\boldsymbol{\mu}}_1$, and $\tilde{M}_1^*(0)\mathbf{e} = \tilde{\boldsymbol{\mu}}_1^*$, respectively. All successive iterates will then have the correct row sum vectors. Convenient choices for initial solutions are $\tilde{\boldsymbol{\mu}}_1\mathbf{g}$ or $m^{-1}\tilde{\boldsymbol{\mu}}_1^*\mathbf{e}'$ respectively. The matrices

$$D_\nu = \sum_{k=0}^{\nu-1} G^k X G^{\nu-k-1}, \quad \text{for } \nu \geqslant 1,$$

may be efficiently computed by noting that

$$D_1 = X, \quad D_\nu = D_{\nu-1}G + G^{\nu-1}X, \quad \text{for } \nu \geqslant 2.$$

Remark *e.* The vectors and matrices which arise in Theorem 3.1.1 have interesting interpretations. The component $(\tilde{\boldsymbol{\mu}}_1)_j$ is the expected number of transitions to reach level i from the state $(i+1,j)$. Similarly, $(\tilde{\boldsymbol{\mu}}_1^*)_j$ is the expected duration of the fundamental period starting in the state $(i+1,j)$. When $G_{jj'} > 0$, the ratio $(\tilde{M}_1)_{jj'}/G_{jj'}$ is the conditional mean number of transitions, given that the fundamental period starts in the state $(i+1,j)$ and ends in the state (i,j'). A corresponding interpretation holds for the ratio $(\tilde{M}_1^*)_{jj'}/G_{jj'}$. By forming the partial derivatives at $z = 1-$, $s = 0+$, taken respectively in z and in s of $\tilde{G}^\nu(z,s)$, we see that the j-th components of the vectors

$$\sum_{k=0}^{\nu-1} G^k \tilde{\boldsymbol{\mu}}_1 = \nu(1-\rho)^{-1}\mathbf{e} + (I - G^\nu)\,[I - A + (\mathbf{e} - \boldsymbol{\beta})\mathbf{g}]^{-1}\mathbf{e}, \quad (3.1.18)$$

$$\sum_{k=0}^{\nu-1} G^k \tilde{\boldsymbol{\mu}}_1^* = \nu(\boldsymbol{\pi}\boldsymbol{\beta}^*)(1-\rho)^{-1}\mathbf{e} + (I - G^\nu)\,[I - A + (\mathbf{e} - \boldsymbol{\beta})\mathbf{g}]^{-1}\boldsymbol{\beta}^*,$$

(3.1.19)

are respectively the expected number of transitions and the expected duration of the first passage time from the state $(i+\nu,j)$ to the level i, $i \geqslant 1$.

The matrices

$$\tilde{M}_1(\nu) = \sum_{k=0}^{\nu-1} G^k \tilde{M}_1 G^{\nu-k-1}, \quad \text{and} \quad \tilde{M}_1^*(\nu) = \sum_{k=0}^{\nu-1} G^k \tilde{M}_1^* G^{\nu-k-1},$$

are respectively the mean matrices of the number of transitions and the duration in real time of the first passage from a level **i** + ν to the level **i**, $i \geq 1$. We notice that the sequence $\{\tilde{M}_1(\nu), \nu \geq 0\}$ is the matrix-convolution product of the sequences $\{G^k \tilde{M}_1, k \geq 0\}$ and $\{G^k, k \geq 0\}$. These two sequences have the limits $\mathbf{eg}\tilde{M}_1$ and $\mathbf{eg}$ respectively. It follows by a direct matrix analogue of Cesaro's limit theorem that

$$\lim_{\nu \to \infty} \nu^{-1} \tilde{M}_1(\nu) = \mathbf{eg}\tilde{M}_1\mathbf{eg} = (1 - \rho)^{-1}\mathbf{eg}, \tag{3.1.20}$$

and similarly

$$\lim_{\nu \to \infty} \nu^{-1} \tilde{M}_1^*(\nu) = (\boldsymbol{\pi}\boldsymbol{\beta}^*)(1 - \rho)^{-1}\mathbf{eg}. \tag{3.1.21}$$

These results show that, in the positive recurrent case, the expected number of transitions and the expected duration in reaching level **i**, $i \geq 1$, from level **i** + ν are asymptotically linear in ν. The corresponding results for the scalar case are trivially proved.

The inverse of the matrix $I - \sum_{\nu=1}^{\infty} A_\nu \sum_{k=0}^{\nu-1} G^k$, which is explicitly given in (3.1.11), has a probabilistic interpretation which is less transparent. That interpretation is given in the next theorem.

Theorem 3.1.2: The (j,r)-element of the matrix in (3.1.11) is the conditional mean number of visits to a state of the form (i',r), $i' > i$, during a first passage from the state $(i+1,j)$ to the level **i**.

Proof: We distinguish the successive transitions during a fundamental period according to the second index r, $1 \leq r \leq m$, of the states visited. Let V_r be the number of times a state of the form (i',r), $i' > i$, is entered. We consider the multivariate generating functions $\tilde{G}_{jj'}(z_1, \ldots, z_m, s) = \tilde{G}_{jj'}(\mathbf{z}, s)$, which are analogous to the $\tilde{G}_{jj'}(z, s)$, introduced in Section 2.2. The random variable V, studied there, is the sum $V_1 + \cdots + V_m$.

Repeating the argument leading to Theorem 2.2.1, we obtain the equation

$$\tilde{G}(\mathbf{z}, s) = \Delta(\mathbf{z}) \sum_{\nu=0}^{\infty} \tilde{A}_\nu(s) \tilde{G}^\nu(\mathbf{z}, s), \tag{3.1.22}$$

where $\Delta(\mathbf{z}) = diag\ (z_1, \ldots, z_m)$. Setting $s = 0$, and differentiating with respect to z_r, we obtain after calculations similar to those leading to (3.1.9) that

$$\left[\frac{\partial \tilde{G}(\mathbf{z},0)}{\partial z_r} \right]_{\mathbf{z}=\mathbf{e}} \mathbf{e} = [I - \sum_{\nu=1}^{\infty} A_\nu \sum_{k=0}^{\nu-1} G^k]^{-1} \mathbf{e}_r,$$

where $\mathbf{e}_r$ is the unit vector with its r-th component equal to one. The right hand side is the r-th column of the inverse, so that the (j,r)-element of the inverse has the stated interpretation. •

Remark *a*. The interest of Theorem 3.1.2 for queueing models is clear. In the $M/SM/1$ queue, for example, it provides us with the expected numbers of services of the various types, dispensed during a fundamental (or busy) period starting with a service of type j.

Remark *b*. In Theorem 3.1.2, the initial state $(i+1,j)$ is included in the counting variable V_j. The final state, in which the set **i** is reached, is not. If we choose to include the final, but not the initial state, the equation (3.1.22) is modified to read

$$\tilde{G}(\mathbf{z},s) = \sum_{\nu=0}^{\infty} \tilde{A}_\nu(s)\, \Delta(\mathbf{z})\, \tilde{G}^\nu(\mathbf{z},s).$$

Remark *c*. A further multivariate generalization of formula (3.1.22) is obtained by including also the cumulative time spent in states of the form (i',r), $i' > i$, for $r = 1, \ldots, m$. Except for leading to other conditional means, the use of such multivariate generalizations appears to be limited and their discussion is marginal to the subject of this book.

Moment Matrices of Higher Order: By further differentiations in formula (3.1.2), we may investigate the moment matrices of higher and mixed order of the durations of the fundamental period in continuous and in discrete time. As we shall see, these calculations lead to matrix equations of the form (3.1.16). Provided that the matrix C is finite, their solution exists by virtue of the argument given in the proof of Theorem 3.1.1. For moments of order higher than two, the matrix C is analytically so complicated that its computation, and therefore that of the corresponding moment matrices, becomes a major task. In most applications, there is fortunately little need to consider moment matrices beyond the second order.

The following is a general approach to the computation of the moment matrices. We first introduce some notation.

$$\tilde{M}_r(i) = \sum_{k=r}^{\infty} k(k-1)\cdots(k-r+1)\ G^{(i)}(k),$$

$$\tilde{M}_r^*(i) = \int_0^{\infty} x^r\ dG^{(i)}(x), \tag{3.1.23}$$

$$\tilde{A}_n^{(r)} = \int_0^{\infty} x^r\ dA_n(x),$$

for $r \geqslant 1$, $i \geqslant 0$. We note that $\tilde{M}_0(i) = \tilde{M}_0^*(i) = G^i$. For $i = 1$, we shall write $\tilde{M}_r$ and $\tilde{M}_r^*$ for $\tilde{M}_r(1)$ and $\tilde{M}_r^*(1)$.

Lemma 3.1.1: The moment matrices for $i \geqslant 2$ are related to those for $i = 1$ by

$$\tilde{M}_r(i) = \sum \frac{r!}{k_1! \cdots k_i!} \tilde{M}_{k_1} \cdots \tilde{M}_{k_i}, \tag{3.1.24}$$

$$\tilde{M}_r^*(i) = \sum \frac{r!}{k_1! \cdots k_i!} \tilde{M}_{k_1}^* \cdots \tilde{M}_{k_i}^*, \tag{3.1.25}$$

where the summations extend over all i-tuples $(k_1, \ldots, k_i)$ satisfying $k_1 \geqslant 0, \ldots, k_i \geqslant 0$, $k_1 + \cdots + k_i = r$.

Proof: By considering the terms obtained in r differentiations of $\tilde{G}^i(z,0)$ and $\tilde{G}^i(1,s)$. •

Theorem 3.1.3: The matrices $\tilde{M}_r$ and $\tilde{M}_r^*$ satisfy the linear equations

$$\tilde{M}_r = C(r) + \sum_{\nu=1}^{\infty} A_\nu \sum_{k=0}^{\nu-1} G^k \tilde{M}_r G^{\nu-k-1}, \tag{3.1.26}$$

$$\tilde{M}_r^* = C^*(r) + \sum_{\nu=1}^{\infty} A_\nu \sum_{k=0}^{\nu-1} G^k \tilde{M}_r^* G^{\nu-k-1}, \tag{3.1.27}$$

where $C(r)$ is a sum of products of matrices $\tilde{M}_j$ with $j < r$ and known matrices, and $C^*(r)$ is similarly expressed in terms of the $\tilde{M}_j^*$ with $j < r$ and known matrices.

Proof: The equations (3.1.26) and (3.1.27) may again be obtained by formal differentiations in (3.1.2), but a direct calculation is somewhat more illuminating. We give the details for (3.1.27).

Starting from the integral equation

$$G(x) = \sum_{\nu=0}^{\infty} \int_0^x A_\nu(x-u)\, dG^{(\nu)}(u),$$

we obtain

$$\tilde{M}_r^* = \int_0^\infty x^r\, dG(x) = \sum_{\nu=0}^{\infty} \int_0^\infty \int_0^x x^r\, dA_\nu(x-u)\, dG^{(\nu)}(u)$$

$$= \sum_{\nu=0}^{\infty} \int_0^\infty \int_0^\infty (u+v)^r\, dA_\nu(u)\, dG^{(\nu)}(v) = \sum_{\tau=0}^{r} \binom{r}{\tau} \sum_{\nu=0}^{\infty} \tilde{A}_\nu^{(r-\tau)} \tilde{M}_\tau^*(\nu).$$

Applying the result of Lemma 3.1.1 and splitting off the terms in $\tau = r$, we obtain (3.1.27) where the matrix $C^*(r)$ is given by

$$C^*(r) = \sum_{\tau=0}^{r} \binom{r}{\tau} \sum_{\nu=0}^{\infty} \tilde{A}_\nu^{(r-\tau)} \tilde{M}_\tau^*(\nu) + \sum_{n=1}^{\infty} A_n \sum \frac{r!}{k_1! \cdots k_n!} \tilde{M}_{k_1}^* \cdots \tilde{M}_{k_n}^*,$$

where the inner summation in the second sum is over all n-tuples $(k_1, \ldots, k_n)$ satisfying $0 \leq k_1 < r, \ldots, 0 \leq k_n < r$, $k_1 + \cdots + k_n = r$.

Upon examining the expression for $C^*(r)$, we see that it is a finite positive linear combination of series of the form $\sum_{\nu=0}^{\infty} \tilde{A}_\nu^{(r-\tau)} \tilde{M}_\tau^*(\nu)$. We may easily show by induction on r that all these matrices are finite if and only if

$$\sum_{\nu=0}^{\infty} \tilde{A}_\nu^{(r)} \tilde{M}_0^*(\nu) = \int_0^\infty x^r\, dA(x),$$

is finite. A corresponding argument shows that the matrix $C(r)$ is finite if and only if $\sum_{\nu=0}^{\infty} \nu^r A_\nu$ converges. •

3.2. THE STEADY-STATE PROBABILITIES OF THE EMBEDDED MARKOV CHAIN

The invariant probability vector $\mathbf{x}$ of the Markov chain $Q(\infty)$ may be partitioned as $\mathbf{x} = [\mathbf{x}_0, \mathbf{x}_1, \mathbf{x}_2, \cdots]$, where the vectors $\mathbf{x}_i$, $i \geq 1$, are of dimension m and the vector $\mathbf{x}_0$, which corresponds to the states in the boundary level 0, is of dimension m_1. In this section, we shall obtain the necessary *and* sufficient conditions for positive recurrence and, by purely probabilistic arguments, express the vectors $\mathbf{x}_0$ and $\mathbf{x}_1$ in terms of quantities already derived.

Let us consider the Markov renewal process $Q(\cdot)$ at its successive visits to the set 0. We shall say that $\xi_n = j$, $1 \leq j \leq m_1$, if the state entered at the n-th visit to 0 is $(0,j)$. The random variable τ_n, $n \geq 2$, is the number of transitions between the $(n-1)$st and the n-th visits to the set 0. Similarly, θ_n is the length of time between the $(n-1)$st and the n-th visits to the level 0. We shall agree that $\tau_0 = \theta_0 = 0$. Initially the Markov renewal process $Q(\cdot)$ will, in general, not be in the set 0, so that τ_1 and θ_1 are defined as the number of transitions and the time required to reach the set 0 for the first time. The random variable ξ_0 may be given an arbitrary value in $\{1, \ldots, m_1\}$, as long as the dependence on the initial conditions of the Markov renewal process $Q(\cdot)$ are brought out in the joint distribution of τ_1 and θ_1.

A key observation is that for every $n \geq 2$, the pairs (τ_1,θ_1), (τ_2,θ_2) $, \ldots,$ (τ_n,θ_n) are conditionally independent, given the random variables ξ_0, $\xi_1, \ldots, \xi_{n+1}$. In particular, the sequences $\{(\xi_n,\tau_n),\ n \geq 0\}$ and $\{(\xi_n,\theta_n),\ n \geq 0\}$ are *Markov renewal sequences,* respectively with the state spaces $\{1, \ldots, m_1\} \times \{0, 1, \cdots\}$ and $\{1, \ldots, m_1\} \times [0,\infty)$. These observations readily follow from the Markov property.

When the Markov renewal process $Q(\cdot)$ is recurrent, both Markov renewal sequences possess proper transition probability matrices. For the purpose of discussing the steady-state probabilities, the specific form of the conditional probabilities $P\{\xi_1 = j', \ \tau_1 = k, \ \theta_1 \leq x \mid \textit{initial conditions}\}$ is immaterial. For $n \geq 2$, the probabilities

$$P\{\xi_n = j', \tau_n = k, \theta_n \leq x \mid \xi_{n-1} = j\},$$

for $1 \leq j, j' \leq m$, $k \geq 1$, $x \geq 0$, are precisely the quantities $K_{jj'}(k;x)$, introduced in Section 2.4. The joint transform $\tilde{K}(z,s)$, defined in (2.4.7), is obtained in Lemma 2.4.2. In discussing the Markov renewal sequence $\{(\xi_n, \tau_n),\ n \geqslant 0\}$, we shall assume that the Markov renewal process is *aperiodic*. This covers all applications known to date and eliminates the need for the minor provisos needed in the periodic case.

It is clear that the recurrence times of the state $(0,j)$ in the Markov chain $Q(\infty)$ and in the Markov renewal process $Q(\cdot)$ are respectively equal to the recurrence times of the state j in the Markov renewal processes $\{(\xi_n, \tau_n)\}$ and $\{(\xi_n, \theta_n)\}$. In order to establish positive recurrence of the Markov chain $Q(\infty)$ and of the Markov renewal process $Q(\cdot)$, it therefore suffices to establish positive recurrence of the corresponding m_1-state Markov renewal processes.

We shall first consider the process $\{(\xi_n, \tau_n)\}$. The corresponding results for $\{(\xi_n, \theta_n)\}$ will be stated at the end of this section. For easy reference, we recall from (2.4.3) and (2.4.8), that the transform matrix $\tilde{K}(z,s)$ is given by

$$\tilde{K}(z,s) = z\tilde{B}_0(s) + \sum_{\nu=1}^{\infty} z\tilde{B}_\nu(s)\tilde{G}^{\nu-1}(z,s) \tag{3.2.1}$$

$$\cdot [I - \sum_{\nu=1}^{\infty} z\tilde{A}_\nu(s)\tilde{G}^{\nu-1}(z,s)]^{-1}\, z\tilde{C}_0(s).$$

From Corollary 2.4.1, we know that the matrix $K = \tilde{K}(1-,0+)$ is an irreducible stochastic matrix. The unique (positive) probability vector κ satisfies the equations

$$\kappa K = \kappa, \qquad \kappa \mathbf{e} = 1. \tag{3.2.2}$$

We now examine the vector of row sum means of the transition probability matrix of the Markov renewal process $\{(\xi_n, \tau_n)\}$. Setting $s = 0$, and differentiating with respect to z in (3.2.1), we obtain for $z = 1-$, that

$$\tilde{\kappa}_1 = \left[\frac{\partial \tilde{K}(z,0)}{\partial z}\right]_{z=1}$$

$$= K\mathbf{e} + \sum_{\nu=2}^{\infty} B_\nu \sum_{k=0}^{\nu-2} G^k \tilde{M}_1 G^{\nu-k-2}\, [I - \sum_{\nu=1}^{\infty} A_\nu G^{\nu-1}]^{-1}\, C_0\mathbf{e}$$

$$+ \sum_{\nu=1}^{\infty} B_\nu G^{\nu-1}\, [I - \sum_{\nu=1}^{\infty} A_\nu G^{\nu-1}]^{-1}$$

$$\cdot \left[\sum_{\nu=1}^{\infty} A_\nu G^{\nu-1} + \sum_{\nu=2}^{\infty} A_\nu \sum_{k=0}^{\nu-2} G^k \tilde{M}_1 G^{\nu-k-2}\right] [I - \sum_{\nu=1}^{\infty} A_\nu G^{\nu-1}]^{-1} C_0 \mathbf{e}$$

$$+ \sum_{\nu=1}^{\infty} B_\nu G^{\nu-1} [I - \sum_{\nu=1}^{\infty} A_\nu G^{\nu-1}]^{-1} C_0 \mathbf{e}.$$

Recalling that $C_0\mathbf{e} = A_0\mathbf{e}$, and noting that

$$[I - \sum_{\nu=1}^{\infty} A_\nu G^{\nu-1}]^{-1} A_0 \mathbf{e} = \mathbf{e},$$

we obtain after routine simplifications that

$$\tilde{\kappa}_1 = \mathbf{e} + \sum_{\nu=2}^{\infty} B_\nu \sum_{k=0}^{\nu-2} G^k \tilde{\mu}_1 \tag{3.2.3}$$

$$+ \sum_{\nu=1}^{\infty} B_\nu G^{\nu-1} [I - \sum_{\nu=1}^{\infty} A_\nu G^{\nu-1}]^{-1} [\mathbf{e} + \sum_{\nu=2}^{\infty} A_\nu \sum_{k=0}^{\nu-2} G^k \tilde{\mu}_1]$$

$$= \psi_2 + \sum_{\nu=1}^{\infty} B_\nu G^{\nu-1} [I - \sum_{\nu=1}^{\infty} A_\nu G^{\nu-1}]^{-1} \psi_1.$$

Formula (3.2.3) is general and does not depend on the irreducibility of G. In the present case, where G is irreducible, the vectors ψ_1 and ψ_2 may be written explicitly by using formula (3.1.12). We obtain

$$\psi_1 = \mathbf{e} + \sum_{\nu=2}^{\infty} A_\nu \sum_{k=0}^{\nu-2} G^k \tilde{\mu}_1 \tag{3.2.4}$$

$$= \mathbf{e} + \sum_{\nu=2}^{\infty} A_\nu [I - G^{\nu-1} + (\nu-1)\mathbf{e}\mathbf{g}] [I - A + (\mathbf{e} - \boldsymbol{\beta})\mathbf{g}]^{-1} \mathbf{e}$$

$$= [I - A_0 - \sum_{\nu=1}^{\infty} A_\nu G^{\nu-1}] [I - A + (\mathbf{e} - \boldsymbol{\beta})\mathbf{g}]^{-1} \mathbf{e} + (1 - \rho)^{-1} A_0 \mathbf{e},$$

$$\psi_2 = \mathbf{e} + [\sum_{\nu=1}^{\infty} B_\nu - \sum_{\nu=1}^{\infty} B_\nu G^{\nu-1}] [I - A + (\mathbf{e} - \boldsymbol{\beta})\mathbf{g}]^{-1} \mathbf{e}$$

$$+ (1 - \rho)^{-1} \sum_{\nu=1}^{\infty} (\nu-1) B_\nu \mathbf{e}. \tag{3.2.5}$$

We see that the vector $\tilde{\kappa}_1$ is finite if and only if $\rho < 1$ and the mean matrix $\sum_{\nu=1}^{\infty} \nu B_\nu$ is finite.

Theorem 3.2.1: When the matrix G is irreducible, the embedded Markov chain $Q(\infty)$ is positive recurrent if and only if $\rho = \pi\beta < 1$, and the matrix $\sum_{\nu=1}^{\infty} \nu B_\nu$ is finite. The vector $\mathbf{x}_0$ is then given by

$$\mathbf{x}_0 = (\kappa\tilde{\kappa}_1)^{-1}\kappa. \tag{3.2.6}$$

Proof: It is clear that the mean recurrence times of states of the form $(0,j)$ are finite if and only if the stated conditions hold. Since positive recurrence is a class property and the Markov chain $Q(\infty)$ is irreducible, the first statement is true. A classical result on positive recurrent Markov renewal processes gives the mean recurrence time of the state $(0,j)$ as $\kappa_j^{-1}(\kappa\tilde{\kappa}_1)$ and the probability $(\mathbf{x}_0)_j$ is clearly the inverse of that quantity. •

The Vector $\mathbf{x}_1$: The vector $\mathbf{x}_1$ may be obtained by a similar probabilistic argument as was used to derive formula (3.2.6). Once the vectors $\mathbf{x}_0$ and $\mathbf{x}_1$ are known, various *moments* of interest may be computed by using explicit matrix formulas. In principle, it is possible to determine also the vectors $\mathbf{x}_i$, $i \geqslant 2$, by studying appropriate first passage times, but the required calculations become extremely involved and impractical for numerical computations. A recursive scheme for the computation of these vectors is discussed at the end of this section.

We consider the embedded Markov chain $Q(\infty)$ at the successive visits to the state $(1,j)$, $1 \leqslant j \leqslant m$, of level 1. The sequence of the times between such visits and of the second indices j of the states visited again defines a Markov renewal sequence of lattice type with m states. We denote the transform matrix of its transition probability matrix by $\tilde{H}(z)$. That matrix is the analogue of $\tilde{K}(z,0)$, but for the level 1. By a first passage argument, similar to that used in Section 2.4, we obtain that

$$\tilde{H}(z) = zC_0(I - zB_0)^{-1} \sum_{\nu=1}^{\infty} zB_\nu \tilde{G}^{\nu-1}(z,0) + \sum_{\nu=1}^{\infty} zA_\nu \tilde{G}^{\nu-1}(z,0). \tag{3.2.7}$$

It is routinely verified that if G is stochastic, so is the matrix $H = \tilde{H}(1-)$. The irreducibility of the Markov chain $Q(\infty)$ implies that

H is also irreducible. The invariant probability vector of H is denoted by $\mathbf{h}$.

The vector $\tilde{\mathbf{h}} = \tilde{H}'(1-)\mathbf{e}$, is given by

$$\tilde{\mathbf{h}} = \boldsymbol{\psi}_1 + C_0(I - B_0)^{-1}\boldsymbol{\psi}_2, \tag{3.2.8}$$

where the vectors $\boldsymbol{\psi}_1$ and $\boldsymbol{\psi}_2$ are as defined in (3.2.4) and (3.2.5).

Theorem 3.2.2: If the Markov chain $Q(\infty)$ is positive recurrent (and the matrix G is irreducible), the vector $\mathbf{x}_1$ is given by

$$\mathbf{x}_1 = (\mathbf{h}\tilde{\mathbf{h}})^{-1}\mathbf{h}. \tag{3.2.9}$$

Proof: By the same argument as for Theorem 3.2.1. •

The following theorem is an analytic accuracy check. It will also serve that purpose in numerical computations.

Theorem 3.2.4: The vectors $\mathbf{x}_0$ and $\mathbf{x}_1$, given by the formulas (3.2.6) and (3.2.7) satisfy

$$\mathbf{x}_0 = \mathbf{x}_0 B_0 + \mathbf{x}_1 C_0. \tag{3.2.10}$$

Proof: Formula (3.2.10) is equivalent to

$$\boldsymbol{\kappa} = (\boldsymbol{\kappa}\tilde{\boldsymbol{\kappa}})(\mathbf{h}\tilde{\mathbf{h}})^{-1}\mathbf{h}C_0(I - B_0)^{-1}.$$

Since $(\boldsymbol{\kappa}\tilde{\boldsymbol{\kappa}})(\mathbf{h}\tilde{\mathbf{h}})^{-1}$ is a (positive) constant, the vector $\mathbf{h}C_0(I - B_0)^{-1}$ must be shown to be a left invariant vector of the matrix K. We have by (2.4.5) and (2.4.9) that

$$\begin{aligned}\mathbf{h}C_0(I - B_0)^{-1}K &= \mathbf{h}C_0(I - B_0)^{-1}B_0 \\ &\quad + \mathbf{h}C_0(I - B_0)^{-1}\sum_{\nu=1}^{\infty} B_\nu G^{\nu-1}[I - \sum_{\nu=1}^{\infty} A_\nu G^{\nu-1}]^{-1}C_0.\end{aligned}$$

The equation $\mathbf{h} = \mathbf{h}H$, is equivalent to

$$\mathbf{h}C_0(I - B_0)^{-1}\sum_{\nu=1}^{\infty} B_\nu G^{\nu-1} = \mathbf{h}[I - \sum_{\nu=1}^{\infty} A_\nu G^{\nu-1}]^{-1}, \tag{3.2.11}$$

so that the preceding two formulas lead to

$$\mathbf{h}C_0(I - B_0)^{-1} = \mathbf{h}C_0(I - B_0)^{-1}K.$$

In order to complete the proof, it suffices to verify that

$$\mathbf{h}\tilde{\mathbf{h}} = (\kappa\tilde{\kappa})\mathbf{h}C_0(I - B_0)^{-1}\mathbf{e} = \mathbf{h}C_0(I - B_0)^{-1}\tilde{\kappa}_1. \tag{3.2.12}$$

By (3.2.3) and (3.2.8), this is equivalent to

$$\mathbf{h}\psi_1 = \mathbf{h}C_0(I - B_0)^{-1}\sum_{\nu=1}^{\infty} B_\nu G^{\nu-1} [I - \sum_{\nu=1}^{\infty} A_\nu G^{\nu-1}]^{-1} \psi_1,$$

which is immediate by formula (3.2.11). •

Remark *a*. In most, but not all, queueing theoretic applications, the components of the vector $\tilde{\kappa}_1$ may be interpreted as the mean number of services dispensed during busy periods starting under various initial conditions. The components of the vector

$$\kappa_1^* = -\left[\frac{\partial \tilde{K}(z,s)}{\partial s}\right]_{z=1,s=0} \tag{3.2.13}$$

are then usually the mean durations of appropriate *busy cycles* of the queue. These interpretations will be made precise in the discussion of specific queueing models in the sequel. For future reference, we record the explicit expression for the vector $\tilde{\kappa}_1^*$. It is obtained by calculations parallel to those leading to (3.2.3).

$$\tilde{\kappa}_1^* = -\sum_{\nu=0}^{\infty} \tilde{B}_\nu'(0+)\mathbf{e} + \sum_{\nu=0}^{\infty} B_\nu \sum_{k=0}^{\nu-2} G^k \tilde{\mu}_1^* \tag{3.2.14}$$

$$+ \sum_{\nu=1}^{\infty} B_\nu G^{\nu-1} [I - \sum_{\nu=1}^{\infty} A_\nu G^{\nu-1}]^{-1}$$

$$\cdot \left[\beta^* + \tilde{A}_0'(0+)\mathbf{e} - \tilde{C}_0'(0+)\mathbf{e} + \sum_{\nu=2}^{\infty} A_\nu \sum_{k=0}^{\nu-2} G^k \tilde{\mu}_1^*\right]$$

$$= \psi_2^* + \sum_{\nu=1}^{\infty} B_\nu G^{\nu-1} [I - \sum_{\nu=1}^{\infty} A_\nu G^{\nu-1}]^{-1}\psi_1^*.$$

When the matrix G is irreducible, the explicit formula (3.1.13) for $\tilde{\mu}_1^*$ may be used to obtain more explicit expressions for ψ_1^* and ψ_2^*. By calculations that are now routine, we obtain

$$\begin{aligned}\psi_1^* &= \beta^* + \tilde{A}_0'(0+)\mathbf{e} - \tilde{C}_0'(0+)\mathbf{e} + \sum_{\nu=2}^{\infty} A_\nu \sum_{k=0}^{\nu-2} G^k \tilde{\mu}_1^* \\ &= (1-\rho)^{-1}(\pi\beta^*)A_0\mathbf{e} + \tilde{A}_0'(0+)\mathbf{e} - \tilde{C}_0'(0+)\mathbf{e} \\ &+ [I - A_0 - \sum_{\nu=1}^{\infty} A_\nu G^{\nu-1}]\,[I - A + (\mathbf{e}-\beta)\mathbf{g}]^{-1}\beta^*,\end{aligned} \tag{3.2.15}$$

$$\begin{aligned}\psi_2^* &= -\sum_{\nu=0}^{\infty} \tilde{B}_\nu'(0+)\mathbf{e} + \sum_{\nu=2}^{\infty} B_\nu \sum_{k=0}^{\nu-2} G^k \tilde{\mu}_1^* \\ &= (1-\rho)^{-1}(\pi\beta^*)\sum_{\nu=1}^{\infty}(\nu-1)B_\nu\mathbf{e} - \sum_{\nu=0}^{\infty} \tilde{B}_\nu'(0+)\mathbf{e} \\ &+ [\sum_{\nu=1}^{\infty} B_\nu - \sum_{\nu=1}^{\infty} B_\nu G^{\nu-1}]\,[I - A + (\mathbf{e}-\beta)\mathbf{g}]^{-1}\beta^*.\end{aligned} \tag{3.2.16}$$

We notice that there are many ingredients common to the expressions for $\tilde{\kappa}_1$ and $\tilde{\kappa}_1^*$. In numerical computations, there is a clear advantage in planning the flow of the algorithm so that the common terms are evaluated only once.

Remark *b*. The Markov renewal process $Q(\cdot)$ is positive recurrent if and only if the vector $\tilde{\kappa}^*$ is finite. Upon examining formula (3.2.14) and noting that $\tilde{\kappa}_1^*$ is expressed as a sum of positive vectors, we see that the *necessary* and *sufficient* conditions for positive recurrence of the Markov renewal process are

$$\begin{aligned}&a. \quad \rho < 1, \\ &b. \quad \sum_{\nu=1}^{\infty} \nu B_\nu \mathbf{e}, \\ &c. \quad \sum_{\nu=0}^{\infty} \nu \tilde{B}_\nu'(0+)\,\mathbf{e} = -\sum_{\nu=0}^{\infty}\int_0^\infty x\, dB_\nu(x)\mathbf{e}, \quad \text{is finite,} \\ &d. \quad \tilde{C}_0'(0+)\,\mathbf{e} = -\int_0^\infty x\, dC_0(x)\mathbf{e}, \quad \text{is finite.}\end{aligned} \tag{3.2.17}$$

With a and b holding, but either c or d not satisfied, we obtain a null-recurrent Markov renewal process with a positive recurrent embedded Markov chain. A simple example of this is given in the discussion of Variant 4 of the $M/G/1$ queue in Section 1.3.

Remark *c*. By proceeding as in Section 3.1, we may in principle calculate the higher moment vectors $\tilde{\boldsymbol{\kappa}}_r$ and $\tilde{\boldsymbol{\kappa}}_r^*$, $r \geqslant 2$, by further differentiations of $\tilde{K}(z,s)$. The complexity of the resulting formulas exceeds that of those for the fundamental period by far. Fortunately these moments are rarely of primary interest.

An Example: The $M/SM/1$ Queue: Its simple boundary behavior makes the $M/SM/1$ queue one of the most tractable models treated in this book. It follows from (2.1.2) that

$$\tilde{K}(z,s) = \frac{\lambda}{\lambda+s}\tilde{G}(z,s), \tag{3.2.18}$$

so that $\boldsymbol{\kappa} = \mathbf{g}$, $\tilde{\boldsymbol{\kappa}}_1 = \tilde{\boldsymbol{\mu}}_1$, and $\tilde{\boldsymbol{\kappa}}_1^* = \lambda^{-1}\mathbf{e} + \tilde{\boldsymbol{\mu}}_1^*$. From (3.2.14) and (3.2.6), we obtain that

$$\mathbf{x}_0 = (1-\rho)\,\mathbf{g}. \tag{3.2.19}$$

Furthermore, since $\boldsymbol{\beta}^* = \boldsymbol{\alpha}$, and $\boldsymbol{\beta} = \lambda\boldsymbol{\alpha}$, it follows that $\mathbf{g}\tilde{\boldsymbol{\mu}}_1^* = \lambda^{-1}\rho(1-\rho)^{-1}$.

The matrix $\tilde{H}(z)$ is given by

$$\tilde{H}(z) = (I - zA_0)^{-1}\sum_{\nu=1}^{\infty} zA_\nu\tilde{G}^{\nu-1}(z,0), \tag{3.2.20}$$

so that the computation of the vector $\mathbf{h}$ is greatly simplified. By using formulas (3.2.9) and (3.2.12), it readily follows that

$$\mathbf{x}_1 = \frac{1-\rho}{\mathbf{h}(I-A_0)^{-1}\mathbf{e} - 1}\,\mathbf{h}. \tag{3.2.21}$$

The Vectors $\mathbf{x}_k$: Until recently, there was no known analogue of Burke's procedure, discussed for the $M/G/1$ queue in Chapter 1, to avoid loss of significance and other numerical problems in the recursive computation of the higher terms of the steady-state probability vector

$\mathbf{x}$. With the vectors $\mathbf{x}_0$ and $\mathbf{x}_1$ computed, the evaluation of the higher terms proceeded by the block Gauss-Seidel solution (or alternative iterative schemes) of high truncations of the steady-state equations. In some cases, this leads to slow convergence and correspondingly high computation times. The following theorem of Ramaswami establishes the matrix analogue of Burke's recursive scheme. Reports on its efficiency in numerical implementation are highly favorable. For ease of notation, we only discuss the case where the matrices B_ν have the same dimension m as the matrices A_ν; the result is easily modified to handle also other forms of boundary behavior.

Theorem 3.2.5: The vectors $\mathbf{x}_i$, $i \geqslant 1$ are given by the recursion formula

$$\mathbf{x}_i = \left[\, \mathbf{x}_0 \overline{B}_i + \sum_{j=1}^{i-1} \mathbf{x}_j \overline{A}_{i+1-j} \,\right](I - \overline{A}_1)^{-1}, \qquad i \geqslant 1, \tag{3.2.22}$$

where

$$\overline{B}_\nu = \sum_{i=\nu}^{\infty} B_i G^{i-\nu}, \quad \text{and} \quad \overline{A}_\nu = \sum_{i=\nu}^{\infty} A_i G^{i-\nu}, \quad \text{for} \quad \nu \geqslant 0.$$

Proof: If we consider the Markov renewal process $Q(\cdot)$ at the epochs of visits to the set of states $\{(n, r) : 0 \leqslant n \leqslant k, 1 \leqslant r \leqslant m\}$, we obtain a new (embedded) Markov renewal process. It is readily seen that the transition probability matrix P_k of its embedded Markov chain is given by the block partitioned matrix

$$P_k = \begin{vmatrix} B_0 & B_1 & B_2 & \cdots & B_{k-1} & \overline{B}_k \\ A_0 & A_1 & A_2 & \cdots & A_{k-1} & \overline{A}_k \\ 0 & A_0 & A_1 & \cdots & A_{k-2} & \overline{A}_{k-1} \\ \cdot & \cdot & \cdot & & \cdot & \cdot \\ \cdot & \cdot & \cdot & & \cdot & \cdot \\ 0 & 0 & 0 & \cdots & A_0 & \overline{A}_1 \end{vmatrix}.$$

A classical elementary result on Markov chains now implies that the vector $\mathbf{y}_k = (\mathbf{x}_0, \cdots, \mathbf{x}_k)$ is an eigenvector of P_k corresponding to the eigenvalue 1. The last set of equations from $\mathbf{y}_k = \mathbf{y}_k P_k$, may be written as

$$\mathbf{x}_k = \mathbf{x}_0 \overline{B}_k + \sum_{i=1}^{k} x_i \overline{A}_{k+1-i}.$$

As shown in the proof of Theorem 2.3.2, the matrix $I - \overline{A}_1$ is invertible. Rearranging the preceding equation and postmultiplying by $(I - \overline{A}_1)^{-1}$ we obtain (3.2.22). •

The implementation of (3.2.22) can be done efficiently by noting that as $i \to \infty$, $\overline{B}_i$, $\overline{A}_i \to 0$. One may choose a large index i, set $\overline{B}_i = \overline{A}_i = 0$ and compute the other required matrices implementing the backward recursion

$$\overline{B}_k = B_k + \overline{B}_{k+1} G, \quad \text{and} \quad \overline{A}_k = A_k + \overline{A}_{k+1} G.$$

The index i is chosen such that the matrices $\sum_{k=i+1}^{\infty} B_k \mathbf{e}$ and $\sum_{k=i+1}^{\infty} A_k \mathbf{e}$ have negligibly small elements. In problems where only a finite number of the matrices A_ν and B_ν differ from zero, the choice of the index is of course obvious. In other cases, the equations $G = A_0 + \overline{A}_1 G$ and $K = B_0 + \overline{B}_1 G$ may be used to check whether the truncation index has been chosen sufficiently large.

3.3. MOMENTS OF THE INVARIANT PROBABILITY VECTOR

The equation $\mathbf{x} = \mathbf{x}\, Q(\infty)$, for the invariant probability vector of the embedded Markov chain may be rewritten as

$$\mathbf{x}_0 = \mathbf{x}_0 B_0 + \mathbf{x}_1 C_0, \tag{3.3.1}$$

$$\mathbf{x}_i = \mathbf{x}_0 B_i + \sum_{\nu=0}^{i} \mathbf{x}_{i-\nu+1} A_\nu, \quad \text{for } i \geqslant 1.$$

The vector generating function $\mathbf{X}(z) = \sum_{i=1}^{\infty} \mathbf{x}_i z^i$, satisfies the equation

$$\mathbf{X}(z)[\, zI - A^*(z)\,] = z\mathbf{x}_0 \sum_{k=1}^{\infty} B_k z^k - z\mathbf{x}_1 A_0, \tag{3.3.2}$$

for all z such that $|z| \leqslant 1$. We shall henceforth denote the matrix $\sum_{k=1}^{\infty} B_k z^k$ by $B(z)$ and the vectors

$$B^{(n)}(1-)\mathbf{e} = \sum_{k=1}^{\infty} k(k-1) \cdots (k-n+1)B_k \mathbf{e},$$

by $\mathbf{b}_n$ for $n \geqslant 1$. We also set $B(1-)\mathbf{e} = \mathbf{b}_0$.

Theorem 3.3.1: When $\rho < 1$, $\mathbf{b}_1$ is finite and A is irreducible, then $\mathbf{X}(1-)$ is explicitly given by

$$\mathbf{X}(1-) = [\mathbf{x}_0 B(1-) - \mathbf{x}_1 A_0](I - A + \mathbf{e}\boldsymbol{\pi})^{-1} + (1 - \mathbf{x}_0\mathbf{e})\boldsymbol{\pi}. \quad (3.3.3)$$

Proof: From (3.3.2), we obtain

$$\mathbf{X}(1-)(I - A) = \mathbf{x}_0 B(1-) - \mathbf{x}_1 A_0,$$

and after adding $\mathbf{X}(1-)\mathbf{e}\boldsymbol{\pi} = (1 - \mathbf{x}_0\mathbf{e})\boldsymbol{\pi}$, to both sides, formula (3.3.3) is immediate, since the matrix $I - A + \mathbf{e}\boldsymbol{\pi}$ is a nonsingular. •

We next investigate the (factorial) moment vectors $\mathbf{X}^{(n)}(1-)$, which, at least for small values of n, are of considerable interest. I have proved in earlier publications on this subject that the vector $\mathbf{X}^{(n)}(1-)$ is finite if and only if the vectors $\mathbf{b}_{n+1}$ and $\boldsymbol{\beta}_{n+1} = A^{*(n+1)}(1-)\mathbf{e}$, are finite. This generalizes the well-known condition for the existence of the n-th moment of the stationary queue length distribution in the $M/G/1$ queue. The proof of this result involves an examination of the derivatives at $z = 1-$ of the Perron-Frobenius eigenvalue $\chi(z)$ and of appropriately normalized corresponding left and right eigenvectors of the matrix $A^*(z)$. That proof also leads to explicit formulas for the vectors $\mathbf{X}^{(n)}(1-)$, but these are unduly complicated. The formal steps of the proof obscure a simple recursive scheme which is ideally suited for numerical computation. The properties of the Perron-Frobenius eigenvalue, used in this proof, are of independent interest and they are discussed in the Appendix.

We believe the recursive scheme to be far more significant than the proof of the moment conditions needed for $\mathbf{X}^{(n)}(1-)$ to be finite. In the discussion that follows, we shall therefore tacitly assume that all moment matrices appearing in the formulas are finite.

Referring to formula (3.3.2), we set

$$\mathbf{U}(z) = z\,\mathbf{x}_0 B(z) - z\,\mathbf{x}_1 A_0, \tag{3.3.4}$$

then clearly

$$\begin{aligned}
\mathbf{U}(1-) &= \mathbf{x}_0 B(1-) - \mathbf{x}_1 A_0,\\
\mathbf{U}'(1-) &= \mathbf{x}_0 B'(1-) + \mathbf{x}_0 B(1-) - \mathbf{x}_1 A_0,\\
\mathbf{U}^{(n)}(1-) &= \mathbf{x}_0 B^{(n)}(1-) + n\,\mathbf{x}_0 B^{(n-1)}(1-), \quad \text{for } n \geqslant 2,
\end{aligned} \tag{3.3.5}$$

$$\begin{aligned}
\mathbf{U}(1-)\mathbf{e} &= 0, && \text{by using equation (3.3.3)},\\
\mathbf{U}'(1-)\mathbf{e} &= \mathbf{x}_0\mathbf{b}_1,\\
\mathbf{U}^{(n)}(1-)\mathbf{e} &= \mathbf{x}_0\mathbf{b}_n + n\,\mathbf{x}_0\mathbf{b}_{n-1}, && \text{for } n \geqslant 2.
\end{aligned}$$

For ease of notation, we shall write

$$(I - A + \mathbf{e}\boldsymbol{\pi})^{-1} = Z, \tag{3.3.6}$$

and note that obviously $\boldsymbol{\pi} Z = \boldsymbol{\pi}$, and $Z\mathbf{e} = \mathbf{e}$. Formula (3.3.3) may be rewritten as

$$\mathbf{X}(1-) = (1 - \mathbf{x}_0\mathbf{e})\boldsymbol{\pi} + \mathbf{U}(1-)Z. \tag{3.3.7}$$

Differentiating n times in (3.3.2) and setting $z = 1-$ yields

$$\begin{aligned}
\mathbf{U}^{(n)}(1-) &= \mathbf{X}^{(n)}(1-)\,(I - A) + n\,\mathbf{X}^{(n-1)}(1-)\,[I - A^{*\prime}(1-)]\\
&\quad + \sum_{\nu=2}^{n} \binom{n}{\nu} \mathbf{X}^{(n-\nu)}(1-) A^{*(\nu)}(1-), \quad \text{for } n \geqslant 1.
\end{aligned} \tag{3.3.8}$$

Using the same device as in the proof of Theorem 3.3.1, we obtain

$$\mathbf{X}^{(n)}(1-) = L_n \boldsymbol{\pi} \tag{3.3.9}$$

$$+\left[\mathbf{U}^{(n)}(1-) + \sum_{\nu=1}^{n} \binom{n}{\nu} \mathbf{X}^{(n-\nu)}(1-)A^{*(\nu)}(1-) - n\mathbf{X}^{(n-1)}(1-)\right] Z,$$

where $L_n = \mathbf{X}^{(n)}(1-)\mathbf{e}$.

It remains to determine the quantities L_n. To that end, we write formula (3.3.8) for $n+1$ and postmultiply by $\mathbf{e}$. (This is the step that causes the difficulties of the formal proof, because $\mathbf{X}^{(n+1)}(1-)$ may not exist.) This yields

$$(n+1)\,L_n = (n+1)\,\mathbf{X}^{(n)}(1-)\boldsymbol{\beta} + \mathbf{U}^{(n+1)}(1-)\mathbf{e} \qquad (3.3.10)$$

$$+\sum_{\nu=2}^{n+1} \binom{n+1}{\nu} \mathbf{X}^{(n+1-\nu)}(1-)\boldsymbol{\beta}_\nu,$$

where $\boldsymbol{\beta}_\nu = A^{*(\nu)}(1-)\mathbf{e}$, for $\nu \geqslant 1$. Postmultiplying in (3.3.9) by $\boldsymbol{\beta}$, leads to

$$\mathbf{X}^{(n)}(1-)\boldsymbol{\beta} = \rho L_n \qquad (3.3.11)$$

$$+\left[\mathbf{U}^{(n)}(1-) + \sum_{\nu=1}^{n} \binom{n}{\nu} \mathbf{X}^{(n-\nu)}(1-)A^{*(\nu)}(1-) - n\mathbf{X}^{(n-1)}(1-)\right] Z\boldsymbol{\beta}.$$

The quantity L_n is (recursively) determined by (3.3.10) and (3.3.11). To write the recursion relations in the most convenient form, we set

$$\theta_n = \left[\mathbf{U}^{(n)}(1-) + \sum_{\nu=1}^{n} \binom{n}{\nu} \mathbf{X}^{(n-\nu)}(1-)A^{*(\nu)}(1-) - n\mathbf{X}^{(n-1)}(1-)\right] Z\boldsymbol{\beta}. \qquad (3.3.12)$$

for $n \geqslant 1$. Replacing $\mathbf{X}^{(n)}(1-)\boldsymbol{\beta}$ by $\rho L_n + \theta_n$ in (3.3.10), we finally obtain

$$L_n = [(n+1)(1-\rho)]^{-1} \Big[(n+1)\theta_n \qquad (3.3.13)$$

$$+ \mathbf{U}^{(n+1)}(1-)\mathbf{e} + \sum_{\nu=2}^{n+1} \binom{n+1}{\nu} \mathbf{X}^{(n+1-\nu)}(1-)\beta_\nu \Bigg], \quad \text{for } n \geqslant 1.$$

An examination of formulas (3.3.7), (3.3.8), (3.3.12) and (3.3.13) shows that the quantities θ_n, L_n and $\mathbf{X}^{(n)}(1-)$ are now recursively computable in the order $\mathbf{X}(1-)$, θ_1, L_1, $\mathbf{X}'(1-)$, θ_2, $L_2, \cdots$. We see that the vector which premultiples $\boldsymbol{\beta}$ in the expression for θ_n also occurs in the formula for $\mathbf{X}^{(n)}(1-)$. For each value of n, that vector is, of course, computed only once. We write the expressions for $n = 1$ and $n = 2$, out in full as these are most useful in applications.

$\mathbf{X}(1-)$ is given by formula (3.3.7) and the recursive scheme yields successively

$$\theta_1 = [\mathbf{U}'(1) + \mathbf{X}(1-)A^{*\prime}(1-) - \mathbf{X}(1-)]\, Z\boldsymbol{\beta},$$

$$L_1 = [2(1-\rho)]^{-1}\, [2\theta_1 + \mathbf{U}''(1)\mathbf{e} + \mathbf{X}(1-)\beta_2],$$

$$\mathbf{X}'(1-) = L_1\boldsymbol{\pi} + [\mathbf{U}'(1) + \mathbf{X}(1-)A^{*\prime}(1-) - \mathbf{X}(1-)]\, Z, \tag{3.3.14}$$

$$\theta_2 = [\mathbf{U}''(1-) + 2\mathbf{X}'(1-)A^{*\prime}(1-) + \mathbf{X}(1-)A^{*\prime\prime}(1-) - 2\mathbf{X}'(1-)]\, Z\boldsymbol{\beta},$$

$$L_2 = [3(1-\rho)]^{-1}\, [3\theta_2 + \mathbf{U}'''(1-)\mathbf{e} + 3\mathbf{X}'(1-)\beta_2 + \mathbf{X}(1-)\beta_3],$$

$$\mathbf{X}''(1-) = L_2\boldsymbol{\pi} + [\mathbf{U}''(1-)$$

$$+ 2\mathbf{X}'(1-)A^{*\prime}(1-) + \mathbf{X}(1-)A^{*\prime\prime}(1-) - 2\mathbf{X}'(1-)]\, Z.$$

Remark *a*. It is only in the rarest of cases that the preceding formulas lead to more transparent expressions for the moment vectors. One such case is the stable $M/SM/1$ queue for which, after lengthy calculations, one obtains the formulas

$$\mathbf{X}(1-) = \boldsymbol{\pi} - (1-\rho)\mathbf{g},$$

$$\theta_1 = (1-\rho)\,[\mathbf{g}Z\boldsymbol{\beta} - \mathbf{g}\boldsymbol{\beta}] + \boldsymbol{\pi} A^{*\prime}(1-)Z\boldsymbol{\beta} - \rho^2,$$

$$L_1 = \mathbf{g}Z\boldsymbol{\beta} + [2(1-\rho)]^{-1}\left[\boldsymbol{\pi}\boldsymbol{\beta}_2 + 2\boldsymbol{\pi} A^{*\prime}(1-)Z\boldsymbol{\beta} - 2\rho^2\right], \tag{3.3.15}$$

$$\mathbf{X}'(1-) = (1-\rho)\mathbf{g}Z - (1-\rho)\mathbf{g} + (\mathbf{g}Z\boldsymbol{\beta})\boldsymbol{\pi} + \boldsymbol{\pi} A^{*\prime}(1-)Z$$

$$+ [2(1-\rho)]^{-1}\left[\boldsymbol{\pi}\boldsymbol{\beta}_2 + 2\boldsymbol{\pi} A^{*\prime}(1-)Z\boldsymbol{\beta} - 2\rho\right]\boldsymbol{\pi},$$

Even for that model, the "explicit" expressions for the second moments are so involved as to be of limited value. The expression for L_1 in (3.3.15) reduces to the familiar expression

$$L_1 = \rho + [\,2(1-\rho)\,]^{-1}\lambda\alpha_2, \tag{3.3.16}$$

for the $M/\,G/\,1$ queue, in all cases where $\boldsymbol{\beta} = \rho\mathbf{e}$, and $\boldsymbol{\beta}_2 = \lambda\alpha_2\mathbf{e}$.

The observation that the right hand side depends only on the first two moments of the service time is an elementary example of an *insensitivity result.* The expressions for L_1 in (3.3.15) clearly shows that the mean queue length in the $M/\,SM/\,1$ queue is *not* insensitive in general. While the second term involves only the first two moments of the semi-Markov matrix of the service times, the vector $\mathbf{g}$ and therefore L_1 depend in a highly implicit manner on the arrival rate and the semi-Markov matrix governing the service times.

Remark *b*. There are several elementary models, such as Bailey's bulk queue described in Section 2.1, for which semi-explicit formulas for the stationary mean queue length are available in the literature. It is interesting to note that in nearly all such cases, the matrix $A^*(z)$ has identical row sums, so that $A^*(z)\,\mathbf{e} = a(z)\,\mathbf{e}$, where $a(z)$ is a scalar probability generating function. We leave it to the initiative of the reader to examine the striking simplifications in the preceding moment formulas under that particular condition. In preparing particular models for algorithmic analysis, such analytic simplifications should obviously be performed beforehand. A few recent publications nevertheless comment on the "complexity" of the present methods while overlooking major simplifications for the models under discussion.

3.4. A CASE WHERE THE MATRIX A IS IRREDUCIBLE, BUT THE MATRIX G IS NOT

A discussion, in full generality, of all cases where the matrix G is reducible requires much additional notation and rather belabored calculations. In this section, the matrix A will remain irreducible, but we shall consider the case where, after relabeling the states if necessary, the matrix G may be cast into the form

$$\begin{vmatrix} G(1) & 0 \\ G(3) & 0 \end{vmatrix}.$$

That case arises when the matrix A_0 has some columns of zeros, as for example in the dam model described in Section 2.1.

Provided that $\rho = \boldsymbol{\pi}\boldsymbol{\beta} \leqslant 1$, the matrix G is stochastic. We shall limit our discussion to the most common case where the matrix $G(1)$ is *irreducible*. If $G(1)$ is further reducible, the analysis proceeds along the same lines, but the full class structure of $G(1)$ needs to be taken into account. The case at hand, the only one of which specific applications have arisen to date, will suffice for purposes of exposition.

The matrices A, A_k, $\tilde{G}(z,s)$ and $\tilde{M}_1$ are accordingly partitioned as

$$A_0 = \begin{vmatrix} A_0(1) & 0 \\ A_0(3) & 0 \end{vmatrix}, \qquad A_k = \begin{vmatrix} A_k(1) & A_k(2) \\ A_k(3) & A_k(4) \end{vmatrix}, \qquad \text{for } k \geqslant 1.$$

$$A = \begin{vmatrix} A(1) & A(2) \\ A(3) & A(4) \end{vmatrix}, \qquad \tilde{G}(z,s) = \begin{vmatrix} \tilde{G}(1;z,s) & 0 \\ \tilde{G}(3;z,s) & 0 \end{vmatrix},$$

$$\tilde{M}_1 = \begin{vmatrix} \tilde{M}_1(1) & 0 \\ \tilde{M}_1(3) & 0 \end{vmatrix}.$$

The matrix $\tilde{M}_1$ is as defined in (3.1.3). The column vector $\tilde{M}_1 \mathbf{e} = \tilde{\boldsymbol{\mu}}_1$, is partitioned in the obvious manner into vectors $\tilde{\boldsymbol{\mu}}_1(1)$ and $\tilde{\boldsymbol{\mu}}_1(3)$. The column vector $\boldsymbol{\beta}$ and the row vector $\boldsymbol{\pi}$ are similarly partitioned into $\boldsymbol{\beta}(1)$, $\boldsymbol{\beta}(3)$, and $\boldsymbol{\pi}(1)$, $\boldsymbol{\pi}(3)$ respectively.

It is clear from the definition of $\boldsymbol{\pi}$ that

$$\boldsymbol{\pi}(1) = \boldsymbol{\pi}(3)A(3)\,[I - A(1)]^{-1}, \tag{3.4.1}$$

and

$$\boldsymbol{\pi}(3) = \boldsymbol{\pi}(1)A(2)\,[I - A(4)]^{-1}.$$

The irreducibility of the matrix A implies that the inverses in (3.4.1) exist. The invariant probability vector of the irreducible stochastic matrix $G(1)$ is denoted by $\mathbf{g}_1$.

Paralleling the development in Section 3.1, we now investigate the matrix $\tilde{M}_1$ and the vector $\tilde{\boldsymbol{\mu}}_1$. The discussion of the matrix $\tilde{M}_1^*$ and of the moment matrices of higher order proceeds along the same lines and will be omitted.

As in Section 3.1, the basic transform equation (3.1.2) leads to the equation

$$[I - \sum_{\nu=1}^{\infty} A_\nu \sum_{k=0}^{\nu-1} G_k]\,\tilde{\boldsymbol{\mu}}_1 = \mathbf{e}. \tag{3.4.2}$$

We define the matrix G° by

$$G^\circ = \begin{vmatrix} \mathbf{e}\mathbf{g}_1 & 0 \\ \mathbf{e}\mathbf{g}_1 & 0 \end{vmatrix}, \tag{3.4.3}$$

where the dimensions of the vectors $\mathbf{e}$ agree with those of the blocks in the partition of G. It is clear that $GG^\circ = G^\circ$, and that the matrix

$$I - G + G^\circ = \begin{vmatrix} I - G(1) + \mathbf{e}\mathbf{g}_1 & 0 \\ \mathbf{e}\mathbf{g}_1 - G(3) & I \end{vmatrix}, \tag{3.4.4}$$

is nonsingular.

By repeating the calculations leading to (3.1.11), we find that $\tilde{\boldsymbol{\mu}}_1$ is given by

$$\tilde{\boldsymbol{\mu}}_1 = (I - G + G^\circ)\,[I - A + G^\circ - \sum_{\nu=1}^{\infty} \nu A_\nu G^\circ]^{-1}\mathbf{e}, \tag{3.4.5}$$

provided that the inverse exists. We note that

$$\sum_{\nu=1}^{\infty} \nu A_\nu G^\circ = \begin{vmatrix} \boldsymbol{\beta}(1)\mathbf{g}_1 & 0 \\ & \\ \boldsymbol{\beta}(3)\mathbf{g}_1 & 0 \end{vmatrix}.$$

The vector $\tilde{\boldsymbol{\mu}}_1$ may be computed directly from (3.4.5) or we may exploit the particular structure of the matrices G and G° to obtain linear systems of lower dimensions. That that end, we write

$$[I - A + G^\circ - \sum_{\nu=1}^{\infty} \nu A_\nu G^\circ]^{-1}\,\mathbf{e} = \mathbf{v}, \tag{3.4.6}$$

and we partition the vector $\mathbf{v}$ into vectors $\mathbf{v}(1)$ and $\mathbf{v}(3)$ in the same manner as the vector $\tilde{\boldsymbol{\mu}}_1$. It is clear that

$$\tilde{\boldsymbol{\mu}}_1(1) = [I - G(1) + \mathbf{e}\mathbf{g}_1]\,\mathbf{v}(1), \tag{3.4.7}$$

$$\tilde{\boldsymbol{\mu}}_1(3) = [\mathbf{e}\mathbf{g}_1 - G(3)]\mathbf{v}(1) + \mathbf{v}(3).$$

From (3.4.6) we obtain after routine calculations that

$$\{ I - A(1) - A(2)\,[I - A(4)]^{-1}A(3) + [\mathbf{e} - \boldsymbol{\beta}(1)]\mathbf{g}_1 \tag{3.4.8}$$

$$+ A(2)[I - A(4)]^{-1}\,[\mathbf{e} - \boldsymbol{\beta}(3)]\mathbf{g}_1 \}\,\mathbf{v}(1) = \mathbf{e} + A(2)\,[I - A(4)]^{-1}\mathbf{e},$$

and

$$\mathbf{v}(3) = [I - A(4)]^{-1}\{\mathbf{e} + A(3)\mathbf{v}(1) - [\mathbf{e} - \beta(3)]\,\mathbf{g}_1\mathbf{v}(1)\}. \tag{3.4.9}$$

If we denote the coefficient matrix in (3.4.8) by C, the

$$\boldsymbol{\pi}(1)C = (1 - \rho)\mathbf{g}_1, \tag{3.4.10}$$

by repeated application of the equations (3.4.1). The system (3.4.8) is therefore singular when $\rho = 1$. By the same proof by contradiction as used in Theorem 3.1.1, we may show that C is nonsingular for $\rho < 1$.

In order to determine $\mathbf{x}_0$ and $\mathbf{x}_1$, we compute the invariant probability vectors $\boldsymbol{\kappa}$ and $\mathbf{h}$ of the irreducible stochastic matrices K and H, as well as the vectors $\tilde{\boldsymbol{\kappa}}_1$ and $\tilde{\mathbf{h}}$. The vectors $\mathbf{x}_0$ and $\mathbf{x}_1$ are then given as in (3.2.6) and (3.2.9). The vectors $\tilde{\boldsymbol{\kappa}}_1$ and $\tilde{\mathbf{h}}$ are expressed in terms of the vectors $\boldsymbol{\psi}_1$ and $\boldsymbol{\psi}_2$ in (3.2.3) and (3.2.8). The same calculations as in Section 3.2, but using the matrix $I - G + G^{\circ}$, yield

$$\boldsymbol{\psi}_1 = [I - A_0 - \sum_{\nu=1}^{\infty} A_\nu G^{\nu-1}]\,[I - A + G^{\circ} - \sum_{\nu=1}^{\infty} \nu A_\nu G^{\circ}]^{-1}\mathbf{e} \tag{3.4.11}$$

$$+ (1 - \rho)^{-1} A_0 \mathbf{e},$$

and

$$\boldsymbol{\psi}_2 = \mathbf{e} + [\sum_{\nu=1}^{\infty} B_\nu - \sum_{\nu=1}^{\infty} B_\nu G^{\nu-1}]\,[I - A + G^{\circ} - \sum_{\nu=1}^{\infty} \nu A_\nu G^{\circ}]^{-1}\mathbf{e} \tag{3.4.12}$$

$$+ (1 - \rho)^{-1} \sum_{\nu=1}^{\infty} (\nu-1) B_\nu \mathbf{e}.$$

The formulas in Section 3.3 apply without modification to the present case as they only require that the matrix A is irreducible.

3.5. A CASE WHERE THE MATRIX A IS REDUCIBLE

When the matrix A is reducible, the necessary and sufficient conditions for the matrix G to be stochastic and the general structure of that matrix are given by Theorem 2.3.3. In this section, we shall discuss the matrix $\tilde{M}_1$ and the related vectors needed in the calculation of the vectors $\mathbf{x}_0$ and $\mathbf{x}_1$. In order to concentrate on the essential ideas, we treat only the simplest case of reducibility. It is also the only one that has arisen in applications to date and its treatment carries over to more complicated cases at the expense of more involved notation.

We assume that the matrix A is of the form

$$A = \begin{vmatrix} A(1) & 0 \\ T(1) & T(0) \end{vmatrix},$$

where the matrix $A(1)$ is irreducible and the matrix $I - T(0)$ is nonsingular. This signifies that the stochastic matrix A has a single irreducible class and a set of transient states which all communicate with the irreducible class. As we shall see, most of the analytic difficulties come from the presence of the transient states. When the matrix A is block diagonal with only irreducible blocks on the diagonal, the treatment is straightforward.

The matrices A_ν, $\nu \geqslant 0$, and the matrix G then have the structures

$$A_\nu = \begin{vmatrix} A_\nu(1) & 0 \\ T_\nu(1) & T_\nu(0) \end{vmatrix}, \qquad G = \begin{vmatrix} G(1) & 0 \\ \hat{G}(1) & \hat{G}(0) \end{vmatrix}.$$

We further assume that the matrix $G(1)$ is *irreducible,* so that the difficulties discussed in Section 3.4 may be avoided. As in Section 3.3, the vector $\pi(1)$ is the invariant probability of the stochastic matrix $A(1)$ and $\beta(1) = \sum_{\nu=1}^{\infty} \nu A_\nu(1)\mathbf{e}$. The matrix $G(1)$ is stochastic if and only if $\rho = \pi(1)\beta(1) \leqslant 1$.

The fundamental equation (3.1.2) leads to

$$\tilde{G}(1;z) = z \sum_{\nu=0}^{\infty} A_\nu(1)\tilde{G}^\nu(1;z), \tag{3.5.1}$$

$$\hat{G}(1;z) = z \sum_{\nu=0}^{\infty} T_\nu(1)\tilde{G}^\nu(1;z) + z \sum_{\nu=0}^{\infty} T_\nu(0) \sum_{k=0}^{\nu-1} \tilde{G}^k(1;z)\hat{G}(1;z)\hat{G}^{\nu-k-1}(0;z), \tag{3.5.2}$$

$$\hat{G}(0;z) = z \sum_{\nu=0}^{\infty} T_\nu(0)\hat{G}^\nu(0;z) \tag{3.5.3}$$

The matrix $\tilde{M}_1$ is of the form

$$\tilde{M}_1 = \begin{vmatrix} \tilde{M}_1(1) & 0 \\ \hat{M}(1) & \hat{M}(0) \end{vmatrix},$$

and $\tilde{M}_1(1)\mathbf{e} = \tilde{\boldsymbol{\mu}}_1(1)$, and $\hat{M}(1)\mathbf{e} + \hat{M}(0)\mathbf{e} = \tilde{\boldsymbol{\mu}}_1(2)$.

It is clear that the discussion of the equation (3.5.1) proceeds exactly as in Section 3.1. With obvious modifications, the matrix $\tilde{M}_1(1)$ is obtained by application of Theorem 3.1.1 and in particular

$$\tilde{\boldsymbol{\mu}}_1(1) = [I - G(1) + \mathbf{e}\mathbf{g}_1]\,[I - A(1) + (\mathbf{e} - \boldsymbol{\beta}(1))\mathbf{g}_1]^{-1}\mathbf{e}, \tag{3.5.4}$$

where $\mathbf{g}_1$ is the invariant probability vector of $G(1)$.

The substochastic matrix $\hat{G}(0)$ is the minimal nonnegative solution of the equation (3.5.3) with $z = 1$. It is to be computed by iterative methods. With the matrices $G(1)$ and $\hat{G}(0)$ known, the equation obtained from (3.5.2) by setting $z = 1$, is a *linear* matrix equation for $\hat{G}(1)$. It is also in a form that is best suited for iterative computations. In order to expedite convergence, it is often convenient first to rewrite that equation as

$$\hat{G}(1) = [I - \sum_{\nu=1}^{\infty} T_\nu(0)G_{\nu-1}(1)]^{-1} \Bigg[\sum_{\nu=0}^{\infty} T_\nu(1)\tilde{G}^\nu(1) \tag{3.5.5}$$

$$+ \sum_{\nu=2}^{\infty} T_\nu(0) \sum_{k=0}^{\nu-2} G^k(1)\hat{G}(1) \quad \hat{G}^{\nu-k-1}(0) \Big],$$

and to perform successive substitutions. The inverse and the first sum inside the braces need to be computed only once. The equality $\hat{G}(1)\mathbf{e} + \hat{G}(0)\mathbf{e} = \mathbf{e}$, provides a numerical accuracy check.

If we let the matrix G° be defined as in formula (3.4.3) and recall from the proof of Theorem 2.3.3 that the matrix $I - \hat{G}(0)$ is nonsingular, then it is clear that the matrix

$$I - G + G^\circ = \begin{vmatrix} I - G(1) + \mathbf{e}\mathbf{g}_1 & 0 \\ \mathbf{e}\mathbf{g}_1 - \hat{G}(1) & I - \hat{G}(0) \end{vmatrix}, \tag{3.5.6}$$

is nonsingular. It is obvious that $GG^\circ = G^\circ$.

By repeating the calculations leading to (3.1.11), we find that

$$\tilde{\boldsymbol{\mu}}_1 = (I - G + G^\circ)\,[I - A + G^\circ - \sum_{\nu=1}^{\infty} \nu A_\nu G^\circ]^{-1}\mathbf{e}, \tag{3.5.7}$$

and upon writing the matrices in the partitioned form, we see that $\tilde{\boldsymbol{\mu}}_1(1)$ is as given in (3.5.4). The vector $\tilde{\boldsymbol{\mu}}_1(3)$ may be computed from

$$\tilde{\boldsymbol{\mu}}_1(3) = [\mathbf{e}\mathbf{g}_1 - \hat{G}(1)]\,[I - A(1) + (\mathbf{e} - \boldsymbol{\beta}(1))\mathbf{g}_1]^{-1}\mathbf{e} \tag{3.5.8}$$

$$+ [I - \hat{G}(0)][I - T(0)]^{-1}$$

$$\cdot \{\mathbf{e} + T(1)[I - A(1) + (\mathbf{e} - \boldsymbol{\beta}(1))\mathbf{g}_1]^{-1}\mathbf{e} - [\mathbf{e} - \boldsymbol{\beta}(3)](1 - \rho)^{-1}\}$$

$$= (1 - \rho)^{-1}\boldsymbol{\beta}(3) + \{\,[I - \hat{G}(0)]\,[I - T(0)]^{-1}T(1) - \hat{G}(1)\}$$

$$\cdot [I - A(1) + (\mathbf{e} - \boldsymbol{\beta}(1))\mathbf{g}_1]^{-1}\mathbf{e},$$

where

$$\beta(3) = \sum_{\nu=1}^{\infty} \nu T_\nu(0)\mathbf{e} + \sum_{\nu=1}^{\infty} \nu T_\nu(1)\mathbf{e}.$$

The vectors ψ_1 and ψ_2, needed in the calculation of $\tilde{\kappa}_1$ and $\tilde{\mathbf{h}}$, may be evaluated by exploiting the structure of G and of the coefficient matrices as in Section 3.4. It is readily verified that ψ_1 and ψ_2 are given by the same expressions as in (3.4.11) and (3.4.12).

Reducibility of the matrix A also affects the calculations discussed in Section 3.3. In order to bring out the structure of the matrix $A^*(z)$, we rewrite the equation (3.3.2) in terms of obvious notation as

$$[\mathbf{U}(1;z),\mathbf{U}(2;z)] = \begin{vmatrix} zI - A^*(1;z) & 0 \\ -T^*(1;z) & zI - T^*(0;z) \end{vmatrix} [\mathbf{X}(1;z),\mathbf{X}(2;z)], \tag{3.5.9}$$

where the right hand side is obtained by partitioning the vector $\mathbf{U}(z)$ of (3.3.4) into row vectors of appropriate dimensions. The vectors $\mathbf{U}^{(n)}(1-)$ are given in (3.3.5) and they are similarly partitioned. We write $Z(1)$ for the inverse of the matrix $I - A(1) + \mathbf{e}\pi(1)$.

From (3.5.9), we obtain

$$\mathbf{X}(1;z)[\, zI - A^*(1;z)\,] = \mathbf{U}(1;z) + \mathbf{X}(2;z)T^*(1;z), \tag{3.5.10}$$

$$\mathbf{X}(2;z)[\, zI - T^*(0;z)\,] = \mathbf{U}(2;z). \tag{3.5.11}$$

Since the matrix $I - T(0) = I - T^*(0;1-)$, is nonsingular, the equation (3.5.11) yields

$$\mathbf{X}(2;1-) = \mathbf{U}(2;1-)\,[I - T(0)]^{-1}. \tag{3.5.12}$$

By letting z tend to $1-$ in (3.5.10) and using the same device as in the proof of Theorem 3.3.1, we obtain

$$\mathbf{X}(1;1-) = [\mathbf{X}(1;1-)\mathbf{e}]\,\pi(1) + [\mathbf{U}(1;1-) + \mathbf{X}(2;1-)T(1)]\,Z(1). \tag{3.5.13}$$

The quantity $\mathbf{X}(1;1-)\mathbf{e}$ is given by

$$\mathbf{X}(1;1-)\mathbf{e} = 1 - \mathbf{x}_0\mathbf{e} - \mathbf{X}(2;1-)\mathbf{e} \tag{3.5.14}$$

$$= 1 - \mathbf{x}_0\mathbf{e} - \mathbf{U}(2;1-)\,[I - T(0)]^{-1}\mathbf{e}.$$

By successive differentiations in (3.5.11), we obtain the recursive expression

$$\mathbf{X}^{(n)}(2;1-) = \left\{ \mathbf{U}^{(n)}(2;1-) + \sum_{\nu=1}^{n} \binom{n}{\nu} \mathbf{X}^{(n-\nu)}(2;1-)T^{*(\nu)}(0;1-) \right\}$$

$$\cdot [I - T(0)]^{-1}, \tag{3.5.15}$$

for the moment vectors $\mathbf{X}^{(n)}(2;1-)$, $n \geqslant 1$.

The calculation of the vectors $\mathbf{X}^{(n)}(1;1-)$, $n \geqslant 1$, proceeds as in Section 3.3. We shall omit the intermediate steps and only state the final recurrence relations.

For $n \geqslant 1$, we have

$$\theta_n = \left\{ \mathbf{U}^{(n)}(1;1-) + \sum_{\nu=0}^{n} \binom{n}{\nu} \mathbf{X}^{(n-\nu)}(2;1-)T^{*(\nu)}(1;1-) \right.$$

$$\left. + \sum_{\nu=1}^{n} \binom{n}{\nu} \mathbf{X}^{(n-\nu)}(1;1-)A^{*(\nu)}(1;1-) \;\; - n\mathbf{X}^{(n-1)}(1;1-) \right\} Z(1)\beta(1),$$

$$L_n(1) = \mathbf{X}^{(n)}(1;1-)\mathbf{e} \tag{3.5.16}$$

$$= [(n+1)(1-\rho)]^{-1} \left\{ (n+1)\theta_n + \sum_{\nu=2}^{n+1} \binom{n+1}{\nu} \mathbf{X}^{(n+1-\nu)}(1;1-)A^{*(\nu)}(1;1-)\mathbf{e} \right.$$

$$\left. + \mathbf{U}^{(n+1)}(1;1-) + \sum_{\nu=0}^{n+1} \binom{n+1}{\nu} \mathbf{X}^{(n+1-\nu)}(2;1-)\;\; T^{*(\nu)}(1;1-)\mathbf{e} \right\},$$

$$\mathbf{X}^{(n)}(1;1-) = L_n(1)\pi(1)$$

$$+ \left\{ \mathbf{U}^{(n)}(1;1-) + \sum_{\nu=1}^{n} \binom{n}{\nu} \mathbf{X}^{(n-\nu)}(2;1-)T^{*(\nu)}(1;1-) \right.$$

$$\left. + \sum_{\nu=1}^{n} \binom{n}{\nu} \mathbf{X}^{(n-\nu)}(1;1-)A^{*(\nu)}(1;1-) - n\mathbf{X}^{(n-1)}(1;1-) \right\} Z(1).$$

The formulas (3.5.16) permit the recursive computation of the quantities $L_n(1)$ and $\mathbf{X}^{(n)}(1;1-)$, for all values of $n \geq 1$, for which the moment expressions appearing in the right hand sides are finite. As in Section 3.3, for the n-th order moments to be finite, the $(n+1)$st moments of the matrix sequences $\{B_\nu\}$ and $\{A_\nu(1)\}$ must exist. There are, however, two additional conditions. We see that the term $\mathbf{X}^{(n+1)}(2;1-)T(1)\mathbf{e}$ appears in the expression for $L_n(1)$. Upon examination of (3.5.14), we see that finiteness of that term is guaranteed in general only if $T^{*(n+1)}(0;1-)$ is finite or equivalently if $\sum_{\nu=1}^{\infty} \nu^{n+1} T_\nu(0)$ converges. With $T^{*(n+1)}(1;1-)$ appearing in the expression for $L_n(1)$, we also require the existence of the $(n+1)$st moment of the matrix sequence $\{T_\nu(1)\}$.

3.6. COMPUTATION OF THE VECTOR $\mathbf{x}$

With the results obtained up to this point, we are now ready to describe an efficient algorithm to compute an adequate number of vectors $\mathbf{x}_i$, $i \geq 2$, in addition to the vectors $\mathbf{x}_0$, $\mathbf{x}_1$, $\mathbf{X}(1-)$, $\mathbf{X}'(1-)$ and $\mathbf{X}''(1-)$. In this section, we shall discuss the algorithm step by step with comments on truncation and stopping criteria, computational organization, internal accuracy checks and the general interpretation of numerical results. We shall assume that the matrix G is *irreducible.* The treatment of reducible cases, as discussed in Sections 3.4 and 3.5, is best performed in conjunction with the exploitation of other special structural features of the models where G is reducible. It will be illustrated by specific examples in the sequel.

Step 1: Set-up Computations: In most applications, the matrix A is explicitly available or easily computable and it is frequently possible to express the vectors $\boldsymbol{\beta}$, $\boldsymbol{\beta}_2$ and, if needed, $\boldsymbol{\beta}_3$ in convenient analytic forms.

The vector $\boldsymbol{\pi}$ may be computed by numerical solution of the equations $\boldsymbol{\pi} = \boldsymbol{\pi} A$, $\boldsymbol{\pi}\mathbf{e} = 1$. A number of algorithms and corresponding library routines are available for this purpose. We shall only describe one procedure, which removes the redundant linear equation in a clever way and evaluates $\boldsymbol{\pi}$ by solving an inhomogeneous system of linear equations. We have routinely implemented this method in a large number of computational projects, but we emphasize that it is only one choice among many and that no general claims for its numerical performance are made.

In the system

$$\sum_{i=1}^{m} \pi_i = 1, \qquad \sum_{i=1}^{m} \pi_i A_{ij} = \pi_j, \quad \text{for } 1 \leqslant j \leqslant m,$$

the first (or any other) equation is multiplied by A_{mj}, $1 \leqslant j \leqslant m-1$, and subtracted from the equation with the corresponding index j. This leads to

$$\sum_{i=1}^{m-1} \pi_i [\delta_{ij} + A_{mj} - A_{ij}] = A_{mj}, \quad \text{for } 1 \leqslant j \leqslant m-1,$$

$$\pi_m = 1 - \sum_{i=1}^{m-1} \pi_i,$$

The inhomogeneous system of the first $m - 1$ equations is shown to have a unique solution, which can be computed by a library routine.

At this point, we check the inequality $\rho = \boldsymbol{\pi\beta} < 1$. In actuality, computation of the vector $\mathbf{x}$ is often very time consuming for values of ρ close to one. A large number of vectors $\mathbf{x}_i$, $i \geqslant 1$, then need to be computed and the numerical results are usually uninformative. What is a value of ρ "close to one" depends on the specific case. We shall return to this point later, but computation of the stationary probabilities is never recommended for $\rho > 0.99$. Even a small value of ρ offers, in general, no assurance that the vectors $\mathbf{x}_i$ decrease rapidly with i.

The computational difficulty of evaluating the matrices A_ν, $\nu \geqslant 0$, depends on the specific application. It ranges from simple cases, where little or no computational effort is required, to cases where that portion requires more processing time and care than all the remaining steps of the algorithm. For some important queueing models, it is even possible to exploit special features so that, at least for the computation of waiting time distributions, the computation and storage of the matrices A_ν may be avoided entirely. Actual procedures will be discussed for several applications in the sequel.

It is commonly necessary to truncate the matrix sequence $\{A_\nu\}$ at a sufficiently high index N. A thorough general analysis of the errors, introduced by this truncation, is not yet available and appears to be difficult. There are, however, sound theoretical reasons for choosing N to be the smallest n for which

$$\max_j \left[\boldsymbol{\beta} - \sum_{\nu=n+1}^{\infty} [\mathbf{e} - \sum_{k=0}^{\nu} A_k \mathbf{e}] \right]_j < \epsilon, \tag{3.6.1}$$

or more conservatively, for which

$$\max_j \left[\boldsymbol{\beta}_2 - 2 \sum_{\nu=n+1}^{\infty} \nu \, [\mathbf{e} - \sum_{k=1}^{\nu} A_k \mathbf{e}] \right]_j < \epsilon. \tag{3.6.2}$$

By virtue of the elementary equalities

$$\sum_{\nu=1}^{\infty} [\, \mathbf{e} - \sum_{k=0}^{\nu} A_k \mathbf{e}\,] = \boldsymbol{\beta}, \qquad 2\sum_{\nu=1}^{\infty} \nu \,[\, \mathbf{e} - \sum_{k=1}^{\nu} A_k \mathbf{e}\,] = \boldsymbol{\beta}_2,$$

the criteria (3.6.1) and (3.6.2) guarantee that the truncated sequence $\{A_\nu\}$ agrees closely with the infinite sequence in its first, respectively second moment. In practice, we usually choose $\epsilon = 10^{-7}$, and reserve the option of a smaller value if subsequent accuracy checks are not satisfactorily met. In order to carry a stochastic sequence $\{A_\nu\}$ to the subsequent steps, we also add the remaining probability mass matrix $A - \sum_{\nu=0}^{N} A_\nu$ to the matrix A_N. This is clearly heuristic, but our experience indicates that this or other procedures for accounting for the missing mass have no significant effect on the numerical results at the level of accuracy that is meaningful in practice.

Step 2: The Matrix G and the Vector $\tilde{\mu}_1$: The definition of G as the minimal nonnegative solution of the equation

$$G = \sum_{\nu=0}^{\infty} A_\nu G^\nu, \tag{3.6.3}$$

suggests computation of the matrix G by successive substitutions starting with the zero matrix. Some authors have, in fact, concluded from our theoretical discussions that the computation should always be carried out in this fashion. That procedure, preferably applied to the equivalent equation

$$G = (I - A_0)^{-1} \sum_{\substack{\nu=0 \\ \nu\neq 1}}^{\infty} A_\nu G^\nu, \tag{3.6.4}$$

has the advantage that the successive iterates increase to G, but it often requires a large number of iterations and the knowledge that, in the recurrent case, the matrix G is stochastic is not used to advantage.

Since G is the unique solution to (3.6.3) in the *compact* set of stochastic matrices of order m, any sequence of iterates $\{G(n)\}$, starting in (3.6.3) or (3.6.4) with a *stochastic* matrix, is a sequence of stochastic matrices converging to G. In extensive experience with the numerical solution of (3.6.3), we have found that approach to be conceptually simple and efficient. We continue iteration until

$$\max_{j,j'} \mid G_{jj'}(n) - [\sum_{\nu=0}^{\infty} A_\nu G^\nu]_{jj'}(n) \mid < \epsilon, \tag{3.6.5}$$

where ϵ is usually set to 10^{-8}. In iterating within the set of stochastic matrices, one should take care to renormalize the iterates $G(n)$ to be stochastic to machine accuracy. This is accomplished by dividing each element of $G(n)$ by the sum of the elements in its row. Without this precaution, accumulation of rounding error may carry the iterates outside the set of stochastic matrices and cause them to diverge.

Much is to be gained by the efficient evaluation of the successive iterates and by good plausible choices of the starting solution. Since in actual computations, the truncated coefficient sequence $\{A_0, \ldots, A_N\}$ is used, the matrix polynomial $\sum_{\nu=0}^{N} A_\nu X^\nu$ is efficiently evaluated by

Horner's rule, that is, by carrying out the matrix operations according to the scheme

$$Y_0 = A_N, \quad Y_\nu = A_{N-\nu} + Y_{\nu-1}X, \quad \text{for } 1 \leqslant \nu \leqslant N. \tag{3.6.6}$$

Clearly $Y_N = \sum_{\nu=0}^{N} A_\nu X^\nu$, and the computation of Y_N requires N matrix multiplications and an equal number of matrix additions.

Without prior information on G, there are no compelling reasons to prefer one stochastic starting solution over another. Convenient choices are the matrices A and $\mathbf{e}\boldsymbol{\pi}$. That second choice has the minor advantage that the first iteration yields

$$A_0 + \sum_{\nu=1}^{\infty} A_\nu \mathbf{e}\boldsymbol{\pi} = \mathbf{e}\boldsymbol{\pi} + A_0(I - \mathbf{e}\boldsymbol{\pi}),$$

so that one iteration is saved simply by starting with that matrix.

In many problems, one or more parameters of a model are varied and the matrix G is to be computed for each set of parameter values. It then commonly occurs that the matrices G do not vary greatly from one case to the next. In such cases, and provided that the order m remains the same throughout, it is plausible to use the preceding G-matrix as the starting solution from the second case on.

The computation of the vector $\mathbf{g}$ proceeds in the same manner as that of the vector $\boldsymbol{\pi}$ in Step 1. In evaluating $\tilde{\boldsymbol{\mu}}_1$, there is a small advantage in first computing and storing the vector

$$\tilde{\mathbf{u}}_1 = [\, I - A + (\mathbf{e} - \boldsymbol{\beta})\mathbf{g}\,]^{-1}\mathbf{e}, \tag{3.6.7}$$

by solving the appropriate system of linear equations. This is convenient in formulas such as (3.2.4) and (3.2.5). The vector $\tilde{\boldsymbol{\mu}}_1$ is then obtained by premultiplying $\tilde{\mathbf{u}}_1$ by $I - G + \mathbf{eg}$. The inner product $\mathbf{g}\tilde{\boldsymbol{\mu}}_1$ and the quantity $(1-\rho)^{-1}$ are now printed as an internal accuracy check. The computed values of these quantities should be in very close agreement. Where it is needed, the vector $\tilde{\boldsymbol{\mu}}_1^*$ is computed in a similar manner.

Step 3: Set-up Computations for the Boundary States: In many applications, it is of interest to modify the matrix $Q(\infty)$ in ways that affect only the boundary matrices B_ν, $\nu \geqslant 0$, and C_0. This is often done through interactive computation and while the remaining computations need to be repeated for each case, there is then no point in repeating Steps 1 and 2. It is therefore convenient to treat the computation of the boundary matrices as a separate step, although all the comments on truncation of the sequence of matrices $\{A_\nu\}$ apply verbatim to that of $\{B_\nu\}$. Finiteness of the required moment matrices $\sum \nu^r B_\nu$ of $\{B_\nu\}$ to assure positive recurrence and the existence of the moments L_1 and L_2 in (3.3.14) is natural to most applications and is assumed to be verified beforehand. In specific models, the boundary matrices often exhibit special structure, which may be exploited to yield further theoretical insight and to expedite the subsequent steps of the algorithm.

Step 4: Computation of $\mathbf{x}_0$ and $\mathbf{x}_1$: The matrices $\sum_{\nu=1}^{\infty} A_\nu G^{\nu-1}$ and $\sum_{\nu=1}^{\infty} B_\nu G^{\nu-1}$ are computed first by use of Horner's rule. Next, we solve the linear equations

$$X = C_0 + \left(\sum_{\nu=1}^{\infty} A_\nu G^{\nu-1} \right) X, \tag{3.6.8}$$

$$Y = \sum_{\nu=1}^{\infty} B_\nu G^{\nu-1} + B_0 Y, \tag{3.6.9}$$

to obtain the main ingredients

$$X = [I - \sum_{\nu=1}^{\infty} A_\nu G^{\nu-1}]^{-1} C_0, \qquad Y = (I - B_0)^{-1} \sum_{\nu=1}^{\infty} B_\nu G^{\nu-1},$$

needed in the computation of the matrices K and H by formulas (3.2.1) and (3.2.8). We note that the matrices X and Y are of dimensions $m \times m_1$ and $m_1 \times m$ respectively and that $X\mathbf{e} = \mathbf{e}$, $Y\mathbf{e} = \mathbf{e}$.

The equations (3.6.8) and (3.6.9) are of a type that is well-suited for iterative solution and the fact that in each case the row sums are equal to one suggests starting with an initial solution that already has that property. The comments on the initial choice in the iterative computation of the matrix G also apply here.

With X and Y computed, K and H are evaluated by

$$K = B_0 + \left(\sum_{\nu=1}^{\infty} B_\nu G^{\nu-1} \right) X,$$

$$H = C_0 Y + \sum_{\nu=1}^{\infty} A_\nu G^{\nu-1}.$$

At this stage, it is prudent to check that the computed matrices K and H are stochastic to within the accuracy that is to be expected from the stopping criteria specified in the iterative solution of (3.6.8) and (3.6.9). One may also make very minor adjustments, say on the order of 10^{-8}, to ensure that the matrices K and H now in memory, are stochastic to within machine accuracy.

The computation of the invariant probability vectors $\boldsymbol{\kappa}$ and $\mathbf{h}$ is now routine and, with the available ingredients, the vectors ψ_1 and ψ_2 are obtained from (3.2.4) and (3.2.5). The vectors

$$[I - \sum_{\nu=1}^{\infty} A_\nu G^{\nu-1}]^{-1}\psi_1, \qquad (I - B_0)^{-1}\psi_2,$$

are evaluated by the (iterative) solutions of systems of linear equations and finally $\tilde{\boldsymbol{\kappa}}_1$ and $\tilde{\mathbf{h}}$ are obtained by substitution in (3.2.3) and (3.2.8).

Using (3.2.6) and (3.2.9), we now obtain the computed vectors $\mathbf{x}_0$ and $\mathbf{x}_1$. The equality (3.2.10) serves as an accuracy check.

Step 5: Computation of the Moment Vectors: We evaluate the vectors $\mathbf{X}(1-)$, $\mathbf{X}'(1-)$ and $\mathbf{X}''(1-)$ by means of the formulas (3.3.3) and (3.3.14). The moment matrices of $\{A_\nu\}$ and $\{B_\nu\}$ are often available in explicit analytic form and need to be computed beforehand. The matrix Z is evaluated by inverting $I - A + \mathbf{e}\boldsymbol{\pi}$, by means of a library routine.

The vectors $\mathbf{X}(1-)$, $\mathbf{X}'(1-)$ and $\mathbf{X}''(1-)$ are important for several reasons. In specific applications, they easily lead to various conditional moments of stationary distributions of interest. In the algorithm, they provide internal accuracy checks and stopping criteria.

Step 6: Iterative Computation of the Vectors $\mathbf{x}_i$, $i \geqslant 2$: Even when the matrix A_0 is nonsingular, it is strongly counter-indicated to compute the vectors $\mathbf{x}_i$, $i \geqslant 2$, by means of the recursion formula

$$\mathbf{x}_{i+1} = [\mathbf{x}_i - \mathbf{x}_0 B_i - \sum_{\nu=1}^{i} \mathbf{x}_{i-\nu+1} A_\nu] A_0^{-1}, \quad \text{for } i \geqslant 2.$$

After a few steps, that recursion typically yields negative or otherwise meaningless numerical values for the vectors $\mathbf{x}_i$. The sources of these errors are two-fold. The subtractions inside the brackets cause loss of significance and the inverse of A_0 typically has positive and negative elements of large magnitude. Errors grow roughly proportionally to θ^{-i}, where θ is the modulus of the smallest eigenvalue of A_0.

In the form

$$\mathbf{x}_2 = (\mathbf{x}_0 B_2 + \mathbf{x}_1 A_2)(I - A_1)^{-1} + \mathbf{x}_3 A_0 (I - A_1)^{-1}, \tag{3.6.10}$$

$$\mathbf{x}_i = (\mathbf{x}_0 B_i + \mathbf{x}_1 A_i)(I - A_1)^{-1}$$

$$+ [\sum_{\nu=2}^{i-1} \mathbf{x}_{i-\nu+1} A_\nu + \mathbf{x}_{i+1} A_0] (I - A_1)^{-1}, \quad \text{for } i \geqslant 3,$$

the equations (3.3.1) are better suited for iterative solution, since the matrix $(I - A_1)^{-1}$ is nonnegative.

The first issue is the truncation of the infinite system. To that end, we compute the variance σ^2 of the probability density $\{\mathbf{x}_i \mathbf{e}, i \geqslant 0\}$ by the formula

$$\sigma^2 = L_2 - L_1(L_1 - 1), \tag{3.6.11}$$

where L_1 and L_2 are given in (3.3.14). As the probability density $\{\mathbf{x}_i \mathbf{e}\}$ is typically highly skewed to the left, most of its probability mass is usually concentrated on $0, \ldots, N$, where $N = [L_1 + 3\sigma + 1]$. For most cases, it suffices to truncate the system (3.6.10) at $i = N$, and to set $\mathbf{x}_{N+1} = 0$ in the equation for $i = N$.

We now compute the vectors

$$\boldsymbol{\phi}_i = (\mathbf{x}_0 B_i + \mathbf{x}_1 A_i)(I - A_1)^{-1}, \quad \text{for } 2 \leqslant i \leqslant N.$$

These vectors are computed only once. The truncated system is then solved iteratively. A number of methods are available for this purpose, but we commonly implement block Gauss-Seidel iteration, which is the iterative scheme

$$\mathbf{x}_2(n+1) = \boldsymbol{\phi}_2 + \mathbf{x}_3(n)A_0(I - A_1)^{-1}, \tag{3.6.12}$$

$$\mathbf{x}_i(n+1) = \boldsymbol{\phi}_i + \Big[\sum_{\nu=2}^{i-1} \mathbf{x}_{i-\nu+1}(n+1)A_\nu + \mathbf{x}_{i+1}(n)A_0\Big](I - A_1)^{-1},$$

$$\text{for } 3 \leqslant i \leqslant N,$$

$$\mathbf{x}_N(n+1) = \boldsymbol{\phi}_N + \sum_{\nu=2}^{N-1} \mathbf{x}_{N-\nu+1}(n+1)A_\nu(I - A_1)^{-1}.$$

Any starting vectors may be chosen, but $\mathbf{x}_i(0) = \boldsymbol{\phi}_i$, $2 \leqslant i \leqslant N$, is a convenient choice.

By careful programming, the next iterate may be written over the preceding one and, in doing so, we keep track of the quantity

$$\max_{2 \leqslant i \leqslant N} \max_{2 \leqslant j \leqslant m} | x_{ij}(n+1) - x_{ij}(n) |,$$

or more conservatively,

$$\sum_{i=2}^{N} \sum_{j=1}^{m} | x_{ij}(n+1) - x_{ij}(n) | .$$

Iteration is continued until the selected norm becomes smaller than a prechosen tolerance ϵ.

The second issue is to decide whether the total probability mass in the truncated sequence is adequate. If it is not, the value of N is to be increased. To that end, compute the vectors $\sum_{i=1}^{N} \mathbf{x}_i$, $\sum_{i=1}^{N} i\mathbf{x}_i$ and $\sum_{i=2}^{N} i(i-1)\mathbf{x}_i$, and compare them to $\mathbf{X}(1-)$, $\mathbf{X}'(1-)$ and $\mathbf{X}''(1-)$ respectively. Depending on the application at hand, we require a close agreement of the first, the first two or all three of the corresponding pairs of vectors. This may be incorporated into the computer code by specifying tolerances on the component-wise absolute differences of these vectors. When N is to be increased (and with N chosen as before and the tolerances properly set, this occurs only rarely), we successively increment N by fixed amounts and we compute the additional vectors $\boldsymbol{\phi}_i$ that are required. The augmented system is again solved by block Gauss-Seidel iteration, starting from the last computed iterate of the smaller system of equations. This process is continued until a satisfactory approximation to the vector $\mathbf{x}$ has been obtained.

While there has been extensive favorable experience with this iterative scheme, there are also many models for which the Gauss-Seidel iteration converges very slowly. When faced with such a case, V. Ramaswami recently discovered the appropriate analogue of Burke's scheme to obtain a stable *recursive* scheme to evaluate the higher terms of the steady state vector $\mathbf{x}$. Ramaswami's method is given in Theorem 3.2.5. In view of its recent discovery, numerical experience with it is still limited to a few cases, but it holds out great promise of numerical efficiency and should be examined as the method of first choice.

Remark *a*. There are a number of different ways of organizing the computations in Step 6, which usually involve a trade-off between memory requirements and processing time. If it is feasible to store the matrices $A_0(I - A_1)^{-1}$ and $A_i(I - A_1)^{-1}$, for $1 \leqslant i \leqslant N-1$, it is possible to eliminate $N-1$ multiplications of an m-vector by a matrix of order m in *each* iteration in (3.6.12). The resulting reduction in computation time is usually substantial.

Remark *b*. Particularly for problems of large size, there is advantage in doing the computations interactively and to report summary information on the state of the algorithm at a number of points in its course. Various accuracy checks, numbers of iterations required and the current values of stopping criteria are reported on a screen, with the bulk of other output diverted to an output file for print-out. In a conversational mode, we may then adjust the tolerances and the numbers of iterations allowed at various steps to suit the requirements of the particular problem.

Remark *c*. The quantities

$$z_i(j) = \frac{x_{ij}}{X_j(1-)}, \quad \text{for } i \geqslant 1,$$

are the stationary conditional probabilities that the Markov chain is in the state (i,j), given that it is in one of the states (i',j), $i' \geqslant 1$. The densities $\{z_i(j), i \geqslant 1\}$, $1 \leqslant j \leqslant m$, are often of interest in studying queues that exhibit wide fluctuations. The first two (factorial) moments of the density $\{z_i(j)\}$ are given by $X_j'(1-)/X_j(1-)$ and $X_j''(1-)/X_j(1-)$, $1 \leqslant j \leqslant m$. In choosing the truncation index N, we have used the unconditional moments L_1 and L_2, which may not be representative of the more detailed behavior of the path functions. In models where the conditional densities $\{z_i(j)\}$ are very different, we may choose the truncation index on the basis of one or more of the mean and variance pairs of the conditional densities. This should, however, be done cautiously as it may lead to very high values of N when the corresponding quantities $X_j(1-)$ are small. This choice is again most useful when it is coded as an option into an interactive program.

NOTES, REFERENCES AND COMMENTS ON CHAPTER 3

Section 3.1

The moment matrices of the fundamental period were first discussed in Neuts [N-028]. Additional explicit expressions for the second moments are given in Neuts [N-034] and a case where the cross-moment matrix of the durations in discrete and continuous time is of interest to a specific application is treated in Lucantoni [L-051].

The probabilistic significance of the matrix in (3.1.11), which is given in Remark *e.*, following the proof of Theorem 3.1.2, was first noted in Neuts [N-042]. In queueing applications, this interpretation shows the relative contributions of various types of services to a fundamental period. The theorem is also useful in specifying the normalizing constants in a multivariate central limit theorem and in analyzing regenerative simulations.

Problem 3.1.1: Consider the following generalization of the classical Gambler's Ruin Problem, as discussed, for example, in Feller [F-010]. A gambler starts with an initial capital of $i > 0$ coins and bets against an infinitely rich adversary. The successive games follow the states of an m-state, irreducible Markov chain with transition probability matrix A. If that Markov chain moves from the state j to the state j', the gambler receives double his wager with probability $p(j,j') > 0$, and loses it with probability $q(j,j') = 1 - p(j,j') > 0$. At every bet, one coin is wagered and the gambler has a net loss or gain of one coin. The game ends when the gambler is ruined. Show that this game may be studied as a Markov chain of $M/G/1$ type with the states 0 and (i,j), where $i > 0$ and $j = 1, \ldots, m$. Determine the necessary and sufficient condition for ruin to occur eventually with probability one. Obtain matrix formulas for the first two moments of the distribution of the time until the gambler is ruined, starting with i coins, $i > 0$, and with the Markov chain in the state j, $j = 1, \ldots, m$. Verify that these matrix expressions are greatly simplified when the initial state is chosen from among $(i,1), \ldots, (i,m)$ according to the vector $\mathbf{g}$, the invariant probability vector of the matrix G. See Neuts [N-053]. •

Problem 3.1.2: With the block-tridiagonal structure of the transition probability matrix in Problem 3.1.1, the matrix sequence $\{G(k), k > 0\}$ for the fundamental period is easily recursively computable. For the case where the moments of the time until ruin are finite, write a well-organized computer code to evaluate up to 200 terms of the matrix sequence $\{G(k)\}$ and allow for matrices A up to order 10. Test your code for the special cases where the $p(j)$ do not depend on j or where the matrix A has identical rows. Use the general results of Problem 3.1.1 to end the computation of the matrix series $\{G(k)\}$ when the mean and variance matrices of the computed sequence are in close agreement with those of the (unbounded) theoretical sequence (Make this precise !), but do not compute more than 200 terms. •

Problem 3.1.3: For the model of Problem 3.1.1, when the moments of the time until ruin are finite, we may appeal to central limit theorems for Markov renewal processes to show that, when the initial capital i is large, the time to ruin is asymptotically normal. Study, if necessary, the central limit theorem of Keilson and Wishart [K-020], which is stated in the Appendix. State the precise result for the model at hand and discuss the computation of the normalizing constants. (The mean is straightforward, but the asymptotic variance requires some work). •

The model of Problems 3.1.1-3 is the subject of the expository paper [N-053]. The evaluation of the asymptotic variance, needed in the central limit theorem, is of particular interest. Similar calculations are needed if the central limit theorem is used for a variety of other first passage times in Markov renewal processes of $M/G/1$ type. It is important to note that the normalizing constants in the central limit theorem involve the matrix G explicitly and are therefore beyond simple calculation. Other central limit theorems, which may be fruitfully applied to the models in this book, are found in Pyke and Schaufele [P-055]. A case with $m = 2$, and special choices of the parameters leads to somewhat explicit generating functions for the absorption time distributions. This was shown in Proudfoot and Lampard [P-046].

Problem 3.1.4: Consider a Markov chain of $M/G/1$ type in which, starting in a state (i,j'), $i > 0$, eventual visits to the boundary states are certain. Using the result of Theorem 3.1.2, derive a matrix formula for the expected number of visits to states of the form (i',j), $i' > 0$, j fixed, during the first passage time from state (i',j), to the boundary level. •

Problem 3.1.5: Consider the following random walk on the nonnegative integers. The states $\{0,1, \ldots, m-1\}$ are absorbing and from a state i with $i > m-1$, the walk moves to the state $i - m$ with probability q or to the state $i + r$ with probability p. Both p and q are positive and $p + q = 1$; m and r are positive integers. Display the transition probability matrix of the Markov chain which describes the state of the random walk and note that it is a highly degenerate case of a Markov chain of $M/G/1$ type in which the matrix A_0 is a diagonal matrix and the matrices A_k are completely determined by the divisibility properties of the parameters m and r.

Show that ultimate absorption occurs after finite expected time if and only if $qm > pr$. If $d = g.c.d.\ (m, r) > 1$, show how the matrix G corresponding to the pair (m, r) may be constructed from the matrix G corresponding to the pair $(m/d, r/d)$ without any additional computation. For the case $d=1$, i. e. when m and r are relatively prime, compute the matrix G efficiently by using the matrix A as a starting solution and taking care to renormalize each successive iterate to remain a stochastic matrix (to machine accuracy). Compute also the vector of mean absorption times from the starting positions $m, m+1, \ldots, 2m-1$, to the set $\{0,1, \ldots, m-1\}$. What is the probabilistic interpretation of the elements of the successive powers of the matrix G?

Suppose that we are interested in computing the conditional probabilities that the walk is absorbed into the state j, $j = 0, \ldots, m-1$, at time n, given the initial state i, and this for many values of i and for an adequate number of values of n for every i, (i.e. so that the absorption time distributions are known up to some small tail probability). How could the insights gained from the general theory of Markov chains of $M/G/1$ type be used to organize this computations efficiently ? Write a computer program to implement the algorithm and study its performance for $m = 5$ and for several values of r. *Hint:* Compute the sequence of matrices $\{G(k)\}$. Examine how the structure of the matrix G depends on the divisibility properties of m and r. Show in particular that G is positive when m and r are relatively prime. •

The absorption times in Problem 3.1.5 can be investigated by a number of elementary methods, but except for special values of m and r the analytically explicit forms of the absorption probabilities are exceedingly complicated. This is primarily due to the effect of the divisibility properties of the parameters m and r and may be seen in Rizzuto and Boullion [R-033], where the absorption time from the state m is studied by combinatorial methods. By the matrix formalism of this book, such difficulties are handled implicitly and arise only in the definition of the matrices A_k.

Problem 3.1.6: Consider the model described in Problem 3.1.5 and assume that $d = 1$. For larger values of m and for cases where the tails of the absorption time densities are long, a direct recursive computation of the sequence of matrices $\{G(k)\}$ requires much storage and substantial computation time. The matrices $G(k)$ are, however, extremely sparse and highly structured. Establish the following properties of the matrices $G(k)$ and use them to write a more belabored computer program in which storage arrays are used sparingly and for which computation times are greatly reduced.

When the matrices $G(k)$, $k > 0$, are successively written down in a common semi-infinite array, show that positive elements occur only on certain diagonal lines, that is lattice points of the form $(h, i+h)$, $h = 1, \ldots, m$. Note that the elements in such a diagonal line may belong to two consecutive matrices $G(k)$. Use the definition of the random walk and the Chinese remainder theorem to determine the indices i for which an element $(1, i+1)$ of the array is positive.

Store only the elements of the positive diagonals as well as all necessary information to identify the matrix $G(k)$ to which particular elements belong. In computing the matrices $G(k)$, note that only matrix multiplications and additions of matrices with positive elements on one sub-diagonal are needed. Develop the necessary multiplication tables to implement arithmetic within this class of matrices and write the code so that only positive elements are accessed and used in the operations to compute the sequence $\{G(k)\}$.

Discuss the necessary modifications to the program to handle also the case where absorption into the set $\{0,1, \ldots, m-1\}$ is not certain or has an infinite expected time. •

Problem 3.1.7: Consider an irreducible, aperiodic Markov chain with a block tridiagonal transition probability matrix of the form

$$P = \begin{vmatrix} B_1 & B_0 & 0 & 0 & \cdots \\ B_2 & A_1 & A_0 & 0 & \cdots \\ 0 & A_2 & A_1 & A_0 & \cdots \\ 0 & 0 & A_2 & A_1 & \cdots \\ & \cdots & & & \cdots \end{vmatrix},$$

where the blocks are square matrices of order m. Markov chains of this type may be studied by the methods of this book, which involve (in the positive recurrent case) the unique stochastic matrix G which satisfies

$$G = A_2 + A_1 G + A_0 G^2.$$

Alternatively, the steady-state probabilities (when they exist) are of matrix-geometric form and then involve the minimal nonnegative solution to the quadratic matrix equation

$$R = A_0 + RA_1 + R^2A_0.$$

Latouche [L-007] has investigated the probabilistically informative relationship between the matrices R and G. That relationship involves a third matrix U, with the following significance: For any $n \geqslant 1$, let $U_{j_1j_2}$ be the conditional probability that, starting in the state (n, j_1), the Markov chain visits the state (n, j_2) at some future time, without moving down to level $\mathbf{n}-1$, and without returning to any state in level $\mathbf{n}$ in between. Succinctly, $U_{j_1j_2}$ is the conditional probability that the chain returns at least once to the level $\mathbf{n}$ before visiting the level $\mathbf{n}-1$, and that, at the first such visit, the chain hits the state (n, j_2).

By using the probabilistic interpretations of the matrices R, G and U, show that, in the positive recurrent case, the following equations are valid:

$$R = A_0(I - U)^{-1}, \qquad G = (I - U)^{-1}A_2,$$

$$R = A_0(I - A_1 - A_0G)^{-1}, \qquad G = (I - A_1 - RA_2)^{-1}A_2,$$

and establish the nonsingularity of all matrices whose inverse is involved. •

The corresponding result for quasi-birth-and-death processes, defined in Chapter 3 of Neuts [N-042], is also found in [L-007].

Problem 3.1.8: Show that the vector

$$\mathbf{v} = [I - A + (\mathbf{e} - \boldsymbol{\beta})\mathbf{g}]^{-1}\mathbf{e},$$

which arises in formula (3.1.12), may be written as

$$\mathbf{v} = (1 - \rho)^{-1}[(1 - \mathbf{g}Z\boldsymbol{\beta})\mathbf{e} + Z\boldsymbol{\beta}],$$

where Z is the inverse of the matrix $I - A + \mathbf{e}\boldsymbol{\pi}$, and therefore that

$$\tilde{\mu}_1 = (1 - \rho)^{-1}[\mathbf{e} + (I - G)Z\beta].$$

State and prove also the corresponding formula for $\tilde{\mu}_1^*$. •

Section 3.2

The derivation of the boundary vectors, here by purely probabilistic arguments based on Markov renewal theory, has been approached in many papers on specific models by applying Rouché's theorem to the determinant of the coefficient matrix in equation (3.3.2). That approach is rarely carried out in full mathematical detail and requires a thorough analysis of the possible branch points of the zeros of that determinant. It also leads to the imposition of a number of technical conditions, which do not appear to be necessary to the problem at hand. A thorough discussion of these matrix analytic difficulties for the $M / SM / 1$ queue was given in Çinlar [C-021,C-022]. The numerical difficulties in obtaining the roots in the unit disk of the appropriate equation and in solving the auxiliary equations for the stationary probabilities of the boundary states are now part of the folklore of this subject. The discussion of the sources of the corresponding difficulties for the matrix-geometric case in Section 1.6 of Neuts [N-042] applies verbatim to the models treated in this book. It will therefore not be repeated here.

It has been noted, for example in Beneš [B-031], that the main source of analytic difficulty in the study of specific stochastic models is their behavior at or near a boundary. This is also recognized, say, in the study of diffusion processes. For the models of $GI / M / 1$ type, treated in Neuts [N-042], the boundary states can, in principle, be handled in a simple manner. This is almost never the case for the models of $M / G / 1$ type to which this book is devoted. It also explains the paucity of variants of the $GI / G / 1$ queue that remain tractable by such a powerful analytic tool as the Wiener-Hopf method. On the other hand, if one is willing to yield on the generality of the arrival process, many such variants become amenable to analysis by the methods proposed here. Specific examples and references are given in the Chapters 5 and 6.

The classical result on the mean recurrence time in an irreducible positive recurrent Markov renewal process is given in Pyke [P-054], Çinlar [C-025] or Hunter [H-076], in addition to several other sources. It is a consequence of the key renewal theorem. This result is the crux of the argument in Theorems 3.2.1 and 3.2.2. The specific results for the $M/SM/1$ queue (with group arrivals) were first discussed by the methods of this book, in Neuts [N-029]. Earlier treatments by transform methods are in Çinlar [C-021] and Neuts [N-011]. Markov renewal processes are the natural embedded processes, not only for many queues, but for a variety of other models of practical importance. We refer to the extensive bibliographies in Teugels [T-055,T-056].

Problem 3.2.1: For the $M/M/c$ queue with group arrivals of Problem 1.4.1, let the initial queue length be $k > c-1$ and denote by $T(k,j)$ the time until for the first time there are j, $1 < j < c+1$, idle servers. Discuss the probability distribution of the random variable $T(k,j)$. *Hint:* A convenient way to do this is to partition the infinitesimal generator into $c \times c$ blocks and to use the results on the fundamental period. Note that the matrix G has a simple special structure. The final form of the distribution of $T(k,j)$ is not particularly simple but much information can be extracted from it. •

Problem 3.2.2: In Problem 3.2.1, suppose that the $M/M/c$ queue is unstable. How could one compute the probability that $T(k,j)$ is finite ? •

Problem 3.2.3: The theory, developed in Chapters 2 and 3, for discrete parameter Markov chains of the canonical form (2.1.9), is directly applicable to regular continuous parameter Markov chains whose infinitesimal generator has the canonical form. Such processes are indeed examples of Markov renewal processes of $M/G/1$ type. The elementary steps needed to translate results from the discrete case to the continuous time case are discussed, for processes of $GI/M/1$ type in Neuts [N-042]. Carry out the necessary steps for infintesimal generators of $M/G/1$ type, with special attention to those integral equations which can be replaced by appropriate differential equations. For quasi-birth-and-death processes, whose generator is block tridiagonal, the details may be found in Chapter 3 of [N-042]. •

Problem 3.2.4: Let P be an infinite stochastic matrix, with the property that for some $K \geqslant 2$, and for all $i > K$, the elements P_{ij} with $1 \leqslant j \leqslant i-2$ vanish. Moreover, assume that P is the transition probability matrix of an *irreducible, positive recurrent and aperiodic* Markov chain. Let B be the $K \times K$ stochastic matrix obtained from the first K rows of P by leaving the first $K-1$ elements in each row unchanged and by replacing the K-th element by the sum of all other elements in that row. We denote by $\mathbf{u}$, the K-vector consisting of the first K terms of the invariant probability vector of P. Prove that $\mathbf{u}$ is proportional to the invariant probability vector $\mathbf{b}$ of B. •

The result in Problem 3.2.4 is not limited to stochastic matrices of $M / G / 1$ type.

Section 3.3

The book on finite Markov chains by Kemeny and Snell [K-042] appears to have given the first systematic treatment of the matrix Z in probability. It has entered Markov renewal theory in Keilson [K-031], Hunter [H-076] and Neuts [N-028,N-034]. Hunter [H-078] has examined its connection with the theory of generalized inverses in depth.

The matrix expressions for the moment vectors of the stationary probability vector were obtained for particular models in Neuts [N-029,N-031] and were discussed in general in Lucantoni and Neuts [L-049]. The recursive scheme, which is well-suited for numerical implementation, is new.

Problem 3.3.1: Other generalized inverses than Z may be used in the matrix calculations which abound in this book. A class of useful alternatives to Z is provided by the following result of Hunter [H-078]. If A is an irreducible stochastic matrix with invariant probability vector $\boldsymbol{\pi}$ and if $\mathbf{u}$ and $\mathbf{v}$ are respectively row and column vectors, then the matrix $I - A + \mathbf{vu}$ is *nonsingular* if and only if neither of the inner products $\boldsymbol{\pi}\mathbf{v}$ and $\mathbf{ue}$ vanishes. Prove this result by contradiction, note that the existence of Z follows and derive the analogues of some of the formulas in Theorem 3.3.1 for arbitrary vectors $\mathbf{u}$ and $\mathbf{v}$. •

Section 3.4

Cases where the matrices A or G are reducible usually involve quite complex physical behavior. This is reflected in the greater detail required in the matrix analytic treatment. It should be stressed, however, that such cases are relatively rare. In a first study, the material in Sections 3.4 and 3.5 may therefore be omitted and studied as the need arises. Several specific models where the matrix A is irreducible, but not the matrix G, have been discussed. One of the earliest is the dam model of Odoom-Lloyd-Ali Khan-Gani [A-017,O-007]. A case where G is upper triangular is discussed in Ramaswami and Lucantoni [R-010].

Problem 3.4.1: For the dam with Markovian inputs of Odoom-Lloyd-Ali Khan-Gani, obtain the generating function of the time until first emptiness, given an initial dam content of i_0 units of water and an arbitrary initial probability vector for the input process. Derive a detailed normal central limit theorem for the time until first emptiness as i_0 tends to infinity. Give the asymptotic variance in form convenient for computation. •

Section 3.5

When the coefficient matrices in the equation (2.2.13) for the matrix G are very sparse or if there are other indications that G may be reducible, it is worthwhile to examine the structure of G in detail before doing any other numerical computations. Often, the most convenient way to do this is to use the computer to see how the positive elements are arranged in a number of successive iterates in (2.2.13). This depends only on the locations of the non-zero elements in the coefficient matrices and not on their magnitudes. The structure of G can therefore be examined by Boolean arithmetic. In connection with a practical situation, described in Problem 2.3.5, we encountered a case where the matrix A_2 was a scalar matrix and the matrices A_0 and A_1 were structured as shown in the following matrices, displayed for a representative order. The non-zero elements are indicated by an asterisk; all remaining elements are zero.

$$A_0 = \left|\begin{array}{ccccccccccccccc}
* & * & \cdot & \cdot & \cdot & \cdot & \cdot & \cdot & \cdot & \cdot & \cdot & \cdot & \cdot & \cdot & \cdot \\
\cdot & * & * & \cdot & \cdot & \cdot & \cdot & \cdot & \cdot & \cdot & \cdot & \cdot & \cdot & \cdot & \cdot \\
\cdot & \cdot & * & * & \cdot & \cdot & \cdot & \cdot & \cdot & \cdot & \cdot & \cdot & \cdot & \cdot & \cdot \\
\cdot & \cdot & \cdot & * & * & \cdot & \cdot & \cdot & \cdot & \cdot & \cdot & \cdot & \cdot & \cdot & \cdot \\
\cdot & \cdot & \cdot & \cdot & * & * & \cdot & \cdot & \cdot & \cdot & \cdot & \cdot & \cdot & \cdot & \cdot \\
\cdot & \cdot & \cdot & \cdot & \cdot & * & * & \cdot & \cdot & \cdot & \cdot & \cdot & \cdot & \cdot & \cdot \\
\cdot & \cdot & \cdot & \cdot & \cdot & \cdot & * & * & \cdot & \cdot & \cdot & \cdot & \cdot & \cdot & \cdot \\
\cdot & \cdot & \cdot & \cdot & \cdot & \cdot & \cdot & * & * & \cdot & \cdot & \cdot & \cdot & \cdot & \cdot \\
\cdot & \cdot & \cdot & \cdot & \cdot & \cdot & \cdot & \cdot & * & * & \cdot & \cdot & \cdot & \cdot & \cdot \\
\cdot & \cdot & \cdot & \cdot & \cdot & \cdot & \cdot & \cdot & \cdot & * & * & \cdot & \cdot & \cdot & \cdot \\
\cdot & \cdot & \cdot & \cdot & \cdot & \cdot & \cdot & \cdot & \cdot & \cdot & \cdot & \cdot & \cdot & \cdot & \cdot \\
\cdot & \cdot & \cdot & \cdot & \cdot & \cdot & \cdot & \cdot & \cdot & \cdot & \cdot & \cdot & \cdot & \cdot & \cdot \\
\cdot & \cdot & \cdot & \cdot & \cdot & \cdot & \cdot & \cdot & \cdot & \cdot & \cdot & \cdot & \cdot & \cdot & \cdot \\
\cdot & \cdot & \cdot & \cdot & \cdot & \cdot & \cdot & \cdot & \cdot & \cdot & \cdot & \cdot & \cdot & \cdot & \cdot \\
\cdot & \cdot & \cdot & \cdot & \cdot & \cdot & \cdot & \cdot & \cdot & \cdot & \cdot & \cdot & \cdot & \cdot & \cdot
\end{array}\right|,$$

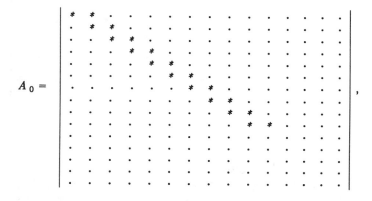

$$A_1 = \begin{vmatrix}
\cdot & \cdot & \cdot & \cdot & \cdot & \cdot & \cdot & \cdot & \cdot & \cdot & \cdot & \cdot & \cdot & \cdot & \cdot \\
\cdot & \cdot & \cdot & \cdot & \cdot & \cdot & \cdot & \cdot & \cdot & \cdot & \cdot & \cdot & \cdot & \cdot & \cdot \\
\cdot & \cdot & \cdot & \cdot & \cdot & \cdot & \cdot & \cdot & \cdot & \cdot & \cdot & \cdot & \cdot & \cdot & \cdot \\
\cdot & \cdot & \cdot & \cdot & \cdot & \cdot & \cdot & \cdot & \cdot & \cdot & \cdot & \cdot & \cdot & \cdot & \cdot \\
\cdot & \cdot & \cdot & \cdot & \cdot & \cdot & \cdot & \cdot & \cdot & \cdot & \cdot & \cdot & \cdot & \cdot & \cdot \\
\cdot & \cdot & \cdot & \cdot & \cdot & \cdot & \cdot & \cdot & \cdot & \cdot & \cdot & \cdot & \cdot & \cdot & \cdot \\
\cdot & \cdot & \cdot & \cdot & \cdot & \cdot & \cdot & \cdot & \cdot & \cdot & \cdot & \cdot & \cdot & \cdot & \cdot \\
\cdot & \cdot & \cdot & \cdot & \cdot & \cdot & \cdot & \cdot & \cdot & \cdot & \cdot & \cdot & \cdot & \cdot & \cdot \\
\cdot & \cdot & \cdot & \cdot & \cdot & \cdot & \cdot & \cdot & \cdot & \cdot & \cdot & \cdot & \cdot & \cdot & \cdot \\
\cdot & \cdot & \cdot & \cdot & \cdot & \cdot & \cdot & \cdot & \cdot & \cdot & * & \cdot & \cdot & \cdot & \cdot \\
\cdot & \cdot & \cdot & \cdot & \cdot & \cdot & \cdot & \cdot & \cdot & \cdot & \cdot & * & \cdot & \cdot & \cdot \\
\cdot & \cdot & \cdot & \cdot & \cdot & \cdot & \cdot & \cdot & \cdot & \cdot & \cdot & \cdot & * & \cdot & \cdot \\
\cdot & \cdot & \cdot & \cdot & \cdot & \cdot & \cdot & \cdot & \cdot & \cdot & \cdot & \cdot & \cdot & * & \cdot \\
\cdot & \cdot & \cdot & \cdot & \cdot & \cdot & \cdot & \cdot & \cdot & \cdot & \cdot & \cdot & \cdot & \cdot & * \\
\cdot & \cdot & \cdot & \cdot & \cdot & \cdot & * & \cdot & \cdot & \cdot & \cdot & \cdot & \cdot & \cdot & \cdot
\end{vmatrix},$$

The matrix A_2 is a scalar matrix and the coefficient matrices of higher index were either zero or scalar. We performed successive substitutions (in Boolean arithmetic) in the equation

$$G = A_0 + (A_1 + A_2 G)G.$$

After a small number of iterations, the following structure for the matrix G emerges. It is preserved in all further iterations.

$$G = \begin{vmatrix}
* & * & * & * & * & * & * & * & * & * & * & \cdot & \cdot & \cdot & \cdot \\
\cdot & * & * & * & * & * & * & * & * & * & * & \cdot & \cdot & \cdot & \cdot \\
\cdot & \cdot & * & * & * & * & * & * & * & * & * & \cdot & \cdot & \cdot & \cdot \\
\cdot & \cdot & \cdot & * & * & * & * & * & * & * & * & \cdot & \cdot & \cdot & \cdot \\
\cdot & \cdot & \cdot & \cdot & * & * & * & * & * & * & * & \cdot & \cdot & \cdot & \cdot \\
\cdot & \cdot & \cdot & \cdot & \cdot & * & * & * & * & * & * & \cdot & \cdot & \cdot & \cdot \\
\cdot & \cdot & \cdot & \cdot & \cdot & * & * & * & * & * & * & \cdot & \cdot & \cdot & \cdot \\
\cdot & \cdot & \cdot & \cdot & \cdot & * & * & * & * & * & * & \cdot & \cdot & \cdot & \cdot \\
\cdot & \cdot & \cdot & \cdot & \cdot & * & * & * & * & * & * & \cdot & \cdot & \cdot & \cdot \\
\cdot & \cdot & \cdot & \cdot & \cdot & * & * & * & * & * & * & \cdot & \cdot & \cdot & \cdot \\
\cdot & \cdot & \cdot & \cdot & \cdot & * & * & * & * & * & * & \cdot & \cdot & \cdot & \cdot \\
\cdot & \cdot & \cdot & \cdot & \cdot & * & * & * & * & * & * & \cdot & \cdot & \cdot & \cdot \\
\cdot & \cdot & \cdot & \cdot & \cdot & * & * & * & * & * & * & \cdot & \cdot & \cdot & \cdot \\
\cdot & \cdot & \cdot & \cdot & \cdot & * & * & * & * & * & * & \cdot & \cdot & \cdot & \cdot \\
\cdot & \cdot & \cdot & \cdot & \cdot & * & * & * & * & * & * & \cdot & \cdot & \cdot & \cdot
\end{vmatrix}.$$

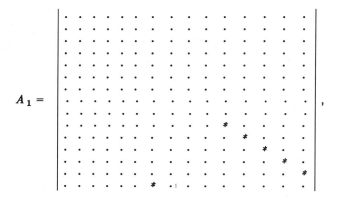

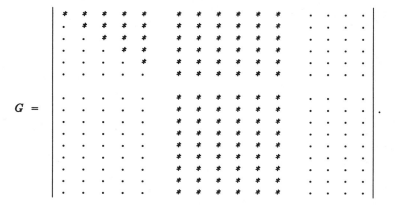

Although that structure agreed with the one we had anticipated on theoretical grounds, it remained interesting to see the rather involved manner in which it came about in the successive iterations. We notice that this case involves both the reducibility treated in Section 3.5 and that of the matrix $G(1)$, treated in Section 3.4. Because of the involved physics of the underlying buffer model, a discussion of the fundamental period for this model is unavoidably complex.

Problem 3.5.1: Discuss the nature of the matrices $K(x)$ and $L(x)$ for the variant of the $M/G/1$ queue described in Problem 2.3.8. Show in particular that the busy period distribution is a mixture of $m+1$ probability distributions, which are the distributions of the durations of the busy periods consisting of $1, \ldots, m$ and "more than m" services. Describe all simplifying features induced by the particular structure of the transition probability matrix for this model. •

Problem 3.5.2: Consider a stable $M/G/1$ queue and study the stationary probability d_k, $k > 0$, that a service is the $k-$th of the busy period during which it occurs. Use the results of Problems 2.3.8 and 3.5.1 to evolve an efficient algorithm for the computation of the probability density $\{d_k\}$. Write a computer program to implement this algorithm for a wide class of service time distribitions, such as discrete or phase type distributions. Discuss the qualitative information on the behavior of the queue which is contained in the computed density. Also consider the conditional means and variances of the queue length at the beginning of services which are the $k-$th in the busy period during which they occur. Add subroutines to the program to evaluate these quantities for k up to some maximum value. Plot and interpret the graphs of the computed functions. •

The investigation proposed in Problem 3.5.2 deals with a sequence of Markov chains embedded in the standard embedded Markov chain at departure epochs. These Markov chains are obtained by considering the $M/G/1$ queue immediately following the successive ends of services which are the first, second $, \ldots,$ $k-$th of the busy periods during which they occur. By the methods of this book, it is possible to obtain detailed results on the corresponding embedded Markov chains for the more general $PH/G/1$ queue. See Ali and Neuts [A-016] and Neuts [N-060]. A related investigation is discussed in Neuts [N-064], where the joint stationary probability density of the queue length and the duration of an excursion of at least k services above a threshold K is used to construct *profile curves* for the $M/G/1$ queue. Profile curves are of interest in designing *a monitor* which alerts the manager of an $M/G/1$ queue, to the occurrence and nature of unusual excursions of the queue.

Problem 3.5.3: The following model may be analyzed by the matrix-geometric approach of [N-042] or by the methods of Section 3.5. In both approaches, there are minor problems due to the reducibility of the matrix A. An $M/M/1$ queue is operated under the $N-$policy.

Whenever the server starts a busy period, he serves at a faster rate for a random length of time, after which the service rate drops to a lower value whenever the current busy period has not yet ended. Assume that the higher rate of service prevails for a length of time which has a phase type distribution. Describe this queue as a quasi-birth-and-death process and examine the structures of the matrices R and G, where R is as defined in [N-042]. Show how both matrices may be efficiently computed. •

Section 3.6

The procedure to compute the invariant probability vector $\boldsymbol{\pi}$, discussed in the text, is due to Paige, Styan and Wachter [P-007]. Various alternatives and generalizations are discussed in Hunter [H-078].

The complexity of the set-up computations for the matrices A_ν and B_ν depends greatly on the specific model. In some cases, a thorough analysis of the model may lead to major simplifications in these and all subsequent algorithmic steps. A superb example of the merits of such a prior analysis may be found in articles by Lucantoni and Ramaswami [L-052,R-012,R-013] on the $GI/PH/c$ queue. There are now several complex models of which thorough studies, with discussions of algorithmic implementation based on the methods of this book, are available in the literature. We mention in particular Lucantoni [L-051], Ramaswami and Lucantoni [R-010], and Chandramouli, Neuts and Ramaswami [C-007]. The promising recursive algorithm for the vectors $\mathbf{x}_i$ is established in Ramaswami [R-016]. A numerical analytic discussion of the non-linear matrix equations for the matrix G may be found in Ramaswami [R-017]. The book by Ortega and Rheinboldt [O-011] is a superb source on the underlying non-linear analysis of such equations.

4

Applications to Queues with Poisson Arrivals

4.1. THE $M/SM/1$ QUEUE AND SOME OF ITS VARIANTS

The general results, established in Chapters 2 and 3, lead to detailed analyses of many queues and other stochastic models with input processes more versatile than the Poisson process. Before we discuss such arrival processes in Chapter 5, we devote this chapter to the examination of a number of familiar queueing models with Poisson arrivals. Among these, the $M/SM/1$ queue introduced as Example A in Section 2.1 stands out by its interest and tractability. In addition to a brief review of the properties of this model, which we have already established, we shall discuss its stationary waiting time distributions and several moment formulas as well as a few of its variants and applications.

The $m \times m$ matrices $A_\nu(\cdot)$ and $B_\nu(\cdot)$, $\nu \geqslant 0$, are as defined in formula (2.1.2). In Section 2.3, it is shown that the matrix G is positive. The $M/SM/1$ queue is stable if and only if $\rho = \lambda \boldsymbol{\pi} \boldsymbol{\alpha} < 1$. The column vector $\boldsymbol{\alpha}$ is the vector of row sum means of the Markov renewal matrix $H(\cdot)$ which governs the service times and $\boldsymbol{\pi}$ is the invariant probability vector of the stochastic matrix $H(\infty) = A$.

The vectors $\mathbf{x}_0$ and $\mathbf{x}_1$ are explicitly given in (3.2.19) and (3.2.21). In particular, $\mathbf{x}_0 = (1-\rho)\mathbf{g}$, is simply a multiple of the invariant probability vector $\mathbf{g}$ of the matrix G. Expressions for the mean queue length L_1 at departure epochs and for several related quantities and vectors are given in the formulas (3.3.15).

The fundamental mean E of the embedded Markov renewal process of the $M/SM/1$ queue is the inner product of the vector $\mathbf{x}$ and the column vector of the row sum means of the transition probability matrix. This second vector is clearly of the form

$$[\lambda^{-1}\mathbf{e} + \boldsymbol{\alpha}, \boldsymbol{\alpha}, \boldsymbol{\alpha}, \cdots]^T,$$

so that

$$E = \mathbf{x}_0(\lambda^{-1}\mathbf{e} + \boldsymbol{\alpha}) + \sum_{i=1}^{\infty} \mathbf{x}_i \boldsymbol{\alpha} = \lambda^{-1}\mathbf{x}_0\mathbf{e} + [\mathbf{x}_0 + \mathbf{X}(1-)]\boldsymbol{\alpha} \qquad (4.1.1)$$

$$= \lambda^{-1}(1 - \rho) + \boldsymbol{\pi}\boldsymbol{\alpha} = \lambda^{-1}.$$

The Queue Length at an Arbitrary Time: The joint stationary probabilities of the queue length and the service type *at an arbitrary time* are obtained by an argument that is entirely similar to that for the $M/G/1$ queue. We introduce the partitioned vector $\mathbf{P} = [\mathbf{P}_0, \mathbf{P}_1, \mathbf{P}_2, \cdots]$, where $(\mathbf{P}_i)_j$, $i \geqslant 1$, $1 \leqslant j \leqslant m$, is the stationary probability that at time t, there are i customers in the system while a service of type j is in course. For $i = 0$, $(\mathbf{P}_0)_j$ is the corresponding probability that the server is idle and will provide a service of type j to the next arriving customer.

We omit the details of the limit argument in which the key renewal theorem is applied to the time-dependent probabilities, as we shall present them for the waiting time distribution. A direct matrix version of the derivations in Section 1.6 (with $N = 1$) leads to the following relations between the probabilities $(\mathbf{P}_i)_j$ and $(\mathbf{x}_i)_j$, $i \geqslant 0$, $1 \leqslant j \leqslant m$.

$$(\mathbf{P}_0)_j = (\mathbf{x}_0)_j,$$

$$(\mathbf{P}_i)_j = \lambda(\mathbf{x}_0)_j \int_0^\infty e^{-\lambda u} \frac{(\lambda u)^{i-1}}{(i-1)!} [1 - (H(u))\mathbf{e})_j] du \tag{4.1.2}$$

$$+ \lambda \sum_{k=1}^{i} (\mathbf{x}_k)_j \int_0^\infty e^{-\lambda u} \frac{(\lambda u)^{i-k}}{(i-k)!} [1 - (H(u))\mathbf{e})_j] du, \quad \text{for } i \geqslant 1.$$

In order to bring out a more transparent relation between the probability vectors $\mathbf{P}$ and $\mathbf{x}$, we write $\Delta^*(z)$ for the diagonal matrix with diagonal elements

$$\Delta^*_{jj}(z) = [h(\lambda - \lambda z)\mathbf{e}]_j,$$

where $h(\lambda - \lambda z)$ is the matrix $h(s)$ of Laplace-Stieltjes transforms of $H(\cdot)$, evaluated at $\lambda - \lambda z$. We shall also write $\Delta(\mathbf{u})$ for a diagonal matrix with the components $u_1, \ldots, u_m$ of a vector $\mathbf{u}$ as its diagonal elements.

Theorem 4.1.1: The vector $\mathbf{P}(z) = \sum_{i=0}^{\infty} \mathbf{P}_i z^i$, is given by

$$\mathbf{P}(z) = \mathbf{x}_0 + [\,\mathbf{X}(z) + z\mathbf{x}_0\,]\,(1 - z)^{-1}\,[I - \Delta^*(z)], \tag{4.1.3}$$

where $\mathbf{X}(z)$ is as defined in Section 3.3. $\mathbf{X}(z)$ satisfies the equation

$$[\mathbf{X}(z) + \mathbf{x}_0]\,[zI - h(\lambda - \lambda z)] = (z - 1)\mathbf{x}_0 h(\lambda - \lambda z). \tag{4.1.4}$$

The equality

$$\mathbf{P}(z)\mathbf{e} = [\mathbf{X}(z) + \mathbf{x}_0]\mathbf{e} = 1 - \rho + \mathbf{X}(z)\mathbf{e}, \tag{4.1.5}$$

holds and the row vectors $\mathbf{P}(1)$ and $\mathbf{P}'(1)$ are given by

$$\mathbf{P}(1) = \mathbf{x}_0 + \lambda\boldsymbol{\pi}\Delta(\boldsymbol{\alpha}) = (1 - \rho)\mathbf{g} + \lambda\boldsymbol{\pi}\Delta(\boldsymbol{\alpha}), \tag{4.1.6}$$

and

$$\mathbf{P}'(1) = [\mathbf{X}'(1) + \mathbf{x}_0]\Delta(\boldsymbol{\alpha}) + \frac{1}{2}\boldsymbol{\pi}\Delta(\boldsymbol{\alpha}_2), \tag{4.1.7}$$

where $\boldsymbol{\alpha}_2$ is the column vector of second moments of the row sums of the semi-Markov matrix $H(\cdot)$. The row vector $\mathbf{X}'(1)$ is as given in (3.3.15).

Proof: Formula (4.1.3) is immediate from (4.1.2) and the equality (4.1.4) is readily obtained from (3.3.2) by substituting the particular forms of the matrices $A^*(z)$ and $\sum_{\nu=1}^{\infty} B_\nu z^\nu$ for the $M/SM/1$ queue.

Postmultiplying by $\mathbf{e}$ in (4.1.4), we obtain

$$z\mathbf{X}(z)\mathbf{e} - \mathbf{X}(z)\Delta^*(z)\mathbf{e} = z\mathbf{x}_0\Delta^*(z)\mathbf{e} - z\mathbf{x}_0\mathbf{e},$$

and with the help of that equality, formula (4.1.5) follows by routine calculations. The remaining formulas follow directly from (4.1.3). •

Formula (4.1.5) shows that the stationary queue length densities $\{\mathbf{P}_i\mathbf{e}\}$ and $\{\mathbf{x}_i\mathbf{e}\}$, respectively at time t and immediately after departures are the same. The conditional density

$$\tilde{q}_i(j) = \frac{(\mathbf{P}_i)_j}{[\mathbf{P}(1)]_j}, \quad \text{for } i \geqslant 0,$$

of the queue length, given that at time t, the service state is j, is different in general from the corresponding conditional density

$$q_i(j) = \frac{(\mathbf{x}_i)_j}{\pi_j}, \quad \text{for } i \geqslant 0,$$

for the queue length at departures.

Since the arrival process is Poisson, the vector $\mathbf{P}$ is also the joint stationary density of the queue length and the service type seen by an arbitrary arriving customer.

The Virtual Waiting Time: Next we shall study the joint probability distribution of *the virtual waiting time and the service type* of a (virtual) customer arriving at time t. The quantity $W_j(x)$, $x \geqslant 0$, $1 \leqslant j \leqslant m$, is the stationary probability that a virtual customer who arrives at time t, waits at most x units of time and receives a service of type j. The row vector $\mathbf{W}(x)$ has components $W_j(x)$, $1 \leqslant j \leqslant m$.

We shall show that the vector distribution $\mathbf{W}(\cdot)$ satisfies a Volterra integral equation which generalizes the classical Pollaczek-Khinchin formula for the $M/G/1$ queue. The proof of this result is the paradigm for the derivation of a number of equations for waiting time distributions for other queueing models to be discussed in the sequel. We shall therefore give a very detailed presentation of all the steps involved so that some later calculations, which proceed along similar lines, may be succinctly presented.

Theorem 4.1.2: For the stable $M/SM/1$ queue, the vector distribution $\mathbf{W}(\cdot)$ satisfies the Volterra integral equation

$$\mathbf{W}(x) = \mathbf{x}_0 + \lambda \int_0^x \mathbf{W}(u)[I - H(x-u)]\,du, \quad \text{for } x \geqslant 0. \tag{4.1.8}$$

Proof: We begin by deriving expressions for the corresponding time dependent distributions $W_j(t;x)$ for the virtual waiting time in the $M/SM/1$ queue with fixed, but not explicitly specified initial conditions. We denote by $M_j(i;t)$, the expected number of visits by the embedded Markov renewal process to the state (i,j), $i \geqslant 0$, $1 \leqslant j \leqslant m$, in $[0,t)$. The explicit form of the Markov renewal matrix $\{M_j(i;t)\}$ will not be needed. It may be studied in the same formal manner as in Section 1.5.

In terms of the functions $M_j(i;t)$, we may write

$$W_j(t;x) = \int_0^t e^{-\lambda(t-u)} dM_j(0;u) \tag{4.1.9}$$

$$+ \sum_{j'=1}^{m} \sum_{j''=1}^{m} \sum_{i=0}^{\infty} \int_{0(u)}^{t} \int_{0(v)}^{t-u} \int_{0(w)}^{x} dM_{j'}(0;u) e^{-\lambda v} \lambda dv \; e^{-\lambda(t-u-v)}$$

$$\cdot \frac{[\lambda(t-u-v)]^i}{i!} dH_{j'j''}(t+w-u-v) \, [H^{(i)}(x-w)]_{j''j}$$

$$+ \sum_{j'=1}^{m} \sum_{j''=1}^{m} \sum_{i=1}^{\infty} \sum_{k=1}^{i} \int_{0(u)}^{t} \int_{0(w)}^{x} dM_{j'}(k;u) e^{-\lambda(t-u)} \frac{[\lambda(t-u)]^{i-k}}{(i-k)!}$$

$$\cdot \, dH_{j'j''}(t+w-u) \, [H^{(i-1)}(x-w)]_{j''j} .$$

Formula (4.1.9) follows by an involved, but straightforward application of the law of total probability. The first term corresponds to the case where the queue is empty at $t = 0$ and the server is ready to dispense a service of type j. The second term corresponds to the case where at time t, there are $i+1$ customers in the queue; the system was empty at the last departure prior to t and the service in course is the first of the subsequent busy period. The case, where there are $i \geqslant 1$ customers at time t and some number k, $1 \leqslant k \leqslant i$, at the last departure prior to t, contributes the third term.

In passing to the limit as $t \to \infty$, we may interchange the summations and limits by virtue of the dominated convergence theorem. Each term is of the form suitable for a direct application of the key renewal theorem. The term-wise limits are non-zero, if and only if the mean recurrence times of (all) the states (i,j) are finite. This is the case if and only if $\rho < 1$. The mean recurrence time of the state (i, j) is then given by $E\, x_{ij}^{-1} = (\lambda x_{ij})^{-1}$.

When $\rho < 1$, it follows from (4.1.9) and the key renewal theorem that (in matrix notation) $\mathbf{W}(x)$ is given by

$$\mathbf{W}(x) = \mathbf{x}_0 + \sum_{i=0}^{\infty} \int_{0(\tau)}^{\infty} d\tau \int_{0(v)}^{\tau} \int_{0(w)}^{x} \lambda \mathbf{x}_0 e^{-\lambda v} \lambda dv \tag{4.1.10}$$

$$\cdot\, e^{-\lambda(\tau - v)} \frac{[\lambda(\tau - v)]^i}{i!} \, dH(\tau + w - v) H^{(i)}(x - w)$$

$$+ \sum_{i=1}^{\infty} \sum_{k=1}^{i} \int_{0(\tau)}^{\infty} d\tau \int_{0(w)}^{x} \lambda \mathbf{x}_k \, e^{-\lambda\tau} \frac{(\lambda\tau)^{(i-k)}}{(i-k)!} \, dH(\tau + w) H^{(i-1)}(x - w).$$

By interchanging the first two integrations, the second term may be written as

$$\lambda \mathbf{x}_0 \sum_{i=0}^{\infty} \int_{0(\tau)}^{\infty} d\tau \int_{0(w)}^{x} e^{-\lambda\tau} \frac{(\lambda\tau)^i}{i!} \, dH(\tau + w) H^{(i)}(x - w).$$

Formula (4.1.10) is most easily simplified by forming the Laplace-Stieltjes transform of both sides. We obtain

$$\mathbf{w}(s) = \int_0^{\infty} e^{-sx} \, d\,\mathbf{W}(x) \tag{4.1.11}$$

$$= \mathbf{x}_0 + \lambda \mathbf{x}_0 \sum_{i=0}^{\infty} \int_{0(\tau)}^{\infty} d\tau \int_{0(x)}^{\infty} \int_{0(w)}^{x} e^{-sx - \lambda\tau} \frac{(\lambda\tau)^i}{i!} \, dH(\tau + w) H^{(i)}(x - w)$$

$$+ \sum_{i=1}^{\infty} \sum_{k=1}^{i} \lambda \mathbf{x}_k \int_{0(\tau)}^{\infty} d\tau \int_{0(x)}^{\infty} \int_{0(w)}^{x} e^{-sx - \lambda\tau} \frac{(\lambda\tau)^{i-k}}{(i-k)!} \, dH(\tau + w) H^{(i-1)}(x - w)$$

$$= \mathbf{x}_0 + \lambda \mathbf{x}_0 \sum_{i=0}^{\infty} \int_{0(\tau)}^{\infty} d\tau \int_{0(w)}^{\infty} e^{-sw - \lambda\tau} \frac{(\lambda\tau)^i}{i!} \, dH(\tau + w) h^i(s)$$

$$+ \sum_{i=1}^{\infty} \sum_{k=1}^{i} \lambda \mathbf{x}_k \int_{0(\tau)}^{\infty} d\tau \int_{0(w)}^{\infty} e^{-sw-\lambda\tau} \frac{(\lambda\tau)^{(i-k)}}{(i-k)!} dH(\tau+w) h^{i-1}(s),$$

where $h(s)$ is the Laplace-Stieltjes transform of the matrix $H(\cdot)$.

The remaining steps are somewhat delicate, due to the non-commutativity of the matrix product. We shall explain every successive step as we proceed.

In terms of the matrix exponential function, we have

$$\sum_{i=0}^{\infty} e^{-\lambda\tau} \frac{(\lambda\tau)^i}{i!} h^i(s) = \exp\{-\lambda\tau[I - h(s)]\},$$

so that after interchanging the two summations in the last term of (4.1.11), we obtain

$$\mathbf{w}(s) = \mathbf{x}_0 + \lambda \mathbf{x}_0 \int_{0(\tau)}^{\infty} d\tau \int_{0(w)}^{\infty} e^{-sw} dH(\tau+w) \exp\{-\lambda\tau[I - h(s)]\}$$

$$+ \lambda \sum_{k=1}^{\infty} \mathbf{x}_k \int_{0(\tau)}^{\infty} d\tau \int_{0(w)}^{\infty} e^{-sw} dH(\tau+w) h^{k-1}(s) \exp\{-\lambda\tau[I - h(s)]\}, \tag{4.1.12}$$

and we note that the matrices $h^{k-1}(s)$ and $\exp\{-\lambda\tau[I - h(s)]\}$ commute.

The double integral in the second term may be written as

$$\int_0^{\infty} dH(v) e^{-sv} \int_0^{v} \exp\{-\tau[(s-\lambda)I - \lambda h(s)]\} d\tau,$$

and similarly for the double integral in the third term. It is clear that

$$\int_0^{v} \exp\{\tau[(s-\lambda)I + \lambda h(s)]\} d\tau \cdot [(s-\lambda)I + \lambda h(s)] \tag{4.1.13}$$

$$= \exp\{+v[(s-\lambda)I + \lambda h(s)]\} - I,$$

since both sides are obviously equal for $v = 0$, and yield the same expression upon differentiation with respect to v. The equality (4.1.13) therefore follows from classical results on linear differential equations with constant coefficients.

If we postmultiply in (4.1.12) by the matrix $(s-\lambda)I + \lambda h(s)$, and use formula (4.1.13) twice, we obtain

$$\mathbf{w}(s)[(s - \lambda)I + \lambda h(s)] = \mathbf{x}_0[(s - \lambda)I + \lambda h(s)] \tag{4.1.14}$$

$$+ \lambda \mathbf{x}_0 \int_0^\infty dH(v)\exp\{-\lambda v[I - h(s)]\} - \lambda \mathbf{x}_0 h(s) - \lambda \sum_{k=1}^\infty \mathbf{x}_k h^k(s)$$

$$+ \lambda \sum_{k=1}^\infty \mathbf{x}_k \int_0^\infty dH(v)\exp\{-\lambda v[I - h(s)]\} h^{k-1}(s).$$

In order to simplify the right hand side of this equation, we turn to the equations (3.3.1), which for the $M/SM/1$ queue read as

$$\mathbf{x}_i = \mathbf{x}_0 A_i + \sum_{\nu=1}^{i+1} \mathbf{x}_\nu A_{i-\nu+1}, \quad \text{for } i \geq 0, \tag{4.1.15}$$

where

$$A_i = \int_0^\infty e^{-\lambda u} \frac{(\lambda u)^i}{i!} dH(u), \quad \text{for } i \geq 0.$$

If in (4.1.15), we postmultiply the equation for the index i by the matrix $h^i(s)$, and sum the resulting equations, we obtain after elementary calculations the equality

$$\sum_{i=0}^\infty \mathbf{x}_i h^i(s) = \mathbf{x}_0 \int_0^\infty dH(v)\exp\{-\lambda v[I - h(s)]\} \tag{4.1.16}$$

$$+ \sum_{i=1}^\infty \mathbf{x}_i \int_0^\infty dH(v)\exp\{-\lambda v[I - h(s)]\} h^{i-1}(s).$$

Upon substitution of (4.1.16) into (4.1.14), it follows that

$$\mathbf{w}(s)[\,(s-\lambda)I+\lambda h(s)\,]=s\,\mathbf{x}_0, \tag{4.1.17}$$

or equivalently

$$\mathbf{w}(s)=\mathbf{x}_0+\lambda\mathbf{w}(s)s^{-1}[I-\lambda h(s)], \tag{4.1.18}$$

which is the transform version of the integral equation (4.1.8). •

Corollary 4.1.1: The vector $\mathbf{W}(\infty)$ is given by

$$\mathbf{W}(\infty)=\boldsymbol{\pi}. \tag{4.1.19}$$

Proof: The vector $\mathbf{w}(0+)=\mathbf{W}(\infty)$, satisfies

$$\mathbf{w}(0+)(I-H)=0, \qquad \mathbf{w}(0+)\mathbf{e}=1,$$

which implies (4.1.19). •

Corollary 4.1.2: For $|\,z\,|\leqslant 1$, we have

$$\mathbf{x}_0+\mathbf{X}(z)=\mathbf{w}(\lambda-\lambda z)h(\lambda-\lambda z). \tag{4.1.20}$$

Proof: Set $s=\lambda(1-z)$ in (4.1.17) and multiply both sides by $h(\lambda-\lambda z)$. The stated equality is obvious upon comparison with (4.1.4), since the stationary probability vector $\mathbf{x}$ of the embedded Markov chain is unique. •

Remark *a*. Formula (4.1.17) is commonly written as

$$\mathbf{w}(s)=s\,\mathbf{x}_0\,[\,(s-\lambda)I+\lambda h(s)\,]^{-1}.$$

Since the components of $\mathbf{w}(s)$ are the Laplace-Stieltjes transforms of probability mass-functions on $[0,\infty)$, the expression in the right hand side can only have *removable* singularities in the half-plane $\mathrm{Re}\,s\geqslant 0$.

By an argument based on Rouché's theorem, one may show that the determinant of the matrix $(s-\lambda)I + \lambda h(s)$ vanishes at $s = 0$ and at exactly $m - 1$ points in Re $s > 0$. The vector $\mathbf{u} = \mathbf{x}_0$ is the unique vector for which the components of the vector $s\mathbf{u}[(s-\lambda)I + \lambda h(s)]^{-1}$ have only removable singularities at the m zeros of the determinant. In our presentation, this argument, which only serves to express $\mathbf{w}(s)$ in a (generally) less useful form, is entirely circumvented.

Remark *b*. For the $M/G/1$ queue, (4.1.8) reduces to the familiar Pollaczek-Khinchin integral equation

$$W(x) = 1 - \rho + \rho W(\cdot) * H^*(x), \quad \text{for } x \geq 0, \tag{4.1.21}$$

where $H^*(x)$ with density $[1 - H(x)]\alpha^{-1}$, is the random modification of the service time distribution $H(\cdot)$. It is worth emphasizing that the matrix $I - H(x)$ is *not* the natural analogue of the mass-function $1 - H(x)$, which arises in the scalar case.

In Markov renewal theory, the analogue of the random modification is the semi-Markov matrix

$$H^*(x) = \Delta^{-1}(\alpha)\int_0^x [H - H(u)]du, \quad \text{for } x \geq 0, \tag{4.1.22}$$

where $\Delta(\alpha)$ is the diagonal matrix with diagonal elements $\alpha_1, \ldots, \alpha_m$. In terms of $H^*(\cdot)$, we may rewrite (4.1.8) as

$$\mathbf{W}(x) = (1-\rho)\mathbf{g} + \lambda \int_0^x \mathbf{W}(u)du\,(I - H) + \rho W(\cdot)\Delta(\hat{\alpha}) * H^*(x), \tag{4.1.23}$$

for $x \geq 0$, where $\hat{\alpha} = (\pi\alpha)^{-1}\alpha$.

Comparing the equations (4.1.21) and (4.1.23), we see that the Pollaczek-Khinchin equation for the $M/SM/1$ queue (and its other matrix generalizations) have a term in the right hand side without counterpart in the scalar case. This explains why the Neumann series solution

$$\mathbf{W}(x) = \sum_{\nu=0}^{\infty} (1-\rho)\lambda^{\nu}\,\mathbf{g}\left[\int_0^x [I - H(u)]du\right]^{(\nu)},$$

of (4.1.8) is neither theoretically interesting, nor suitable for numerical computation.

Remark *c*. The same calculations as in the proof of Theorem 4.1.2 may be used to obtain transform formulas for the joint stationary distribution of the queue length and the residual service time as well as of the joint distribution of the queue length and the waiting time at an arbitrary epoch.

Let $\phi_j(i;x)$, $i \geq 1$, $x \geq 0$, $1 \leq j \leq m$, be the stationary probability that an arriving customer finds i customers in the system, the residual service time is at most x and the next service will be of type j. Similarly, let $W_j(i;x)$, $i \geq 0$, $x \geq 0$, $x \geq 0$, $1 \leq j \leq m$, be the stationary probability that an arriving customer finds i customers in the system and has to wait a time at most x before receiving a service of type j. The row vectors $\mathbf{\Phi}(z;s)$ and $\mathbf{w}(z;s)$ have the components

$$\Phi_j(z;s) = \sum_{i=1}^{\infty} z^i \int_0^{\infty} e^{-sx}\, d\phi_j(i;x),$$

$$w_j(z;s) = \sum_{i=1}^{\infty} z^i \int_0^{\infty} e^{-sx}\, dW_j(i;x), \quad \text{for } 1 \leq j \leq m,$$

and are given by

$$\mathbf{\Phi}(z;s) = \lambda(1-\rho)z(z-1)\,\mathbf{g}[zI - h(\lambda - \lambda z)] \cdot [h(\lambda - \lambda z) - h(s)]\,(s - \lambda + \lambda z)^{-1}, \tag{4.1.24}$$

and

$$\mathbf{w}(z;s)\,[sI - \lambda I + \lambda z h(s)] = s\,\mathbf{x}_0 - \lambda(1-z)\sum_{i=0}^{\infty} \mathbf{x}_i h^i(s) z^i. \tag{4.1.25}$$

These formulas are primarily useful in deriving expressions for the correlations between the queue length and respectively, the residual service and waiting times.

Moment Calculations: By using the formulas (4.1.17) and (4.1.20), we may relate the moments of $\mathbf{W}(\cdot)$ to the moments of the queue length discussed in Section 3.3. We consider the moment vectors

$$\mathbf{W}_n = \int_0^\infty x^n \, d\,\mathbf{W}(x), \tag{4.1.26}$$

and the moments matrices

$$H_n = \int_0^\infty x^n \, dH(x), \tag{4.1.27}$$

and we set $\mathbf{W}_n \mathbf{e} = W_n^*$. Since $A^*(z) = h(\lambda - \lambda z)$, the matrix $A^{*(n)}(1-) = \lambda^n H_n$, for $n \geqslant 0$. The matrices $A^{*(n)}(1-)$ are needed in the evaluation of the vectors $\mathbf{X}^{(n)}(1-)$ and the n-th factorial moments L_n, discussed in Section 3.3.

By differentiating n times in (4.1.20), setting $z = 1-$, and post-multiplying by $\mathbf{e}$, we obtain

$$L_n = \lambda^n W_n^* + \lambda^n \sum_{\nu=0}^{n-1} \binom{n}{\nu} \mathbf{W}_\nu H_{n-\nu} \mathbf{e}. \tag{4.1.28}$$

In particular, for $n = 1$, we have

$$\lambda W_1^* = L_1 - \rho. \tag{4.1.29}$$

A first differentiation in (4.1.17) leads to

$$\lambda \mathbf{W}_1 (I - H) + \boldsymbol{\pi}(I - \lambda H_1) = \mathbf{x}_0,$$

and by adding $\lambda \mathbf{W}_1 \mathbf{e} \boldsymbol{\pi} = \lambda W_1^* \boldsymbol{\pi}$, to both sides of that equation, we obtain

$$\lambda \mathbf{W}_1 = \mathbf{x}_0 Z + \lambda \boldsymbol{\pi} H_1 Z + (L_1 - \rho - 1)\boldsymbol{\pi}, \tag{4.1.30}$$

by using (4.1.29). The matrix $Z = (I - H + \mathbf{e}\boldsymbol{\pi})^{-1}$, is the same matrix as in formulas (3.3.15). A comparison of (4.1.30) with the expression for $\mathbf{X}'(1-)$ in (3.3.15) shows, after elementary manipulations, that

$$\lambda \mathbf{W}_1 = \mathbf{X}'(1-) + (1-\rho)\mathbf{g} - \boldsymbol{\pi} = \mathbf{X}'(1-) - \mathbf{X}(1-). \qquad (4.1.31)$$

The formula (4.1.31) is an interesting vector analogue of the Little-type formula (4.1.29). It is clear that (4.1.29) also follows from (4.1.31) by forming the inner product with $\mathbf{e}$ on both sides of the equalities.

Further differentiations in (4.1.17) show that, for $n \geqslant 2$,

$$\lambda \mathbf{W}_n (I - H) = -\, n\, \mathbf{W}_{n-1} + \lambda \sum_{\nu=0}^{n-1} \binom{n}{\nu} \mathbf{W}_\nu H_{n-\nu},$$

and by using the same device as in the case $n = 1$, we obtain

$$\lambda \mathbf{W}_n = [-\, n\, \mathbf{W}_{n-1} + \lambda \sum_{\nu=0}^{n-1} \binom{n}{\nu} \mathbf{W}_\nu H_{n-\nu}]\, Z + \lambda W_n^* \boldsymbol{\pi}. \qquad (4.1.32)$$

We see that the formulas (4.1.28) and (4.1.32) yield a simple recursive scheme for the evaluation of the vectors $\mathbf{W}_n$. Since the quantities L_n are needed, it is most efficient to attach a step to the recursive procedure given by the formulas (3.3.7), (3.3.8), (3.3.12) and (3.3.13), so that with the moment vectors for the queue length, we also evaluate the vectors $\mathbf{W}_n$. In doing so, we may combine a number of arithmetic operations for greater computational efficiency. The details are left to the initiative of the interested reader.

The Expected Make-Up of the Queue: The stationary $M/SM/1$ queue, viewed either at departure epochs or at an arbitrary time, is made up of customers of m possible types. It is of interest to quantify the contribution of each customer type to the expected queue length L_1 and to the expected waiting time W_1^*. Let $L_{1,j}$ and $L_{1,j}^*$ be the expected number of customers of type j in the stationary queue, respectively after departures and at an arbitrary time. Similarly $W_{1,j}^*$ is the expected total remaining service time of all customers of type j in the stationary queue at an arbitrary time. In Theorem 4.1.3, we shall obtain explicit expressions for the row vectors $\mathbf{L}_1 =$

$(L_{1,1}, \ldots, L_{1,m})$, $\mathbf{L}_1^* = L_{1,1}^*, \ldots, L_{1,m}^*$ and $\mathbf{W}_1^* = (W_{1,1}^*, \ldots, W_{1,m}^*)$. Before starting the necessary calculations, we derive some preliminary results which are also of independent interest. They deal with several conditional means of random variables related to the stationary version of the $M/SM/1$ queue.

Let $\psi_{jj'}(i;x)$ be the stationary probability that at time t, there are $i \geq 1$ customers in the system, that the service in course is of type j, $1 \leq j \leq m$, that this service ends no later than time $t+x$ and is followed by a service of type j', $1 \leq j' \leq m$. By the usual argument based on the key renewal theorem, we obtain

$$\psi_{jj'}(i;x) = \lambda(\mathbf{x}_0)_j \int_0^\infty e^{-\lambda\tau} \frac{(\lambda\tau)^{i-1}}{(i-1)!} [H_{jj'}(\tau+x) - H_{jj'}(\tau)] d\tau \quad (4.1.33)$$

$$+ \sum_{k=1}^{i} \lambda(\mathbf{x}_k)_j \int_0^\infty e^{-\lambda\tau} \frac{(\lambda\tau)^{i-k}}{(i-k)!} [H_{jj'}(\tau+x) - H_{jj'}(\tau)] d\tau .$$

For future use, we record the following formulas, which are obtained from (4.1.33) by routine calculations. The moment matrices H_1 and H_2 are as defined in (4.1.27).

$$\psi_{jj'}(i) = \psi_{jj'}(i,\infty) = \lambda(\mathbf{x}_0)_j \int_0^\infty e^{-\lambda\tau} \frac{(\lambda\tau)^{i-1}}{(i-1)!} [H_{jj'} - H_{jj'}(\tau)] d\tau$$

$$(4.1.34)$$

$$+ \sum_{k=1}^{i} \lambda(\mathbf{x}_k)_j \int_0^\infty e^{-\lambda\tau} \frac{(\lambda\tau)^{i-k}}{(i-k)!} [H_{jj'} - H_{jj'}(\tau)] d\tau .$$

$$\sum_{i=1}^{\infty} \psi_{jj'}(i) = \lambda\pi_j (H_1)_{jj'}. \quad (4.1.35)$$

$$\int_0^\infty x \, d\psi_{jj'}(i) = \lambda(\mathbf{x}_0)_j \int_0^\infty \int_0^\infty e^{-\lambda\tau} \frac{(\lambda\tau)^{i-1}}{(i-1)!} x \, dH_{jj'}(\tau+x) d\tau$$

$$(4.1.36)$$

$$+ \sum_{k=1}^{i} \lambda(\mathbf{x}_k)_j \int_0^\infty \int_0^\infty e^{-\lambda\tau} \frac{(\lambda\tau)^{i-k}}{(i-k)!} x \, dH_{jj'}(\tau+x) d\tau$$

$$= \lambda(\mathbf{x}_0)_j \int_0^\infty dH_{jj'}(v) \int_0^v e^{-\lambda\tau} \frac{(\lambda\tau)^{i-1}}{(i-1)!} (v-\tau) d\tau$$

$$+ \sum_{k=1}^{i} \lambda(\mathbf{x}_k)_j \int_0^\infty dH_{jj'}(v) \int_0^v e^{-\lambda\tau} \frac{(\lambda\tau)^{i-k}}{(i-k)!} (v-\tau) d\tau .$$

$$\sum_{i=1}^{\infty} \int_0^\infty x \, d\psi_{jj'}(i;x) = \frac{1}{2}\lambda\pi_j (H_2)_{jj'}. \tag{4.1.37}$$

$$\sum_{i=1}^{\infty} (i-1)\psi_{jj'}(i) = \frac{1}{2}\lambda^2\pi_j (H_2)_{jj'} + \lambda[\mathbf{X}'(1-) - \mathbf{X}(1-)]_j (H_1)_{jj'}. \tag{4.1.38}$$

Next, let $V_{j'j''}(j;n)$, $n \geqslant 1$, be the conditional expected number of services of type j in n consecutive services with the service type at the end of the n-th service being j'', given that the first service is of type j'. Similarly $\tilde{V}_{j'j''}(j;n)$, $n \geqslant 1$, is the conditional expected time devoted to all services of type j in n consecutive services with the service type at the end of the n-th service being j'', given that the first service is of type j'.

We denote by $\Delta(\mathbf{e}_j)$, the $m \times m$ matrix with $\Delta_{jj}(\mathbf{e}_j) = 1$, and all its other elements zero. By considering the contributions of each of the n services to the expected value, we obtain that the matrices $V(j;n) = \{V_{j'j''}(j;n)\}$ and $\tilde{V}(j;n) = \{\tilde{V}_{j'j''}(j;n)\}$ are given by

$$V(j;n) = \sum_{\nu=1}^{n} H^{\nu-1}\Delta(\mathbf{e}_j)H^{n-\nu+1}, \tag{4.1.39}$$

and

$$\tilde{V}(j;n) = \sum_{\nu=1}^{n} H^{\nu-1}\Delta(\mathbf{e}_j)H_1H^{n-\nu}. \tag{4.1.40}$$

Theorem 4.1.3: The vectors $\mathbf{L}_1$, $\mathbf{L}_1^*$ and $\mathbf{W}_1^*$ are given by

$$\mathbf{L}_1 = (L_1 + 1)\boldsymbol{\pi} - \sum_{i=0}^{\infty} \mathbf{x}_i H^i Z, \tag{4.1.41}$$

$$\mathbf{L}_1^* = \lambda\boldsymbol{\pi}\Delta(\boldsymbol{\alpha}) + (\mathbf{x}_0 + \lambda\boldsymbol{\pi} H_1)Z + c\,\boldsymbol{\pi}, \tag{4.1.42}$$

$$\mathbf{W}_1^* = \frac{1}{2}\lambda\boldsymbol{\pi}\Delta(\boldsymbol{\alpha}_2) + (\mathbf{x}_0 + \lambda\boldsymbol{\pi} H_1)ZH_1 + c\,\boldsymbol{\pi} H_1, \tag{4.1.43}$$

where $Z = (I - H + \mathbf{e}\boldsymbol{\pi})^{-1}$, and the constant c is defined as

$$c = \lambda \mathbf{g} Z\boldsymbol{\alpha} + (1-\rho)^{-1}\left[\,\lambda^2\boldsymbol{\pi} H_1 Z\boldsymbol{\alpha} + \frac{1}{2}\lambda^2\boldsymbol{\pi}\boldsymbol{\alpha}_2 - 1\,\right].$$

Proof: It is clear that

$$L_{1,j} = \sum_{i=1}^{\infty} \mathbf{x}_i V(j;i)\mathbf{e} = \sum_{i=1}^{\infty} \mathbf{x}_i \sum_{\nu=1}^{i} H^{\nu-1}\Delta(\mathbf{e}_j)\mathbf{e},$$

from which it immediately follows that

$$\mathbf{L}_1 = \sum_{i=1}^{\infty} \mathbf{x}_i \sum_{\nu=1}^{i} H^{\nu-1} = \sum_{i=1}^{\infty} \mathbf{x}_i (I - H^i + i\mathbf{e}\boldsymbol{\pi})Z$$

$$= [\,\mathbf{X}(1-) - \sum_{i=1}^{\infty} \mathbf{x}_i H^i + \mathbf{X}'(1-)\mathbf{e}\boldsymbol{\pi}\,]Z$$

$$= (L_1 + 1)\boldsymbol{\pi} - \sum_{i=0}^{\infty} \mathbf{x}_i H^i Z.$$

Since $Z\mathbf{e} = \mathbf{e}$, it is obvious that $\mathbf{L}_1\mathbf{e} = L_1$. The proofs of the formulas (4.1.42) and (4.1.43) are considerably more involved. We begin by observing that there are three sets of alternatives, which contribute to the expected value $L_{1,j}^*$. These are respectively the cases where there is only one customer in the queue and he receives a service of type j; where there are two or more customers in the queue and the one in service is of type j, and finally, the case where the customer in service is not of type j. Clearly, in that last case, there can be a contribution to $L_{1,j}^*$, only if there are two or more customers in the system. These alternatives contribute the three terms in the formula

$$L_{1,j}^{*} = \sum_{j_1=1}^{m} \psi_{jj_1}(1) + \sum_{j_1=1}^{m} \sum_{j_2=1}^{m} \sum_{i=2}^{\infty} \psi_{jj_1}(i)\, [1 + V_{j_1j_2}(j\,;i-1)]$$

$$+ \sum_{j_3 \neq j} \sum_{j_1=1}^{m} \sum_{j_2=1}^{m} \sum_{i=2}^{\infty} \psi_{j_3j_1}(i)\, V_{j_1j_2}(j\,;i-1)$$

$$= \sum_{i=1}^{\infty} [\psi(i)\mathbf{e}]_j + \sum_{i=2}^{\infty} \sum_{j_3=1}^{m} [\psi(i)\, V(j\,;i-1)\mathbf{e}]_{j_3}.$$

By using formula (4.1.35) to evaluate the first term and substituting the expressions in (4.1.34) and (4.1.39) into the second term, we see that in matrix notation

$$\mathbf{L}_1^{*} = \lambda\boldsymbol{\pi}\Delta(\boldsymbol{\alpha}) + \lambda\mathbf{x}_0 \sum_{i=2}^{\infty} \int_0^{\infty} e^{-\lambda\tau} \frac{(\lambda\tau)^{i-1}}{(i-1)!} [H - H(\tau)]\, d\tau \sum_{\nu=1}^{i-1} H^{\nu-1}$$

$$+ \lambda \sum_{i=2}^{\infty} \sum_{k=1}^{i} \mathbf{x}_k \int_0^{\infty} e^{-\lambda\tau} \frac{(\lambda\tau)^{i-k}}{(i-k)!} [H - H(\tau)]\, d\tau \sum_{\nu=1}^{i-1} H^{\nu-1}$$

$$= \lambda\boldsymbol{\pi}\Delta(\boldsymbol{\alpha})$$

$$+ \lambda\mathbf{x}_0 \sum_{i=1}^{\infty} \int_0^{\infty} e^{-\lambda\tau} \frac{(\lambda\tau)^{i-1}}{(i-1)!} [H - H(\tau)]\, d\tau (I - H^{i-1} + (i-1)\mathbf{e}\boldsymbol{\pi})HZ$$

$$+ \lambda \sum_{i=1}^{\infty} \sum_{k=1}^{i} \mathbf{x}_k \int_0^{\infty} e^{-\lambda\tau} \frac{(\lambda\tau)^{i-k}}{(i-k)!} [H - H(\tau)]\, d\tau$$

$$\cdot (I - H^{i-1} + (i-1)\mathbf{e}\boldsymbol{\pi})HZ.$$

Noting that

$$\int_0^\infty [H - H(\tau)]d\tau = H_1,$$

and using similar elementary simplifications, the preceding expression for $\mathbf{L}_1^*$ may be rewritten as

$$\mathbf{L}_1^* = \lambda\boldsymbol{\pi}\Delta(\boldsymbol{\alpha}) + \lambda\boldsymbol{\pi}H_1Z + \frac{1}{2}\lambda^2(\boldsymbol{\pi}\boldsymbol{\alpha}_2)\boldsymbol{\pi}$$

$$+ \lambda[\mathbf{X}'(1-) - \mathbf{X}(1-)]\boldsymbol{\alpha}\boldsymbol{\pi} - \lambda\mathbf{x}_0\int_0^\infty dH(v)\int_0^v \exp[-\lambda\tau(I - H)]\,d\tau Z$$

$$- \lambda\sum_{k=1}^\infty \mathbf{x}_k\int_0^\infty dH(v)\int_0^v \exp[-\lambda\tau(I - H)]d\tau H^{k-1}Z.$$

We note that the matrices H and Z commute, but that, in general, neither H nor Z commute with H_1. The evaluation of the last two terms in the preceding expression requires some care. Since $I - H$ is singular, the inner integrals are calculated by the following steps.

$$\int_0^v \exp[-\lambda\tau(I - H)]d\tau \cdot (I - H + \mathbf{e}\boldsymbol{\pi}) \tag{4.1.44}$$

$$= \sum_{\nu=0}^\infty \int_0^v e^{-\lambda\tau}\frac{(\lambda\tau)^\nu}{\nu!}d\tau\, H^\nu(I - H + \mathbf{e}\boldsymbol{\pi})$$

$$= v\mathbf{e}\boldsymbol{\pi} + \frac{1}{\lambda}\{I - \exp[-\lambda\tau(I - H)]\},$$

by elementary partial integrations. It follows that

$$\int_0^v \exp[-\lambda\tau(I - H)]d\tau = v\mathbf{e}\boldsymbol{\pi} + \frac{1}{\lambda}\{I - \exp[-\lambda\tau(I - H)]\}Z.$$

By using this formula, it follows after simple calculations that

$$\mathbf{x}_0 \int_0^\infty dH(v) \int_0^v \exp[-\lambda\tau(I - H)] d\tau$$

$$+ \sum_{k=1}^\infty \mathbf{x}_k \int_0^\infty dH(v) \int_0^v \exp[-\lambda\tau(I - H)] d\tau H^{k-1}$$

$$= (\boldsymbol{\pi}\boldsymbol{\alpha})\boldsymbol{\pi} + \frac{1}{\lambda}\mathbf{x}_0 HZ + \frac{1}{\lambda}\{\sum_{k=1}^\infty \mathbf{x}_k H^k$$

$$- \mathbf{x}_0 \int_0^\infty dH(v) \exp[-\lambda v(I - H)] \tag{4.1.45}$$

$$- \sum_{k=1}^\infty \mathbf{x}_k \int_0^\infty dH(v) \int_0^v \exp[-\lambda v(I - H)] H^{k-1} \} Z.$$

If we postmultiply the equation of index i in (4.1.15) by H^i and sum over i, we see that the terms inside the braces in (4.1.45) simplify to $-\mathbf{x}_0$. The right hand side of (4.1.45) therefore reduces to

$$(\boldsymbol{\pi}\boldsymbol{\alpha})\boldsymbol{\pi} - \frac{1}{\lambda}\mathbf{x}_0(I - H)Z = \frac{1}{\lambda}\rho\boldsymbol{\pi} - \frac{1}{\lambda}\mathbf{x}_0 + \frac{1}{\lambda}(1 - \rho)\boldsymbol{\pi} = \frac{1}{\lambda}(\boldsymbol{\pi} - \mathbf{x}_0).$$

Upon substitution in the expression for $\mathbf{L}_1^*$, we obtain that

$$\mathbf{L}_1^* = \lambda\boldsymbol{\pi}\Delta(\boldsymbol{\alpha}) + (\mathbf{x}_0 + \lambda\boldsymbol{\pi}H_1)Z$$

$$+ [\frac{1}{2}\lambda^2\boldsymbol{\pi}\boldsymbol{\alpha}_2 + \lambda\mathbf{X}'(1-)\boldsymbol{\alpha} - 1 - \rho + \lambda\mathbf{x}_0\boldsymbol{\alpha}]\boldsymbol{\pi}.$$

The constant c, which is the coefficient of $\boldsymbol{\pi}$, is simplified by using the expression for $\mathbf{X}'(1-)$ in (3.3.15). We obtain the form given in the statement of the theorem.

Formulas (4.1.33) and (4.1.36) and a decomposition as used for $L_{1,j}^*$ lead to the following expression for $\mathbf{W}_1^*$,

$$\mathbf{W}_1^* = \frac{1}{2}\lambda\boldsymbol{\pi}\Delta(\boldsymbol{\alpha}_2) + \lambda\mathbf{x}_0 \sum_{i=2}^{\infty} \int_0^{\infty} e^{-\lambda\tau} \frac{(\lambda\tau)^{i-1}}{(i-1)!} [H - H(\tau)] d\tau \sum_{\nu=1}^{i-1} H^{\nu-1} H_1$$

$$+ \lambda \sum_{i=2}^{\infty} \sum_{k=1}^{i} \mathbf{x}_k \int_0^{\infty} e^{-\lambda\tau} \frac{(\lambda\tau)^{i-k}}{(i-k)!} [H - H(\tau)] d\tau \sum_{\nu=1}^{i-1} H^{\nu-1} H_1.$$

This expression can be simplified in exactly the same manner as before to yield formula (4.1.43), but a more insightful way is to compare the preceding formula to the corresponding expression for $\mathbf{L}_1^*$ and to note that

$$\mathbf{W}_1^* = \frac{1}{2}\lambda\boldsymbol{\pi}\Delta(\boldsymbol{\alpha}) + [\, \mathbf{L}_1^* - \lambda\boldsymbol{\pi}\Delta(\boldsymbol{\alpha}) \,] H_1. \qquad (4.1.46)$$

That equality immediately leads to (4.1.43). As accuracy checks, we postmultiply by $\mathbf{e}$ in (4.1.42) and (4.1.43) to obtain, after routine calculations, that $\mathbf{L}_1^*\mathbf{e} = L_1$, and $\lambda\mathbf{W}_1^*\mathbf{e} = L_1 - \rho$. Formula (4.1.46) is a vector analogue of Little's formula, which is the preceding equality. •

Applications and Variants of the $M/\,SM/\,1$ Queue: A number of applications, such as a queue in which the customer types vary in cyclic order, are immediate cases of the $M/\,SM/\,1$ queue. As for the $M/\,G/\,1$ queue, many interesting variants involve modified matrices $B_\nu(\cdot)$. The formal expressions for the stationary distributions and their moments are then different from those for the $M/\,SM/\,1$ queue, but the crucial steps in their derivations are the same as in the preceding discussion.

We shall describe a number of variants for the purpose of bringing out the essential structural features shared with the $M/\,SM/\,1$ queue, but we shall not state the formulas specific to these. It is unlikely that a particular application would fit any one of our examples in all details. It is therefore more important that the reader by able to adapt the preceding derivations as needed, than to look up a number of special formulas. Such derivations of specific formulas for these and other variants should be both illuminating and instructive.

Example A : Change-over times: This model was the earliest to motivate the study of the $M/SM/1$ queue. The successive types of customers, in a queue with Poisson input of rate λ, form an m-state Markov chain with the irreducible transition probability matrix P. We assume that $P_{jj} > 0$, for $1 \leqslant j \leqslant m$. Customers are served by a single server in order of arrival.

Whenever the server switches to a different type of customer, a certain length of time is expended to set up the service mechanism for that type of customer. We model this by ascribing a different service time distribution $\tilde{H}_j(\cdot)$ to the first customer in each run of customers of type j. All subsequent customers of type j during the same run have the service time distribution $\hat{H}_j(\cdot)$. Service times are conditionally independent, given the successive customer types.

Without the set-up times, the queue with m types of customers (with the types forming a Markov chain) is an $M/SM/1$ queue with the matrix $H(\cdot)$ defined by

$$H_{jj'}(x) = \hat{H}_j(x) P_{jj'}, \quad \text{for } 1 \leqslant j, j' \leqslant m.$$

The model with the set-up times is also an $M/SM/1$ queue provided we distinguish between the first and the subsequent customers of each run. A customer of type j who follows a customer of a different type will be said to be of type j^* and we list the new $2m$ customer types in the order $1, \ldots, m, 1^*, \ldots, m^*$. The transition probability matrix $H(\cdot)$ for this case is then given by

$$H(x) = \begin{vmatrix} \tilde{H}(x) & 0 \\ 0 & \hat{H}(x) \end{vmatrix} \begin{vmatrix} \Delta & P - \Delta \\ \Delta & P - \Delta \end{vmatrix},$$

where $\tilde{H}(x)$, $\hat{H}(x)$ and Δ are diagonal matrices respectively with the diagonal elements $\tilde{H}_j(x)$, $\hat{H}_j(x)$ and P_{jj}, $1 \leqslant j \leqslant m$. The results on the $M/SM/1$ queue may be directly implemented to assess the influence of the change-over times on the stationary distributions of the queue.

Example B: A Service Unit with a Buffer of Parts: Semi-Markovian service times frequently arise from a feature of the service mechanism which imposes a type on the services of otherwise identical customers. This is illustrated by the following model which has many variants of interest.

Suppose that a server is equipped with a reservoir of parts and that each service consumes up to K parts. The buffer is of capacity $m \geqslant k \geqslant 1$. The numbers of parts required by successive services are independent and identically distributed. The probability that a service requires k parts is p_k, $1 \leqslant k \leqslant K$, and we assume that $p_K > 0$.

If at the beginning of a service, the buffer contains fewer than K parts, it is replenished up to its capacity m. This replenishment requires a certain amount of time, which is added to the subsequent service time. Ordinary service times have the probability distribution $D(\cdot)$ and those augmented by a replenishment period, the distribution $R(\cdot)$.

With Poisson arrivals and conditional independence of the service times, we obtain an $M/SM/1$ queue with the type of a service defined as the number of parts in the buffer after the preceding service completion.

In order to illustrate its structure, we display the matrix $H(\cdot)$ for the representative values $K = 3$ and $m = 7$.

	0	1	2	3	4	5	6
0	0	0	0	0	$p_3R(x)$	$p_2R(x)$	$p_1R(x)$
1	0	0	0	0	$p_3R(x)$	$p_2R(x)$	$p_1R(x)$
2	0	0	0	0	$p_3R(x)$	$p_2R(x)$	$p_1R(x)$
3	$p_3D(x)$	$p_2D(x)$	$p_1D(x)$	0	0	0	0
4	0	$p_3D(x)$	$p_2D(x)$	$p_1D(x)$	0	0	0
5	0	0	$p_3D(x)$	$p_2D(x)$	$p_1D(x)$	0	0
6	0	0	0	$p_3D(x)$	$p_2D(x)$	$p_1D(x)$	0

A variant with modified boundary matrices arises if the buffer is also replenished whenever the queue becomes empty and this regardless of the number of parts remaining in the buffer. The corresponding replenishment period may then possibly extend beyond the period of emptiness.

Example C: A Random Environment Model: Consider an $M/G/1$ queue evolving in a random environment which is described by an m-state, irreducible Markov process with the infinitesimal generator D. The durations of the successive service times are conditionally independent, given the states of the environment at their beginnings. All services starting with the Markov process D in the state j have the service time distribution $H_j(\cdot)$, $1 \leqslant j \leqslant m$.

This model differs from an ordinary $M/SM/1$ queue only in the transitions from the empty state, since the state of the environment process may change during the idle period. The type of a service is conveniently defined as the state of the environment at its beginning.

The transition probability matrix $H(\cdot)$ may be concisely written as

$$H(x) = \int_0^x d\Delta[\mathbf{H}(u)] \exp(Du), \quad \text{for } x \geqslant 0, \tag{4.1.47}$$

where $\Delta[\mathbf{H}(u)]$ is a diagonal matrix with diagonal elements $H_1(u), \ldots, H_m(u)$. The matrices $A_\nu(x)$, $\nu \geqslant 0$, are as shown in (2.1.2), while the corresponding matrices $B_\nu(x)$ are now given by

$$B_\nu(x) = \int_0^\infty \exp(Du)\, e^{-\lambda u} \lambda A_\nu(x-u)\, du. \tag{4.1.48}$$

We note that $\rho = \lambda \pi \alpha$, where α_j is the mean of $H_j(\cdot)$ and $\alpha = (\alpha_1, \ldots, \alpha_m)$. The computation of the vector π can be quite involved as the matrix $H = H(\infty)$ is given by

$$H = \int_0^\infty d\Delta[\mathbf{H}(u)] \exp(Du), \quad \text{for } x \geqslant 0, \tag{4.1.49}$$

so that we need, in principle, to integrate a system of linear differential equations to obtain $\exp(Du)$, for $u \geqslant 0$, and to evaluate m^2 integrals numerically before we can solve the linear equations $\pi H = \pi$, $\pi \mathbf{e} = 1$, for the vector π. The most efficient way to do that appears to be to rewrite $\exp(Du)$ as $\exp(-\theta u)\, \exp[(\theta I + D)u]$, and to expand the matrix exponential in its Maclaurin series (as in the elementary uniformization method.) The matrix H can so be written as a series of nonnegative matrix terms. This is an example of a simple queueing model for which even the verification of the equilibrium condition requires substantial computation.

4.2. BULK SERVICE QUEUES

The literature on bulk service queues with Poisson input is extensive, but highly repetitive. The reason for the recurrence of particular analytic steps lies in the fact that most queues with bulk service have embedded Markov renewal processes of $M/G/1$ type with a common form of the transition probability matrix $Q(\cdot)$. The bulk service model, which we have selected for detailed discussion in this book, is known as *the $M/G/1$ queue with a general bulk service rule.* It is described under Example B in Section 2.1, where we have also pointed out the special form of the transition probability matrix $Q(\cdot)$ away from the m boundary states. We shall spare the reader the details of the many variants of this and related models. These differ from the model to be discussed in details of the definition of the boundary matrices $B_\nu(\cdot)$, $\nu \geqslant 0$, and their analysis requires few new ideas. However, the derivation of the waiting time distributions for the $M/G/1$ queue with a general bulk service rule is quite challenging. Without the matrix-analytic methods of this book those distributions would appear to be intractable.

We refer to Section 2.1 for the definition of the elements of the matrix $Q(\cdot)$. In Section 2.3, we have pointed out that the stochastic matrix A is a *circulant* and that the vector π is therefore given by $m^{-1}\mathbf{e}'$. The quantity ρ is then given by

$$\rho = \frac{1}{m}\lambda\alpha_m,$$

where α_m is the mean service time of a group of size m.

We shall now express the matrix $A^*(z) = \sum_{\nu=0}^{\infty} A_\nu z^\nu$, in a convenient form. Let $\phi(z)$ be the probability generation function

$$\phi(z) = \sum_{\nu=0}^{\infty} a_\nu z^\nu = h_m(\lambda - \lambda z), \tag{4.2.1}$$

and let the functions $\phi_j(z)$, $0 \leqslant j \leqslant m-1$, be defined by

$$\phi_j(z) = \sum_{\nu=0}^{\infty} a_{m\nu+j}\, z^{m\nu+j}. \tag{4.2.2}$$

The functions $\phi_j(z), 0 \leqslant j \leqslant m-1$, are related to $\phi(z)$ in an interesting and useful way. We shall write $\mathbf{\Phi}(z)$ and $\hat{\mathbf{\Phi}}(z)$ for the column vectors with the components $\phi_j(z)$ and $\phi(\omega^j z)$, $0 \leqslant j \leqslant m-1$, respectively. The constant ω is defined as

$$\omega = \exp\left(\frac{2\pi i}{m}\right) = \cos\left(\frac{2\pi}{m}\right) + i\sin\left(\frac{2\pi}{m}\right).$$

Theorem 4.2.1: The vectors $\mathbf{\Phi}(z)$ and $\hat{\mathbf{\Phi}}(z)$ are related by

$$\mathbf{\Phi}(z) = \Omega^{-1}\hat{\mathbf{\Phi}}(z), \tag{4.2.3}$$

where Ω is the Vandermonde matrix with elements $\Omega_{jj'} = \omega^{jj'}$, for $0 \leqslant j, j' \leqslant m-1$.

Proof: For $0 \leqslant j' \leqslant m-1$, we have

$$\phi(\omega^{j'} z) = \sum_{\nu=0}^{\infty} a_\nu \omega^{j'\nu} z^\nu$$

$$= \sum_{j=0}^{m-1} \omega^{jj'} \sum_{\nu=0}^{\infty} a_{m\nu+j} z^{m\nu+j} = \sum_{j=0}^{m-1} \omega^{jj'} \phi_j(z),$$

or equivalently

$$\hat{\mathbf{\Phi}}(z) = \Omega\, \mathbf{\Phi}(z).$$

Since the quantities $1, \omega, \omega^2, \ldots, \omega^{m-1}$ are the distinct m-th roots of unity, the matrix Ω is nonsingular and this implies (4.2.3).

It is readily seen that the elements of the matrix $A^*(z)$ are given by

$$A^*_{jj'}(z) = z^{\frac{j-j'}{m}} \phi_{j'-j}(z^{\frac{1}{m}}), \qquad \text{for } 0 \leqslant j \leqslant j' \leqslant m-1, \tag{4.2.4}$$

$$= z^{\frac{j-j'}{m}} \phi_{m+j'-j}(z^{\frac{1}{m}}), \quad \text{for } 0 \leqslant j' < j \leqslant m-1.$$

After routine differentiations, we obtain the following expressions for the components $\beta_0, \ldots, \beta_{m-1}$ of the vector $\boldsymbol{\beta} = \sum_{\nu=1}^{\infty} \nu A_\nu \mathbf{e}$. We have

$$\beta_0 = \frac{1}{m}\lambda\alpha_m - \frac{1}{m}\sum_{\nu=1}^{m-1} \nu\phi_\nu(1), \tag{4.2.5}$$

$$\beta_j = \beta_0 + \sum_{\nu=m-j}^{m-1} \nu\phi_\nu(1), \qquad \text{for } 1 \leqslant j \leqslant m-1. \qquad \bullet$$

For the stable bulk service queue, the matrix G is computed by solving the standard equation

$$G = \sum_{\nu=0}^{\infty} A_\nu G^\nu, \tag{4.2.6}$$

for its unique stochastic solution. The matrix G is positive and has no special structural features. The computations of the invariant probability vector $\mathbf{g}$ and of the mean vector $\tilde{\boldsymbol{\mu}}_1$ proceed as in the general case, discussed in Section 3.6.

The elements $b_{ij} = b_{ij}(\infty)$, $0 \leqslant i \leqslant m-1$, $j \geqslant 0$, of the first m rows of the stochastic matrix $\tilde{Q}(\infty)$ have the probability generating functions

$$\sum_{j=0}^{\infty} b_{ij} z^j = h_N(\lambda - \lambda z), \quad \text{for } 0 \leqslant i < N, \tag{4.2.7}$$

$$= h_i(\lambda - \lambda z), \quad \text{for } N \leqslant i \leqslant m-1.$$

Since the probability distributions $H_i(\cdot)$, $N \leqslant i \leqslant m-1$, have finite means, the condition $\rho < 1$ is both necessary and sufficient for the positive recurrence of the embedded Markov renewal process. When the service time distributions depend on the group sizes, there are no appreciable simplifications in the computations of the vectors $\mathbf{x}_0$ and $\mathbf{x}_1$ over the general case. These vectors are then evaluated by the procedure of Section 3.6.

Several treatments of bulk queues have exclusively dealt with the case where the service time distribution is the same for all group sizes i, $N \leqslant i \leqslant m$. In that case, there are substantial simplifications also in the present approach. It is then clear that the first $m + 1$ rows of the matrix $\tilde{Q}(\infty)$ are identical. The results for this particular case are given in the following theorem.

Theorem 4.2.2: When the service time distributions do not depend on the group sizes, the vector $\mathbf{x}_0$ is given by

$$\mathbf{x}_0 = (\tilde{\mu}_{1,0})^{-1}\boldsymbol{\gamma},$$

where the vector $\boldsymbol{\gamma}$ is the first row of the matrix G and $\tilde{\mu}_{1,0}$ is the first component of the vector $\tilde{\boldsymbol{\mu}}_1$.

The matrix H is given by

$$H = [\, I + A_0 + (1 - \boldsymbol{\theta}\mathbf{e})^{-1} A_0 \mathbf{e}\boldsymbol{\theta}\,] \sum_{\nu=1}^{\infty} A_\nu G^{\nu-1},$$

where $\boldsymbol{\theta}$ is the first row of the matrix A_0. The vector $\mathbf{h}$ is the invariant probability vector of H. The vector $\boldsymbol{\psi}_1$ is as given in formula (3.2.4) and $\boldsymbol{\psi}_2 = \psi_{1,0}\mathbf{e}$, where $\psi_{1,0}$ is the first component of $\boldsymbol{\psi}_1$. The vector $\tilde{\mathbf{h}}$ is then given by

$$\tilde{\mathbf{h}} = \boldsymbol{\psi}_1 + (1 - \boldsymbol{\theta}\mathbf{e})^{-1}\psi_{1,0} A_0 \mathbf{e},$$

and the vector $\mathbf{x}_1$ is evaluated as

$$\mathbf{x}_1 = (\mathbf{h}\tilde{\mathbf{h}})^{-1}\mathbf{h}.$$

Proof: Upon inspection of formula (3.2.1), we see that in the present case the matrix $\tilde{K}(z,0)$ has m identical rows equal to the first row of the matrix $\tilde{G}(z,0)$. This implies that $K = \mathbf{e}\boldsymbol{\gamma}$, $\boldsymbol{\kappa} = \boldsymbol{\gamma}$ and $\tilde{\boldsymbol{\kappa}}_1 = \tilde{\mu}_{1,0}\mathbf{e}$. The stated expression for $\mathbf{x}_0$ is now immediate.

The matrix B_0 may be written as $B_0 = \mathbf{e}\boldsymbol{\theta}$, so that

$$(I - B_0)^{-1} = I + (1 - \boldsymbol{\theta}\mathbf{e})^{-1}\mathbf{e}\boldsymbol{\theta},$$

and clearly $C_0 = A_0$. The particular form of the vector ψ_2 is clear upon inspection of the formula (3.2.5) and follows from the special form of the matrices B_ν for the present case. The remaining formulas then follow by routine substitutions into the general formulas. •

Remark *a*. We see that in the case of Theorem 4.2.2, the vector $\mathbf{x}_0$ is obtained without additional computations.

Remark *b*. Theorem 4.2.2 applies in particular to Bailey's bulk service queue, described in Example B of Section 2.1.

Remark *c*. Since the matrix A_0 is obviously nonsingular, we may deduce from the equation

$$\mathbf{x}_0 = \mathbf{x}_0 B_0 + \mathbf{x}_1 A_0,$$

that

$$\mathbf{x}_1 = (\tilde{\mu}_{1,0})^{-1}(\boldsymbol{\gamma} - \boldsymbol{\theta})A_0^{-1},$$

but for large values of m, this formula may suffer from loss of significance. For the same reason, the direct recursive computation of the vectors $\mathbf{x}_i$ for $i \geqslant 2$, is not advisable, though Ramaswami's recursion of Theorem 3.2.5 may be readily adapted for use in this model.

The particular form of the matrices A_ν, $\nu \geqslant 0$, is primarily useful in a treatment by probability generating functions. To illustrate this, we also write the invariant probability vector of $\tilde{Q}(\infty)$ as $[u_0, u_1, \cdots]$. The vectors $\mathbf{x}_i$, are then given by $\mathbf{x}_i = [u_{mi}, u_{mi+1}, \ldots, u_{mi+m-1}]$, for $i \geqslant 0$. The steady-state equations may be written as

$$u_i = \sum_{\nu=0}^{m-1} u_\nu b_{\nu i} + \sum_{\nu=0}^{i} u_{m+i-\nu} a_\nu, \quad \text{for } i \geqslant 0. \tag{4.2.8}$$

The scalar probability generating function $U(z) = \sum_{i=0}^{\infty} u_i z^i$, then satisfies the equation

$$[z^m - h_m(\lambda - \lambda z)]U(z) = \sum_{\nu=0}^{N} u_\nu [z^m h_N(\lambda - \lambda z) - z^\nu h_m(\lambda - \lambda z)] \tag{4.2.9}$$

$$+\sum_{\nu=N+1}^{m-1} u_\nu \left[z^m h_\nu(\lambda - \lambda z) - z^\nu h_m(\lambda - \lambda z) \right]$$

which is primarily useful in the derivation of the moments of the stationary queue length after service completions. We see that with the vector $\mathbf{x}_0$ known, formula (4.2.9) fully determines the probability generating function $U(z)$.

By routine differentiations, (4.2.9) leads to

$$2m(1-\rho)U'(1-) = \lambda^2\alpha_m'' - m(m-1) + \sum_{\nu=0}^{m-1} [m(m-1) - \nu(\nu-1)]u_\nu$$

$$+ \lambda \sum_{\nu=0}^{N} [\lambda(\alpha_N'' - \alpha_m'') + 2(m\alpha_N - \nu\alpha_m)]u_\nu \qquad (4.2.10)$$

$$+ \lambda \sum_{\nu=N+1}^{m-1} [\lambda(\alpha_\nu'' - \alpha_m'') + 2(m\alpha_\nu - \nu\alpha_m)]u_\nu,$$

from which the mean queue length $U'(1-)$ is immediately obtained. The quantities α_j and α_j'' are respectively the mean and the second moment of the service time distribution $H_j(\cdot)$, $N \leqslant j \leqslant m$.

The mean and higher moments of the queue length may also be obtained from the matrix formulas derived in Section 3.3. In doing so, one should remember the different state description which is induced by the canonical partitioning of the matrix $\tilde{Q}(\cdot)$. The state (i,j), $i \geqslant 0$, $0 \leqslant j \leqslant m-1$, corresponds to a queue length $im + j$. In terms of the vectors $\mathbf{X}(1-)$ and $\mathbf{X}'(1-)$ of the formulas (3.3.7) and (3.3.14), the mean queue length $U'(1-)$ is given by

$$U'(1-) = m\mathbf{X}'(1-)\mathbf{e} + \sum_{\nu=0}^{m-1} \nu\,[u_\nu + X_\nu(1-)]$$

$$= mL_1 + \sum_{\nu=0}^{m-1} \nu\,[u_\nu + X_\nu(1-)].$$

We see that L_1 is the expected number of "complete" groups of size m

remaining in the queue after a departure. From the explicit form of the elements of the transition probability matrix $\tilde{Q}(\cdot)$ in Section 2.1, we easily see that the row sum means of $\tilde{Q}(\cdot)$ are given

$$\frac{N}{\lambda} + \alpha_N, \frac{N-1}{\lambda} + \alpha_N, \ldots, \alpha_N, \alpha_{N+1}, \ldots, \alpha_{m-1},$$

for the first m rows, and by α_m for all other rows.

It follows that the *fundamental mean* E of the embedded Markov renewal process $\tilde{Q}(\cdot)$ is given by

$$E = \sum_{\nu=0}^{N} u_\nu \left(\frac{N-\nu}{\lambda} + \alpha_N \right) + \sum_{\nu=N+1}^{m-1} u_\nu \alpha_\nu + \alpha_\nu (1 - \sum_{\nu=0}^{m-1} u_\nu). \quad (4.2.11)$$

The quantity E^{-1} is therefore the *rate* at which completed batches leave the stationary queue.

Theorem 4.2.3: We have the equality

$$[N \sum_{\nu=0}^{N} u_\nu + \sum_{\nu=N+1}^{m-1} \nu u_\nu + m(1 - \sum_{\nu=0}^{m-1} u_\nu)] E^{-1} = \lambda. \quad (4.2.12)$$

Proof: By differentiating in (4.2.9) and setting $z = 1$, we obtain

$$m(1 - \sum_{\nu=0}^{m-1} u_\nu) + \sum_{\nu=N+1}^{m-1} \nu u_\nu$$

$$= \lambda \{\alpha_m (1 - \sum_{\nu=0}^{m-1} u_\nu) + \alpha_N \sum_{\nu=0}^{N} u_\nu + \sum_{\nu=N+1}^{m-1} u_\nu \alpha_\nu - \frac{1}{\lambda} \sum_{\nu=0}^{N} \nu u_\nu\},$$

and by adding $N(u_0 + \cdots + u_N)$ to both sides, (4.2.12) follows.

The expression inside the brackets in (4.2.12) is the expected size of the group completing service at an arbitrary departure epoch. Formula (4.2.12) therefore says that the stationary departure rate of individual customers is equal to their arrival rate λ.

Next we consider the *time averages* for the $M/G/1$ queue with a general bulk service rule. By y_j, $0 \leqslant j \leqslant N-1$, we denote the probability that at an arbitrary time the server is idle and there are j customers present in the stationary version of the queue. Similarly y_{ij}, $N \leqslant j \leqslant m$, $i \geqslant j$, is the corresponding probability that there are i customers in the system and a group of size j is being served. •

Theorem 4.2.4: The probabilities y_j, $0 \leqslant j \leqslant N-1$, and y_{ij}, for $N \leqslant j \leqslant m$, and $i \geqslant j$, are related to the sequence $\{u_\nu\}$ by

$$y_i = (\lambda E)^{-1} \sum_{\nu=0}^{i} u_\nu, \quad \text{for } 0 \leqslant i \leqslant N-1,$$

$$Y_N(z) = \sum_{i=N}^{\infty} y_{iN} z^i = E^{-1} z^N \sum_{\nu=0}^{N} u_\nu \frac{1 - h_N(\lambda - \lambda z)}{\lambda(1-z)}, \tag{4.2.13}$$

$$Y_j(z) = \sum_{i=j}^{\infty} y_{ij} z^i = E^{-1} u_j z^j \frac{1 - h_j(\lambda - \lambda z)}{\lambda(1-z)}, \quad \text{for } N < j < m,$$

$$Y_m(z) = \sum_{i=m}^{\infty} y_{im} z^i = E^{-1} \left[U(z) - \sum_{\nu=0}^{m-1} u_\nu z^\nu\right] \frac{1 - h_m(\lambda - \lambda z)}{\lambda(1-z)}.$$

Proof: The *time dependent* probabilities of the events under consideration may be expressed in terms of the Markov renewal matrix $M(t)$ of the embedded Markov renewal process $\tilde{Q}(\cdot)$. The resulting formulas are obtained by routine applications of the law of total probability. As an example, we write the expression for $y_{iN}(t)$, $i \geqslant N$. The other expressions are similar but simpler.

$$y_{iN}(t) = \sum_{\nu=0}^{N-1} \int_0^t dM_\nu(u) \int_0^{t-u} e^{-\lambda v} \frac{(\lambda v)^{N-\nu-1}}{(N-\nu-1)!} \lambda dv$$

$$\cdot\, e^{-\lambda(t-u-v)} \frac{[\lambda(t-u-v)]^{i-N}}{(i-N)!} [1 - H_N(t-u-v)]$$

$$+ \int_0^t dM_N(u)\, e^{-\lambda(t-u)} \frac{[\lambda(t-u)]^{i-N}}{(i-N)!} [1 - H_N(t-u)],$$

where $M_\nu(u)$ is the expected number of times in $(0,u]$ that the queue length following a departure is equal to ν.

By applying the key renewal theorem to this and other expressions for the time dependent probabilities, we obtain after routine simplifications that

$$y_i = (\lambda E)^{-1} \sum_{\nu=0}^{i} u_\nu, \quad \text{for } 0 \leqslant i \leqslant N-1,$$

$$y_{iN} = E^{-1} \sum_{\nu=0}^{N} u_\nu \int_0^\infty e^{-\lambda u} \frac{(\lambda u)^{i-N}}{(i-N)!} [1 - H_N(u)] du, \quad \text{for } i \geqslant N,$$

$$y_{ij} = E^{-1} u_j \int_0^\infty e^{-\lambda u} \frac{(\lambda u)^{i-j}}{(i-j)!} [1 - H_j(u)] du, \quad \text{for } i \leqslant j, \quad N < j < m,$$

$$y_{im} = E^{-1} \sum_{\nu=0}^{\infty} u_\nu \int_0^\infty e^{-\lambda u} \frac{(\lambda u)^{i-\nu}}{(i-\nu)!} [1 - H_m(u)] du, \quad \text{for } i \geqslant m.$$

The expressions for the generating functions in (4.2.13) follow by straightforward calculations. •

Corollary 4.2.1: In the stationary version of the queue, the fraction of time the server is *idle* is given by

$$\sum_{i=0}^{N-1} y_i = (\lambda E)^{-1} \sum_{i=0}^{N-1} (N-\nu) u_\nu.$$

The fractions of time devoted to serving groups of the various sizes $N, \ldots, m$ are given by

$$E^{-1} \alpha_N \sum_{\nu=0}^{N} u_\nu, \qquad \text{for } j = N,$$

$$E^{-1} \alpha_j u_j, \qquad \text{for } N < j < m,$$

$$E^{-1}\alpha_m \, [1 - \sum_{\nu=0}^{m-1} u_\nu], \qquad \text{for } j = m.$$

It is clear from (4.2.11) that the sum of all this quantities is one.

Proof: By setting $z = 1-$ in the probability generating functions given in (4.2.13). •

Remark: It is further clear from the way we have written the formulas (4.2.13) that *no numerical integrations are needed* to compute the probabilities y_{ij} once the quantities u_ν are known. Using the expressions for $Y_m(z)$ as an example, we write

$$Y_m(z) = E^{-1}\alpha_m [\, U(z) - \sum_{\nu=0}^{m-1} u_\nu z^\nu \,] \, \frac{1 - h_m(\lambda - \lambda z)}{\lambda\alpha_m(1-z)},$$

which shows that the sequence $\{y_{im}, i \geqslant m\}$ is obtained by forming the convolution of the sequence $\{u_\nu, \nu \geqslant m\}$ and the probability density $\{a'_{\nu m}\}$, with

$$a'_{\nu m} = (\lambda\alpha_m)^{-1} [\, 1 - \sum_{r=0}^{\nu} a_r \,], \qquad \nu \geqslant 0,$$

and thereupon multiplying each term of that convolution by the constant $E^{-1}\alpha_m$.

The computation of the lower order moments of the queue length at an arbitrary time in terms of the corresponding moments of $\{u_\nu\}$ is a routine matter, which is left to the initiative of the reader.

The Stationary Waiting Time Distribution: We assume that customers are served in the order of their arrival and we study the stationary waiting time distribution of a (real or virtual) customer arriving to the queue. For the following reasons the derivation of this probability distribution is somewhat complicated.

The waiting time process is, in general, *not Markovian.* The duration of a customer's wait also depends on the number of arrivals after he joins the queue. We shall therefore have to express the waiting time distribution in terms of m auxiliary functions $R_\nu(\cdot)$, $0 \leqslant \nu \leqslant m - 1$. Tractable equations for these functions will be obtained by using Markovian arguments. In the derivations, we shall also need to perform certain manipulations similar to those in the proof of Theorem 4.1.2.

These steps would be completely obscured if we chose to work only with the scalar transforms. It will therefore be to our advantage to reintroduce the matrix formalism at the appropriate point in the derivations.

Let $W_j(x)$, $N \leqslant j \leqslant m$, $x \geqslant 0$, be the stationary probability that an arriving customer C waits at most a time x and is served as a member of a group of size j. We shall first relate the probability mass-functions $W_j(\cdot)$ to m auxiliary functions $R_\nu(\cdot)$, $0 \leqslant \nu \leqslant m-1$, defined as follows.

Suppose that there are i customers in the system when the customer C arrives and define the random variables T and J as follows. If $0 \leqslant i \leqslant N-1$, set $T = 0$ and $J = i$. If $i \geqslant N$ and a group is size r is in service, set $J = i - r - m\left[\frac{i-r}{m}\right]$, and let T be the time until the first departure after the arrival of C at which *all complete groups ahead of C have departed.* The probability mass-functions $R_\nu(\cdot)$ are defined by

$$R_\nu(x) = P\{J = \nu,\, T \leqslant x\}, \quad 0 \leqslant \nu \leqslant m-1, \quad x \geqslant 0.$$

Lemma 4.2.1: The mass-functions $W_j(\cdot)$, $N \leqslant j \leqslant m$, and $R_\nu(\cdot)$, $0 \leqslant j \leqslant m-1$, are related by the formulas:

$$W_N(x) = \sum_{\nu=0}^{N-1} \int_0^x e^{-\lambda u} \frac{(\lambda u)^{N-\nu-1}}{(N-\nu-1)!}\, dR_\nu(u)$$

$$+ \sum_{k=0}^{N-2} \sum_{\nu=0}^{k} \int_0^x e^{-\lambda u} \frac{(\lambda u)^{k-\nu}}{(k-\nu)!}\, dR_\nu(u) \int_0^{x-u} e^{-\lambda v} \frac{(\lambda v)^{N-k-2}}{(N-k-2)!}\, \lambda dv,$$

$$W_j(x) = \sum_{\nu=0}^{j-1} \int_0^x e^{-\lambda u} \frac{(\lambda u)^{j-\nu-1}}{(j-\nu-1)!}\, dR_\nu(u), \quad \text{for } m > j > N, \tag{4.2.14}$$

$$W_m(x) = \sum_{\nu=0}^{m-1} \int_0^x \sum_{r=m-\nu-1}^{\infty} e^{-\lambda u} \frac{(\lambda u)^r}{r!}\, dR_\nu(u)$$

$$= \sum_{\nu=0}^{m-1} R_\nu(x) - \sum_{\nu=0}^{m-2} \sum_{r=0}^{m-\nu-2} \int_0^x e^{-\lambda u} \frac{(\lambda u)^r}{r!}\, dR_\nu(u).$$

Proof: To see how the expression for $W_N(x)$ is obtained, we distinguish between two cases. In addition to the ν customers remaining when all complete groups are served, there could be exactly $N-\nu-1$ arrivals during the service time of the complete groups. Together with the customer C, they will make up the group of size N to be served next. This case contributes the first term.

The second term corresponds to the case where ν customers remain ahead of C at time u and $k-\nu$ additional customers have arrived during the time length u, with $k \leqslant N-2$. Including the customer C, there are now $k+1 \leqslant N-1$ customers in the queue. The server remains idle until $N-k-1$ additional customers have arrived and then starts serving a group of size N. The arguments leading to the other expressions in (4.2.14) are similar, but simpler. •

Setting

$$\phi_{\nu,j} = \int_0^\infty e^{-\lambda u} \frac{(\lambda u)^j}{j!}\, dR_\nu(u), \quad \text{for } j \geqslant 0, \quad 0 \leqslant \nu \leqslant m-1,$$

we see that

$$W_N(\infty) = \sum_{k=0}^{N-1} \sum_{\nu=0}^{k} \phi_{\nu,k-\nu}, \tag{4.2.15}$$

$$W_j(\infty) = \sum_{k=0}^{j-1} \phi_{k,j-k-1}, \quad \text{for } N < j < m,$$

$$W_m(\infty) = \sum_{k=0}^{m-1} R_k(\infty) - \sum_{\nu=0}^{m-2} \sum_{r=0}^{m-\nu-2} \phi_{\nu,r},$$

from which it easily follows that

$$\sum_{j=N}^{m} W_j(\infty) = \sum_{k=0}^{m-1} R_k(\infty). \tag{4.2.16}$$

The quantities $W_j(\infty)$, $N \leqslant j \leqslant m$, are the stationary probabilities of the various sizes of the group in which the arriving customer C is eventually served.

Next, we shall prove that the vector $\mathbf{R}(x)$, with components $R_\nu(x)$, $0 \leqslant \nu \leqslant m-1$, satisfies a system of Volterra integral equations which is similar to the Pollaczek-Khinchin equation. Before doing so, we need to introduce some notation and establish two lemmas.

From formula (4.2.12), it follows that

$$R_j(0+) = (\lambda E)^{-1} \sum_{\nu=0}^{j} u_\nu, \quad \text{for } 0 \leqslant j \leqslant N-1, \tag{4.2.17}$$

$$= 0, \qquad \text{for } N \leqslant j \leqslant m-1.$$

The vector $\mathbf{R}(0+)$ is therefore known. We now denote by $\tilde{\Delta}(x)$, a diagonal matrix with

$$\tilde{\Delta}_{jj}(x) = H_N(x), \quad \text{for } 0 \leqslant j \leqslant N, \tag{4.2.18}$$

$$= H_j(x), \quad \text{for } N < j < m.$$

Let us set $\phi_\nu = e^{-\lambda\tau} \dfrac{(\lambda\tau)^\nu}{\nu!}$, for $\nu \geqslant 0$, and define the $m \times m$ matrices $T_k(\tau)$, $k \geqslant 0$, by

$$T_0(\tau) = \begin{vmatrix} \phi_0 & \phi_1 & \phi_2 & \cdots & \phi_{m-1} \\ 0 & \phi_0 & \phi_1 & \cdots & \phi_{m-2} \\ 0 & 0 & \phi_0 & \cdots & \phi_{m-3} \\ & \cdots & & \cdots & \\ 0 & 0 & 0 & \cdots & \phi_0 \end{vmatrix},$$

and for $k \geqslant 1$,

$$T_k(\tau) = \begin{vmatrix} \phi_{km} & \phi_{km+1} & \phi_{km+2} & \cdots & \phi_{km+m-1} \\ \phi_{km-1} & \phi_{km} & \phi_{km+1} & \cdots & \phi_{km+m-2} \\ \phi_{km-2} & \phi_{km-1} & \phi_{km} & \cdots & \phi_{km+m-3} \\ & \cdots & & \cdots & \\ \phi_{km-m+1} & \phi_{km-m+2} & \phi_{km-m+3} & \cdots & \phi_{km} \end{vmatrix}.$$

The $m \times m$ matrices $S_k(\tau)$ have m identical rows given by the first row of the corresponding $T_k(\tau)$. The matrix $F(z)$ of order m is defined by

$$F(z) = \begin{vmatrix} 0 & 1 & 0 & \cdots & 0 \\ 0 & 0 & 1 & \cdots & 0 \\ & \cdots & & \cdots & \\ 0 & 0 & 0 & \cdots & 1 \\ z & 0 & 0 & \cdots & 0 \end{vmatrix},$$

and F_1 is a matrix of order m with all elements in the first column equal to one and all others equal to zero.

Lemma 4.2.2: The matrix $\psi(\tau,z) = \sum_{k=0}^{\infty} T_k(\tau)z^k$, is given for $|z| \leqslant 1$, and $\tau \geqslant 0$, by

$$\psi(\tau,z) = \exp\{-\lambda\tau[I - F(z)]\}. \tag{4.2.19}$$

The matrix $\hat{\psi}(\tau,z) = \sum_{k=0}^{\infty} S_k(\tau)z^k$, is given by

$$\hat{\psi}(\tau,z) = F_1\,\psi(\tau,z). \tag{4.2.20}$$

Proof: Clearly the matrix $\psi(\tau;z)$ is given by

$$\begin{vmatrix} \psi_0(\tau;z) & \psi_1(\tau;z) & \psi_2(\tau;z) & \cdots & \psi_{m-1}(\tau;z) \\ z\psi_{m-1}(\tau;z) & \psi_0(\tau;z) & \psi_1(\tau;z) & \cdots & \psi_{m-2}(\tau;z) \\ z\psi_{m-2}(\tau;z) & z\psi_{m-1}(\tau;z) & \psi_0(\tau;z) & \cdots & \psi_{m-3}(\tau;z) \\ & \cdots & & \cdots & \\ z\psi_1(\tau;z) & z\psi_2(\tau;z) & z\psi_3(\tau;z) & \cdots & \psi_0(\tau;z) \end{vmatrix},$$

where

$$\psi_j(\tau;z) = \sum_{\nu=0}^{\infty} e^{-\lambda\tau}\frac{(\lambda\tau)^{\nu m+j}}{(\nu m+j)!}z^{\nu}, \quad \text{for } 0 \leqslant j \leqslant m-1.$$

Obviously, we have the differential equations

$$\frac{\partial}{\partial \tau} \psi_j(\tau;z) = -\lambda\psi_j(\tau;z) + \lambda\psi_{j-1}(\tau;z), \quad \text{for } 1 \leqslant j \leqslant m-1.$$

$$\frac{\partial}{\partial \tau} \psi_0(\tau;z) = -\lambda\psi_0(\tau;z) + \lambda z\,\psi_{m-1}(\tau;z),$$

so that

$$\frac{\partial}{\partial \tau} \psi(\tau;z) = \psi(\tau;z)\,[-\lambda I + \lambda F(z)],$$

and since $\psi(0;z) = I$, (4.2.19) follows. The formula (4.2.20) is obvious from the definition of the matrices $S_k(\tau)$.

The matrices A_k and B_k, $k \geqslant 0$, in the transition probability matrix $Q(\infty)$ of the present model may be written as

$$A_k = \int_0^\infty T_k(\tau)\,dH_m(\tau), \qquad B_k = \int_0^\infty S_k(\tau)\,d\tilde{\Delta}(\tau),$$

where $\tilde{\Delta}(\cdot)$ is as defined in (4.2.18). •

Lemma 4.2.3: For $s \geqslant 0$, we have the equality

$$\sum_{i=0}^\infty \mathbf{x}_i h_m^i(s) = \mathbf{x}_0 F_1 \int_0^\infty \exp\{-\lambda\tau[I - F(h_m(s))]\}\,d\tilde{\Delta}(\tau) \qquad (4.2.21)$$

$$+ \sum_{\nu=1}^\infty \mathbf{x}_\nu h_m^{\nu-1}(s) \int_0^\infty \exp\{-\lambda\tau[I - F(h_m(s))]\}\,dH_m(\tau).$$

Proof: By routine calculations, starting with the steady-state equations and using (4.2.19) and (4.2.20). •

Theorem 4.2.5: The vector $\mathbf{R}(x)$ with components $R_j(x)$, $0 \leqslant j \leqslant m-1$, satisfies the matrix-Volterra integral equation

$$\mathbf{R}(x) = \mathbf{b}(x) + \lambda \int_0^x \mathbf{R}(u) \{ I - F[H_m(x-u)] \} du \tag{4.2.22}$$

where the components of the vector $\mathbf{b}(x)$ are given by

$$b_0(x) = R_0(0+) - E^{-1}(\mathbf{ue}) \int_0^x H_N(u) du,$$

$$b_j(x) = R_j(0+), \quad \text{for } 1 \leqslant j \leqslant N-1,$$

$$b_N(x) = E^{-1}x \sum_{\nu=0}^{N} u_\nu,$$

$$b_j(x) = E^{-1}x\ u_j, \quad \text{for } N+1 \leqslant j \leqslant m-1.$$

The vector $\mathbf{u} = \mathbf{x}_0$, has as its components the stationary probabilities u_j that at a service completion, there remain j, $0 \leqslant j \leqslant m-1$, customers in the system.

Proof: By the standard argument relating the time dependent version of $R_j(x)$ to the Markov renewal matrix and using the key renewal theorem to justify passage to the limit, we obtain that for $0 \leqslant j \leqslant N-1$,

$$R_j(x) = (\lambda E)^{-1} \sum_{\nu=0}^{j} u_\nu$$

$$+ \sum_{\nu=0}^{N-1} E^{-1} u_\nu \int_0^\infty d\tau \int_0^\tau e^{-\lambda v} \frac{(\lambda v)^{N-\nu-1}}{(N-\nu-1)!} \lambda dv \int_0^x dH_N(\tau + w - v)$$

$$\cdot \sum_{r=0}^{\infty} e^{-\lambda(\tau - v)} \frac{[\lambda(\tau - v)]^{mr+j}}{(mr+j)!} H_m^{(r)}(x-w)$$

$$+ \sum_{\nu=N}^{m-1} E^{-1} u_\nu \int_0^\infty d\tau \int_0^x \sum_{r=0}^\infty e^{-\lambda\tau} \frac{(\lambda\tau)^{mr+j}}{(mr+j)!} \, dH_\nu(\tau+w) H_m^{(r)}(x-w)$$

$$+ \sum_{k=0}^{m-1} \sum_{\nu=1}^\infty E^{-1} u_{m\nu+k} \sum_{\substack{mr+j \geqslant \\ m\nu+k}} \int_0^\infty d\tau \int_0^x e^{-\lambda\tau}$$

$$\cdot \frac{(\lambda\tau)^{m(r-\nu)+j-k}}{(mr-m\nu+j-k)!} \, dH_\nu(\tau+w) H_m^{(r-1)}(x-w).$$

The second term is routinely simplified to

$$\sum_{\nu=0}^{N-1} E^{-1} u_\nu \int_0^\infty d\tau \int_0^x e^{-\lambda\tau} \frac{(\lambda\tau)^{mr+j}}{(mr+j)!} \, dH_N(\tau+w) H_m^{(r)}(x-w)$$

The expression for $R_j(x)$, $N \leqslant j \leqslant m-1$, is entirely similar, except that the first term is omitted.

Rewritten in matrix notation, using the matrices $S_k(\tau)$ and $T_k(\tau)$, we obtain $\mathbf{R}(x)$ in the form

$$\mathbf{R}(x) = \mathbf{R}(0) + E^{-1}\mathbf{x}_0 \sum_{r=0}^\infty \int_0^\infty d\tau \int_0^x S_r(\tau) d\tilde{\Delta}(\tau+w) H_m^{(r)}(x-w)$$

$$+ E^{-1} \sum_{r=1}^\infty \sum_{\nu=1}^r \mathbf{x}_\nu \int_0^\infty d\tau \int_0^x T_{r-\nu}(\tau) dH_m(\tau+w) H_m^{(r-1)}(x-w).$$

In order to simplify this expression further, it is convenient to evaluate the Laplace-Stieltjes transform

$$\mathbf{R}^*(s) = \int_0^\infty e^{-sx} d\mathbf{R}(x).$$

We successively obtain the equalities

$$\mathbf{R}^*(s) = \mathbf{R}(0) + E^{-1}\mathbf{x}_0 \sum_{r=0}^{\infty} \int_0^{\infty} \int_0^{\infty} \int_0^{x} e^{-sx} S_r(\tau) d\tau \, d\tilde{\Delta}(\tau+w) \, H_m^{(r)}(x-w)$$

$$+ E^{-1}\sum_{r=1}^{\infty} \sum_{\nu=1}^{r} \mathbf{x}_\nu \int_0^{\infty} \int_0^{\infty} \int_0^{x} e^{-sx} T_{r-\nu}(\tau) d\tau \;\; dH_m(\tau+w) dH_m^{(r)}(x-w)$$

$$= \mathbf{R}(0) + E^{-1}\mathbf{x}_0 \int_0^{\infty} \int_0^{v} e^{-sv} \sum_{r=0}^{\infty} S_r(\tau) e^{s\tau} h_m^r(s) d\tau \, d\tilde{\Delta}(v)$$

$$+ E^{-1}\sum_{\nu=1}^{\infty} \sum_{r=0}^{\infty} \mathbf{x}_\nu \int_0^{\infty} \int_0^{v} e^{-sv} T_r(\tau) e^{s\tau} h_m^r(s) \; d\tau \, dH_m(v) \; h_m^{\nu-1}(s)$$

$$= \mathbf{R}(0) + E^{-1}\mathbf{x}_0 \int_0^{\infty} e^{-sv} \int_0^{v} F_1 \exp\{s\tau I - \lambda\tau I + \lambda\tau F[h_m(s)]\} d\tau \, d\tilde{\Delta}(v)$$

$$+ E^{-1}\sum_{\nu=1}^{\infty} \mathbf{x}_\nu \int_0^{\infty} e^{-sv} \int_0^{v} \exp\{s\tau I - \lambda\tau I + \lambda\tau F[h_m(s)]\} d\tau \, dH_m(v) h_m^{\nu-1}(s).$$

Next we observe that

$$\int_0^{v} \exp\{ s\tau I - \lambda\tau I + \lambda\tau F[h_m(s)] \} d\tau \cdot \{ sI - \lambda I + \lambda F[h_m(s)] \}$$

$$= \exp\{ svI - \lambda vI + \lambda vF[h_m(s)] \} - I,$$

so that

$$[\mathbf{R}^*(s) - \mathbf{R}(0)] \cdot \{ sI - \lambda I + \lambda F[h_m(s)] \}$$

$$= E^{-1}\mathbf{x}_0 F_1 \int_0^\infty \exp\{-\lambda v I + \lambda v F[h_m(s)]\}\, d\tilde{\Delta}(v)$$

$$- E^{-1}\mathbf{x}_0 F_1 \int_0^\infty e^{-sv}\, d\tilde{\Delta}(v) - E^{-1}\sum_{\nu=1}^\infty \mathbf{x}_\nu h_m^\nu(s)$$

$$+ E^{-1}\sum_{\nu=1}^\infty \mathbf{x}_\nu h_m^{\nu-1}(s) \int_0^\infty \exp\{-\lambda v I + \lambda v F[h_m(s)]\}\, dH_m(v).$$

Now using formula (4.2.21) and simplifying, we obtain that

$$[\mathbf{R}^*(s) - \mathbf{R}(0)] \cdot \{sI - \lambda I + \lambda F[h_m(s)]\}$$

$$= E^{-1}\mathbf{x}_0 - E^{-1}\mathbf{x}_0 F_1 \int_0^\infty e^{-sv}\, d\tilde{\Delta}(v).$$

The vector $\mathbf{x}_0 F_1$ is equal to $(\mathbf{x}_0\mathbf{e}) \cdot (1,0,\ldots,0)$, and $\mathbf{R}(0)$ is given in formula (4.2.17). Upon substitution, we obtain

$$\mathbf{R}^*(s)\{sI - \lambda I + \lambda F[h_m(s)]\} \tag{4.2.24}$$

$$= s\,\mathbf{R}(0) + E^{-1}\{-\mathbf{ue}h_N(s), 0, \ldots, 0, u_0 + \cdots u_N, u_{N+1}, \ldots, u_{m+1}\},$$

or equivalently

$$\mathbf{R}^*(s) = \lambda s^{-1}\mathbf{R}^*(s)\{I - F[h_m(s)]\} + \mathbf{R}(0) \tag{4.2.25}$$

$$+ E^{-1}(\mathbf{ue})s^{-1}[1 - h_N(s)](1, 0, \ldots, 0)$$

$$+ E^{-1}s^{-1}\{-\mathbf{ue}, 0, \ldots, 0, u_0 + \cdots u_N, u_{N+1}, \ldots, u_{m-1}\}.$$

Upon inversion, we obtain the integral equation (4.2.22). The row vector $\mathbf{R}(\infty) = \mathbf{R}^*(0+)$, may be explicitly obtained as follows. By setting $s = 0+$, in the equation (4.2.24), we get the equations

$$-\lambda R_0^*(0+) + \lambda R_{m-1}^*(0+) = -E^{-1}(\mathbf{ue}),$$

$$R_j^*(0+) = R_0^*(0+), \qquad \text{for } 1 \leqslant j \leqslant N-1,$$

$$\lambda R_{N-1}^*(0+) - \lambda R_N^*(0+) = E^{-1}\sum_{\nu=0}^{N} u_\nu,$$

$$\lambda R_{j-1}^*(0+) - \lambda R_j^*(0+) = E^{-1}u_j, \qquad \text{for } N+1 \leqslant j \leqslant m-1.$$

Since $\mathbf{R}^*(0+)\mathbf{e} = 1$, it follows that

$$R_j^*(0+) = R_0^*(0+), \qquad \text{for } 1 \leqslant j \leqslant N-1,$$

$$R_j^*(0+) = R_0^*(0+) - (\lambda E)^{-1}\sum_{\nu=0}^{j} u_\nu, \qquad \text{for } N \leqslant j \leqslant m-1,$$

and

$$R_0^*(0+) = \frac{1}{m}\{1 + (\lambda E)^{-1}[(m-N)\mathbf{ue} - \sum_{\nu=1}^{m-N-1} \nu u_{N+\nu}]\} = (\lambda E)^{-1}.$$

This final simplification is obtained by application of formula (4.2.12). We note that clearly, the components $R_j^*(0+)$, are decreasing for $N-1 \leqslant j \leqslant m-1$, and that $R_{m-1}^*(0+) = (\lambda E)^{-1}(1 - \mathbf{ue})$, is obviously a positive quantity. •

For purposes of numerical computation, it is useful to note that the matrix-Volterra integral equation (4.2.22) may be rewritten as a single scalar integral equation

$$R_0(x) = \lambda\int_0^x R_0(u) - \lambda\int_0^x R_{m-1}(u)H_m(x-u)du + b_0(x),$$

and $m - 1$ linear differential equations

$$R_j'(x) = \lambda\,[\,R_j(x) - R_{j-1}(x)\,] + b_j'(x), \quad \text{for } 1 \leqslant j \leqslant m-1,$$

with initial values $R_j(0+)$ given by (4.2.17).

As the probability mass-functions $R_j(x)$, $0 \leqslant j \leqslant m-1$, are evaluated for increasing values of $x \geqslant 0$, the mass-functions $W_j(x)$, $N \leqslant j \leqslant m-1$, are simultaneously computed from the formula (4.2.14). That computation may be organized in a variety of ways. One of these is to replace the equations (4.2.14) by the equivalent system of linear differential equations

$$W_N'(x) = \sum_{\nu=0}^{N-1} e^{-\lambda x}\frac{(\lambda x)^{N-\nu-1}}{(N-\nu-1)!}R_\nu'(x) + \lambda\sum_{\nu=0}^{N-2} e^{-\lambda x}\frac{(\lambda x)^{N-\nu-2}}{(N-\nu-2)!}R_\nu(x),$$

$$W_j'(x) = \sum_{\nu=0}^{j-1} e^{-\lambda x}\frac{(\lambda x)^{j-\nu-1}}{(j-\nu-1)!}R_\nu'(x), \quad \text{for } N < j < m, \tag{4.2.26}$$

$$W_m'(x) = \sum_{\nu=0}^{m-1} R_\nu'(x) - \sum_{\nu=0}^{m-2}\sum_{r=0}^{m-\nu-2} e^{-\lambda x}\frac{(\lambda x)^r}{r!}R_\nu'(x),$$

with the initial conditions $W_j(0) = R_{j-1}(0)$, for $N \leqslant j \leqslant m-1$, and

$$W_m(0) = \sum_{\nu=0}^{m-1} R_\nu(0) = \sum_{\nu=0}^{m-1}(N-\nu)u_\nu.$$

The equations (4.2.26) are obtained from (4.2.14) by elementary calculations.

NOTES, REFERENCES AND COMMENTS ON CHAPTER 4

Section 4.1

The specific results in this section were first discussed in Neuts [N-029] for the $M / SM / 1$ queue (with group arrivals). Earlier treatments by transform methods are in Çinlar [C-021] and Neuts [N-011]. In spite of its versatility as a queueing model, the $M / SM / 1$ queue has not been extensively used in practical modelling. For applications suggested by computer operations, we refer to Gaver [G-014], Hofri [H-057,H-058], Levy [L-026] and Neuts [N-031]. Alternative theoretical approaches may be found in de Smit [D-036], McNickle [M-022], Pestalozzi [P-019], and Takács [T-033].

In discussing queues with Poisson arrivals, we need not distinguish between the stationary distributions of queue lengths and waiting times *at arrivals* and *at arbitrary times,* because of the general result that *"Poisson arrivals see time averages".* See Wolff [W-019] and in particular, the book by Franken, König, Arndt and Schmidt [F-032].

The mean queue length and waiting time in the $M / G / 1$ queue depends on the distribution of the service times *only* through the first two moments of the latter. This *insensitivity result* is considered to be of major importance and has been extended to other models. The formulas for the corresponding means for the $M / SM / 1$ queue show that such tractable results appear to be limited to elementary queueing models. Similarly, the relation $L = \lambda W$, *(Little's formula),* has been established by many different arguments and was generalized in varying ways. The expressions for the $M / SM / 1$ model suggest that challenging problems often lie in finding at least one side of that equality.

To see that the matrix $H^*(x)$ in equation (4.1.22) is the natural analogue of the random modification of a probability distribution of a nonnegative random variable, we refer to the construction of the stationary version of a positive recurrent, finite-state Markov renewal process in Pyke [P-053,P-054]. See also Çinlar [C-025].

Problem 4.1.1: Consider the stationary probability that, at time t, a service of type j is in course and that the residual service time of the customer in service does not exceed x. How is that probability related to the semi-Markov matrix $H^*(\cdot)$? •

Problem 4.1.2: In Chapter 1, we present a detailed analysis of the $M / G / 1$ queue with the N-policy. Discuss the modifications needed to obtain the corresponding results for the $M / SM / 1$ queue operating under the N – policy. Is there a factorization similar to that in formula (1.6.8) for the vector $\mathbf{X}(z)$? Does the relation between the stationary waiting time distributions $W_N(\cdot)$ and $W_1(\cdot)$, noted in Chapter 1, have a natural generalization to the vectors $\mathbf{W}_N(\cdot)$ for the $M / SM / 1$ queue with the N – policy? •

Problem 4.1.3: The $M / SM / 1$ queue, even with two customer types, is instructive in numerically demonstrating the profound effect of the mild dependence between service times, modelled by semi-Markovian service times. Consider the case $m = 2$, and, for convenience, let the matrix $H = H(\infty)$, be positive. Fix the stationary probability vector $\boldsymbol{\pi}$, and note that this is equivalent to keeping the ratio

$$H_{12}(H_{12} + H_{21})^{-1},$$

constant. Notice that there is a unique choice of H_{11}, for which the successive customer types are independent. That choice corresponds to an $M / G / 1$ queue with two customer types, in which the types of successive customers are determined by Bernoulli trials with probabilities π_1 and π_2. Appropriately specifying the four probability distributions

$$H_{jj'}(x)\,[H_{jj'}]^{-1},$$

enables us to model set-up times whenever service changes from one type to the other. Discuss the (rather modest) analytic simplifications which arise in that particular case.

An illuminating numerical study consists in varying the probability H_{12}, while keeping the sojourn time distributions and the vector $\boldsymbol{\pi}$ fixed. In doing so, one obtains sequences of services generated by a two-state Markov chain. The runs of services of each type can now be quite different from those obtained from Bernoulli trials. By keeping the vector $\boldsymbol{\pi}$ fixed, the overall fraction of services of each type remains the same, but arrivals of a each customer type occur in longer or shorter runs.

Choose service time distributions from a simple parametric family, such as Erlang distributions of various orders, so that only a minimum of set-up calculations are required. By easy modifications of the parameters of the model, one can now generate a wealth of numerical examples, each requiring only a small computation time. We encourage the reader to interpret the numerical results for each of these examples in detail and to propose as many inferences about the physical behavior of the queue as possible. Select some perceived behavior of the queue for further study by examining, through additional computer runs, how the noted effects are affected by well-understood parameter changes. Insightful though they can be, inferences from numerical results are not easily made and, for a variety of reasons, are only rarely discussed in the literature. Examples of the type of qualitative results to look for are given in Chapter 5 of [N-042], in Latouche [L-006], Ramaswami and Latouche [R-015], and in Neuts [N-059]. •

Many results on the $M/SM/1$ queue may be adapted to the variant where the rate of the Poisson arrival process depends on the type of service in course (with a different value when the queue is empty). The fundamental period for that model is discussed in Purdue [P-050]. The following problem calls for the investigation of other stationary distributions.

Problem 4.1.4: Extend the theory of the $M/SM/1$ queue to the model where, conditional on the type of service in course (or on the queue being empty), the arrival rate assumes a value depending on that type. Discuss, in particular, the generalization of Theorem 4.1.2. There are now $m+1$ arrival rates; λ_0, which corresponds to an empty queue and λ_j, $1 \leqslant j \leqslant m$, the Poisson arrival rates during services of type j. Study the quantities $W_j(r;x)$, $x \geqslant 0$, $1 \leqslant j \leqslant m$, $0 \leqslant r \leqslant m$, that a virtual customer who arrives at time t finds the arrival rate λ_r prevailing, has to wait at most a length of time x and receives a service of type j. Set up equations which are the analogues of (4.1.10) and carry out the simplification of the Laplace-Stieltjes transforms as far as possible. Obtain also the joint stationary distribution of the waiting time of an actual arrival and of the type of service received. •

Problem 4.1.5: Note that the variant of the $M/G/1$ queue in which m types of service are performed on successive customers *in cyclic order*, is a special case of the $M/SM/1$ queue. Let the service time distribution of the $j-$th customer in each cycle of length m be $H_j(\cdot)$. Discuss the simplifications in the analytic results in Section 4.1 for that model. Observe further that by letting the probability distributions $H_j(\cdot)$, $1 \leqslant j \leqslant m-1$, tend to the degenerate distribution $U(\cdot)$, we obtain results for the variant of the $M/G/1$ queue in which customers are served in batches of (fixed) size m. Compare these results with those in Takács [T-006] and Osone and Fujisawa [O-012]. •

A direct formulation of the queue with services in batches of size m as an $M/SM/1$ model is possible, but requires that customers whose rank number (in the arrival process) is not a multiple of m be assigned a service time of length zero. Because of the instantaneous states, which are now introduced in the Markov renewal process, that approach poses mild problems of interpretation, which the reader is encouraged to think through.

Problem 4.1.6: In Problem 1.2.2, a variant of the $M/G/1$ queue in which the server can fail, but only while busy, is considered. Examine that same model, but suppose that the state of the service mechanism is described by an irreducible m-state continuous-parameter Markov chain with infinitesimal generator R. Note that the Markov chain R is "alive" only when the server is busy. During sojourns in the state j of that Markov chain, the server has a constant hazard rate θ_j, which only depends on j. When a failure occurs with the service mechanism in the state j, the customer in service is lost and the server enters a repair period with distribution $C_j(\cdot)$ of finite mean. The state of the service mechanism does not change during repair periods and the successive repair periods are conditionally independent, given the states of the Markov process R. Show that this model is an $M/SM/1$ queue and determine the transition probability matrix $H(\cdot)$. Establish the equilibrium condition. In addition to all the properties of this model which follow directly from the general theory, many particular questions may be asked. For example, consider the queue in steady-state immediately after a departure. How can one determine the expected number ξ_j of customers present at that time, who will be lost due to a failure of the server occurring in its environmental state j? What is the expected time τ_j devoted to repairs occurring with the server in its environmental state j? •

Problem 4.1.7: For the model described in Problems 1.2.2 and 4.1.6, suppose instead that the times between failures of the service mechanism have a PH-distribution with irreducible representation $(\boldsymbol{\alpha}, T)$. Again, the phase of the service mechanism does not change during idle periods of the queue. At the end of the repair periods, the phase of the server is re-initialized according to the probability vector $\boldsymbol{\alpha}$. Show that this model is again an $M/SM/1$ queue and determine its transition probability matrix $H(\cdot)$. •

Problem 4.1.8: In Section 1.8, the relationship between the $M/G/1$ queue and ideas from branching processes is discussed. Carry out the same analysis for the $M/SM/1$ queue, defining functional iterates of power series with matrix coefficients (carefully!). Note that *formally* most of the steps remain valid. Study the steady-state probability vector of the analogue of the embedded Markov chain, discussed in Section 1.8. Now think through the formidable computational effort that would be needed to carry over the rather routine algorithm mentioned at the end of Section 1.8. This difference in computational difficulty is due to the fact that the matrix functional iterates, needed here, do *not* have a major simplifying property of their scalar versions. Identify that property. •

At one time, we considered the relation between queues and branching processes a major unifying idea. The observation which underlies this problem led us to discount the utility of that idea for algorithmic work. The difficulty in implementing the computations corresponding to deceptively easy iterations of transforms accounts for the formidable computational problems of branching processes. These have not yet received much attention.

Section 4.2

The bulk service queue, treated in this section, has a long history and many special cases have been treated, often by repetitive methods. See Borthakur [B-073], Borthakur and Medhi [B-072], Chaudhry [C-013], Chaudhry and Templeton [C-017,C-018], Curry and Feldman [C-087], Easton and Chaudhry [E-001], Loris-Teghem [L-030-L-032,L-041], Medhi [M-023-M-025,M-026] and Neuts [N-012-N-014,N-032,N-038]. Nice applications of the bulk service queue arise in statistical quality control tests in which the items to be tested arrive according to a Poisson process, rather than all being present at the beginning of testing. Such an application to *group testing* is discussed in Neuts and Chandramouli [N-065], with special attention to the trade-off between larger group sizes and time lost due to retesting of groups containing flawed items. Transportation problems offer another rich area of application, which is being investigated by Powell and his coauthors [P-029-P-034]. For a combination of group arrivals, bulk service and vacations, see Chatterjee and Mukkerjee [C-010].

Problem 4.2.1: Discuss the simplifications in the analysis of the bulk service queue when all service time distributions are assumed to be *of phase type*. •

Bulk service models with service time distributions of phase type are amenable, even with renewal input, to the methods of analysis in Neuts [N-042]. The main challenges of that approach lie in the careful definition of the embedded Markov renewal process and in the exploitation of its special structure to expedite numerical computation. Several examples of such an analysis are briefly discussed in Section 4.2c of [N-042].

Problem 4.2.2: This model, which is a generalization of Bailey's bulk queue, combines features of the queue with semi-Markovian services and bulk service. Suppose that a shuttle of capacity m arrives at the epochs of a renewal process. The numbers of available spaces at successive arrivals of the shuttle form an irreducible Markov chain with state space $\{0,1, \ldots, m\}$ and transition probability matrix P. Customers (passengers) arrive at the stop according to a Poisson process of rate λ and waiting customers may board the shuttle provided they can find a free space. Set up the Markov renewal process of $M/G/1$ type, embedded immediately after arrivals of the shuttle and discuss the stationary distributions of the queue length at these epochs and at arbitrary times. Discuss that probability that a customer will wait no longer than x and will depart in a shuttle carrying j customers, where $1 \leqslant j \leqslant m$. Write a computer program to implement your algorithmic solution for the case where the interdeparture times of the shuttle are constant. •

This model again has many variants, which we encourage the reader to formulate and to examine to what extent they may be analyzed by the methods of this book. It should be noted that at an arrival of a shuttle without free spaces, no waiting customers are removed. There are various ways of handling this case in the analysis.

5

Versatile Arrival Processes

5.1. THE *PH*–DISTRIBUTIONS AND THE *PH*–RENEWAL PROCESSES

The theory of Markov renewal processes of $M/G/1$ type is applicable to many queues and other models, whose input processes retain certain Markovian features without being Poisson or even renewal processes. In this chapter, we shall develop the formalism to handle a versatile class of such processes. The price of that versatility is an elaborate notation, which may hide the essential simplicity of the underlying ideas as well as the substantial simplifications present in most specific applications. In presenting this material, we shall therefore proceed in a leisurely manner from the simplest cases to the more complex. We shall also discuss a number of applications in detail. These are of independent interest, but also serve to illustrate the utility of the formalism to be developed when used in conjunction with the results established in the earlier chapters.

A continuous probability distribution $F(\cdot)$ on $[0,\infty)$ is *of phase type* (*PH*–distribution) if it is the distribution of the time until absorption in a finite-state Markov process with a single absorbing state, that is, there exists a probability vector $(\boldsymbol{\alpha}, \alpha_{m+1})$ and an infinitesimal generator of the form

$$Q = \begin{vmatrix} T & \mathbf{T}^\circ \\ 0 & 0 \end{vmatrix},$$

such that

$$F(x) = 1 - \boldsymbol{\alpha}\exp(Tx)\mathbf{e}, \quad \text{for } x \geqslant 0. \tag{5.1.1}$$

The $m \times m$ matrix T is nonsingular, has negative diagonal elements and nonnegative off-diagonal elements. The vector $\mathbf{T}^\circ$ is nonnegative and satisfies $T\mathbf{e} + \mathbf{T}^\circ = 0$. The pair $(\boldsymbol{\alpha}, T)$ is called a *representation* of $F(\cdot)$. The *PH*–distribution $F(\cdot)$ has a point mass α_{m+1} at 0 and a density

$$F'(x) = -\alpha\exp(Tx)T\mathbf{e} = \alpha\exp(Tx)\mathbf{T}^{\circ}, \tag{5.1.2}$$

on $(0,\infty)$. The Laplace-Stieltjes transform $f(s)$ of $F(\cdot)$ is given by

$$f(s) = \alpha_{m+1} + \alpha(sI - T)^{-1}\mathbf{T}^{\circ}, \quad \text{for Re } s \geqslant 0, \tag{5.1.3}$$

and its moments λ_{ν}', $\nu \geqslant 1$, are all finite and given by

$$\lambda_{\nu}' = (-1)^{\nu}\nu!\,\alpha T^{-\nu}\mathbf{e}. \tag{5.1.4}$$

Standard, but very special examples of $PH-$ distributions are the *hyperexponential* distributions

$$F(x) = \sum_{\nu=1}^{m} \alpha_{\nu}(1 - e^{-\lambda_{\nu}x}),$$

which may be represented by $\alpha = (\alpha_1, \ldots, \alpha_m)$, $\alpha_{m+1} = 0$, and $T = -\,diag\,(\lambda_1, \ldots, \lambda_m)$, and the (mixed) *Erlang* distributions

$$F(x) = \sum_{\nu=1}^{m} p_{\nu}E_{\nu}(\lambda;x),$$

which may be represented by $\alpha = (p_m, p_{m-1}, \ldots, p_1)$, $\alpha_{m+1} = 0$, and

$$T = \begin{vmatrix} -\lambda & \lambda & 0 & \cdots & 0 & 0 & 0 \\ 0 & -\lambda & \lambda & \cdots & 0 & 0 & 0 \\ & \cdots & & & \cdots & & \\ 0 & 0 & 0 & \cdots & 0 & -\lambda & \lambda \\ 0 & 0 & 0 & \cdots & 0 & 0 & -\lambda \end{vmatrix}.$$

The infinitesimal generator

$$Q^* = T + (1 - \alpha_{m+1})^{-1}\mathbf{T}^\circ\alpha, \tag{5.1.5}$$

plays an important role in what follows. In order to see its significance, we make the absorbing state $m + 1$ an instantaneous return state. That is, upon absorption in the Markov process Q, we instantaneously restart the process by selecting a new initial state (independently of the past), using the same probability vector (α, α_{m+1}). This resetting is repeated indefinitely. It is obvious that the epochs of successive visits to the "instantaneous" state $m + 1$ form a renewal process with the underlying life time distribution $F(\cdot)$. Equivalently, we may view the process as generated by the phase type distribution $F_1(\cdot)$ with representation $\{(1-\alpha_{m+1})^{-1}\alpha, T\}$ and add independent, identically distributed numbers $N_0, N_1, \cdots$ of arrivals at the successive epochs of resettings. The random variables N_k are geometrically distributed on $\nu \geqslant 1$ with $p = 1 - \alpha_{m+1}$.

Upon defining the path functions of the restarted process to be right continuous, we obtain an m-state Markov process whose infinitesimal generator is Q^*. It is elementary to show that the matrix Q^* is reducible if and only if there are superfluous phases in the representation (α, T). These can be deleted without any consequence, so that, without loss of generality, we may require that Q^* is *irreducible*. The corresponding pair (α, T) is then called an *irreducible* representation. We henceforth assume that all given representations of PH-distributions are irreducible.

The case of a geometrically distributed number of "arrivals" at each resetting epoch will turn out to be a particular case of a later construction. In discussing the *renewal process of phase type* (PH-renewal process), we may, for ease of notation, set $\alpha_{m+1} = 0$, The stationary probability vector π of Q^* then satisfies

$$\pi(T + \mathbf{T}^\circ\alpha) = 0, \qquad \pi\mathbf{e} = 1, \tag{5.1.6}$$

from which we easily deduce that

$$\pi = \lambda_1'^{-1}\alpha(-T)^{-1}, \qquad \pi\mathbf{T}^\circ = \lambda_1'^{-1}. \tag{5.1.7}$$

Theorem 5.1.1: The probability distribution

$$F^*(x) = \lambda_1'^{-1}\int_0^x [1 - F(u)]\, du, \tag{5.1.8}$$

is a PH – distribution with the representation (π, T).

Proof: We have that

$$F^*(x) = \lambda_1'^{-1}\int_0^x \alpha\exp(Tu)\mathbf{e}\,du = \lambda_1'^{-1}\alpha T^{-1}[\exp(Tx) - I]\mathbf{e}$$

$$= 1 - \pi\exp(Tu)\mathbf{e}. \quad \bullet$$

We note that the moments of $F^*(\cdot)$ are given by $\phi_\nu' = (\nu+1)^{-1}\lambda_1'^{-1}\lambda_{\nu+1}'$, for $\nu \geqslant 1$.

By choosing the initial conditions for the Markov process Q^* in various ways, we can generate several related PH – renewal processes. By choosing the initial state according to the probability vector α, we obtain the *ordinary* PH – renewal process. If it is chosen according to π, the time until the first resetting has the distribution $F^*(\cdot)$ and we then obtain the *stationary* PH – renewal process. These are the most important two cases, but it is clear that by determining the initial state of Q^* by the probability vector $\tilde{\alpha}$, we may give the first interval the PH – distribution with representation $(\tilde{\alpha}, T)$, where $\tilde{\alpha}$ is arbitrary. For brevity, we shall now refer to the epochs at which resetting occurs as *renewals*.

The Counting Process: The matrices $\{P(k;t),\ k \geqslant 0\}$, which are introduced next are basic to the study of the counting random variables of the PH – renewal process. By $P_{jj'}(k;t)$, $1 \leqslant j, j' \leqslant m$, $k \geqslant 0$, $t \geqslant 0$, we denote the conditional probability that the Markov process Q^* is in the phase j' at time t and that k renewals occur in $(0,t)$, given that the process Q^* was started in the phase j at time 0.

The matrices $P(k;t) = \{P_{jj'}(k;t)\}$, then satisfy the Chapman-Kolmogorov differential equations

$$P'(0;t) = P(0;t)T, \tag{5.1.9}$$

$$P'(k;t) = P(k;t)T + P(k-1;t)\mathbf{T}^\circ\alpha, \quad \text{for } k \geqslant 1,$$

and also

$$P'(0;t) = TP(0;t), \tag{5.1.10}$$

$$P'(k;t) = TP(k;t) + \mathbf{T}^{\circ}\boldsymbol{\alpha}P(k-1;t), \quad \text{for } k \geqslant 1,$$

The initial conditions are given by $P(0;0) = I$, $P(k;0) = 0$, $k \geqslant 1$. From either of these systems, we may easily derive the matrix generating function

$$P^*(z;t) = \sum_{k=0}^{\infty} P(k;t)z^k = \exp\{ (T + z\mathbf{T}^{\circ}\boldsymbol{\alpha})t \}, \quad \text{for } |z| \leqslant 1. \quad (5.1.11)$$

By using elementary properties, we may show that with an irreducible representation $(\boldsymbol{\alpha},T)$, the matrices $P(k;t)$, $k \geqslant 1$ are *positive* for $t > 0$. This is not, in general, the case for the matrix $P(0;t) = \exp(Tt)$. In the case of the Erlang distribution, for example, T and therefore also the matrices $P(0;t)$ are upper-triangular. The matrix $P(0;t)$ is, however, nonsingular for $t \geqslant 0$, and therefore cannot have either vanishing rows or columns.

For the Poisson process with its simplest representation $\alpha = 1$, $T = -\lambda$, the matrices $P(k;t)$ reduce to the scalar functions $P(k;t) = e^{-\lambda t}\frac{(\lambda t)^k}{k!}$, $k \geqslant 0$. It is rarely useful to derive explicit analytic expressions for the matrices $P(k;t)$ for the few other cases where this is feasible. It is important to note that any PH-distribution has infinitely many representations and, in particular, there are infinitely many pairs (α,T) for which the corresponding PH-distribution $F(\cdot)$ is *exponential.* Such pairs are useful in testing general algorithms involving PH-distributions. In such cases, numerical results should agree with those obtained from the tractable formulas, which often hold under exponential assumptions.

The following result is frequently useful in theoretical considerations. It also shows that for all $t \geqslant 0$, the moment matrices of the matrix sequence $\{P(k;t)\}$ are finite.

Theorem 5.1.2: The eigenvalue with largest real part of the matrix T is real and negative. Denoting that eigenvalue by $-\sigma < 0$, and assuming that the representation $(\boldsymbol{\alpha},T)$ is *irreducible,* we have that

a. The Laplace-Stieltjes transform $f(s)$ in (5.1.3) converges for $\operatorname{Re} s > -\sigma$, and diverges for $s = -\sigma$.

b. The matrix series $P^*(z)$ in (5.1.11) converges for all z.

Proof: Let the abscissa of convergence of $f(s)$ be $-\tau < 0$. If the representation $(\boldsymbol{\alpha},T)$ is irreducible, the matrix $T + z\mathbf{T}^{\circ}\boldsymbol{\alpha}$ is an irreducible, stable matrix for $0 < z < 1$. For $z = 1$, that matrix is irreducible and has 0 as a simple eigenvalue. The matrix $T + z\mathbf{T}^\circ\boldsymbol{\alpha}$ therefore has (by the Perron-Frobenius theorem) a simple eigenvalue $-\sigma(z) < 0$, which is its eigenvalue of maximum real part. That eigenvalue decreases to $-\sigma$ as $z \to 0+$. Let $\mathbf{u}(z), 0 < z \leqslant 1$, be the corresponding left eigenvector of $T + z\mathbf{T}^\circ\boldsymbol{\alpha}$, then $\mathbf{u}(z)$ may be chosen to be a positive vector and may be normalized by setting $\mathbf{u}(z)\mathbf{e} = 1$.

The equation

$$\mathbf{u}(z)\,[T + z\mathbf{T}^\circ\alpha] = -\,\sigma(z)\mathbf{u}(z),$$

readily leads to

$$\mathbf{u}(z) = z\,[\mathbf{u}(z)\mathbf{T}^\circ]\,\alpha[-\,\sigma(z)I - T]^{-1},$$

and since $\mathbf{u}(z)$ is positive, $\mathbf{u}(z)\mathbf{T}^\circ$ does not vanish. Postmultiplying by $\mathbf{T}^\circ$, we see that $-\sigma(z)$ is a root of the equation $f(s) = z^{-1}$.

On the interval $(-\tau,\infty)$, $f(s)$ is a strictly decreasing convex function of s, which tends to infinity as $s \to -\tau+$, and clearly $f(0) = 1$. By consideration of the graph of $f(s)$, it follows that the equation $f(s) = z^{-1}$, has a unique root $-\sigma(z)$ in $(-\tau,0)$ and that $\sigma(z) \to \tau$, as $z \to 0+$. This implies that $\sigma = \tau$.

The matrix exponential series $\exp(A) = \sum_{\nu=0}^{\infty} \frac{A^\nu}{\nu!}$, converges for all finite matrices A. The probability generating function in (5.1.11) is therefore an entire matrix function of z and its derivatives of all orders at $z = 1-$ are finite matrices. •

Remark *a*. The non-restrictive technical condition that the representation (α,T) is *irreducible* is needed to avoid cases such as, for example, $\alpha = (0,1)$, and

$$T = \begin{vmatrix} -1 & 0 \\ 0 & -2 \end{vmatrix},$$

for which $\sigma = 1$, but $\tau = 2$.

Remark *b*. It readily follows from

$$\mathbf{u}(z)P^*(z;t) = \mathbf{u}(z)e^{-\sigma(z)t},$$

and

$$\mathbf{u}(z)\int_0^\infty P^*(z;t)e^{-st}\,dt = \mathbf{u}(z)[\,sI - (T + z\mathbf{T}^\circ\boldsymbol{\alpha})\,]^{-1}$$

$$= [\,s + \sigma(z)\,]^{-1}\mathbf{u}(z), \quad \text{for } 0 \leqslant z \leqslant 1,$$

that the Perron-Frobenius eigenvalue of $P^*(z;t)$ is given by $\exp[-\sigma(z)t]$, $0 \leqslant z \leqslant 1$, and that the matrix $sI - (T + z\mathbf{T}^\circ\boldsymbol{\alpha})$ is non-singular for Re $s > -\sigma(z)$, and $0 \leqslant z \leqslant 1$.

Next, we consider the factorial moment matrices

$$V_n(t) = \sum_{k=n}^{\infty} \frac{k!}{(k-n)!} P(k;t), \quad \text{for } n \geqslant 0, \tag{5.1.12}$$

of the sequence $\{P(k;t)\}$. By using the differential equations (5.1.9) and (5.1.10), we may verify that the matrices $V_n(t)$, $n \geqslant 0$, $t \geqslant 0$, satisfy the differential equations

$$V_0'(t) = V_0(t)Q^*, \tag{5.1.13}$$

$$V_n'(t) = V_n(t)Q^* + nV_{n-1}(t)\mathbf{T}^\circ\boldsymbol{\alpha}, \quad \text{for } n \geqslant 1,$$

and

$$V_0'(t) = Q^*V_0(t), \tag{5.1.14}$$

$$V_n'(t) = Q^*V_n(t) + n\mathbf{T}^\circ\boldsymbol{\alpha}V_{n-1}(t), \quad \text{for } n \geqslant 1,$$

with the initial conditions $V_0(0) = I$, $V_n(0) = 0$, for $n \geqslant 1$. It is often necessary to compute the matrices $V_n(t)$, $t \geqslant 0$, for small values of n. In most cases, this requires the numerical integration of the equations (5.1.13) or (5.1.14). It is clear that

$$V_0(t) = \exp(Q^* t) = P^*(1;t).$$

The following results are frequently needed in matrix analytic calculations involving PH-renewal processes.

Theorem 5.1.3: For any irreducible infinitesimal generator Q with stationary probability vector $\boldsymbol{\pi}$, the matrix $\mathbf{e}\boldsymbol{\pi} - Q$ is nonsingular and

$$\int_0^t \exp(Qu)du = \mathbf{e}\boldsymbol{\pi} + [I - \exp(Qt)]\,(\mathbf{e}\boldsymbol{\pi} - Q)^{-1}. \tag{5.1.15}$$

For $Q^* = T + \mathbf{T}^\circ\boldsymbol{\alpha}$, we have in addition that

$$V_1(t)\mathbf{e} = \lambda_1'^{-1} t\,\mathbf{e} + \lambda_1'^{-1}\,[I - \exp(Q^* t)]\; T^{-1}\mathbf{e}. \tag{5.1.16}$$

Proof: If $\mathbf{v}(\mathbf{e}\boldsymbol{\pi} - Q) = 0$, then, since $Q\mathbf{e} = 0$, it follows that $\mathbf{ve} = 0$, and therefore $\mathbf{v}Q = 0$. Since Q is irreducible, this implies that $\mathbf{v} = k\boldsymbol{\pi}$, for some constant k. Finally, $\mathbf{ve} = k = 0$, so that $\mathbf{v} = 0$. The matrix $\mathbf{e}\boldsymbol{\pi} - Q$ is therefore nonsingular.

Formula (5.1.15) is proved by noting that

$$\int_0^t \exp(Qu)du\;(\mathbf{e}\boldsymbol{\pi} - Q) = \mathbf{e}\boldsymbol{\pi}t + I - \exp(Qt),$$

and since $\boldsymbol{\pi}(\mathbf{e}\boldsymbol{\pi} - Q)^{-1} = \boldsymbol{\pi}$, the stated expression follows. It readily follows from (5.1.13) that

$$V_1(t)\mathbf{e} = \int_0^t V_0(u)du\,\mathbf{T}^\circ$$

$$= (\boldsymbol{\pi}\mathbf{T}^\circ)t\,\mathbf{e} + [I - \exp(Q^* t)]\,(\mathbf{e}\boldsymbol{\pi} - Q^*)^{-1}\mathbf{T}^\circ.$$

By (5.1.7), $\boldsymbol{\pi}\mathbf{T}^\circ = \lambda_1'^{-1}$. If we set $(\mathbf{e}\boldsymbol{\pi} - Q^*)^{-1}\mathbf{T}^\circ = \mathbf{v}$, then

$$(\boldsymbol{\pi}\mathbf{v})\mathbf{e} - T\mathbf{v} - (\boldsymbol{\alpha}\mathbf{v})\mathbf{T}^{\circ} = \mathbf{T}^{\circ},$$

and upon premultiplication by $\boldsymbol{\pi}$, we obtain that $\boldsymbol{\pi}\mathbf{v} = \lambda_1'^{-1}$. Furthermore, it follows that

$$\mathbf{v} = \lambda_1'^{-1}T^{-1}\mathbf{e} - (1 + \boldsymbol{\alpha}\mathbf{v})\mathbf{e},$$

and upon substitution, we obtain (5.1.16). •

Remark *a*. We note that

$$\boldsymbol{\alpha} V_1(t)\mathbf{e}$$

$$= \lambda_1'^{-1}t + \lambda_1'^{-1}\boldsymbol{\alpha}(I - \mathbf{e}\boldsymbol{\pi})T^{-1}\mathbf{e} + \lambda_1'^{-1}\boldsymbol{\alpha}[\,\mathbf{e}\boldsymbol{\pi} - \exp(Q^*t)\,]T^{-1}\mathbf{e}$$

$$= \lambda_1'^{-1}t + \frac{\lambda_2'}{2\lambda_1'^2} - 1 + \lambda_1'^{-1}\boldsymbol{\alpha}[\,\mathbf{e}\boldsymbol{\pi} - \exp(Q^*t)\,]T^{-1}\mathbf{e},$$

is the renewal function of the (ordinary) PH-renewal process. Since $\lim_{t\to\infty} \exp(Q^*t) = \mathbf{e}\boldsymbol{\pi}$, the last term tends to zero as $t \to \infty$. The other terms give the familiar expression for the linear asymptote of the renewal function. The relation $\boldsymbol{\pi} V_1(t)\mathbf{e} = \lambda_1'^{-1}t$, gives the linear renewal function for the stationary PH-renewal process.

Remark *b*. By elementary manipulations, we may deduce the integral formulas

$$V_1(t) = \int_0^t du_1 V_0(u_1)\,\mathbf{T}^{\circ}\boldsymbol{\alpha}\; V_0(t-u_1),$$

$$V_n(t) = n!\int_0^t du_n \int_0^{u_n} du_{n-1} \cdots \int_0^{u_2} du_1 V_0(u_1)\,\mathbf{T}^{\circ}\boldsymbol{\alpha}\; V_0(u_2-u_1)\,\mathbf{T}^{\circ}\boldsymbol{\alpha} \cdots$$

$$\cdot\, V_0(u_n - u_{n-1})\,\mathbf{T}^{\circ}\boldsymbol{\alpha}\; V_0(t-u_n), \quad \text{for } n \geqslant 2.$$

By repeating the manipulations in the proof of Theorem 5.1.3, we may in principle obtain explicit expressions for $V_n(t)\mathbf{e}$, similar to, but much more complicated than formula (5.1.16). These permit us to study the time dependence of the variance and higher moments of the counting process of a PH-renewal process. An explicit expression for the variance of $N(t)$ is obtained as a special case of the result in Theorem 5.4.1.

Discrete PH-Distributions - Closure Properties: There is an exactly parallel development of the notion of a *discrete PH-distribution.* We consider an $(m+1)$-state Markov chain P with a single absorbing state $m+1$ and m transient states $\{1, \ldots, m\}$. The initial probability vector is given by $(\boldsymbol{\alpha}, \alpha_{m+1})$ and the matrix P is written as

$$P = \begin{vmatrix} T & \mathbf{T}^{\circ} \\ 0 & 1 \end{vmatrix},$$

where T is a substochastic matrix such that $I - T$ is nonsingular and $T\mathbf{e} + \mathbf{T}^{\circ} = \mathbf{e}$.

The discrete density $\{p_k, k \geqslant 0\}$ of the time until absorption is given by

$$p_0 = \alpha_{m+1}, \qquad p_k = \boldsymbol{\alpha} T^{k-1} \mathbf{T}^{\circ}, \quad \text{for } k \geqslant 1.$$

The density $\{p_k, k \geqslant 0\}$ is a *discrete PH*-density.

It is easy to show that both classes of the continuous PH-distributions and discrete PH-densities are *closed* under finite mixtures and finite convolution products and representations for the resulting PH-distributions may be constructed.

In addition to the closure result proved in Theorem 5.1.1, we shall state (without proof) one further closure theorem which deals with mixtures of the form

$$G(x) = \sum_{k=0}^{\infty} p_k F^{(k)}(x), \tag{5.1.17}$$

where $\{p_k\}$ is a discrete PH-density with (irreducible) representation

(β,S) where β and S are of dimension n. The probability distribution $F(x)$ is either a continuous or discrete PH-distribution. We denote its (irreducible) representation by (α,T), but recall, of course, that T has a different significance in the continuous and discrete cases.

In order to write the representation of $G(\cdot)$ and also for several purposes in the sequel, we shall need the *Kronecker product* of matrices. For any two matrices L and M of dimensions $k_1 \times k_2$ and $k_1' \times k_2'$ respectively, their Kronecker product $L \otimes M$ is the matrix of dimensions $k_1k_1' \times k_2k_2'$, written in partitioned form as

$$\begin{vmatrix} L_{11}M & L_{12}M & \cdots & L_{1k_2}M \\ \cdots & & & \cdots \\ L_{k_11}M & L_{k_12}M & \cdots & L_{k_1k_2}M \end{vmatrix}.$$

The following is a useful property, of frequent use in calculations with Kronecker products. If L, M, U and V are rectangular matrices such that the ordinary matrix products LU and MV are defined, then

$$(L \otimes M)(U \otimes V) = LU \otimes MV. \tag{5.1.18}$$

Theorem 5.1.4: The representation of the PH-distributions $G(\cdot)$ in (5.1.17) is given by the pair (γ,L), where

$$\gamma = \alpha \otimes \beta(I - \alpha_{m+1}S)^{-1}, \tag{5.1.19}$$

$$L = T \otimes I + \mathbf{T}^{\circ}\alpha \otimes (I - \alpha_{m+1}S)^{-1}S.$$

A useful simple case of Theorem 5.1.4 arises when the mixing density $\{p_\nu\}$ is *geometric*. The probability distribution

$$G(x) = \sum_{\nu=0}^{\infty} (1-p)p^{\nu} F^{(\nu)}(x),$$

then has the representation

$$\gamma = (1 - \alpha_{m+1}p)^{-1} p\,\alpha, \qquad L = T + p(1 - \alpha_{m+1}p)^{-1}\mathbf{T}^{\circ}\alpha.$$

For the sake of illustration, we give a few applications of Theorem 5.1.4. The first of these yields a simple representation for the stationary waiting time distribution $W(\cdot)$ for the stable $M/\ PH/\ 1$ queue with arrival rate λ, traffic intensity ρ and a phase type service time distribution $F(\cdot)$ with representation (α, T), $\alpha_{m+1} = 0$.

It is well-known that

$$W(x) = \sum_{\nu=0}^{\infty} (1 - \rho)\rho^{\nu} F^{*(\nu)}(x), \tag{5.1.20}$$

where $F^{*}(\cdot)$ is the probability distribution defined in (5.1.8). It then readily follows from Theorems 5.1.1 and 5.1.4 that $W(\cdot)$ is the $PH-$distribution with representation

$$\gamma = \rho\pi, \qquad L = T + \rho\mathbf{T}^{\circ}\pi. \tag{5.1.21}$$

The second application deals with the *mixed Erlang distributions*

$$G(x) = \sum_{\nu=1}^{\infty} p_{\nu} E(\lambda; x), \tag{5.1.22}$$

which have frequently been considered in stochastic models. Under the mild restriction that $\{p_{\nu}\}$ is a discrete $PH-$density, the probability distribution $G(\cdot)$ is of phase type. By Theorem 5.1.4, its representation is given by

$$\gamma = \beta, \qquad L = \lambda(S - I). \tag{5.1.23}$$

Important Applications of $PH-$distributions: We conclude this overview of the basic properties of $PH-$distributions by discussing two of their most far reaching applications. The first is based on the fact that the $PH-$distributions are ***dense*** in the set of all probability distributions on $[0,\infty)$. This is of primarily theoretical utility. The second, important to algorithmic research, lies in the avoidance of many cumbersome numerical integrations through the use of the formalism of PH – distributions.

The proof of the denseness of the set of PH – distributions is elementary. It is obvious that any probability distribution on $[0,\infty)$ may be arbitrarily closely and uniformly approximated by a discrete distribution with *finite support*. Such a distribution is clearly a finite mixture of degenerate distributions. Any degenerate distribution at $x = a > 0$, is the uniform limit of a sequence of Erlang distributions with mean a and increasing orders. Any probability distribution $F(\cdot)$ on $[0,\infty)$ may therefore by obtained as the uniform limit of a sequence of probability distributions, each of which consists of a finite mixture of Erlang distributions and possibly a jump at 0. Such distributions are clearly of phase type. While the merits of this general result *as a practical approximation theorem* are limited, it has the following theoretical application, which will only be described here in an informal manner.

Suppose that a stochastic model involves one or more general probability distributions $F_j(\cdot)$, $1 \leqslant j \leqslant N$, on $[0,\infty)$ and that we wish to evaluate a *continuous* functional $\Phi[F_1(\cdot), \ldots, F_N(\cdot)]$ of these probability distributions. If an expression for $\Phi(\cdot)$ can be found for the case where $F_1(\cdot), \ldots, F_N(\cdot)$ are PH – distributions and *if that expression does not explicitly depend on the special formalism of PH – distributions,* then that expression is also valid for *arbitrary* distributions $F_1(\cdot), \ldots, F_N(\cdot)$.

A formal proof of this result is lengthy, primarily due to the necessary notation and the definitions of the required topological notions. The main idea, however, is the classical uniqueness of the continuous extension of a function which is continuous on a dense subset of a topological space. We shall illustrate the utility of this result by examples in the sequel.

The use of the formalism of PH – distributions in avoiding or simplifying numerical integrations is illustrated by two important examples.

Example 1: In a number of cases, we are led to the evaluation of integrals of the form

$$A_k = \int_0^\infty P(k;u)\,dH(u), \quad \text{for } k \geqslant 0, \tag{5.1.24}$$

or

$$\tilde{A}_k = \int_0^\infty P(k;u)\,[1 - H(u)]\,du \quad \text{for } k \geqslant 0, \tag{5.1.25}$$

where the matrices $P(k;\cdot)$, $k \geq 0$, satisfy the equations (5.1.9) and (5.1.10) and $H(\cdot)$ is a PH-distribution with the (irreducible) representation $(\boldsymbol{\beta},S)$, where $\boldsymbol{\beta}$ and S are of dimension n. Without loss of generality, we may set $\beta_{n+1} = 0$.

Theorem 5.1.5: The matrices A_k and $\tilde{A}_k$, $k \geq 0$, are given by

$$A_0 = -(I \otimes \boldsymbol{\beta})(T \otimes I + I \otimes S)^{-1}(I \otimes \mathbf{S}^\circ), \tag{5.1.26}$$

$$A_k = DC^{k-1}E, \quad \text{for } k \geq 1,$$

and

$$\tilde{A}_0 = -(I \otimes \boldsymbol{\beta})(T \otimes I + I \otimes S)^{-1}(I \otimes \mathbf{e}), \tag{5.1.27}$$

$$\tilde{A}_k = DC^{k-1}E_0, \quad \text{for } k \geq 1,$$

where the matrices D, C, E, and E_0 are given by

$$D = -(I \otimes \boldsymbol{\beta})(T \otimes I + I \otimes S)^{-1}(\mathbf{T}^\circ \otimes I), \tag{5.1.28}$$

$$C = -(\boldsymbol{\alpha} \otimes I)(T \otimes I + I \otimes S)^{-1}(\mathbf{T}^\circ \otimes I),$$

$$E = -(\boldsymbol{\alpha} \otimes I)(T \otimes I + I \otimes S)^{-1}(I \otimes \mathbf{S}^\circ),$$

$$E_0 = -(\boldsymbol{\alpha} \otimes I)(T \otimes I + I \otimes S)^{-1}(I \otimes \mathbf{e}).$$

The matrices D and C are of dimensions $m \times n$ and $n \times n$ respectively. The matrices E and E_0 are $n \times m$.

Proof: By using the product formula (5.1.18), we may write

$$A_0 = \int_0^\infty \exp(Tu)\boldsymbol{\beta}\exp(Su)\mathbf{S}^\circ \, du$$

$$= (I \otimes \beta) \int_0^\infty \exp(Tu) \otimes \exp(Su)du \; (I \otimes \mathbf{S}^\circ).$$

The matrix $V(u) = \exp(Tu) \otimes \exp(Su)$, clearly satisfies

$$V'(u) = V(u)(T \otimes I + I \otimes S), \quad \text{for } u \geqslant 0,$$

with $V(0) = I \otimes I$, so that it may also be written as

$$V(u) = \exp[\,(T \otimes I + I \otimes S)u\,], \quad \text{for } u \geqslant 0.$$

Since all the eigenvalues of T and of S lie in the left half plane, the same is true of the eigenvalues of the matrix

$$T \oplus S = T \otimes I + I \otimes S,$$

which is therefore nonsingular. The stated expression for A_0 is now immediate. The matrix $T \oplus S$ is called the *Kronecker sum* of the matrices T and S.

For $k \geqslant 1$, we write A_k as

$$A_k = (I \otimes \beta) \int_0^\infty P(k;u) \otimes \exp(Su)du \; (I \otimes \mathbf{S}^\circ),$$

and by partial integration and using the differential equations (5.1.9), we obtain that

$$\int_0^\infty P(k;u) \otimes \exp(Su)\, du = -\int_0^\infty P(k;u)T \otimes \exp(Su)S^{-1}\, du$$

$$-\int_0^\infty P(k-1;u)\mathbf{T}^\circ\alpha \otimes \exp(Su)S^{-1}\, du,$$

so that

$$\int_0^\infty P(k;u) \otimes \exp(Su)\, du$$

$$= -\int_0^\infty P(k-1;u) \otimes \exp(Su)du$$

$$\cdot (\mathbf{T}^\circ \otimes I)(\boldsymbol{\alpha} \otimes I)(T \otimes I + I \otimes S)^{-1}$$

$$= (T \otimes I + I \otimes S)^{-1}(\mathbf{T}^\circ \otimes I)$$

$$\cdot C^{k-1} \cdot (\boldsymbol{\alpha} \otimes I)(T \otimes I + I \otimes S)^{-1}.$$

Upon substitution, we obtain the stated expression for A_k, $k \geqslant 1$. Since $1 - H(u) = \boldsymbol{\beta}\exp(Su)\mathbf{e}$, the formulas (5.1.27) are obtained by replacing $\mathbf{S}^\circ$ by $\mathbf{e}$ in the preceding derivations. •

Remark *a*. In algorithmic implementations of Theorem 5.1.5, we first compute the matrices

$$U = -(I \otimes \boldsymbol{\beta})\,(T \otimes I + I \otimes S)^{-1},$$

and

$$V = -(\boldsymbol{\alpha} \otimes I)\,(T \otimes I + I \otimes S)^{-1},$$

by solving the linear systems of equations

$$I \otimes \boldsymbol{\beta} = -\,[U_1, \ldots, U_m]\,(T \otimes I + I \otimes S), \tag{5.1.29}$$

$$\boldsymbol{\alpha} \otimes I = -\,[V_1, \ldots, V_m]\,(T \otimes I + I \otimes S),$$

which are well-suited for solution by block Gauss-Seidel iteration. The matrices U_j and V_j, $1 \leqslant j \leqslant m$, are of dimensions $m \times n$ and $n \times n$ respectively. With the matrices U and V known, the evaluation of all other matrices is straightforward.

Remark *b*. The matrices C, D and E have interesting probabilistic interpretations, which account for the geometric form of A_k for $k \geqslant 1$. Let the length of the interval $(0,X)$ have the PH-distribution

$H(\cdot)$, with representation (β, S). We call the corresponding absorbing Markov process the *killing* process. The $PH-$renewal process with the representation (α, T) is called the *arrival* process.

The element $D_{j\nu'}$, $1 \leqslant j \leqslant m$, $1 \leqslant \nu' \leqslant n$, of D is the conditional probability that, starting the arrival process in the phase j, there is a first arrival prior to X and the killing process is in the phase ν' at the time of the first arrival. The element $C_{\nu\nu'}$, $1 \leqslant \nu, \nu' \leqslant n$, is the conditional probability that, starting from an arrival with the killing process in the phase ν, the next arrival occurs when the killing process is in the (non-absorbing) state ν'. The element $E_{\nu j'}$, $1 \leqslant \nu \leqslant n$, $1 \leqslant j' \leqslant m$, is the conditional probability that, starting from an arrival with the killing process in the phase ν, the killing process is absorbed prior to the next arrival and at time X, the arrival phase is j'.

Example 2: The convolution $F * H(\cdot)$ of an arbitrary probability distribution $F(\cdot)$ on $[0,\infty)$ and of a phase type distribution $H(\cdot)$ with representation (β, S) may be computed by solving a simple system of linear differential equations. The simple manipulation, involved in showing this, is useful in several other contexts.

Theorem 5.1.6: The convolution $G(x) = F * H(x)$, may be evaluated by solving the differential equation

$$\mathbf{y}'(x) = \mathbf{y}(x)S + F(x)\beta, \quad \text{for } x > 0, \tag{5.1.30}$$

with $\mathbf{y}(0) = 0$, and forming

$$G(x) = \beta_{n+1}F(0+) + \mathbf{y}(x)\mathbf{S}^{\circ}, \quad \text{for } x \geqslant 0. \tag{5.1.31}$$

Proof: Clearly

$$G(x) = \beta_{n+1}F(0+) + \int_0^x \beta \exp[S(x-u)]\mathbf{S}^{\circ} F(u)du.$$

Rewriting

$$\mathbf{y}(x) = \beta \int_0^x \exp[S(x-u)] F(u)du,$$

as

$$\mathbf{y}(x)\exp(-Sx) = \beta\int_0^x \exp[-Su)]\, F(u)du\,,$$

and differentiating, we obtain the equation (5.1.30), since $\exp(-Su)$ is nonsingular. Equation (5.1.31) is now obvious. •

5.2. THE *PH* / *G* / 1 QUEUE

The $PH/G/1$ queues form the subset of the $GI/G/1$ queueing models in which the input stream is a PH-renewal process. The interarrival time distribution $F(\cdot)$ has the irreducible representation (α,T) with m phases and moments λ_j', $j \geqslant 1$, given by formula (5.1.4). The service time distribution is denoted by $H(\cdot)$. It has a finite mean μ_1' and its higher moments are denoted by μ_j', $j \geqslant 2$, provided they exist. The model $PH/PH/1$, where also $H(\cdot)$ is of phase type, is particularly tractable. For that model, the use of the formalism of the PH-distributions, developed in Section 5.1, is highly useful in the derivation of explicit matrix formulas. When also $H(\cdot)$ is a PH-distribution, it will have the representation (β,S) involving n phases.

We consider the standard Markov renewal process, embedded at departure epochs. The state (i,j), $i \geqslant 0$, $1 \leqslant j \leqslant m$, signifies that there are i customers left behind in the system and that the arrival process is in the phase j at the service completion. It is readily verified that the transition probability matrix $Q(\cdot)$ of the embedded Markov renewal process is of the form (2.1.9) with $C_0(x) = A_0(x)$, and that

$$A_k(x) = \int_0^x P(k;u)dH(u), \tag{5.2.1}$$

$$B_k(x) = \int_0^x \exp[T(x-u)]\,\mathbf{T}^\circ\alpha\, A_k(u)\, du\,, \quad \text{for } k \geqslant 0, \quad x \geqslant 0.$$

All these blocks are square matrices of order m. The matrices $P(k;\cdot)$, $k \geqslant 0$, are as defined in Section 5.1. It is important to note that

$$B_k = B_k(\infty) = -T^{-1}\mathbf{T}^{\circ}\boldsymbol{\alpha}A_k = \mathbf{e}\boldsymbol{\alpha}A_k, \quad \text{for } k \geqslant 0. \tag{5.2.2}$$

so that the matrices B_k have identical rows. This is a consequence of the fact that the beginnings of busy periods are regeneration points. It is also the source of major analytical simplifications.

The following theorem summarizes the main properties of the embedded Markov chain.

Theorem 5.2.1: The embedded Markov chain is positive recurrent if and only if

$$\rho = \mu_1'\lambda_1'^{-1} < 1. \tag{5.2.3}$$

In that case, the matrix G is the unique stochastic solution of the equation

$$G = \sum_{k=0}^{\infty} A_k G^k, \tag{5.2.4}$$

and the matrix G is positive. The vector $\tilde{\boldsymbol{\mu}}_1$ is given by

$$\tilde{\boldsymbol{\mu}}_1 = (I - G + \mathbf{eg})\,[I - A + (\mathbf{e} - \hat{\boldsymbol{\beta}})\mathbf{g}]^{-1}\mathbf{e}, \tag{5.2.5}$$

where

$$\hat{\boldsymbol{\beta}} = \rho\mathbf{e} + \lambda_1'^{-1}(I - A)T^{-1}\mathbf{e}. \tag{5.2.6}$$

The particular form of the vector $\hat{\boldsymbol{\beta}}$ leads to the simplified expression

$$\tilde{\boldsymbol{\mu}}_1 = (1 - \rho)^{-1}[\,\mathbf{e} + \lambda_1'^{-1}(I - G)T^{-1}\mathbf{e}\,]. \tag{5.2.7}$$

The vector $\mathbf{x}_0 = \theta\,\boldsymbol{\alpha}G$, where the constant $\theta = \mathbf{x}_0\mathbf{e}$, is given by

$$\theta = (\boldsymbol{\alpha}\tilde{\boldsymbol{\mu}}_1)^{-1} = (1 - \rho)\lambda_1'[\,\boldsymbol{\alpha}G(-T^{-1}\mathbf{e})\,]^{-1}. \tag{5.2.8}$$

The vector $\mathbf{x}_1$ is explicitly given by

$$\mathbf{x}_1 = \theta\{\alpha[I - \sum_{k=1}^{\infty} A_k G^{k-1}]^{-1} - \alpha\}. \tag{5.2.9}$$

Proof: Since

$$A = \int_0^{\infty} \exp(Q^* u)\, dH(u),$$

the probability vector $\boldsymbol{\pi}$ in formula (5.1.7) obviously satisfies $\boldsymbol{\pi} A = \boldsymbol{\pi}$. It also readily follows from (5.1.16) that the vector $\hat{\boldsymbol{\beta}} = \sum_{k=1}^{\infty} k\, A_k \mathbf{e}$, is given by (5.2.6). The inner product $\boldsymbol{\pi}\hat{\boldsymbol{\beta}} = \rho$, so that the matrix G is stochastic if and only if $\rho \leqslant 1$.

The matrix G is positive since A_0 does not have vanishing columns and the matrices A_k, $k \geqslant 1$, are positive. When $\rho < 1$, the expression (5.2.5) for the vector $\tilde{\boldsymbol{\mu}}_1$ is obtained from formula (3.1.12). The component $(\tilde{\boldsymbol{\mu}}_1)_j$, $1 \leqslant j \leqslant m$, is the expected number of services during a busy period starting with the arrival process in the phase j. The inner product $\boldsymbol{\alpha}\tilde{\boldsymbol{\mu}}_1$ is the expected number of services completed during a busy period.

In order to obtain the simplified formula (5.2.7), we calculate the vector

$$\mathbf{v} = [I - A + (\mathbf{e} - \hat{\boldsymbol{\beta}})\mathbf{g}]^{-1}\mathbf{e},$$

by solving the linear system

$$(I - A)\mathbf{v} + (\mathbf{g}\mathbf{v})(\mathbf{e} - \hat{\boldsymbol{\beta}}) = \mathbf{e}.$$

Premultiplication by $\boldsymbol{\pi}$ yields that $\mathbf{g}\mathbf{v} = (1 - \rho)^{-1}$, so that by using (5.2.6) we obtain

$$(I - A)\mathbf{v} = \lambda_1'^{-1}(1 - \rho)^{-1}(I - A)T^{-1}\mathbf{e},$$

which implies that for some constant k,

$$\mathbf{v} = k\mathbf{e} + \lambda_1'^{-1}(1 - \rho)^{-1}T^{-1}\mathbf{e}.$$

Since **gv** is known, the constant k is explicitly given by

$$k = (1-\rho)^{-1}[\,1 - \lambda_1'^{-1}\mathbf{g}T^{-1}\mathbf{e}\,] > 0.$$

Upon substituting the explicit expression for **v** into the formula (5.2.5), we obtain (5.2.7).

For the $PH/G/1$ queue, the matrix $\tilde{K}(z,0)$, obtained by setting $s = 0$ in (3.2.1), reduces to

$$\tilde{K}(z,0) = \mathbf{e}\alpha\,\tilde{G}(z,0),$$

so that

$$\kappa = \alpha G, \qquad \tilde{\kappa}_1 = (\alpha\tilde{\mu}_1)\mathbf{e}.$$

The stated expression for $\mathbf{x}_0$ now readily follows from (3.2.6) and (5.2.7). The expression for $\mathbf{x}_1$ may be obtained by particularizing the general formulas in Section 3.2 or by noting that the matrix $\tilde{H}(z)$ of (3.2.7) is here given by

$$\tilde{H}(z) = \{I + (1-\gamma z)^{-1}z\,A_0\,\mathbf{e}\alpha\}\sum_{k=1}^{\infty} zA_k\,\tilde{G}^{k-1}(z,0),$$

where $\gamma = \alpha A_0\mathbf{e}$.

In order to find the invariant probability vector **h** of $\tilde{H}(1)$, we note that the equation $\mathbf{h}\tilde{H}(1) = \mathbf{h}$, is equivalent to

$$\mathbf{h}[I - \sum_{k=1}^{\infty} A_k\,G^{k-1}] = (1-\gamma)^{-1}\mathbf{h}\,A_0\,\mathbf{e}\,\alpha\sum_{k=1}^{\infty} A_k\,G^{k-1}.$$

Since $\mathbf{he} = 1$, this leads to

$$\mathbf{h} = \{\,\alpha[I - \sum_{k=1}^{\infty} A_k\,G^{k-1}]^{-1}\mathbf{e} - 1\,\}^{-1}\{\,\alpha[I - \sum_{k=1}^{\infty} A_k\,G^{k-1}]^{-1} - \alpha\,\}.$$

The calculation of the constant $\mathbf{h}\tilde{H}'(1)\mathbf{e} = \mathbf{h}\tilde{\mathbf{h}}$, is somewhat belabored, but after routine simplifications, we obtain that

$$\mathbf{h}\tilde{\mathbf{h}} = \{ \boldsymbol{\alpha}[I - \sum_{k=1}^{\infty} A_k G^{k-1}]^{-1}\mathbf{e} - 1 \}^{-1} \lambda_1'^{-1}(1 - \rho)^{-1}\boldsymbol{\alpha} G(-T^{-1}\mathbf{e}),$$

so that, by appealing to (3.2.9), the vector $\mathbf{x}_1$ is given by (5.2.9). •

The simplifications which are obtained in the results of Theorem 5.2.1 for the $PH/PH/1$ model, are noteworthy. They are described in the following corollary and result from the particular form of the matrices A_k, $k \geqslant 0$, given in Theorem 5.1.5.

Corollary 5.2.1: For the stable $PH/PH/1$ queue, the matrix G may be obtained by computing the minimal nonnegative solution (G,U) to the system of equations

$$G = A_0 + DUG, \qquad (5.2.10)$$
$$U = E + CUG,$$

where U is a matrix of dimensions $n \times m$.

The vector $\mathbf{x}_1$ is then given by

$$\mathbf{x}_1 = \theta[\boldsymbol{\alpha}(I - DU)^{-1} - \boldsymbol{\alpha}]. \qquad (5.2.11)$$

Proof: By setting $U = \sum_{k=0}^{\infty} C^k EG^k$, the equation (5.2.4) readily leads to the equivalent system (5.2.10). It is also immediate that $\sum_{k=1}^{\infty} A_k G^{k-1} = DU$, so that (5.2.11) follows from (5.2.9). •

Remark: We have that $U\mathbf{e} = (I-C)^{-1}E\mathbf{e}$, so that the n-vector $U\mathbf{e}$ may be computed beforehand once and for all. As a starting solution, we choose a pair (G°, U°), consisting of an arbitrary (positive) stochastic matrix G° and an arbitrary positive $n \times m$ matrix U° with $U^\circ\mathbf{e} = U\mathbf{e}$. Upon performing successive substitutions in (5.2.10), the successive iterates then all have the correct row sums. Only three matrix multiplications per iteration are required.

For the stable $PH/G/1$ queue, the vector

$$(\mathbf{x}_0\mathbf{e})^{-1}\mathbf{x}_0 = \boldsymbol{\alpha} G, \tag{5.2.12}$$

is the vector of the conditional probabilities of the various arrival phases at the end of an arbitrary busy period. It is therefore clear that the successive idle periods, which are independent and identically distributed, have the common PH-distribution with representation $(\boldsymbol{\alpha} G, T)$ and mean $E(I) = \boldsymbol{\alpha} G(-T^{-1}\mathbf{e})$. The formula (5.2.8) may therefore be rewritten as

$$\theta = (1 - \rho)\, \lambda_1{}'\, [E(I)]^{-1}. \tag{5.2.13}$$

The moment relation (5.2.13) does not depend on the formalism of the PH-distributions and, by continuity, is therefore also valid for the stable $GI/G/1$ queue. θ is the fraction of services in the stationary version of the queue, which terminate with the server becoming idle. For the general $GI/G/1$ queue, no tractable expressions for the distribution or the moments of the idle time are available.

Stationary Queue Length Densities: The generating function

$$\mathbf{X}(z) = \sum_{i=0}^{\infty} \mathbf{x}_i z^i, \tag{5.2.14}$$

is given by

$$\mathbf{X}(z)[\, zI - A^*(z)\,] = \theta \boldsymbol{\alpha}[\, zI - G\,]\, A^*(z), \tag{5.2.15}$$

where

$$A^*(z) = \int_0^{\infty} \exp[(T + z\mathbf{T}^\circ\boldsymbol{\alpha})u]\, dH(u). \tag{5.2.16}$$

Formula (5.2.15) is routinely derived from the steady-state equations for the embedded Markov chain and is particularly useful in the derivation of moment formulas. By calculations similar to those in the proof of Theorem 3.3.1, we obtain that

$$\mathbf{X}(1) = \boldsymbol{\pi} - \theta(I - G)AZ, \tag{5.2.17}$$

where $Z = (I - A + \mathbf{e}\boldsymbol{\pi})^{-1}$. The components of $X(1)$ are the conditional probabilities of the arrival phases at the end of an arbitrary service.

The Queue Length at an Arbitrary Time: Next we consider the stationary joint density of the queue length and the phase of the arrival process *at an arbitrary time.* Let the j-th component, $1 \leqslant j \leqslant m$, of the vector $\mathbf{y}_i$, $i \geqslant 0$, be the stationary probability that at an arbitrary time t, there are i customers in the system and the phase of the arrival process is j. The following theorem relates the vectors $\mathbf{y}_i$, $i \geqslant 0$, and $\mathbf{x}_i$, $i \geqslant 0$, in a very simple way.

Theorem 5.2.2: In the stable $PH/G/1$ queue, the vectors $\mathbf{y}_i$, $i \geqslant 0$, are given by

$$\mathbf{y}_0 = \lambda_1'^{-1}\mathbf{x}_0(-T^{-1}) = \lambda_1'^{-1}\theta\,\alpha G(-T^{-1}), \tag{5.2.18}$$

$$\mathbf{y}_i = (\mathbf{x}_{i-1}\mathbf{e})\boldsymbol{\pi} + \lambda_1'^{-1}(\mathbf{x}_i - \mathbf{x}_{i-1})(-T^{-1}), \quad \text{for } i \geqslant 1.$$

The unconditional stationary queue length density $\{\phi_i\}$ *at arrivals* is the same as the unconditional stationary queue length density $\{\mathbf{x}_i\mathbf{e}\}$ *following departures.* We also have the equality

$$\sum_{i=0}^{\infty} \mathbf{y}_i = \boldsymbol{\pi}. \tag{5.2.19}$$

Proof: The fundamental mean E of the embedded Markov renewal process of the $PH/G/1$ queue is given by

$$E = \mathbf{x}_0(\mu_1'\mathbf{e} - T^{-1}\mathbf{e}) + [X(1) - \mathbf{x}_0]\,\mu_1'\mathbf{e}$$

$$= \mu_1' + \mathbf{x}_0(-T^{-1}\mathbf{e}) = \mu_1' + \lambda_1'(1 - \rho) = \lambda_1',$$

by virtue of formula (5.2.8). The departure rate E^{-1} from the stationary queue is therefore, as is to be expected, equal to its arrival rate $\lambda_1'^{-1}$.

By the classical argument based on the key renewal theorem, the vectors $\mathbf{y}_i$, $i \geqslant 0$, are related to the vectors $\mathbf{x}_i$, $i \geqslant 0$, by

$$\mathbf{y}_0 = \lambda_1'^{-1}\theta\,\boldsymbol{\alpha}\,G(-T^{-1}),$$

$$\mathbf{y}_i = \lambda_1'^{-1}\theta\,\boldsymbol{\alpha}\int_0^\infty P(i-1;u)\,[1-H(u)]du$$

$$+\lambda_1'^{-1}\sum_{\nu=1}^{i}\mathbf{x}_\nu\int_0^\infty P(i-\nu;u)\,[1-H(u)]du, \quad \text{for } i \geqslant 1.$$

Upon forming the generating function

$$\mathbf{Y}(z) = \sum_{i=0}^{\infty}\mathbf{y}_i z^i,$$

we obtain that for $|z| < 1$,

$$\mathbf{Y}(z) = \mathbf{y}_0 + \lambda_1'^{-1}z\,\theta\,\boldsymbol{\alpha}\int_0^\infty \exp[(T + z\mathbf{T}^\circ\boldsymbol{\alpha})u]\,[1-H(u)]du$$

$$+\lambda_1'^{-1}[\mathbf{X}(z)-\mathbf{x}_0]\int_0^\infty \exp[(T + z\mathbf{T}^\circ\boldsymbol{\alpha})u]\,[1-H(u)]du$$

$$= \mathbf{y}_0 + \lambda_1'^{-1}[\mathbf{X}(z) + \theta\,\boldsymbol{\alpha}(zI - G)]\,[A^*(z) - I]\,(T + z\mathbf{T}^\circ\boldsymbol{\alpha})^{-1}.$$

We now postmultiply by the matrix $T + z\mathbf{T}^\circ\boldsymbol{\alpha}$, replace $\mathbf{y}_0$ by its explicit form and we notice that, by virtue of formula (5.2.15), the terms

$$\mathbf{X}(z)\,A^*(z) + \theta\,\boldsymbol{\alpha}\,[zI - G]\,A^*(z)$$

may be replaced by $z\mathbf{X}(z)$. After routine simplifications, we obtain that

$$\mathbf{Y}(z)(T + z\mathbf{T}^\circ\alpha) = \lambda_1'^{-1}(z - 1)\mathbf{X}(z). \tag{5.2.20}$$

Setting $z = 1$, and recalling that $\mathbf{Y}(1)\mathbf{e} = 1$, it follows that $\mathbf{Y}(1) = \boldsymbol{\pi}$. The stationary queue length density $\{\phi_i\}$ at arrivals is clearly given by

$$\phi_i = [\mathbf{Y}(1)\mathbf{T}^\circ]^{-1}\mathbf{y}_i\mathbf{T}^\circ = \lambda_1'\mathbf{y}_i\mathbf{T}^\circ, \quad \text{for } i \geqslant 0,$$

and has the probability generating function

$$\phi(z) = \lambda_1'\mathbf{Y}(z)\mathbf{T}^\circ.$$

Postmultiplying by $\mathbf{e}$ in (5.2.20), it follows that $\phi(z) = \mathbf{X}(z)\mathbf{e}$, so that the same stationary queue length densities prevail at arrivals and following departures. Finally, by postmultiplying by T^{-1}, the equation (5.2.20) may be rewritten as

$$\mathbf{Y}(z) = z\phi(z)\boldsymbol{\pi} + \lambda_1'^{-1}(1 - z)\mathbf{X}(z)(-T^{-1}),$$

and upon equating the coefficients of z^i, the equalities (5.2.18) are obtained. •

Stationary Waiting Time Distributions: In studying the stationary *waiting time distributions* for the stable $PH/G/1$ queue, it is convenient to define the vector $\mathbf{W}(\cdot)$, where $W_j(x)$, $1 \leqslant j \leqslant m$, is the probability that a *virtual* customer arriving at an arbitrary time t has to wait at most a length of time x and finds the arrival process in the phase j at time t. The vector $\mathbf{W}(\cdot)$ satisfies a generalized Pollaczek-Khinchin equation, as is shown in the following theorem.

Theorem 5.2.3: In the stable $PH/G/1$ queue, the vector $\mathbf{W}(\cdot)$ satisfies the integral equation

$$\mathbf{W}(x) = \mathbf{y}_0 - \int_0^x \mathbf{W}(u)du\,(T + \mathbf{T}^\circ\alpha) \tag{5.2.21}$$

$$+ \rho \int_0^x \frac{1 - H(x-u)}{\mu_1'} \lambda_1' \mathbf{W}(u) \mathbf{T}^\circ du \, \boldsymbol{\alpha},$$

for $x \geqslant 0$. The vector $\mathbf{W}(\infty) = \boldsymbol{\pi}$, and the stationary waiting time distribution of an *arriving* customer is given by

$$W_A(x) = \lambda_1' \mathbf{W}(x) \mathbf{T}^\circ, \quad \text{for } x \geqslant 0. \tag{5.2.22}$$

Proof: The classical argument, based on the key renewal theorem, yields that the vector $\mathbf{W}(\cdot)$, is related to the vectors $\mathbf{x}_i$, $i \geqslant 0$, by the equation

$$\mathbf{W}(x) = \mathbf{y}_0 + \lambda_1'^{-1} \theta \, \boldsymbol{\alpha} \sum_{\nu=0}^{\infty} \int_0^{\infty} \int_0^x P(\nu;t) dH(t+v) dt \; H^{(\nu)}(x-v) \tag{5.2.23}$$

$$+ \lambda_1'^{-1} \sum_{i=1}^{\infty} \sum_{\nu=0}^{\infty} \mathbf{x}_i \int_0^{\infty} \int_0^x P(\nu;t) dH(t+v) dt \; H^{(i+\nu-1)}(x-v).$$

Upon calculation of the Laplace-Stieltjes transform

$$\mathbf{W}^*(s) = \int_0^{\infty} e^{-sx} d\mathbf{W}(x),$$

the equation (5.2.23) is seen to be equivalent to

$$\mathbf{W}^*(s) \left[sI + T + h(s) \mathbf{T}^\circ \boldsymbol{\alpha} \right] = s \, \mathbf{y}_0, \tag{5.2.24}$$

where $h(s)$ is the Laplace-Stieltjes transform of the service time distribution $H(\cdot)$.

By setting $s = 0$, we readily deduce that $\mathbf{W}^*(0) = \boldsymbol{\pi}$, or equivalently that $\mathbf{W}(\infty) = \boldsymbol{\pi}$. Since $\boldsymbol{\pi} \mathbf{T}^\circ = \lambda_1'^{-1}$, it is clear that the Laplace-Stieltjes transform of the stationary distribution of the waiting time *at an arrival epoch* is given by

$$W_A^*(s) = \lambda_1' \mathbf{W}^*(s) \mathbf{T}^\circ.$$

The equation (5.2.24) may be rewritten in the form

$$\mathbf{W}^*(s) = \mathbf{y}_0 - s^{-1}\mathbf{W}^*(s)(T + \mathbf{T}^\circ\boldsymbol{\alpha}) + \rho\,\frac{1 - h(s)}{s\,\mu_1'}\,\lambda_1'\mathbf{W}^*(s)\mathbf{T}^\circ\boldsymbol{\alpha},$$

which is equivalent to the equation (5.2.21). •

Remark *a*. Theorem 5.2.3 is very rich in consequences and leads, particularly for the $PH/PH/1$ queue, to algorithms of exquisite simplicity. We begin by noting that equation (5.2.21) is indeed a natural generalization of the Pollaczek-Khinchin equation for the stationary waiting time distribution in the $M/G/1$ queue. When $T = -\lambda$, $\mathbf{T}^\circ = \lambda$, and $\boldsymbol{\alpha} = 1$, it reduces to the familiar equation

$$W(x) = 1 - \rho + \rho\int_0^\infty \frac{1 - H(x-u)}{\mu_1'}\,W(u)\,du, \quad \text{for } x \geqslant 0.$$

Remark *b*. Upon postmultiplication by $\mathbf{e}$, the equation (5.2.21) yields

$$\mathbf{W}(x)\mathbf{e} = 1 - \rho + \rho\int_0^\infty \frac{1 - H(x-u)}{\mu_1'}\,W_A(u)\,du, \quad \text{for } x \geqslant 0, \qquad (5.2.25)$$

which is a well-known relation between the stationary distribution $\mathbf{W}(x)\mathbf{e}$ of the *virtual* waiting time and the stationary distribution $W_A(x)$ of the waiting time *at an arrival*. Formula (5.2.25) does not depend on the PH-form of the interarrival time distribution and therefore also holds for the $GI/G/1$ queue, as was indeed shown by different methods. The equation (5.2.25), by itself, clearly does not suffice to determine either $\mathbf{W}(x)\mathbf{e}$ or $W_A(x)$.

Remark *c*. For the stable $PH/PH/1$ queue, we may write the equation (5.2.21) as

$$\mathbf{W}(x) = \mathbf{y}_0 - \int_0^x \mathbf{W}(u)\,du\,(T + \mathbf{T}^\circ\boldsymbol{\alpha}) + \mathbf{V}(x)\mathbf{e}\boldsymbol{\alpha}, \qquad (5.2.26)$$

where

$$\mathbf{V}(x) = \int_0^x \mathbf{W}(u)\mathbf{T}^\circ \cdot \boldsymbol{\beta}\exp[S(x-u)]\,du, \quad \text{for } x \geqslant 0, \tag{5.2.27}$$

The following result is then of major algorithmic interest.

Theorem 5.2.4: The $(n+m)-$vector $[\mathbf{V}(x),\mathbf{W}(x)]$ is the solution to the system of linear differential equations

$$[\mathbf{V}'(x),\mathbf{W}'(x)] = [\mathbf{V}(x),\mathbf{W}(x)] \begin{vmatrix} S & -\mathbf{S}^\circ\boldsymbol{\alpha} \\ \mathbf{T}^\circ\boldsymbol{\beta} & -T \end{vmatrix}, \tag{5.2.28}$$

for $x \geqslant 0$, with initial conditions $\mathbf{V}(0) = 0$, $\mathbf{W}(0) = \mathbf{y}_0$.

The stationary distributions of the virtual waiting time, of the waiting time of an arriving customer and of the total time in system of an arriving customer are then respectively given by $\mathbf{W}(x)\mathbf{e}$, $\lambda_1{}'\mathbf{W}(x)\mathbf{T}^\circ$ and $\lambda_1{}'\mathbf{V}(x)\mathbf{S}^\circ$.

Proof: Equation (5.2.27) may be written as

$$\mathbf{V}(x)\exp(-Sx) = \int_0^x \mathbf{W}(u)\mathbf{T}^\circ \cdot \boldsymbol{\beta}\exp(-Su)\,du, \quad \text{for } x \geqslant 0,$$

from which, upon differentiation, we obtain

$$\mathbf{V}'(x) = \mathbf{V}(x)S + \mathbf{W}(x)\mathbf{T}^\circ\boldsymbol{\beta}, \quad \text{for } x \geqslant 0. \tag{5.2.29}$$

Postmultiplication by $\mathbf{e}$ further yields that

$$\mathbf{V}'(x)\mathbf{e} = -\mathbf{V}(x)\mathbf{S}^\circ + \mathbf{W}(x)\mathbf{T}^\circ. \tag{5.2.30}$$

Differentiation in (5.2.26) yields

$$\mathbf{W}'(x) = -\mathbf{W}(x)(T + \mathbf{T}^\circ\boldsymbol{\alpha}) + \mathbf{V}'(x)\mathbf{e}\boldsymbol{\alpha},$$

so that by (5.2.29),

$$\mathbf{W}'(x) = -\mathbf{W}(x)T - \mathbf{V}(x)\mathbf{S}^\circ\boldsymbol{\alpha}, \quad \text{for } x \geqslant 0. \tag{5.2.31}$$

The equations (5.2.29) and (5.2.31) may clearly be rewritten as (5.2.28) and the initial conditions $\mathbf{V}(0) = 0$, $\mathbf{W}(0) = \mathbf{y}_0$, are evident. The interpretations of the inner products $\mathbf{W}(x)\mathbf{e}$ and $\lambda_1{}'\mathbf{W}(x)\mathbf{T}^\circ$ are obvious and

$$\lambda_1{}'\mathbf{V}(x)\mathbf{S}^\circ = \int_0^\infty W_A(u)\boldsymbol{\beta}\exp[S(x-u)]\mathbf{S}^\circ du,$$

is the convolution of the stationary waiting time distribution $W_A(\cdot)$ of an arriving customer and the distribution of his subsequent service time. •

Corollary 5.2.2: We have that

$$\mathbf{V}(\infty) = \rho\boldsymbol{\psi}, \tag{5.2.32}$$

where $\boldsymbol{\psi} = \mu_1{}'^{-1}\boldsymbol{\beta}(-S^{-1})$, is the invariant vector of the irreducible generator $S + \mathbf{S}^\circ\boldsymbol{\beta}$.

Proof: It follows from (5.2.27) that

$$\mathbf{V}(\infty) = \mathbf{W}(\infty)\mathbf{T}^\circ\int_0^\infty \boldsymbol{\beta}\exp(Su)du = \lambda_1{}'^{-1}\boldsymbol{\beta}(-S^{-1}) = \rho\boldsymbol{\psi}. \qquad \bullet$$

Moments: The important (generalized) Pollaczek-Khinchin equation (5.2.21) may also be used to derive recursive expressions for the *moments* of the stationary virtual waiting time and of the stationary waiting time at arrivals.

For either of the embedded processes, the n-th moment of the waiting time involves the moments up to order $n+1$ of the interarrival time, the service time and the idle time distributions. For the $PH/G/1$ queue, the n-th moments of the waiting times therefore exist if and only if the service times have a finite moment of order $n+1$. As we shall see, the analytic form of the recurrence relation does not depend on the formalism of the PH-distributions and is therefore, by continuity, also valid for the $GI/G/1$ queue, at least under the proviso of the existence of the moments of order $n+1$ of the interarrival, service and idle times.

The recurrence relations may be written in a simple mnemonic form, provided we make the following *operational* conventions. Let $\tilde{W}$ and W_A denote random variables with the same distributions as the virtual waiting time and the waiting time at arrivals. The random variable S has the probability distribution $H(\cdot)$ with moments $\mu_n{}'$, $n \geqslant 0$. The random variables U and V have the moments $\psi_n{}'$ and $\phi_n{}'$ respectively, where

$$\psi_n{}' = [(n+1)\lambda_1{}']^{-1}\lambda_{n+1}{}', \quad \phi_n{}' = [(n+1)\mu_1{}']^{-1}\mu_{n+1}{}', \quad (5.2.33)$$

for $n \geqslant 0$. These are clearly the n-th moments of the probability distributions with the densities $\lambda_1{}'^{-1}[1 - F(x)]$, and $\mu_1{}'^{-1}[1 - H(x)]$. The n-th moment of the random variable J is defined by

$$E(J^n) = [(n+1)E(I)]^{-1}E(I^{n+1}), \quad \text{for } n \geqslant 0, \qquad (5.2.34)$$

where I has the stationary distribution of the idle time. In the following mnemonic formula, the random variables W_A, S, U, V and J are to be treated as *mutually independent*. We stress, however, that no relation between *random variables* related to the $GI/G/1$ queue is thereby intended.

Theorem 5.2.5: The moments of the stationary waiting time distributions may be recursively calculated from the relations

$$E(\tilde{W}^n) = \rho E(W_A + V)^n, \quad \text{for } n \geqslant 1, \qquad (5.2.35)$$

and

$$E(W_A + S - U)^n - \rho E(W_A + V)^n = (-1)^n(1-\rho)E(J^n), \quad (5.2.36)$$

for $n \geqslant 0$.

Proof: Postmultiplying by $\mathbf{e}$ in formula (5.2.24), we obtain

$$\mathbf{W}^*(s)\mathbf{e} = 1 - \rho + \rho W_A^*(s)\frac{1 - h(s)}{\mu_1{}'},$$

which is the transform version of the relation (5.2.25). Differentiating n times, we see that

$$E(\tilde{W}^n) = \rho \sum_{\nu=0}^{n} \binom{n}{\nu} E(W_A^{n-\nu})\phi_\nu', \quad \text{for } n \geq 1,$$

which may be symbolically written as (5.2.35).

Differentiating n times in (5.2.24), we see that for $n \geq 1$,

$$\mathbf{W}^{*\,(n)}(s)(sI + T) + n\mathbf{W}^{*\,(n-1)}(s)$$

$$= \delta_{1n}\mathbf{y}_0 - \lambda_1'^{-1}\sum_{\nu=0}^{n} \binom{n}{\nu} h^{(\nu)}(s)\, W_A^{*\,(n-\nu)}(s)\boldsymbol{\alpha}.$$

Setting $s = 0$, and recalling that $\mathbf{W}^*(0) = \boldsymbol{\pi}$, it follows that

$$\mathbf{W}^{*\,\prime}(0)T = -\boldsymbol{\pi} + \mathbf{y}_0 + \lambda_1'^{-1}[\mu_1' + E(W_A)]\boldsymbol{\alpha},$$

and for $2 \leq r \leq n$,

$$\mathbf{W}^{*\,(r)}(0)T = -r\mathbf{W}^{*\,(r-1)}(0) - (-1)^r\lambda_1'^{-1}\sum_{\nu=0}^{r} \binom{r}{\nu} \mu_\nu'\, E(W_A^{r-\nu})\boldsymbol{\alpha}.$$

We now postmultiply the equation corresponding to the index r by the matrix $(-1)^{n-r}(n!/r!)T^{-(n-r+1)}$ and sum the resulting equations for r ranging from one to n to obtain

$$\mathbf{W}^{*\,(n)}(0) = (-1)^n n!\,\boldsymbol{\pi}T^{-n} - (-1)^n n!\,\mathbf{y}_0T^{-n}$$

$$+ (-1)^{n+1}\lambda_1'^{-1}\sum_{r=1}^{n} \frac{n!}{r!} \sum_{\nu=0}^{r} \binom{r}{\nu} \mu_\nu'\, E(W_A^{r-\nu})\,\boldsymbol{\alpha}T^{-(n-r+1)}.$$

Finally, we postmultiply by $\mathbf{e}$ and notice that

$$\psi_n' = (-1)^n n!\,\boldsymbol{\pi}T^{-n}\mathbf{e}, \qquad (-1)^n n!\,\mathbf{y}_0T^{-n}\mathbf{e} = (1-\rho)E(J^n).$$

It now follows routinely that

$$E(\tilde{W}^n) = (-1)^n \psi_n{}' - (-1)^n (1-\rho)E(J^n)$$

$$+ \sum_{r=1}^{n} (-1)^{n-r} \sum_{\nu=0}^{r} \frac{n!}{\nu!(r-\nu)!(n-r)} \mu_\nu{}' \, E(W_A^{r-\nu})\psi_{n-r}{}'$$

$$= (-1)^{n+1}(1-\rho)E(J^n) + E(W_A + S - U)^n,$$

and by (5.2.35) this may be written as (5.2.36). •

Remark *a.* The explicit analytic expressions for $E(W_A^n)$ and $E(\tilde{W}^n)$, obtained by expansion in (5.2.35) and (5.2.36), are very involved, except for $n = 1$ or 2. The formulas (5.2.35) and (5.2.36) nevertheless remain useful for numerical computation. We illustrate their use by considering the first two moments in detail.

For $n = 1$, we obtain that

$$(1-\rho)E(W_A) = \frac{\lambda_2{}'}{2\lambda_1{}'} + \rho \frac{\mu_2{}'}{2\mu_1{}'} - \mu_1{}' - (1-\rho)E(J),$$

and

$$E(\tilde{W}) = \rho E(W_A) + \rho \frac{\mu_2{}'}{2\mu_1{}'}.$$

In terms of the coefficients of variation

$$C_A = \left(\frac{\lambda_2{}'}{\lambda_1{}'^2} - 1\right)^{\frac{1}{2}}, \qquad C_S = \left(\frac{\mu_2{}'}{\mu_1{}'^2} - 1\right)^{\frac{1}{2}},$$

of the interarrival and service times, these expressions may be rewritten in the familiar forms

$$(1-\rho)E(W_A) = \frac{1}{2}\mu_1{}' \, [\rho^{-1}C_A^2 + \rho C_S^2 + \rho^{-1}(1-\rho)^2] \qquad (5.2.37)$$

$$- (1 - \rho)E(J),$$

and

$$(1 - \rho)E(\tilde{W}) = \frac{1}{2}\mu_1{}' \, [C_A^2 + \rho C_S^2 + (1 - \rho) - 2(1 - \rho)\lambda_1{}'^{-1}E(J)]. \tag{5.2.38}$$

For $n = 2$, Theorem 5.2.5 yields that

$$\begin{aligned}(1 - \rho)E(W_A^2) &= 2E(W_A)[\, \rho E(V) + E(U) - E(S)\,] \\ &+ [\, \rho E(V^2) - E(S^2) - E(U^2) + 2E(U)E(S)\,] \\ &+ (1 - \rho)E(J^2),\end{aligned}$$

and

$$E(\tilde{W}^2) = \rho E(W_A^2) + 2\rho E(V)E(W_A) + \rho E(V^2).$$

We notice that the coefficient $\rho E(V) + E(U) - E(S)$ may be replaced by $(1 - \rho)\ [E(W_A) + E(J)]$, and we also express the noncentral moments of the service and interarrival times in terms of the central moments in order to obtain formulas that are analogous to (5.2.37) and (5.2.38).

After routine, but laborious calculations one obtains the following expressions:

$$\begin{aligned}&(1 - \rho)E(W_A^2) \\ &= 2(1 - \rho)E(W_A)[E(W_A) + E(J)] + (1 - \rho)E(J^2) \\ &+ \frac{1}{3}\mu_1{}'^2\{\, \rho D_S^3 - \rho^{-2}D_A^3 - 3(1 - \rho)(C_S^2 + \rho^{-2}C_A^2) - \rho^{-2}(1 - \rho)^3\,\},\end{aligned} \tag{5.2.39}$$

and

$$(1 - \rho)E(\tilde{W}^2) \tag{5.2.40}$$

$$= 2\rho(1-\rho)E(W_A)\,[E(W_A) + E(J) + \frac{1}{2}\mu_1'(C_S^2+1)]$$

$$+ \rho(1-\rho)E(J^2)$$

$$+ \frac{1}{3}\mu_1'^2\{\rho D_S^3 - \rho^{-1}D_A^3 - 3\rho^{-1}(1-\rho)C_A^2 + \rho^{-1}(1-\rho)(2\rho-1)\},$$

where $D_A^3 = \lambda_1'^{-3}\lambda_3$, and $D_S^3 = \mu_1'^{-3}\mu_3$, are the scaled central moments of the interarrival and service time distributions.

For the case of Poisson arrivals, i. e. for the ordinary $M/G/1$ queue, the first two moments are given by the simpler expressions

$$(1-\rho)E(W_A) = (1-\rho)E(\tilde{W}) = \frac{1}{2}\mu_1'\rho(C_S^2+1), \qquad (5.2.41)$$

and

$$(1-\rho)E(W_A^2) = (1-\rho)E(\tilde{W}^2) \qquad (5.2.42)$$

$$= \frac{1}{6}\mu_1'^2\,[2\rho D_S^3 + 3\rho^2(1-\rho)^{-1}(C_S^2+1)^2 + 6\rho C_S^2 + 2\rho].$$

Remark *b*. For the $PH/G/1$ queue, the moments of the idle time are explicitly known in terms of the matrix G. They are given by

$$E(I^n) = (-1)^n n!\,\boldsymbol{\alpha} G T^{-n}\mathbf{e}, \qquad (5.2.43)$$

since the idle time I has the $PH-$ distribution with representation $(\boldsymbol{\alpha} G, T)$.

5.3. EXAMPLES OF OTHER MARKOVIAN ARRIVAL PROCESSES

In this section, we develop a number of examples of arrival processes generated by means of continuous-parameter Markov processes. These will serve in the first place to motivate the study of the *versatile Markovian arrival process* to be defined in the next section. Several of the point processes, used here as examples, have been discussed separately in connection with varied applications of stochastic models. In discussing them, we shall establish some particular properties of independent interest.

A. **Arrivals at the Transition Epochs of a Markov Process.**

Consider an m-state Markov process with the irreducible infinitesimal generator Q. We obtain a number of point processes by allowing arrivals to occur whenever the Markov process makes transitions into one or more specified states. In some cases, we allow group arrivals and the probability densities of the group sizes may then depend on the specific transitions occurring in the underlying Markov process Q.

We first consider the case where a single arrival occurs whenever the Markov process enters a specified state i, $1 \leqslant i \leqslant m$.

Theorem 5.3.1: The point process obtained by placing a single arrival at each visit to a state i is a PH-renewal process.

Proof: It is well-known that the successive visits to a state i in an irreducible Markov process form a renewal process. To show that it is a PH-renewal process, we may choose $i = 1$. We may then write the matrix Q as

$$Q = \left| \begin{matrix} -a & a\boldsymbol{\beta} \\ \mathbf{S}^{\circ} & S \end{matrix} \right| ,$$

where $a > 0$ and $\boldsymbol{\beta}$ is a probability vector. Without essential loss of generality, we also start the Markov process Q in the state 1. The Laplace-Stieltjes transform of the distribution of the time between visits to state 1 is then given by

$$f(s) = \frac{a}{s+a}\, \boldsymbol{\beta}(sI - S)^{-1}\mathbf{S}^{\circ}, \tag{5.3.1}$$

which is also the Laplace-Stieltjes transform of the PH-distribution $F(\cdot)$ with representation $\boldsymbol{\alpha} = (1, 0, \ldots, 0)$ and

$$T = \begin{vmatrix} -a & a\beta \\ 0 & S \end{vmatrix},$$

We note that $T + \mathbf{T}^{\circ}\alpha = Q$. In order to construct the stationary version of the PH-renewal process, it therefore suffices to choose the initial state of the Markov process according to the stationary probability vector π of the matrix Q. Other choices of the initial conditions for the Markov process result in PH-renewal processes with different PH-distributions for the time until the first arrival.

The point process obtained by placing single arrivals at the visits to states in a set $E \subseteq \{1,2, \ldots, m\}$, where E contains more than one state is generally not a renewal process. It is a rather particular *semi-Markovian* arrival process. To see this, we partition the matrix Q, possibly after relabeling of the states, as

$$Q = \begin{vmatrix} Q(1,1) & Q(1,2) \\ Q(2,1) & Q(2,2) \end{vmatrix},$$

where the row and column indices of $Q(1,1)$ correspond to the $m_1 < m$ states in the set E. The case $m_1 = m$ is trivial.

Without loss of generality, we set $J_0 = i \in E$, and $X_0 = \tau_0 = 0$. For $n \geqslant 1$, the τ_n are the times of the successive arrivals and $X_n = \tau_n - \tau_{n-1}$. We define the random variables J_n, $n \geqslant 1$, with values in $\{1, \ldots, m_1\}$ to be the states of the Markov chain Q at the epochs $\tau_n +$.

It is an easy consequence of the Markov property that the sequence $\{(J_n, X_n), \ n \geqslant 0\}$ is a Markov renewal sequence. The elements $f_{ii'}(s)$, $1 \leqslant i, i' \leqslant m_1$, of the Laplace-Stieltjes transform of its transition probability matrix are given by

$$f_{ii'}(s) = (1 - \delta_{ii'}) [s - Q_{ii}(1,1)]^{-1} Q_{ii'}(1,1)$$

$$+ [s - Q_{ii}(1,1)]^{-1} \{Q(1,2) [sI - Q(2,2)]^{-1} Q(2,1)\}_{ii'}.$$

It is easily verified that all the functions $[f_{ii'}(0)]^{-1} f_{ii'}(s)$,

$1 \leqslant i, i' \leqslant m_1$, are Laplace-Stieltjes transforms of $PH-$distributions.

A further generalization of the preceding point process is obtained by placing arrivals at those epochs where transitions take place from a state i to a state j, where the pair (i,j), $i \neq j$, belongs to a specified subset E_1 of the set $\{1, \ldots, m\} \times \{1, \ldots, m\}$. This case may formally be reduced to the preceding one by enlarging the state space of the underlying Markov chain. As we shall see in Section 5.4, there is no need to do so explicitly, at least not for algorithmic analyses. We describe a specific model where such a point process arises.

The queueing model $M/M/c/c+K$, the $M/M/c$ queue with K waiting spaces is described by the Markov process with generator Q of order $m = c+K+1$, with the states $0,1, \ldots, c+K$. The matrix Q has elements

$$Q_{i,i+1} = \lambda, \qquad \text{for } 0 \leqslant i \leqslant c+K-1,$$

$$Q_{i,i-1} = \mu \min(i,c), \qquad \text{for } 1 \leqslant i \leqslant c+K,$$

$$Q_{i,i} = -Q_{i,i-1} - Q_{i,i+1},$$

$$Q_{i,j} = 0, \qquad \text{for } |i-j| > 1.$$

The output process of the $M/M/c/c+K$ queue is obtained by ascribing an "event", in this case a departure, to the epochs of transitions from i to $i-1$, for $1 \leqslant i \leqslant c+K$, and only to those epochs. For example, in the analysis of a tandem queue, of which the first unit is an $M/M/c/c+K$ queue, the arrival process to the second queue may be described in this manner. The tandem queue is thereby replaced, at least conceptually, by a single queue with a somewhat complex arrival process. Constructions of this type are frequently used in the approximate analysis of networks of queues.

It is clear that, in the models already described, all single arrivals may be replaced by group arrivals. In doing so, it is assumed that the sizes of the successive groups depend only on the specific transition $i \rightarrow j$, which occurs in the Markov process Q at the arrival epoch and are conditionally independent. Under this proviso, which is stated formally in Section 5.4, we obtain a versatile class of bunched arrival processes in which arrivals are triggered by transitions in an underlying

Markov process Q.

B. The Markov-Modulated Poisson Process.

A *doubly stochastic Poisson process* is a generalization of the Poisson process in which the rate $\lambda^*(t)$ is a nonnegative-valued stochastic process. The *Markov-modulated Poisson process* is the elementary, but useful case of a doubly stochastic Poisson process in which the arrival rate is given by $\lambda^*[J(t)]$, where $J(t)$, $t \geqslant 0$, is an m-state irreducible Markov process. The arrival rate can therefore take on only m values $\lambda_1, \ldots, \lambda_m$. It is equal to λ_j whenever the Markov process is in the state j, $1 \leqslant j \leqslant m$. The Markov-modulated Poisson process is fully parametrized by specifying the initial probability vector α and the infinitesimal generator Q of the (modulating) Markov process and the vector $\lambda = (\lambda_1, \ldots, \lambda_m)$ of arrival rates. It is notationally useful to define the diagonal matrix Λ with $\Lambda_{jj} = \lambda_j$, $1 \leqslant j \leqslant m$. For brevity, we shall often refer to the Markov-modulated Poisson process with generator Q and rate vector λ as a (Q,Λ)*-source*. The most useful choices of the initial probability vector α will be discussed in the sequel.

Let us now consider the epochs of successive arrivals in a Markov-modulated Poisson process and assume (without essential loss of generality) that $t = 0$ is an arrival epoch. We denote by J_n, $n \geqslant 0$, the states of the underlying Markov process at the times of the successive arrivals (J_0 is the state at $t = 0$). We set $X_0 = 0$, and let X_n, $n \geqslant 1$, be the time between the $(n-1)$st and the n-th arrivals. In the following theorem, the Markov-modulated Poisson process is shown to be a *semi-Markovian* arrival process of a particular type.

Theorem 5.3.2: The sequence $\{(J_n, X_n), n \geqslant 0\}$ is a Markov renewal sequence with the transition probability matrix

$$F(x) = \int_0^x \exp[(Q - \Lambda)u]\,du\,\Lambda \tag{5.3.2}$$

$$= \{I - \exp[(Q - \Lambda)x]\}(\Lambda - Q)^{-1}\Lambda = F(\infty) - \exp[(Q - \Lambda)x]F(\infty).$$

Proof: Let $P_{jj'}(t)$ be the conditional probability that the Markov process Q is in the state j' at time t and that no arrivals occur in $(0,t)$, given that the state at $t = 0$ is j. The probabilities $P_{jj'}(t)$ then satisfy the Chapman-Kolmogorov equations

$$P_{jj'}'(t) = \sum_{\nu=1}^{m} P_{j\nu}(t)Q_{\nu j'} - P_{jj'}(t)\lambda_{j'},$$

for $t \geqslant 0$, $1 \leqslant j, j' \leqslant m$, with the initial conditions $P_{jj'}(0) = \delta_{jj'}$. This system may be written in matrix notation as

$$P'(t) = P(t)(Q - \Lambda), \qquad P(0) = I,$$

and therefore has the solution

$$P(t) = \exp[(Q - \Lambda)t], \quad \text{for } t \geqslant 0.$$

By the Markov property, the elementary conditional probability that the first n arrivals occur respectively at the times u_1, $u_1 + u_2$, $\ldots$, $u_1 + \cdots + u_n$, given the initial state j, is given by

$$P_{jj_1}(u_1)\lambda_{j_1}P_{j_1j_2}(u_2)\lambda_{j_2} \cdots P_{j_{n-1}j_n}(u_n)\lambda_{j_n}du_1 \cdots du_n.$$

This readily implies that the sequence $\{(J_n, X_n), n \geqslant 0\}$ satisfies the defining properties of a Markov renewal sequence, whose transition probability matrix $F(\cdot)$ is given by

$$F(x) = \int_0^x \exp[(Q - \Lambda)u]du \ \Lambda.$$

The alternative expressions in (5.3.2) follow readily from the nonsingularity of the matrix $Q - \Lambda$, which is proved by the same argument by contradiction as used in the proof of Theorem 5.1.3. The uninteresting trivial case $\Lambda = 0$, is of course to be excluded. •

The matrix $F(\infty) = (\Lambda - Q)^{-1}\Lambda$, is clearly stochastic. When all λ_j are positive, it is also irreducible. Let π be the stationary probability vector of Q. The vector $\tilde{\pi}$ is defined by

$$\tilde{\pi} = (\pi\lambda)^{-1}\pi\Lambda, \tag{5.3.3}$$

and is clearly a probability vector. The component $\tilde{\pi}_j$ is positive if and only if $\lambda_j > 0$. One may also routinely verify that

$$\tilde{\boldsymbol{\pi}} = \tilde{\boldsymbol{\pi}} F(\infty). \tag{5.3.4}$$

If the time epoch $t = 0$ is chosen to correspond to an arrival epoch and if $\tilde{\boldsymbol{\pi}}$ is the initial probability vector of the Markov process Q, we obtain the Markov-modulated Poisson process started at an "arbitrary" arrival epoch. This version of the point process is said to be *interval-stationary.*

The *environment-stationary* version of the Markov-modulated Poisson process is obtained by choosing $\boldsymbol{\pi}$ as the initial probability vector of the Markov process Q. In the description as a Markov renewal sequence, this corresponds to setting $P\{J_0 = j\} = \pi_j$, for $1 \leqslant j \leqslant m$ and by defining the matrix with elements

$$P\{X_1 \leqslant x, J_1 = j' \mid J_0 = j\}$$

to be equal to $F(x)$. The origin of time is now not an arrival epoch, but is chosen so that the environmental Markov process Q is stationary.

Finally, we may construct the *time-stationary* version of the Markov-modulated Poisson process by choosing the initial conditions of the Markov renewal sequence so as to obtain its stationary version. This is a classical construction, which we carry out first for the case where all λ_j are positive.

The Laplace-Stieltjes transform $f(s)$ of the matrix $F(\cdot)$ is given by

$$f(s) = (sI + \Lambda - Q)^{-1}\Lambda, \tag{5.3.5}$$

so that the vector $\boldsymbol{\phi}^* = -f'(0+)\mathbf{e}$, of the row sum means is given by

$$\boldsymbol{\phi}^* = (\Lambda - Q)^{-2}\Lambda\mathbf{e} = (\Lambda - Q)^{-1}\mathbf{e}.$$

We clearly have that

$$\tilde{\boldsymbol{\pi}}\boldsymbol{\phi}^* = (\boldsymbol{\pi}\boldsymbol{\lambda})^{-1}\boldsymbol{\pi}\Lambda(\Lambda - Q)^{-1}\mathbf{e} = (\boldsymbol{\pi}\boldsymbol{\lambda})^{-1}.$$

To obtain the stationary version of the Markov renewal sequence, we define $\psi_j = P\{J_0 = j\}$ and $F^*_{jj'}(x) = P\{X_1 \leq x, J_1 = j' \mid J_0 = j\}$, for $1 \leq j, j' \leq m$, $x \geq 0$, respectively by

$$\psi_j = (\tilde{\pi}\phi^*)^{-1}\tilde{\pi}_j\phi_j^* = \pi_j\lambda_j\phi_j^*,$$

and

$$F^*_{jj'}(x) = \phi_j^{*\,-1}\int_0^x [F_{jj'}(\infty) - F_{jj'}(u)]du.$$

Writing $\Delta(\phi^*)$ for the matrix $diag\{\phi_j^*, 1 \leq j \leq m\}$, these formulas may, by virtue of (5.3.2), be written as

$$\psi = \pi\Lambda\Delta(\phi^*), \tag{5.3.6}$$

and

$$F^*(x) = \Delta^{-1}(\phi^*)\int_0^x [F(\infty) - F(u)]du \tag{5.3.7}$$

$$= \Delta^{-1}(\phi^*)(\Lambda - Q)^{-1}F(x).$$

When some, but not all, of the rates λ_j are zero, the matrix $F(x)$ has zero columns corresponding to those (and only those) indices j for which $\lambda_j = 0$. By virtue of the irreducibility of Q, the vector ϕ^* remains strictly positive. We notice that the zero components of the vector ψ in (5.3.6) also correspond exactly to those indices j for which λ_j vanishes. In constructing the stationary version of the Markov renewal process when some λ_j are zero, we may either delete the corresponding states (which are never visited) or we may preserve the notation of the formulas (5.3.6) and (5.3.7) with the observation that the states j for which $\lambda_j = 0$, are redundant. This second alternative is clearly notationally more convenient and will therefore be chosen.

The following theorem shows that the point processes of the environment and time stationary versions are stochastically equivalent.

Theorem 5.3.3: For every $n \geqslant 1$, the joint distributions of the random variables $X_1, \ldots, X_n, J_n$, for the environment and time stationary versions agree and are given by

$$P\{X_1 \leqslant x_1, \ldots, X_n \leqslant x_n, J_n = j\} = \{\pi F(x_1) \cdots F(x_n)\}_j, \quad (5.3.8)$$

for $x_1 \geqslant 0, \ldots, x_n \geqslant 0, 1 \leqslant j \leqslant m$.

Proof: The expression in the right hand of (5.3.8) is clearly valid for the environment-stationary version. For the time-stationary version the corresponding probability is given by the j-th component of the vector

$$\psi F^*(x_1)F(x_2) \cdots F(x_n),$$

but by virtue of (5.3.6) and (5.3.7) and after obvious simplifications, this agrees with the stated expression. •

Theorem 5.3.3 has the important consequence that any features of the Markov-modulated Poisson process, which involve only the joint distributions of the interarrival intervals, may be studied by using the convenient formalism for the environment-stationary version.

A Markov-modulated Poisson process is only in very special cases a *renewal process*. We briefly discuss these cases which are fully characterized by more general results of Kingman for doubly stochastic Poisson processes.

Kingman has established that a stationary doubly stochastic Poisson process is a renewal process if and only if its rate function $\lambda^*(\cdot)$ takes on the values $\lambda > 0$ and zero alternatingly on the successive intervals of a stationary *alternating renewal process,* in which in addition the intervals where the rate λ prevails have an *exponential* distribution (say, with parameter θ). Doubly stochastic Poisson processes which are also renewal processes, are therefore characterized by two positive parameters λ and θ and by a general probability distribution $H(\cdot)$ of finite mean μ_1' on $(0,\infty)$. The environment process is the stationary alternating renewal process with underlying distributions $H(\cdot)$ and $1 - \exp(-\theta x)$. Its initial conditions are therefore as follows. The time origin falls with probability $p = (1 + \theta \mu_1')^{-1}$ during an interval where the rate λ prevails and with probability $1 - p$ during an interval where the rate is zero. The (conditional) distribution of the first interval is exponential in the first case and is given by

$$\mu_1'^{-1} \int_0^x [1 - H(u)]du ,$$

in the second case. The distributions of the intervals after the first are alternatingly $H(\cdot)$ and $1 - \exp(-\theta x)$.

Let the probability distribution, underlying the renewal process so obtained, be denoted by $A(\cdot)$ and its mean by α_1'. The following result expresses $A(\cdot)$ in terms of the parameters λ, θ and $H(\cdot)$ of the special doubly stochastic Poisson process.

Theorem 5.3.4: The probability distribution $A(\cdot)$ is the unique solution to the Volterra integral equation

$$A(x) = \lambda(\lambda + \theta)^{-1}[1 - e^{-(\lambda+\theta)x}] \tag{5.3.9}$$

$$+ \theta \int_0^x A(u) \int_0^{x-u} e^{-(\lambda+\theta)(x-u-v)}\, dH(v)\,du ,$$

for $x \geqslant 0$. Its mean is given by $\alpha_1' = \lambda^{-1}(1 + \theta\mu_1')$. The probability distribution $A(\cdot)$ is absolutely continuous on $[0,\infty)$.

Proof: It is most expeditious to calculate the Laplace-Stieltjes transform $a^*(s)$ of the distribution of the time until the first arrival in the doubly stochastic Poisson process. $a^*(s)$ is related to the Laplace-Stieltjes transform $a(s)$ of $A(\cdot)$ by $a^*(s) = (s\alpha_1')^{-1}[1 - a(s)]$.

By considering the interval during which the first arrival occurs and applying the law of total probability, we obtain that

$$a^*(s) = (1 + \theta\mu_1')^{-1} \sum_{\nu=0}^{\infty} \left[\frac{\theta h(s)}{s+\lambda+\theta}\right]^{\nu} \frac{\lambda}{s+\lambda+\theta}$$

$$+ \theta\mu_1'(1 + \theta\mu_1')^{-1} \frac{1 - h(s)}{s\mu_1'} \sum_{\nu=0}^{\infty} \left[\frac{\theta h(s)}{s+\lambda+\theta}\right]^{\nu} \frac{\lambda}{s+\lambda+\theta}$$

$$= (1 + \theta\mu_1')^{-1} \frac{s+\lambda+\theta-\theta h(s)}{s+\theta-\theta h(s)} \frac{1}{s}.$$

Upon equating the two expressions for $sa^*(s)$ and letting $s \to \infty$ in the resulting equation, we obtain that $\alpha_1' = \lambda^{-1}(1 + \theta \mu_1')$, and also that

$$a(s) = \frac{\lambda}{s+\lambda+\theta} + \frac{\theta}{s+\lambda+\theta} h(s)a(s),$$

which is the transform version of the integral equation (5.3.9). Since $a(s)$ may also be written as

$$a(s) = \frac{\lambda}{s+\lambda+\theta}\left[1 - \frac{\theta}{s+\lambda+\theta} h(s)\right]^{-1},$$

we see that $A(\cdot)$ is the convolution of two mass-functions, one of which is absolutely continuous on $[0,\infty)$. It is therefore clear that $A(\cdot)$ also has a density. •

By virtue of the preceding discussion, every Markov-modulated Poisson process, for which exactly one of the rates λ_j is positive, is a renewal process. For definiteness, let $\lambda_1 = \lambda$ be positive and $\lambda_j = 0$, for $2 \leqslant j \leqslant m$. We also write the matrix Q as

$$Q = \begin{vmatrix} -\theta & \theta \beta \\ \mathbf{S}^\circ & S \end{vmatrix},$$

It is then easy to see that the probability distribution $A(\cdot)$ is the PH-distribution with the representation $\gamma = (1,0)$ and

$$L = \begin{vmatrix} -\theta - \lambda & \theta \beta \\ \mathbf{S}^\circ & S \end{vmatrix},$$

It is readily verified that $H(\cdot)$ is then the PH-distribution with representation (β, S). The probability distribution $A(\cdot)$ may be computed from the expression

$$A(x) = 1 - v_0(x) - \mathbf{v}(x)\mathbf{e}, \quad \text{for } x \geqslant 0,$$

where the vector $[v_0(x), \mathbf{v}(x)]$ is the solution to the differential equations

$$v_0'(x) = -v_0(x)(\theta + \lambda) + \mathbf{v}(x)\mathbf{S}^\circ,$$

$$\mathbf{v}'(x) = \theta\, v_0(x)\beta + \mathbf{v}(x)S,$$

for $x \geqslant 0$, with initial conditions $v_0(0) = 1$, $\mathbf{v}(0) = 0$.

A particularly simple case arises when $S = -\sigma$. This corresponds to the *interrupted Poisson process,* in which the arrival rate is λ or 0 on alternating independent intervals of exponential distributions with parameters θ and σ. The transform $a(s)$ is then given by

$$a(s) = \frac{\lambda(s+\sigma)}{s^2+(\lambda+\theta+\sigma)s+\lambda\sigma} = p\,\frac{\eta_1}{s+\eta_1} + (1-p)\,\frac{\eta_2}{s+\eta_2},$$

where $\eta_1 > \lambda > \eta_2 > 0$, and $p = (\lambda-\eta_2)(\eta_1-\eta_2)^{-1}$. The quantities $-\eta_1$ and $-\eta_2$ are the zeros of the denominator. We see that $A(\cdot)$ is the *hyperexponential distribution* with parameters η_1, η_2 and p.

Conversely, we see for any bona fide hyperexponential distribution with parameters η_1, η_2 and p, the quantities

$$\lambda = p\eta_1 + (1-p)\eta_2,$$

$$\theta = p(1-p)(\eta_1-\eta_2)^2\,[p\eta_1 + (1-p)\eta_2]^{-1},$$

$$\sigma = \eta_1\eta_2\,[p\eta_1 + (1-p)\eta_2]^{-1},$$

are valid parameters for an interrupted Poisson process whose distribution $A(\cdot)$ is the given hyperexponential distribution. It follows that the interrupted Poisson and the hyperexponential renewal processes are stochastically equivalent point processes.

It is possible to give examples of Markov-modulated Poisson processes for which $\lambda_j = \lambda > 0$, for $j \in E$, and $\lambda_j = 0$, for $j \in E^c$, which are renewal processes and in which the E and E^c are non-empty subsets of the state space $\{1, \ldots, m\}$ of Q. In addition, the set E

contains *more than one* state. In constructing such examples, the matrix Q must be chosen in such a manner that in the time-stationary version the successive sojourn times in the sets E and E^c are mutually independent and that the sojourn time distribution in the set E is exponential. While it is easy to write these conditions formally, they cannot, in general, be simplified to characterize the corresponding matrix Q in a tractable manner.

The following theorem shows that the *superposition* of independent Markov-modulated Poisson processes is itself a Markov-modulated Poisson process.

Theorem 5.3.5: The superposition of $N \geqslant 2$ independent Markov-modulated Poisson processes with parameters $Q(\nu)$, $\Lambda(\nu)$, $1 \leqslant \nu \leqslant N$, is itself a (Q, Λ)–source with Q and Λ given by the Kronecker sums

$$Q = Q(1) \oplus Q(2) \oplus \cdots \oplus Q(N),$$

and

$$\Lambda = \Lambda(1) \oplus \Lambda(2) \oplus \cdots \oplus \Lambda(N).$$

Proof: We recall that if $Q(1), \ldots, Q(N)$ are square matrices of orders $m(1), \ldots, m(N)$ respectively, their Kronecker sum is the matrix of order $m(1)m(2) \cdots m(N)$, given by

$$Q(1) \otimes I_{m(2) \cdots m(N)} + I_{m(1)} \otimes Q(2) \otimes I_{m(3) \cdots m(N)}$$
$$+ \cdots + I_{m(1) \cdots m(N-1)} \otimes Q(N).$$

It suffices to prove the theorem for $N = 2$. The product of the independent Markov processes $Q(1)$ and $Q(2)$ is itself a Markov process on the state space $\{1, \ldots, m(1)\} \times \{1, \ldots, m(2)\}$. With the states of the product listed in lexicographic order, its infinitesimal generator Q is given by

$$Q = Q(1) \otimes I_{m(2)} + I_{m(1)} \otimes Q(2) = Q(1) \oplus Q(2).$$

Now consider any sojourn time in the state (j_1,j_2) of the product process. During such an interval, the arrivals in the two component processes are independent Poisson processes of rates $\lambda_{j_1}(1)$ and $\lambda_{j_2}(2)$ respectively. Their superposition is therefore a Poisson process of rate $\lambda_{j_1}(1) + \lambda_{j_2}(2)$. The fact, that the successive sojourn times in a Markov process are exponentially distributed, allows us to treat the successive sojourn times as conditionally independent and to appeal to the superposition theorem of Poisson processes separately for each pair (j_1,j_2).

It is now clear that the superposition is itself a Markov-modulated Poisson process on the Markov chain Q and with

$$\Lambda = \Lambda(1) \otimes I_{m(2)} + I_{m(1)} \otimes \Lambda(2) = \Lambda(1) \oplus \Lambda(2). \qquad \bullet$$

Remark: The rapid increase in the dimension of the matrix Q severely limits the utility of Theorem 5.3.5 for practical modelling, although it has a number of theoretical applications. When the superimposed processes have the same parameters, the superposition can be described with matrices Q and Λ of a smaller dimension, but with a more involved analytic description. We shall not pursue this elementary matter further, but present only the following example which is useful in communications modelling.

The superposition of $N \geqslant 2$ independent (Q,Λ)-sources with identical parameters

$$Q = \begin{vmatrix} -\sigma_1 & \sigma_1 \\ \sigma_2 & -\sigma_2 \end{vmatrix}, \qquad \Lambda = \begin{vmatrix} \lambda_1 & 0 \\ 0 & \lambda_2 \end{vmatrix},$$

may be described as the (Q_N,Λ_N)-source where Q_N and Λ_N are matrices of dimension $N+1$ given by

$$(Q_N)_{i,i} = -i\sigma_1 - (N-i)\sigma_2, \qquad \text{for } 0 \leqslant i \leqslant N,$$

$$(Q_N)_{i,i-1} = i\sigma_1, \qquad \text{for } 1 \leqslant i \leqslant N,$$

$$(Q_N)_{i,i+1} = (N-i)\sigma_2, \qquad \text{for } 0 \leqslant i \leqslant N-1,$$

the other elements are zero, and

$$\Lambda = diag\{\, i\lambda_1 + (N-i)\lambda_2, \quad 0 \leqslant i \leqslant N \,\}.$$

5.4. THE VERSATILE MARKOVIAN POINT PROCESS

The point processes, already described in this chapter, are particular cases of a *versatile Markovian point process* to be defined in this section. In its most general form, the process has a very large number of parameters. For the cases arising in typical applications, most of these parameters take on particular simple values, which lead to useful simplifications of details in the algorithmic analysis. Conceptualizing the matrix-analytic treatment of the most general case is not more complicated than that of most of its special cases. The versatile Markovian point process may therefore be used to give a *unified treatment* of queueing models with a wide variety of arrival processes. Many such models have been treated by ad hoc methods in the extensive literature on queues.

We start by considering a PH-renewal process for which the interrenewal time distribution $F(\cdot)$ has the irreducible representation (α, T) of dimension m. We assume that $\alpha_{m+1} = 0$, and we construct the irreducible m-state Markov process with the irreducible generator $Q^* = T + \mathbf{T}^\circ\alpha$, as in Section 5.1.

Next, the Markov process Q^* is used as the environment process for a Markov-modulated Poisson process with rates $\lambda_1, \ldots, \lambda_m$. As in Section 5.3, $\Lambda = diag\ (\lambda_1, \ldots, \lambda_m)$. In the (Q^*, Λ)-source, there are now three types of "events." These are:

a. The Markov-modulated Poisson arrivals, which are of rate λ_j during sojourns in the phase j.

b. The transitions from a phase j to a phase j', $j' \neq j$, which do not occur at a renewal epoch of the PH-renewal process. The elementary probability of such a transition in $(t, t+dt)$ is equal to $T_{jj'}dt$, for $j \neq j'$. Such a transition is called a (j,j')-*transition.*

c. The transitions from a phase j to a phase j' at a renewal in the PH-renewal process. Such transitions involve an instantaneous absorption and restarting of the Markov process and their elementary probability of occurrence in $(t, t+dt)$ is $T_j^o \alpha_{j'} dt$. For such transitions, it is possible that $j' = j$. For brevity, we shall refer to them as $(j, j')^*$-transitions.

We now consider the labeled point process of the times τ_n, $n \geq 1$ of the successive events and of corresponding labels L_n, $n \geq 1$. The labels L_n can take on $2m^2$ possible values, to wit, j, $1 \leq j \leq m$, (j, j'), $j \neq j'$, $1 \leq j, j' \leq m$, or $(j, j')^*$, $1 \leq j, j' \leq m$.

Finally, we associate with each event in this labeled point process a nonnegative integer ν_n which is interpreted as the *number of arrivals* at that point. We assume that for every $N \geq 2$, the group sizes ν_n, $1 \leq n \leq N$, are conditionally independent, given the random variables τ_n and L_n, $1 \leq n \leq N$. Furthermore, we assume that the (conditional) density of ν_n depends only L_n. The size of each group arrival therefore depends *only* on the type of transition which occurs at its epoch of arrival.

The conditional probability densities of the group sizes are defined as follows:

a. At an event in the Markov-modulated Poisson process, which occurs during a sojourn in the phase j, the group size density is $\{p_j(k), k \geq 0\}$. The probability generating function of $\{p_j(k)\}$ is denoted by $\phi_j(z)$ and we may assume, without loss of generality, that $\phi_j(0) = 0$, for $1 \leq j \leq m$. The matrix $\Delta[\phi(z)]$ is defined as the diagonal matrix $diag[\phi_1(z), \ldots, \phi_m(z)]$.

b. At a (j, j')–transition, there are group arrivals with probability density $\{q_{jj'}(k), k \geq 0\}$. The corresponding probability generating function is denoted by $\psi_{jj'}(z)$, for $j \neq j'$. It is notationally convenient to define $\psi_{jj}(z) = 1$, for $1 \leq j \leq m$ and to denote the matrix with elements $\psi_{jj'}(z)$ by $\Psi(z)$.

c. At a $(j, j')^*$-transition, the group size density is given by $\{r_{jj'}(k), k \geq 0\}$. The corresponding probability generating functions $\Phi_{jj'}(z)$, $1 \leq j, j' \leq m$, are the elements of the $m \times m$ matrix $\Phi(z)$.

The *versatile Markovian point process* is the labeled point process with group arrivals which is so obtained. It is parametrized by the representation (α, T) of $F(\cdot)$, by the matrices Λ, $\Delta[\phi(z)]$, $\Psi(z)$, $\Phi(z)$ and by the initial probability vector **a** of the Markov process Q^*.

The random variables $N(t)$ and $J(t)$, $t \geq 0$, with $N(0) = 0$, are respectively defined as the number of arrivals in $(0, t]$ in the versatile

Markovian point process and as the state of the Markov process Q^* at time t. The *counting process* $\{N(t), t \geqslant 0\}$ of the versatile Markovian point process is of basic importance to the sequel.

It is straightforward to verify that the process $\{N(t), J(t), t \geqslant 0\}$ is a Markov process with the state space $\{k \geqslant 0\} \times \{1, \ldots, m\}$. Moreover, the increments of the process $\{N(t), t \geqslant 0\}$ are conditionally independent, given the process $\{J(t), t \geqslant 0\}$. Specifically for every choice of $t_0 = 0 < t_1 < \cdots < t_n$, and $n \geqslant 2$, the random variables $N(t_1)$, $N(t_2) - N(t_1)$, . . . , $N(t_n) - N(t_{n-1})$ are conditionally independent, given the random variables $J(t_r)$, $0 \leqslant r \leqslant n$.

We shall now consider the probabilities

$$P_{jj'}(k;t) = P\{N(t) = k, J(t) = j' \mid N(0) = 0, J(0) = j\},$$

for $k \geqslant 0$, $t \geqslant 0$, $1 \leqslant j, j' \leqslant m$, and obtain a system of difference-differential equations for the matrices $P(k;t) = \{P_{jj'}(k;t)\}$. These equations generalize the equations (5.1.9) for the PH-renewal process.

By considering all possible occurrences in the interval $(t, t+dt)$, we obtain the (forward) Chapman-Kolmogorov equations

$$P_{jj'}'(k;t) = P_{jj'}(k;t)\,(T_{j'j'} - \lambda_{j'}) + \sum_{\nu=0}^{k} P_{jj'}(\nu;t)\lambda_{j'}p_{j'}(k-\nu)$$

$$+ \sum_{\substack{h=1 \\ h \neq j'}}^{m} \sum_{\nu=0}^{k} P_{jh}(\nu;t)T_{hj'}q_{hj'}(k-\nu) + \sum_{h=1}^{m} \sum_{\nu=0}^{k} P_{jh}(\nu;t)T_h^{\circ}\alpha_{j'}r_{hj'}(k-\nu),$$

for $k \geqslant 0$, $t \geqslant 0$, $1 \leqslant j, j' \leqslant m$.

In matrix notation, we have

$$P'(k;t) = -P(k;t)\Lambda + \sum_{\nu=0}^{k} P(\nu;t)\Lambda\,\Delta[\mathbf{p}(k-\nu)] \tag{5.4.1}$$

$$+ \sum_{\nu=0}^{k} P(\nu;t)[T \circ q(k-\nu)] + \sum_{\nu=0}^{k} P(\nu;t)\,[\mathbf{T}^{\circ}\boldsymbol{\alpha} \circ r(k-\nu)].$$

The matrix $\Delta[\mathbf{p}(\nu)] = diag\{p_1(\nu), \ldots, p_m(\nu)]$, and the matrices $q(\nu)$ and $r(\nu)$ have elements $q_{jj'}(\nu)$ and $r_{jj'}(\nu)$ respectively. The symbol $\circ$

denotes the *Schur* or *element-wise* product of two matrices.

From (5.4.1), it is immediate that the matrix generating function

$$P^*(z;t) = \sum_{k=0}^{\infty} P(k;t)z^k, \quad \text{for } |z| \leqslant 1,$$

satisfies the differential equation

$$\frac{\partial P^*(z,t)}{\partial t} = P^*(z;t)R(z), \qquad P^*(z;0) = I, \tag{5.4.2}$$

where

$$R(z) = \Lambda\Delta[\phi(z)] - \Lambda + T \circ \Psi(z) + \mathbf{T}^\circ\boldsymbol{\alpha} \circ \Phi(z). \tag{5.4.3}$$

It therefore follows that

$$P^*(z;t) = \exp[R(z)t], \quad \text{for } t \geqslant 0. \tag{5.4.4}$$

We see that $\Psi(1) = \Phi(1) = E$, where $E_{jj'} = 1$, so that $P^*(1;t) = \exp(Q^*t)$, as is to be expected.

The initial probability vector $\mathbf{a}$ of the Markov process Q^* may be chosen to generate particular versions of the versatile Markovian point process. This choice is usually clear in each application. The choices $\mathbf{a} = \boldsymbol{\alpha}$ and $\mathbf{a} = \boldsymbol{\pi}$ are particularly useful. In the first, the underlying $PH-$renewal process is started with a renewal at $t = 0$; the second choice yields the stationary version of the $PH-$renewal process and therefore *also* of the versatile Markovian point process.

Moments: The calculation of the (factorial) moment matrices $V_n(t)$, which are defined as in formula (5.1.12) proceeds as in Section 5.1. Successive differentiations with respect to z in (5.4.4) yield the equations

$$V_0'(t) = V_0(t)Q^*, \tag{5.4.5}$$

$$V_n'(t) = V_n(t)Q^* + \sum_{\nu=0}^{n-1} \binom{n}{\nu} V_\nu(t)R^{(n-\nu)}(1), \quad \text{for } n \geqslant 1.$$

It is clear from (5.4.3) that for $\nu \geqslant 1$, the elements of the matrices $R^{(\nu)}(1)$ are linear combinations with nonnegative coefficients of the $\nu-$th factorial moments of the various group size densities. In order for the matrix $V_n(t)$ to be finite for $t > 0$, it is therefore necessary and sufficient that the group size densities have moments of order n.

By repeating the same calculations as in Theorem 5.1.3, we obtain that for $t \geqslant 0$

$$V_1(t)\mathbf{e} = [\boldsymbol{\pi} R'(1)\mathbf{e}]t\mathbf{e} + [I - \exp(Q^*t)]\,(\mathbf{e}\boldsymbol{\pi} - Q^*)^{-1}R'(1)\mathbf{e} \qquad (5.4.6)$$

$$= [\boldsymbol{\pi} R'(1)\mathbf{e}]t\mathbf{e} + [I - \exp(Q^*t)]\,\{\,[\boldsymbol{\pi} R'(1)\mathbf{e}]T^{-1}\mathbf{e} - T^{-1}R'(1)\mathbf{e}\,\}.$$

The constant $\xi^* = \boldsymbol{\pi} R'(1)\mathbf{e}$, is called the *fundamental rate* of the versatile Markovian point process. It plays an important role in all applications. Since

$$\boldsymbol{\pi} V_1(t)\mathbf{e} = \xi^* t, \quad \text{for } t \geqslant 0,$$

we see that ξ^* is the *expected number of arrivals per unit of time* in the stationary version of the versatile Markovian point process. By recalling from formula (5.1.7) that $\boldsymbol{\pi} = \lambda_1'^{-1}\boldsymbol{\alpha}(-T)^{-1}$, we may also write ξ^* as

$$\xi^* = \lambda_1'^{-1}\,\boldsymbol{\alpha}(-T^{-1})R'(1)\mathbf{e}, \qquad (5.4.7)$$

and it is easily seen that $-\boldsymbol{\alpha} T^{-1}R'(1)\mathbf{e}$ is the expected number of arrivals during an interrenewal interval of the underlying $PH-$renewal process. The quantity ξ^* is therefore also the *ratio of the number of arrivals during an interval of the PH−renewal process* and the *expected length of such an interval.*

Since $\exp(Q^*t) \to \mathbf{e}\boldsymbol{\pi}$, as $t \to \infty$, we readily obtain from (5.4.6) the following asymptotic formula:

$$V_1(t)\mathbf{e} = \xi^* t\mathbf{e} + \{\,\xi^* T^{-1}\mathbf{e} - T^{-1}R'(1)\mathbf{e} \qquad (5.4.8)$$

$$+ \frac{\lambda_2'}{2\lambda_1'}\xi^*\mathbf{e} - [\boldsymbol{\pi} T^{-1}R'(1)\mathbf{e}]\,\mathbf{e}\} + o(1),$$

as $t \to \infty$. Premultiplying by α and using (5.4.7) , we obtain

$$\alpha V_1(t)\mathbf{e} = \xi^* t + \frac{\lambda_2'}{2\lambda_1'}\xi^* - [\pi T^{-1}R'(1)\mathbf{e}] + o(1)$$

$$= \xi^* t + \frac{\lambda_2'}{2\lambda_1'}\xi^* \mathbf{e} + \frac{1}{\lambda_1'}\alpha T^{-2}R'(1)\mathbf{e} + o(1),$$

as $t \to \infty$. This is the generalization to the counting process $N(t)$ of the familiar linear asymptote for the renewal function of an ordinary PH-renewal process. Similar explicit formulas may be obtained for the second moments, but their derivations and the resulting analytic expressions are considerably more complicated. We shall only derive the analytic form of the *variance* of $N(t)$ for the stationary version of the versatile Markovian point process.

Theorem 5.4.1: For the stationary version of the versatile Markovian point process, the variance $\sigma^2(t) = E[N(t) - \xi^* t]^2$, of the number $N(t)$ of arrivals in $(0,t]$ is given by

$$\sigma^2(t) = [\ \xi_2^* + \xi^* - 2\xi^{*2} + 2\pi R'(1)(\mathbf{e}\pi - Q^*)^{-1}R'(1)\mathbf{e}\ ]t \tag{5.4.9}$$

$$- 2\pi R'(1)[I - \exp(Q^* t)]\ (\mathbf{e}\pi - Q^*)^{-2}R'(1)\mathbf{e}$$

$$= [\ \xi_2^* + \xi^* + \xi^{*2}\frac{\lambda_2'}{\lambda_1'} + 2\xi^* \pi R'(1)T^{-1}\mathbf{e}$$

$$- 2\pi R'(1)T^{-1}R'(1)\mathbf{e} + 2\xi^* \pi T^{-1}R'(1)\mathbf{e}\ ]t$$

$$+ 2\pi R'(1)[I - \exp(Q^* t)]\ \{\ [\pi T^{-1}R'(1)\mathbf{e} + \xi^* \frac{\lambda_2'}{2\lambda_1'}]T^{-1}\mathbf{e}$$

$$+ \xi^* T^{-2}\mathbf{e} - T^{-2}R'(1)\mathbf{e}\ \},$$

where $\xi_2^* = \pi R''(1)\mathbf{e}$. λ_1' and λ_2' are the first two moments of the interrenewal time distribution of the PH-renewal process.

Proof: It is clear that

$$\sigma^2(t) = \pi V_2(t)\mathbf{e} + \xi^* t - \xi^{*2} t^2, \quad \text{for } t \geqslant 0,$$

so that it suffices to calculate $\pi V_2(t)\mathbf{e}$.

We first observe that the differential equation

$$V_1'(t) = V_1(t)Q^* + \exp(Q^* t), \quad \text{for } t \geqslant 0,$$

with $V_1(0) = 0$, leads after routine manipulations to the integral formula

$$V_1(t) = \int_0^t \exp[Q^*(t-u)]\, R'(1) \exp(Q^* u)\, du. \tag{5.4.10}$$

The differential equation for $n = 2$ in (5.4.5) yields, upon premultiplication by π and postmultiplication by $\mathbf{e}$ that

$$\pi V_2'(t)\mathbf{e} = 2\pi V_1(t) R'(1)\mathbf{e} + \pi R''(1)\mathbf{e} \tag{5.4.11}$$

$$= \xi_2^* + 2\pi R'(1) \int_0^t \exp(Q^* u)\, du \;\; R'(1)\mathbf{e},$$

by virtue of (5.4.10). By using the integration formula (5.1.15), it follows that

$$\pi V_2'(t)\mathbf{e} = \xi_2^* + 2\xi^{*2} t + 2\pi R'(1)[I - \exp(Q^* t)]\, (\mathbf{e}\pi - Q^*)^{-1} R'(1)\mathbf{e}.$$

Upon integrating both sides of this equality between 0 and t, and using formula (5.1.15) once more, we get

$$\pi V_2(t)\mathbf{e} = \xi_2^* t + \xi^{*2} t^2 - 2\xi^{*2} t + 2t\,\pi R'(1)(\mathbf{e}\pi - Q^*)^{-1} R'(1)\mathbf{e}$$

$$- 2\pi R'(1)[I - \exp(Q^* t)] (\mathbf{e}\pi - Q^*)^{-2} R'(1)\mathbf{e},$$

from which the first expression in (5.4.9) is obtained.

The second expression is obtained by using the particular form $T + \mathbf{T}^\circ \alpha$ of the matrix Q^* to solve the system of linear equations

$$(\mathbf{e}\pi - T - \mathbf{T}^\circ \alpha)\mathbf{v} = R'(1)\mathbf{e}, \tag{5.4.12}$$

$$(\mathbf{e}\pi - T - \mathbf{T}^\circ \alpha)\mathbf{w} = \mathbf{v},$$

whose solutions are the vectors

$$\mathbf{v} = (\mathbf{e}\pi - Q^*)^{-1} R'(1)\mathbf{e}, \quad \text{and} \quad \mathbf{w} = (\mathbf{e}\pi - Q^*)^{-2} R'(1)\mathbf{e}.$$

It is clear that $\pi\mathbf{v} = \pi\mathbf{w} = \xi^*$, and since T is nonsingular, it follows that

$$\mathbf{v} = \xi^* T^{-1}\mathbf{e} - T^{-1} R'(1)\mathbf{e} + (\alpha\mathbf{v})\mathbf{e},$$

$$\mathbf{w} = \xi^* T^{-1}\mathbf{e} - T^{-1}\mathbf{v} + (\alpha\mathbf{w})\mathbf{e}$$

$$= \xi^* T^{-1}\mathbf{e} - \xi^* T^{-2}\mathbf{e} + T^{-2} R'(1)\mathbf{e} - (\alpha\mathbf{v}) T^{-1}\mathbf{e} + (\alpha\mathbf{w})\mathbf{e}.$$

Premultiplication by π in the expression for $\mathbf{v}$ yields that

$$\alpha\mathbf{v} = \xi^* + \xi^* \frac{\lambda_2'}{2\lambda_1'} + \pi T^{-1} R'(1)\mathbf{e}.$$

Substitution of $\mathbf{w}$ shows that the term corresponding to $(\alpha\mathbf{w})\mathbf{e}$ cancels, so that we do not need the inner product $\alpha\mathbf{w}$ explicitly. The second expression for $\sigma^2(t)$ in (5.4.9) is now obtained by straightforward but laborious substitutions and simplifications. •

Remarks: The second form of the variance $\sigma^2(t)$ is particularly useful if we wish to bring out the role of the underlying PH-renewal process. In other cases, where the PH-renewal process is less significant, the first formula is more convenient. We illustrate this by some examples.

For the *PH – renewal process* itself, i.e., when arrivals occur only at renewal epochs, the matrix $R(z) = z\mathbf{T}^{\circ}\boldsymbol{\alpha}$, so that $R'(1) = \mathbf{T}^{\circ}\boldsymbol{\alpha}$, $R''(1) = 0$. The second expression for $\sigma^2(t)$ then simplifies to

$$\sigma^2(t) = \frac{\lambda_2' - \lambda_1'^2}{\lambda_1'^2}t + 2\alpha[I - \exp(Q^* t)]\,[\lambda_1'^{-2}T^{-2}\mathbf{e} + \frac{1}{2}\lambda_1'^{-3}\lambda_2' T^{-1}\mathbf{e}],$$

for $t \geqslant 0$. Since $\exp(Q^* t) \to \mathbf{e}\pi$, as $t \to \infty$, this also leads after some calculations to the well-known asymptotic formula

$$\sigma^2(t) = \frac{\sigma^2}{\lambda_1'^2}t + \frac{1}{6} + \frac{\sigma^4}{2\lambda_1'^4} - \frac{\lambda_3}{3\lambda_1'^3} + o(1), \quad \text{as } t \to \infty.$$

σ^2 and λ_3^* are respectively the variance and the third central moment of the interrenewal time distribution $F(\cdot)$.

For the *Markov-modulated Poisson process,* we may consider the successive visits to any one of the states of the Markov process Q as the epochs of the underlying PH – renewal process, although this is purely an artifact. We see that the matrix $T + \mathbf{T}^{\circ}\boldsymbol{\alpha}$ is then none other than the matrix Q, so that the matrix $R(z)$ is given by

$$R(z) = \Lambda(z - 1) + Q,$$

and therefore $R'(1) = \Lambda$, $R''(1) = 0$.

The first expression for $\sigma^2(t)$ now reduces to

$$\sigma^2(t) = [\xi^* - 2\xi^{*\,2} + 2\pi\Lambda(\mathbf{e}\pi - Q)^{-1}\lambda]\,t$$
$$- 2\pi\Lambda[I - \exp(Qt)]\,(\mathbf{e}\pi - Q)^{-2}\lambda,$$

for $t \geqslant 0$, where $\xi^* = \pi\lambda$. As $t \to \infty$, we have that

$$\sigma^2(t) = [\xi^* - 2\xi^{*\,2} + 2\pi\Lambda(\mathbf{e}\pi - Q)^{-1}\lambda]\,t$$
$$- 2[\pi\Lambda(\mathbf{e}\pi - Q)^{-2}\lambda - \xi^{*\,2}] + o(1).$$

For $\lambda = \xi^* \mathbf{e}$, the Markov-modulated Poisson process has the same counting function $N(t)$ as a Poisson process of rate ξ^*. We see that then $\sigma^2(t) = \xi^* t$, as it should be. The asymptotic formula also suggests that the differences

$$\theta_1 = \boldsymbol{\pi}\Lambda(\mathbf{e}\boldsymbol{\pi} - Q)^{-1}\boldsymbol{\lambda} - \xi^{*2}, \quad \text{and} \quad \theta_0 = \boldsymbol{\pi}\Lambda(\mathbf{e}\boldsymbol{\pi} - Q)^{-2}\boldsymbol{\lambda} - \xi^{*2},$$

may be used as measures of the "deviation" of a Markov-modulated Poisson process from the Poisson process with the same fundamental rate. For $m = 2$ and Q explicitly given by

$$Q = \begin{vmatrix} -\sigma_1 & \sigma_1 \\ \sigma_2 & -\sigma_2 \end{vmatrix},$$

the quantities θ_0 and θ_1 are expressed in terms of the parameters λ_1, λ_2, σ_1 and σ_2 by

$$\theta_1 = (\sigma_1 + \sigma_2)\,\theta_0 = (\sigma_1 + \sigma_2)^{-3}\sigma_1\sigma_2(\lambda_1 - \lambda_2)^2.$$

Theorem 5.4.2: For the stationary version of the versatile Markovian point process, the *covariance* $C(t,t';t_1)$ of the random variables $N(t)$ and $N(t+t_1+t') - N(t+t_1)$, with $t > 0$, $t_1 \geqslant 0$ and $t' > 0$, is given by

$$C(t,t';t_1) = \boldsymbol{\pi}R'(1)[I - \exp(Q^*t)]\exp(Q^*t_1) \tag{5.4.13}$$

$$\cdot [I - \exp(Q^*t')]\,(\mathbf{e}\boldsymbol{\pi} - Q^*)^{-2}R'(1)\mathbf{e}.$$

Proof: The conditional independence of the increments of the counting process $\{N(t),\ t \geqslant 0\}$ implies that for the stationary process

$$E\{z_1^{N(t)}z_2^{N(t+t_1+t') - N(t+t_1)}\} = \boldsymbol{\pi}\exp[R(z_1)t]\exp(Q^*t_1)\exp[R(z_2)t']\mathbf{e}.$$

Differentiating once each with respect to z_1 and z_2 and setting $z_1 =$

$z_2 = 1$, we obtain that

$$C(t,t';t_1) = \pi V_1(t) \exp(Q^* t_1) V_1(t')\mathbf{e} - \xi^{*2} tt'.$$

By using (5.4.6) and the corresponding formula

$$\pi V_1(t) = \xi^* t \pi + \pi R'(1)[I - \exp(Q^* t)] (\mathbf{e}\pi - Q^*)^{-1},$$

and noting that the matrices $(\mathbf{e}\pi - Q^*)^{-1}$ and $\exp(Q^* u)$ commute, formula (5.4.13) follows by simple matrix calculations. •

We conclude this section with a technical lemma needed in the study of queues with a versatile Markovian input process.

Lemma 5.4.1: Except in the trivial case where $R(z) = Q^*$, $0 \leqslant z \leqslant 1$, the matrix $R(0)$ is *stable*. All its eigenvalues have negative real parts and the inverse

$$[sI - R(0)]^{-1} = \int_0^\infty e^{-su} P(0;u)\,du,$$

exists for all s with $\text{Re}\, s \geqslant 0$.

Proof: Obviously $R_{jj}(0) < 0$, $1 \leqslant j \leqslant m$, $R_{jj'}(0) \geqslant 0$, $j \neq j$, and $[R(0)\mathbf{e}]_j \leqslant 0$. The matrix $R(0)$ is therefore a semi-stable matrix and the real parts of its eigenvalues are nonpositive. If $R(0)$ has an eigenvalue with vanishing real part, then $P(0;t) = \exp[R(0)t]$, has an eigenvalue of modulus one. Since $P(0;t)$, $t \geqslant 0$, is obviously substochastic, this would in turn imply that the Perron-Frobenius eigenvalue of $P(0;t)$ is one. However, we have that $P(0;t) \leqslant P^*(1;t) = \exp(Q^* t)$, that the matrix Q^* is irreducible and $P^*(1;t)$ is positive. A corollary of the Perron-Frobenius theorem now yields that $P(0;t) = \exp(Q^* t)$, for all $t \geqslant 0$, and therefore $R(0) = Q^*$. Clearly, in all other cases, the eigenvalues of $R(0)$ all have negative real parts and $sI - R(0)$ is nonsingular for $\text{Re}\, s \geqslant 0$. •

Remark: $R(0) = Q^*$ corresponds to the trivial case where, with probability one, $N(t) = 0$, for all $t \geqslant 0$. In all other cases, the lemma guarantees that the time until the first arrival is a proper random variable.

5.5. SOME QUEUES WITH VERSATILE MARKOVIAN INPUT

The formal analogy of the versatile Markovian process to the homogeneous Poisson process suggests that it is possible to study any queueing model, that is tractable for Poisson input, *also* for the general case of a versatile Markovian input process. The details of such a generalization are, of course, quite involved and rely heavily on the matrix-analytic methods treated in this book. In this section, the analyses of the *single server queue* with general independent services times and of the *c − server queue with constant services times* will be carried out for a versatile Markovian input process. These generalize the classical $M/G/1$ and $M/D/c$ models.

A. THE SINGLE SERVER QUEUE

The single server queue with independent, identically distributed service times of distribution $H(\cdot)$ with mean μ_1' and a versatile Markovian arrival process with parameters as defined in Section 5.4, is a far-reaching generalization of the $M/G/1$ queue. By considering the queue lengths and the phases of the arrival process immediately after departure epochs, it is clear that this model has an embedded Markov renewal sequence with transition probability matrix $\tilde{Q}(\cdot)$, given for $x \geq 0$, by

$$\tilde{Q}(x) = \begin{vmatrix} B_0(x) & B_1(x) & B_2(x) & B_3(x) & B_4(x) & \cdots \\ A_0(x) & A_1(x) & A_2(x) & A_3(x) & A_4(x) & \cdots \\ 0 & A_0(x) & A_1(x) & A_2(x) & A_3(x) & \cdots \\ 0 & 0 & A_0(x) & A_1(x) & A_2(x) & \cdots \\ \cdot & \cdot & \cdot & \cdot & \cdot & \\ \cdot & \cdot & \cdot & \cdot & \cdot & \end{vmatrix}. \quad (5.5.1)$$

The $m \times m$ matrices $A_k(\cdot)$ and $B_k(\cdot)$, $k \geq 0$, are given by

$$A_k(x) = \int_0^x P(k;t)\,dH(t), \quad \text{for } k \geq 0, \qquad (5.5.2)$$

$$B_k(x) = \sum_{\nu=0}^{k} \int_0^x U(k-\nu+1;x-t)P(\nu;t)\,dH(t)$$

$$= \sum_{\nu=0}^{k} U(k-\nu+1;\cdot) * A(\nu;x), \quad \text{for } k \geqslant 0. \tag{5.5.3}$$

The matrices $P(k;t)$ are the solutions to the matrix-differential equations (5.4.1) and the matrices $U(k;t)$, $k \geqslant 1$, $t \geqslant 0$, are defined by

$$U(k;t) = \int_0^t P(0;u)\,du\,\{\Delta(\lambda)\Delta[\mathbf{p}(k)] + T \circ q(k) + \mathbf{T}^{\circ}\alpha \circ r(k)\}. \tag{5.5.4}$$

It follows from (5.4.4) and (5.5.2) that the transform matrix

$$A^*(z,s) = \sum_{k=0}^{\infty} \tilde{A}_k(s) z^k,$$

is given by

$$A^*(z,s) = \int_0^{\infty} \exp[R(z)t]e^{-st}\,dH(t). \tag{5.5.5}$$

This implies that

$$A = A^*(1,0) = \int_0^{\infty} \exp(Q^* t)\,dH(t), \tag{5.5.6}$$

so that the vector π of formula (5.1.7) is also the invariant probability vector of the positive stochastic matrix A.

By virtue of formula (5.4.6), the vector β defined in (2.3.5) is explicitly given by

$$\beta = \xi^* \mu_1{}'\mathbf{e} + (I - A)\,[\xi^* T^{-1}\mathbf{e} - T^{-1}R'(1)\mathbf{e}], \tag{5.5.7}$$

so that

$$\rho = \pi\beta = \xi^* \mu_1'. \tag{5.5.8}$$

The vector β, and therefore also ρ, is finite if and only if the (realizable) group size densities of the arrival process all have finite means. We shall henceforth assume that this is the case.

From (5.5.3) and (5.5.4), we see that the transform matrix

$$B^*(z,s) = \sum_{k=0}^{\infty} \tilde{B}_k(s) z^k,$$

is given by

$$B^*(z,s) = U^*(z,s) A^*(z,s), \tag{5.5.9}$$

where

$$U^*(z,s) = \sum_{k=0}^{\infty} \tilde{U}(k,s) z^k = [sI - R(0)]^{-1} [R(z) - R(0)]. \tag{5.5.10}$$

Lemma 5.4.1 guarantees that, for all non-trivial cases of the arrival process, the inverse in (5.5.10) exists.

It readily follows from the special form of $B^*(z,s)$ that the conditions b, c and d, in (3.2.17) are always satisfied. The embedded Markov renewal process is therefore *positive recurrent,* or equivalently, the queue is *stable* if and only if $\rho < 1$.

Many results for the single server queue with versatile Markovian input may now be obtained by invoking general theorems established in Chapters 2 and 3. We shall state these results only briefly, so that our discussion may dwell on those results that require separate arguments.

All results on the matrix $\tilde{G}(z,s)$, proved in Chapters 2 and 3 are immediately applicable. Since the matrix A_0 has positive diagonal elements and the matrices A_k, $k \geqslant 1$, are positive, it is clear that the matrix $G = \tilde{G}(1,0)$, is positive.

The particular form of the matrices $B_k(x)$, $k \geqslant 0$, implies that the matrix $\tilde{K}(z,s)$ of Lemma 2.4.2 is given by

$$\tilde{K}(z,s) = z \sum_{k=0}^{\infty} \tilde{B}_k(s) \tilde{G}^k(z,s) = U^*[\tilde{G}(z,s), s] \tag{5.5.11}$$

$$= [sI - R(0)]^{-1}\{R[\tilde{G}(z,s)] - R(0)\}.$$

The matrix $\tilde{K}(z,s)$ is the transform of the duration and the number of customers served in a ***busy cycle*** of the queue, account taken of the phases of the beginning and the end of the busy cycle.

We stress that for this queueing model the idle period and the subsequent busy period are, in general, not (conditionally) independent. A direct application of the law of total probability yields the corresponding transform matrix $\tilde{K}^*(z,s)$ for the duration of and the number served during a ***busy period,*** as follows.

$$\tilde{K}^*(z,s) = \sum_{k=1}^{\infty} \tilde{U}(k;0)\tilde{G}^k(z,s) = U^*[\tilde{G}(z,s),0] \tag{5.5.12}$$

$$= [-R(0)]^{-1}\{R[\tilde{G}(z,s)] - R(0)\}.$$

Similarly, the distribution of the duration of the *idle period* and the arrival phase at the start of the subsequent busy period, conditioned on the arrival phase at the end of the preceding busy period, has the transform matrix

$$U^*(1,s) = [sI - R(0)]^{-1}[Q^* - R(0)]. \tag{5.5.13}$$

From (5.5.13) it follows that the vector of row sum means of the semi-Markov matrix $U(x) = \sum_{k=1}^{\infty} U(k;x)$ with transform $U^*(1,s)$, is given by

$$\boldsymbol{\phi} = [-R(0)]^{-1}\mathbf{e}. \tag{5.5.14}$$

The components of the vector $\boldsymbol{\phi}$ are the conditional mean durations of idle periods starting in the various phases of the arrival process.

Let the column vector $\mathbf{v}$ be defined as

$$\mathbf{v} = [I - A + (\mathbf{e} - \boldsymbol{\beta})\mathbf{g}]^{-1}\mathbf{e}, \tag{5.5.15}$$

then the vectors $\tilde{\boldsymbol{\mu}}_1$ and $\tilde{\boldsymbol{\mu}}_1^*$ of (3.1.12) and (3.1.13) are given by

$$\tilde{\boldsymbol{\mu}}_1 = (I - G + \mathbf{eg})\mathbf{v}, \qquad \tilde{\boldsymbol{\mu}}_1^* = \mu_1'\tilde{\boldsymbol{\mu}}_1, \tag{5.5.16}$$

and the vectors $\tilde{\boldsymbol{\kappa}}_1$ of (3.2.3) and $\tilde{\boldsymbol{\kappa}}_1^*$ of (3.2.13) are given by

$$\tilde{\boldsymbol{\kappa}}_1 = [U^*(1,0) - U^*(G,0) - R^{-1}(0)R'(1)\mathbf{eg}]\,\mathbf{v}, \tag{5.5.17}$$

and

$$\tilde{\boldsymbol{\kappa}}_1^* = \boldsymbol{\phi} + \mu_1'\tilde{\boldsymbol{\kappa}}_1. \tag{5.5.18}$$

The components of the vectors $\tilde{\boldsymbol{\kappa}}_1$ and $\tilde{\boldsymbol{\kappa}}_1^*$ are respectively the conditional mean numbers served during and the conditional mean durations of busy cycles starting in the various phases of the arrival process. The formulas (5.5.17) and (5.5.18) may be obtained by direct calculations starting with (5.5.11), or by particularizing the formulas (3.2.3) and (3.2.14) to the present model. We note that the term $\mu_1'\tilde{\boldsymbol{\kappa}}_1$ in (5.5.18) gives the corresponding mean durations for the busy periods.

There are no appreciable simplifications in the calculation of the matrix $K = \tilde{K}(1,0)$. We see from (5.5.11) that

$$K = U^*(G,0), \tag{5.5.19}$$

a matrix which also arises in formula (5.5.17). When $\rho < 1$, we know that K is a stochastic matrix, which is also positive. We denote its invariant probability vector by $\boldsymbol{\kappa}$, as in formula (3.2.2).

The particular form of the matrices B_k, $k \geqslant 0$, leads to some simplifications in the form of the matrix $\tilde{H}(z)$ of formula (3.2.7). As these expressions correspond to $s = 0$, we shall suppress that variable throughout. In particular, the matrix $U(k;\infty) = \tilde{U}(k;0)$, will be written as $U(k)$ for $k \geqslant 1$.

Lemma 5.5.1: The matrix $\tilde{H}(z)$ of formula (3.2.7) may here be written as

$$\tilde{H}(z) = [I - zA_0U(1)]^{-1}\{\sum_{\nu=1}^{\infty} zA_\nu\tilde{G}^{\nu-1}(z) \tag{5.5.20}$$

$$+ A_0 \sum_{\nu=2}^{\infty} U(\nu) \tilde{G}^{\nu-1}(z)\}.$$

Proof: By virtue of (5.5.3), we have that

$$z \sum_{\nu=1}^{\infty} B_\nu \tilde{G}^{\nu-1}(z) = z \sum_{\nu=1}^{\infty} \sum_{k=0}^{\nu} U(\nu-k+1) A_k \tilde{G}^{\nu-1}(z)$$

$$= U(1)\, z \sum_{\nu=1}^{\infty} A_\nu \tilde{G}^{\nu-1}(z) + z \sum_{k=0}^{\infty} \sum_{\nu=0}^{\infty} U(\nu+2) A_k \tilde{G}^{k+\nu}(z)$$

$$= U(1)\, z \sum_{\nu=1}^{\infty} A_\nu \tilde{G}^{\nu-1}(z) + \sum_{\nu=2}^{\infty} U(\nu) \tilde{G}^{\nu-1}(z).$$

Furthermore, $B_0 = U(1) A_0$, so that

$$zA_0 [I - zU(1)A_0]^{-1} U(1) + I = [I - zU(1)A_0]^{-1}.$$

Substitution into (3.2.7) now leads to

$$\tilde{H}(z) = [I - zA_0 U(1)]^{-1} \sum_{\nu=1}^{\infty} zA_\nu \tilde{G}^{\nu-1}(z)$$

$$+ A_0 [I - zU(1)A_0]^{-1} \sum_{\nu=2}^{\infty} U(\nu) \tilde{G}^{\nu-1}(z).$$

which may clearly be written as in (5.5.20). •

From formula (5.5.20), we obtain after lengthy, but routine calculations that the vector $\tilde{\mathbf{h}}$ is given by

$$\tilde{\mathbf{h}} = (I - H)\mathbf{v} - \mathbf{e} + [I - A_0 U(1)]^{-1}\{ \mathbf{e} - A_0 \mathbf{e} \qquad (5.5.21)$$

$$+ A_0 [- R(0)]^{-1} [R(1)\mathbf{v} + (1-\rho)^{-1} R'(1)\mathbf{e}] \},$$

where $\mathbf{v}$ is as in formula (5.5.15).

The vector $\mathbf{h}$ is the invariant probability vector of the stochastic matrix

$$H = \tilde{H}(1) = [I - A_0 U(1)]^{-1} [\sum_{\nu=1}^{\infty} A_\nu G^{\nu-1} + A_0 \sum_{\nu=2}^{\infty} U(\nu) G^{\nu-1}]. \quad (5.5.22)$$

It is clear from (5.5.21), that

$$\mathbf{h}\tilde{\mathbf{h}} = -1 + \mathbf{h}[I - A_0 U(1)]^{-1}\{ \mathbf{e} - A_0 \mathbf{e} \quad (5.5.23)$$

$$+ A_0 [- R(0)]^{-1} [R(1)\mathbf{v} + (1-\rho)^{-1} R'(1)\mathbf{e}] \}.$$

With the help of (5.5.22) and (5.5.23), the vector $\mathbf{h}$ and the constant $\mathbf{h}\tilde{\mathbf{h}}$ may be evaluated by using ingredients already required in earlier computations.

The Stationary Queue Length Densities: The results on the stationary density of the queue length and the arrival phase at departure epochs are summarized in the following theorem. The stationary density is written as the sequence of m-vectors $\{\mathbf{x}_i, i \geqslant 0\}$.

Theorem 5.5.1: When $\rho < 1$, the vectors $\mathbf{x}_0$ and $\mathbf{x}_1$ are given by

$$\mathbf{x}_0 = (\boldsymbol{\kappa}\tilde{\boldsymbol{\kappa}}_1)^{-1}\boldsymbol{\kappa}, \qquad \mathbf{x}_1 = (\mathbf{h}\tilde{\mathbf{h}})^{-1}\mathbf{h}. \quad (5.5.24)$$

The generating function

$$\mathbf{X}(z) = \sum_{i=0}^{\infty} \mathbf{x}_i z^i,$$

is given by

$$\mathbf{X}(z)[zI - A^*(z,0)] = \mathbf{x}_0 [-R(0)]^{-1} R(z) A^*(z,0), \quad (5.5.25)$$

and for $z = 1$, we have

$$\mathbf{X}(1) = \sum_{i=0}^{\infty} \mathbf{x}_i = \mathbf{x}_0 [-R(0)]^{-1} Q^* (Z - I) + \boldsymbol{\pi}, \tag{5.5.26}$$

where $Z = (I - A + \mathbf{e}\boldsymbol{\pi})^{-1}$. Furthermore, we have the equality

$$\xi^* \mathbf{x}_0 [-R(0)]^{-1} \mathbf{e} = 1 - \rho. \tag{5.5.27}$$

Proof: The formulas (5.5.24) are obtained by applying Theorems 3.2.1 and 3.2.2. The expression in (5.5.25) follows from the steady-state equations or by particularizing (3.3.2). The same calculations as in the proof of Theorem 3.3.1 lead from (5.5.25) to the expression for $\mathbf{X}(1)$.

In order to obtain (5.5.27), we differentiate in (5.5.25), set $z = 1$, and postmultiply by $\mathbf{e}$. This yields that

$$1 - \mathbf{X}(1)\boldsymbol{\beta} = \mathbf{x}_0 [-R(0)]^{-1} [R'(1)\mathbf{e} + Q^* \boldsymbol{\beta}].$$

Postmultiplying by $\boldsymbol{\beta}$ in (5.5.26) and substituting, it follows that

$$\mathbf{x}_0 [-R(0)]^{-1} [R'(1)\mathbf{e} + Q^* Z \boldsymbol{\beta}] = 1 - \rho.$$

If we can show that

$$R'(1)\mathbf{e} + Q^* Z \boldsymbol{\beta} = \xi^* \mathbf{e}, \tag{5.5.28}$$

the equality (5.5.27) will be proved. To that end, we note that clearly

$$R(z)A^*(z,0) = A^*(z,0)R(z).$$

Differentiation, setting $z = 1$, and postmultiplication by $\mathbf{e}$ yields

$$Q^* \boldsymbol{\beta} = -(I - A)R'(1)\mathbf{e} = -Z^{-1}R'(1)\mathbf{e} + \xi^* \mathbf{e},$$

since $\xi^* = \boldsymbol{\pi} R'(1)\mathbf{e}$. Finally, since $Z\mathbf{e} = \mathbf{e}$, and Q^* and Z commute, the equality (5.5.28) follows. •

The stationary joint density of the queue length and the phase of the arrival process *at an arbitrary time* is represented by the sequence of m-vectors $\{\mathbf{y}_i, i \geqslant 0\}$ with generating function $\mathbf{Y}(z) = \sum_{i=0}^{\infty} \mathbf{y}_i z^i$. The remarkably simple relation between $\mathbf{Y}(z)$ and $\mathbf{X}(z)$ is given in the following theorem.

Theorem 5.5.2: The generating function $\mathbf{Y}(z)$ is given by

$$\mathbf{Y}(z) = \xi^*(z - 1)\mathbf{X}(z)R^{-1}(z), \quad \text{for } 0 \leqslant z < 1, \tag{5.5.29}$$

and $\mathbf{Y}(1) = \boldsymbol{\pi}$. The probability $\mathbf{y}_0\mathbf{e}$ that the server is idle at an arbitrary time is equal to $1 - \rho$.

Proof: The classical argument based on the key renewal theorem leads to expressions for the vectors $\mathbf{y}_i$ in terms of the vectors $\mathbf{x}_i$, $i \geqslant 0$. After some direct simplifications, using the particular form of the matrices B_k, $k \geqslant 0$, these may be written as

$$\mathbf{y}_0 = \xi^* \mathbf{x}_0 [- R(0)]^{-1}, \tag{5.5.30}$$

$$\mathbf{y}_i = \xi^* \sum_{\nu=1}^{i} [\mathbf{x}_0 U(\nu) + \mathbf{x}_\nu] \int_0^\infty P(i-\nu;u)[1 - H(u)]\,du, \quad \text{for } i \geqslant 1.$$

From (5.5.27), it is clear that $\mathbf{y}_0\mathbf{e} = 1 - \rho$. Upon taking generating functions and noting that for $0 \leqslant z < 1$,

$$\int_0^\infty P^*(z;u)[1 - H(u)]\,du = [A^*(z,0) - I]\,[R(z)]^{-1},$$

by virtue of (5.4.4), we obtain (5.5.29).

Writing (5.5.29) as

$$\mathbf{Y}(z)R(z) = \xi^*(z - 1)\mathbf{X}(z), \quad \text{for } 0 \leqslant z < 1, \tag{5.5.31}$$

and setting $z = 1-$, it follows that $\mathbf{Y}(1-)Q^* = 0$. Since $\mathbf{Y}(1-)$ must be a probability vector, we obtain $\mathbf{Y}(1-) = \boldsymbol{\pi}$. •

Moment Vectors: The vectors $\mathbf{X}^{(n)}(1-)$, $n \geqslant 1$, and the factorial moments $L_n = \mathbf{X}^{(n)}(1-)\mathbf{e}$, $n \geqslant 1$, may be computed from equation (5.5.25) by adapting the recursive scheme proposed in the equations (3.3.11) and (3.3.13). The simple form of the right hand side in (5.5.25) induces minor simplifications in the formulas (3.3.5), but the essential recurrence is as complex as in most applications of the matrix formalism. Substantial simplifications are obtained only for special cases. The existence of the $n-$th moment of the queue length requires the existence of moments of order $n+1$ of the service time distribution and of all (realizable) group size densities in the arrival process.

Formula (5.5.31) may be used to relate the moment vectors $\mathbf{Y}^{(n)}(1-)$ and the factorial moments $L_n^* = \mathbf{Y}^{(n)}(1-)\mathbf{e}$, $n \geqslant 1$, to the corresponding quantities for the queue length at departure epochs. The resulting formulas may be implemented with little additional computations. We shall show the essential manipulations for the first two moments only.

Three successive differentiations in (5.5.31), followed by setting $z = 1$, lead to

$$\mathbf{Y}'(1)Q^* = \xi^*\mathbf{X}(1) - \boldsymbol{\pi}R'(1),$$

$$\mathbf{Y}''(1)Q^* = 2\xi^*\mathbf{X}'(1) - 2\mathbf{Y}'(1)R'(1), \tag{5.5.32}$$

$$\mathbf{Y}'''(1)Q^* = 3\xi^*\mathbf{X}''(1) - 3\mathbf{Y}''(1)R'(1) - 2\mathbf{Y}'(1)R''(1).$$

Adding $\mathbf{Y}'(1)\mathbf{e}\boldsymbol{\pi} = L_1^*\boldsymbol{\pi}$, to both sides of the first equation and recalling from Theorem 5.1.3 that the matrix $\mathbf{e}\boldsymbol{\pi} - Q^*$ is nonsingular, we obtain

$$\mathbf{Y}'(1) = L_1^*\boldsymbol{\pi} + [\boldsymbol{\pi}R'(1) - \xi^*\mathbf{X}(1)]\,(\mathbf{e}\boldsymbol{\pi} - Q^*)^{-1}, \tag{5.5.33}$$

and similarly

$$\mathbf{Y}''(1) = L_2^*\boldsymbol{\pi} + 2[\mathbf{Y}'(1)R'(1) - \xi^*\mathbf{X}'(1)]\,(\mathbf{e}\boldsymbol{\pi} - Q^*)^{-1}. \tag{5.5.34}$$

Postmultiplying by $\mathbf{e}$ in the last two of the equations (5.5.32) leads to

$$\mathbf{Y}'(1)R'(1)\mathbf{e} = \xi^* L_1,$$

$$3\mathbf{Y}''(1)R'(1)\mathbf{e} = 3\xi^* L_2 - 2\mathbf{Y}'(1)R''(1)\mathbf{e},$$

and by postmultiplying in (5.5.33) and (5.5.34) by $R'(1)\mathbf{e}$, we obtain further expressions for $\mathbf{Y}'(1)R'(1)\mathbf{e}$ and $\mathbf{Y}''(1)R'(1)\mathbf{e}$. Equating the corresponding expressions, it follows that

$$L_1^* = L_1 - \xi^{*\,-1}\,[\pi R'(1) - \xi^* \mathbf{X}(1)]\,(\mathbf{e}\pi - Q^*)^{-1}R'(1)\mathbf{e}, \qquad (5.5.35)$$

$$L_2^* = L_2 - 2\xi^{*\,-1}\,[\mathbf{Y}'(1)R'(1) - \xi^* \mathbf{X}'(1)]\,(\mathbf{e}\pi - Q^*)^{-1}R'(1)\mathbf{e},$$

In the course of the proof of Theorem 5.4.1, we have shown that

$$\mathbf{v} = (\mathbf{e}\pi - Q^*)^{-1}R'(1)\mathbf{e} = T^{-1}[\xi^*\mathbf{e} - R'(1)\mathbf{e}] + C\mathbf{e},$$

where C is a known constant which cancels out upon substitution in the formulas (5.5.35). Substitution of this expression into (5.5.35) results in some further cancellations, but does not appreciably simplify these formulas.

The Virtual Waiting Time: We denote by $W_j(x)$, $1 \leqslant j \leqslant m$, $x \geqslant 0$, the stationary probability that a (virtual) customer, arriving to the queue at an arbitrary time, finds the arrival process in the phase j and has to wait at most a time x. The service discipline is first-come, first-served. The vector $\mathbf{W}(x)$ has components $W_j(x)$, $1 \leqslant j \leqslant m$, and has the Laplace-Stieltjes transform $\mathbf{w}(s)$.

Theorem 5.5.3: The vector $\mathbf{W}(\cdot)$ satisfies the generalized Pollaczek-Khinchin integral equation

$$\mathbf{W}(x) = \mathbf{y}_0 + \int_0^x \mathbf{W}(x-u)\,d\,\Theta(u), \quad \text{for } x \geqslant 0, \qquad (5.5.36)$$

where the kernel matrix $\Theta(\cdot)$ is defined by

$$\Theta(u) = \int_0^u \left[\Delta(\lambda)\Delta\left(\mathbf{e} - \sum_{k=0}^{\infty} \mathbf{p}(k)H^{(k)}(v)\right)\right. \qquad (5.5.37)$$

$$- T \circ \sum_{k=0}^{\infty} q(k) H^{(k)}(v) - \mathbf{T}^{\circ}\boldsymbol{\alpha} \circ \sum_{k=0}^{\infty} r(k) H^{(k)}(v) \Big] dv,$$

for $u \geqslant 0$. The Laplace-Stieltjes transform $\Theta^*(s)$ of $\Theta(\cdot)$ is given by

$$\Theta^*(s) = s^{-1} R[h(s)], \tag{5.5.38}$$

where $h(s)$ is the Laplace-Stieltjes transform of the service time distribution $H(\cdot)$. The vector $\mathbf{y}_0$ is as given in formula (5.5.30). The vector $\mathbf{W}(\infty)$ is given by

$$\mathbf{W}(\infty) = \boldsymbol{\pi}. \tag{5.5.39}$$

Proof: The arriving customer does not have to wait, if and only if the queue is empty at time t. It is therefore obvious that $\mathbf{W}(0+) = \mathbf{y}_0$. The vector $\mathbf{W}(x) - \mathbf{W}(0+)$, $x > 0$, is obtained by the classical argument based on the law of total probability and the key renewal theorem. By carefully enumerating all possible configurations of the queue length and the arrival phase at time t and the corresponding waiting times experienced by the virtual customer, we obtain that

$$\mathbf{W}(x) - \mathbf{y}_0 = \xi^* \sum_{\nu=1}^{\infty} \sum_{k=0}^{\infty} \int_{0_{(t)}}^{\infty} \int_{0_{(w)}}^{x} \mathbf{x}_\nu P(k;t) dt \; dH(t+w) H^{(k+\nu-1)}(x-w)$$

$$+ \xi^* \mathbf{x}_0 \sum_{\nu=1}^{\infty} \sum_{k=0}^{\infty} \int_{0_{(t)}}^{\infty} \int_{0_{(y)}}^{t} \int_{0_{(w)}}^{x} dt \; dU(\nu; t-y) P(k;y) dy dH(y+w) H^{(k+\nu-1)}(x-w).$$

The first term corresponds to the case where the virtual customer arrives after the first service of the current busy period has terminated. The second term corresponds to the case where the service in course at time t is the first of the current busy period.

Upon evaluating the Laplace-Stieltjes transform of the preceding expressions, we obtain after routine interchanges of integrations that

$$\mathbf{w}(s) = \mathbf{y}_0 + \xi^* \sum_{\nu=1}^{\infty} \mathbf{x}_\nu \int_0^{\infty} \int_0^{\infty} \exp\{R[h(s)]t\} \, dt \; e^{-sw} dH(t+w) \; h^{\nu-1}(s)$$

$$+ \xi^* \mathbf{x}_0 \sum_{\nu=1}^{\infty} U(\nu) \int_0^{\infty} \int_0^{\infty} \exp\{R[h(s)]t\}\, dt\; e^{-sw}\, dH(t+w)\; h^{\nu-1}(s).$$

In simplifying this expression, we proceed as in the proof of Theorem 4.1.2. The integral appearing in the last two terms is written as

$$I(s) = \int_0^{\infty} \int_0^{\infty} \exp\{R[h(s)]t\} e^{-sw}\, dt\; dH(t+w)$$

$$= \int_0^{\infty} \int_0^{v} \exp\{R[h(s)]w\} e^{-sw}\, dw\; \exp\{R[h(s)]v\} dH(v),$$

so that

$$I(s)\{sI + R[h(s)]\} = \int_0^{\infty} \exp\{R[h(s)]v\} dH(v) - h(s)I$$

$$= A^*[h(s),0] - h(s)I,$$

by virtue of formula (5.5.5). It now follows that

$$\mathbf{w}(s)\{sI + R[h(s)]\} = s\,\mathbf{y}_0 + \mathbf{y}_0 R[h(s)]$$

$$+ \xi^*[h(s)]^{-1}\{\mathbf{x}_0 U^*[h(s),0] + \mathbf{X}[h(s)] - \mathbf{x}_0\}\{A^*[h(s),0] - h(s)I\}.$$

Replacing $U^*[h(s),0]$ by the expression obtained from (5.5.10), $\mathbf{y}_0$ by $\xi^* \mathbf{x}_0[-R(0)]^{-1}$, and using formula (5.5.25) with $z = h(s)$, we obtain

$$\mathbf{w}(s)\{sI + R[h(s)]\} = s\,\mathbf{y}_0, \tag{5.5.40}$$

which is the transform version of the integral equation (5.5.36). Upon setting $s = 0$, in (5.5.40), it follows that $\mathbf{w}(0)Q^* = 0$, so that $\mathbf{w}(0) =$

$\mathbf{W}(\infty) = \boldsymbol{\pi}$. •

Moment Formulas: The computation of the moment vectors $(-1)^n \mathbf{w}^{(n)}(0)$ and of the moments $W_n^* = (-1)^n \mathbf{w}^{(n)}(0)\mathbf{e}$, $n \geq 1$, of the virtual waiting time proceeds routinely from formula (5.5.40). The intermediate manipulations are the same as used to obtain the formulas (5.5.33-35). We record the results for the first two moments. The quantities μ_j', $j \geq 1$, are the non-central moments of the service time distribution $H(\cdot)$.

$$-\mathbf{w}'(0) = (W_1^* - 1)\boldsymbol{\pi} + [\, \mathbf{y}_0 + \mu_1'\boldsymbol{\pi} R'(1)\,]\,(\mathbf{e}\boldsymbol{\pi} - Q^*)^{-1}, \tag{5.5.41}$$

$$\mathbf{w}''(0) = W_2^*\boldsymbol{\pi} + [\, 2\mu_1'\mathbf{w}'(0)R'(1) - 2\mathbf{w}'(0) - \mu_1'^2\boldsymbol{\pi} R''(1)$$

$$- \mu_2'\boldsymbol{\pi} R'(1)]\,(\mathbf{e}\boldsymbol{\pi} - Q^*)^{-1},$$

where

$$W_1^* = (1-\rho)^{-1}\{\, \mu_1'[\mathbf{y}_0 + \mu_1'\boldsymbol{\pi} R'(1)]\,(\mathbf{e}\boldsymbol{\pi} - Q^*)^{-1}R'(1)\mathbf{e} \tag{5.5.42}$$

$$- \rho + \frac{1}{2}\mu_2'\xi^* + \frac{1}{2}\mu_1'^2\boldsymbol{\pi} R''(1)\mathbf{e}\,\},$$

and

$$3(1-\rho)W_2^* = (W_1^* - 1)\,[\mu_2'\xi^* + 3\mu_1'^2\boldsymbol{\pi} R''(1)\mathbf{e}] \tag{5.5.43}$$

$$+ [\mathbf{y}_0 + \mu_1'\boldsymbol{\pi} R'(1)]\,(\mathbf{e}\boldsymbol{\pi} - Q^*)^{-1}\,[\mu_2' R'(1)\mathbf{e} + 3\mu_1'^2 R''(1)\mathbf{e}]$$

$$+ 3\mu_1'\,[2\mu_1'\mathbf{w}'(0)R'(1) - 2\mathbf{w}'(0) - \mu_1'^2\boldsymbol{\pi} R''(1)$$

$$- \mu_2'\boldsymbol{\pi} R'(1)]\,(\mathbf{e}\boldsymbol{\pi} - Q^*)^{-1}\,R'(1)\mathbf{e}$$

$$+ \mu_3'\xi^* + 3\mu_1'\mu_2'\boldsymbol{\pi} R''(1)\mathbf{e} + \mu_1'^3\boldsymbol{\pi} R'''(1)\mathbf{e}.$$

An Example: The generality of the versatile Markovian point process accounts for the complexity of the matrix formulas which appear in the analysis of the single server queue with that input process. Few applications involve all the possible types of group arrivals allowed by the general formulation. For specific models, it is clearly indicated to exploit the analytic and algorithmic simplification that are present. The much simpler results for the *PH / G / 1* queue, discussed in Section 5.2, may also be obtained as particular cases of results in the present section.

As an additional and useful example, we shall now consider the case of a Markov-modulated Poisson process as input to a single server. The matrix $R(z)$ is now given by

$$R(z) = (z - 1)\Lambda + Q, \tag{5.5.44}$$

which leads to

$$R'(1) = \Lambda, \qquad R^{(n)}(1) = 0, \quad \text{for } n \geqslant 2,$$

$$[-R(0)]^{-1} = (\Lambda - Q)^{-1}, \qquad \xi^* = \pi\lambda.$$

From (5.5.5) and (5.5.10), we obtain

$$A^*(z,s) = h[sI + (1 - z)\Lambda - Q],$$

$$U^*(z,s) = z[sI + \Lambda - Q]^{-1}\Lambda,$$

so that $\tilde{U}(k;s) = 0$, for $k \geqslant 2$.

The matrices A_k, $k \geqslant 0$, and the matrix G must be computed numerically. This is inherent to the matrix methods of this book and there is no further simplification here, unless further conditions are imposed on the service time distribution.

The matrices $\tilde{K}(z,s)$ of (5.5.11) and $\tilde{H}(z)$ of (5.5.20) are greatly simplified. We have

$$\tilde{K}(z,s) = [sI + \Lambda - Q]^{-1}\Lambda\tilde{G}(z,s),$$

and

$$\tilde{H}(z) = [I - zA_0(\Lambda - Q)\Lambda]^{-1} \sum_{\nu=1}^{\infty} zA_\nu \tilde{G}^{\nu-1}(z,0).$$

It follows that the matrices K and H are given by

$$K = (\Lambda - Q)^{-1}\Lambda G,$$

and

$$H = [I - A_0(\Lambda - Q)^{-1}\Lambda]^{-1} \sum_{\nu=1}^{\infty} A_\nu G^{\nu-1}.$$

Their invariant probability vectors $\boldsymbol{\kappa}$ and $\mathbf{h}$ are numerically computed. There are useful simplifications in the expressions for $\tilde{\boldsymbol{\kappa}}_1$ and $\mathbf{h}\tilde{\mathbf{h}}$. We see that

$$\tilde{\boldsymbol{\kappa}}_1 = \{ (\Lambda - Q)^{-1}\Lambda - (\Lambda - Q)^{-1}\Lambda G + (\Lambda - Q)^{-1}\Lambda\, \mathbf{eg}\}\, \mathbf{v}$$

$$= (\Lambda - Q)^{-1}\Lambda\tilde{\boldsymbol{\mu}}_1.$$

From formula (5.5.23), we obtain that

$$\mathbf{h}\tilde{\mathbf{h}} = -1 + \mathbf{h}\,[I - A_0(\Lambda - Q)^{-1}\Lambda]^{-1}\{ \mathbf{e} - A_0\mathbf{e}$$

$$+ A_0(\Lambda - Q)^{-1}\,[Q\mathbf{v} + (1 - \rho)^{-1}\Lambda\mathbf{e}]\, \}.$$

Since $(\Lambda - Q)^{-1}\Lambda$ is a stochastic matrix, the expression inside the braces simplifies to

$$\mathbf{e} + \rho(1 - \rho)^{-1}A_0\mathbf{e} - A_0\mathbf{v} + A_0(\Lambda - Q)^{-1}\Lambda\mathbf{v}.$$

By further elementary matrix manipulations, we ultimately obtain

$$\mathbf{h}\tilde{\mathbf{h}} = \mathbf{h}\,[I - A_0(\Lambda - Q)^{-1}\Lambda]^{-1}\,[\mathbf{v} - A_0\mathbf{v} + (1-\rho)^{-1}\mathbf{e}] - (1-\rho)^{-1} - \mathbf{h}\mathbf{v},$$

which is a computationally convenient form.

The evaluation of the vectors $\mathbf{x}_0$ and $\mathbf{x}_1$ is now immediate by formula (5.5.24). Formula (5.5.25) becomes here

$$\mathbf{X}(z)\,[zI - A^*(z)] = z\,\mathbf{u}_0 - \mathbf{x}_0 A^*(z),$$

where $A^*(z) = A^*(z,0)$, and $\mathbf{u}_0 = \mathbf{x}_0(\Lambda - Q)^{-1}\Lambda$. Because of the simple form of the right hand side, there are minor simplifications in the computation of the moment of the queue length at departures. We shall display the results for the first two moments.

We recall that

$$\boldsymbol{\beta}_\nu = A^{*\,(\nu)}(1)\mathbf{e}, \quad \text{for } \nu \geqslant 1,$$

and we set

$$\mathbf{U}(z) = z\,\mathbf{u}_0 - \mathbf{x}_0 A^*(z),$$

We clearly have that

$$\mathbf{U}(1) = \mathbf{u}_0 - \mathbf{x}_0 A\,, \qquad \mathbf{U}(1)\mathbf{e} = 0,$$

$$\mathbf{U}'(1) = \mathbf{u}_0 - \mathbf{x}_0 A^{*\,\prime}(1), \qquad \mathbf{U}'(1)\mathbf{e} = \mathbf{x}_0\mathbf{e} - \mathbf{x}_0\boldsymbol{\beta},$$

$$\mathbf{U}''(1) = -\,\mathbf{x}_0 A^{*\,\prime\prime}(1), \qquad \mathbf{U}''(1)\mathbf{e} = -\,\mathbf{x}_0\boldsymbol{\beta}_2,$$

$$\mathbf{U}'''(1) = -\,\mathbf{x}_0 A^{*\,\prime\prime\prime}(1), \qquad \mathbf{U}'''(1)\mathbf{e} = -\,\mathbf{x}_0\boldsymbol{\beta}_3.$$

By implementing the recursive scheme described in Section 3.3, we may evaluate the first two moments by successive computation of the following expressions.

$$\mathbf{X}(1) = (\mathbf{u}_0 - \mathbf{x}_0)(Z - I) + \boldsymbol{\pi},$$

$$\theta_1 = [\mathbf{U}'(1) + \mathbf{X}(1)A^{*\prime}(1) - \mathbf{X}(1)]\, Z\boldsymbol{\beta},$$

$$L_1 = [2(1-\rho)]^{-1}\, [\mathbf{X}(1)\beta_2 + \mathbf{U}''(1)\mathbf{e} + 2\theta_1],$$

$$\mathbf{X}'(1) = L_1\boldsymbol{\pi} + [\mathbf{U}'(1) + \mathbf{X}(1)A^{*\prime}(1) - \mathbf{X}(1)]\, Z,$$

$$\theta_2 = [\mathbf{U}''(1) + \mathbf{X}(1)A^{*\prime\prime}(1) + 2\mathbf{X}'(1)A^{*\prime}(1) - 2\mathbf{X}'(1)]Z\boldsymbol{\beta},$$

$$L_2 = [3(1-\rho)]^{-1}\, [3\mathbf{X}'(1)\beta_2 + \mathbf{X}(1)\beta_3 + \mathbf{U}'''(1)\mathbf{e} + 3\theta_2],$$

$$\mathbf{X}''(1) = L_2\boldsymbol{\pi} + [\mathbf{U}''(1) + \mathbf{X}(1)A^{*\prime\prime}(1) + 2\mathbf{X}'(1)A^{*\prime}(1) - 2\mathbf{X}'(1)]\, Z.$$

The computational effort in implementing these formulas does not primarily come from the required matrix calculations. These can be very efficiently organized and, at least for moderate values of m, do not require much processing time. The main effort comes from the computation of the matrices A, $A^{*\prime}(1)$, $A^{*\prime\prime}(1)$ and $A^{*\prime\prime\prime}(1)$ and their row sum vectors. These require the numerical integration of the differential equations in (5.4.5), followed by the evaluation of matrix integrals of the form $\int_0^\infty V_n(t)\,dH(t)$. Such belabored set-up computations appear to be unavoidable in all queues with a non-Poisson arrival process and a *general* service time distribution. As for the $PH/G/1$ queue, we may avoid all numerical integrations if we restrict the service time distribution to be *of phase type*. The matrices $\{A_k,\, k \geq 0\}$ and the related moment matrices may then all be computed by means of matrix-iterative formulas similar to those given in Theorem 5.1.5 for the $PH/PH/1$ queue. We shall leave their derivation to the initiative of the reader.

The simplifications, due to the Markov-modulated Poisson process, in the formulas (5.5.29-35) are both obvious and minor. The Pollaczek-Khinchin integral equation becomes

$$\mathbf{W}(x) = \mathbf{y}_0 + \int_0^x \mathbf{W}(x-u)[\,\Lambda[1 - H(u)] - Q\,]\, du \qquad (5.5.45)$$

$$= \mathbf{y}_0 - \int_0^x \mathbf{W}(u)du \; Q + \int_0^x \mathbf{W}(x-u)[1 - H(u)] \, du \; \Lambda,$$

for $x \geqslant 0$. As a Volterra integral equation of the second kind, it is of a type that is well-suited for numerical solution.

The algorithmic utility of PH – distributions may be demonstrated also in the computation of the vector $\mathbf{W}(x)$, $x \geqslant 0$. When the service time distribution $H(\cdot)$ is of phase type and has the representation (γ, S) with n phases, the waiting time and several related distributions may be evaluated by the integration of a highly structured system of $mn + m$ linear differential equations with constant coefficients. This result is analogous to that in Theorem 5.2.4.

Theorem 5.5.4: In the stable single server queue with Markov-modulated Poisson arrivals and a service time distribution $H(\cdot)$ of phase type, the vector $\mathbf{W}(x)$, $x \geqslant 0$, may be computed by solving the differential equations

$$\mathbf{W}'(x) = \mathbf{W}(x)(\Lambda - Q) - \mathbf{V}(x)(I \otimes \mathbf{S}^\circ), \tag{5.5.46}$$

$$\mathbf{V}'(x) = \mathbf{V}(x)(I \otimes S) + \mathbf{W}(x)\Lambda \otimes \gamma,$$

with initial conditions $\mathbf{W}(0) = \mathbf{y}_0$, $\mathbf{V}(0) = 0$.

The conditional waiting time distribution $\tilde{W}_j(x)$ of a customer who arrives when the Markov process is in the state j, is given by

$$\tilde{W}_j(x) = \pi_j^{-1} W_j(x), \quad \text{for } x \geqslant 0. \tag{5.5.47}$$

When the mn – vector $\mathbf{V}(x)$ is partitioned into m n – vectors $\mathbf{V}_\nu(x)$, $1 \leqslant \nu \leqslant m$, then $\pi_\nu^{-1}\mathbf{V}_\nu(x)\mathbf{S}^\circ$, $x \geqslant 0$, is the probability distribution of the time-in-system of a customer who arrives while the Markov process Q is in the state ν.

Proof: When $H(\cdot)$ is of phase type with representation (γ, S), the equation (5.5.45) may be written as

$$\mathbf{W}(x) = \mathbf{y}_0 + \int_0^x \mathbf{W}(u)du \; Q + \int_0^x \mathbf{W}(u)\Lambda \; \gamma \exp[S(x-u)]\mathbf{e}] \, du \tag{5.5.48}$$

$$= \mathbf{y}_0 + \int_0^x \mathbf{W}(u)du \; Q + \int_0^x \mathbf{W}(u)\Lambda \otimes \gamma \exp[S(x-u)]du \; (I \otimes \mathbf{e}).$$

Setting

$$\mathbf{V}(x) = \int_0^x \mathbf{W}(u)\Lambda \otimes \gamma \exp[S(x-u)]du \quad \text{for } x \geqslant 0, \tag{5.5.49}$$

it follows that

$$\mathbf{V}(x)[I \otimes \exp(-Sx)] = \int_0^x \mathbf{W}(u)\Lambda \otimes \gamma \exp(-Su)du.$$

Upon differentiation, we obtain

$$\mathbf{V}'(x) = \mathbf{V}(x)(I \otimes S) + \mathbf{W}(x)\Lambda \otimes \gamma, \quad \text{for } x \geqslant 0, \tag{5.5.50}$$

which is the second equation in (5.5.46). Writing (5.5.48) as

$$\mathbf{W}(x) = \mathbf{y}_0 - \int_0^x \mathbf{W}(u)du \; Q + \mathbf{V}(x)\,(I \otimes \mathbf{e}),$$

and differentiating, we get

$$\mathbf{W}'(x) = -\mathbf{W}(x)Q + \mathbf{V}'(x)(I \otimes \mathbf{e}). \tag{5.5.51}$$

However, from (5.5.50) it follows that

$$\mathbf{V}'(x)(I \otimes \mathbf{e}) = -\mathbf{V}(x)(I \otimes \mathbf{S}^{\circ}) + \mathbf{W}(x)\Lambda,$$

and substituting into (5.5.51), we obtain the first equation in (5.5.46).
•

Remark: Once the vector $\mathbf{y}_0$ is computed, the system (5.5.46) is ideally suited for numerical integration. Although its coefficient matrix is formally of dimension $mn + m$, it may be conveniently rewritten as

$$\mathbf{W}'(x) = \mathbf{W}(x)(\Lambda - Q) - [\mathbf{V}_1(x)\mathbf{S}^\circ, \ldots, \mathbf{V}_m(x)\mathbf{S}^\circ], \qquad (5.5.52)$$

$$\mathbf{V}_\nu'(x) = \mathbf{V}_\nu(x)S + \lambda_\nu W_\nu(x)\gamma, \quad \text{for } 1 \leqslant \nu \leqslant m,$$

so that no computations with matrices of order higher than $\max(m,n)$ are involved.

B. THE c-SERVER QUEUE WITH CONSTANT SERVICE TIMES

The model to be discussed in this section is noteworthy in that a Markov chain of $M/G/1$ type arises, not as an embedded chain, but as the description of a stationary queue at an arbitrary time. It is also, in a broadly generalized form, related to one of the oldest results in the theory of queues. As was shown by Crommelin in 1932, the stationary density of the queue length at an arbitrary time in the $M/D/c$ queue is the invariant probability vector of the same Markov chain as arises in *Bailey's bulk service queue* (Section 2.1 - Example B) with

$$a_k = e^{-\lambda a} \frac{(\lambda a)^k}{k!}, \quad \text{for } k \geqslant 0.$$

The parameters λ and a are the Poisson arrival rate and the duration of each service. The $M/D/c$ queue is stable if and only if $\lambda a c^{-1} = \rho < 1$. We shall show that these results may be generalized to any c-server queue with constant service times (of length a) and a *versatile Markovian point process* as input.

Let $P(k;a)$, $k \geqslant 0$, be the matrices of formula (5.4.1) evaluated at $t = a$. The elements $P_{jj'}(k;a)$ are the conditional probabilities that k arrivals occur during $(0,a)$ in the versatile Markovian point process and that the Markov process Q^* is in the state j' at time a, given that it was in the state j, $1 \leqslant j, j' \leqslant m$, at time 0.

The stochastic matrix P is defined by

$$P = \begin{vmatrix} P(0;a) & P(1;a) & P(2;a) & P(3;a) & P(4;a) & \cdots \\ P(0;a) & P(1;a) & P(2;a) & P(3;a) & P(4;a) & \cdots \\ \cdot & \cdot & \cdot & \cdot & \cdot & \\ \cdot & \cdot & \cdot & \cdot & \cdot & \\ P(0;a) & P(1;a) & P(2;a) & P(3;a) & P(4;a) & \cdots \\ 0 & P(0;a) & P(1;a) & P(2;a) & P(3;a) & \cdots \\ 0 & 0 & P(0;a) & P(1;a) & P(2;a) & \cdots \\ 0 & 0 & 0 & P(0;a) & P(1;a) & \cdots \\ \cdot & \cdot & \cdot & \cdot & \cdot & \\ \cdot & \cdot & \cdot & \cdot & \cdot & \end{vmatrix}, \quad (5.5.53)$$

where the first $c + 1$ rows of blocks are identical.

The vector $\mathbf{x}(\tau)$, $\tau \geqslant 0$, has components $x_{ij}(\tau)$, $i \geqslant 0$, $1 \leqslant j \leqslant m$, listed in lexicographic order on i and j. The quantity $x_{ij}(\tau)$ is the steady-state probability that at time τ, there are i customers in the system and the process Q^* is in the state j.

A key observation is that any customers *in service* at time τ and *only those* leave the system during the time interval $(\tau, \tau+a]$. The queue lengths $\xi(\tau)$ and $\xi(\tau+a)$ are therefore related by

$$\xi(\tau+a) = [\xi(\tau) - c]^+ + \nu(a), \quad (5.5.54)$$

where $\nu(a)$ is the number of arrivals during $(\tau, \tau+a]$. It readily follows from (5.5.54) that

$$\mathbf{x}(\tau+a) = \mathbf{x}(\tau)P. \quad (5.5.55)$$

The vector $\mathbf{x}(\tau)$ should not depend on τ, and, if it exists, is the invariant probability vector $\mathbf{x}$ of the stochastic matrix P.

The matrix P is partitioned into blocks of order cm as was done for the scalar case displayed in (2.1.3). The first row of blocks has elements B_k, $k \geqslant 0$. Each matrix B_k consists of c identical rows of $m \times m$ matrices. The blocks in the other rows are denoted by A_k.

By (5.4.4), it is clear that

$$P^*(z;a) = \sum_{k=0}^{\infty} P(k;a)z^k = \exp[R(z)a], \quad (5.5.56)$$

and hence $P^*(1;a) = \exp(Q^*a)$. The vector π of formula (5.1.7) is therefore also the invariant probability vector of $P^*(1;a)$.

As in the scalar case, discussed in Section 2.3, the matrix $A = \sum_{k=0}^{\infty} A_k$, has the *block-circulant* structure

$$\begin{vmatrix} \tilde{A}_0 & \tilde{A}_1 & \tilde{A}_2 & \cdots & \tilde{A}_{c-1} \\ \tilde{A}_{c-1} & \tilde{A}_0 & \tilde{A}_1 & \cdots & \tilde{A}_{c-2} \\ \tilde{A}_{c-2} & \tilde{A}_{c-1} & \tilde{A}_0 & \cdots & \tilde{A}_{c-3} \\ \cdots & & \cdots & & \cdots \\ \tilde{A}_1 & \tilde{A}_2 & \tilde{A}_3 & \cdots & \tilde{A}_0 \end{vmatrix},$$

where

$$\tilde{A}_r = \sum_{\nu=0}^{\infty} P(c\nu+r;a), \quad \text{for } 0 \leqslant r \leqslant c-1.$$

and

$$\sum_{r=0}^{c-1} \tilde{A}_r = P^*(1;a).$$

This readily implies that the invariant probability vector π^* of A is given by

$$\pi^* = c^{-1}[\pi, \pi, \ldots, \pi]. \tag{5.5.57}$$

Lemma 5.5.2: The vector $\beta^* = \sum_{k=1}^{\infty} kA_k \mathbf{e}$ is finite and the inner product $\rho = \pi^* \beta^*$, is given by

$$\rho = c^{-1}\xi^* a, \tag{5.5.58}$$

where $\xi^* = \pi R'(1)\mathbf{e}$, is the fundamental arrival rate of the versatile Markovian point process. The stationary probability vector $\mathbf{x}$ of P exists if and only if $\rho < 1$.

Proof: We partition the column vector β^* into c m-vectors $[\beta_0, \beta_1, \ldots, \beta_{c-1}]^T$. The structure of the matrices A_k, $k \geqslant 0$, then implies that for $0 \leqslant j \leqslant c-1$,

$$\beta_j = \sum_{h=1}^{c-1} \sum_{k=1}^{\infty} kP(kc+h;a)\mathbf{e} + \sum_{h=1}^{c-1} \sum_{k=0}^{\infty} P(kc+h;a)\mathbf{e},$$

and upon summation over j, we obtain

$$\sum_{j=0}^{c-1} \beta_j + \sum_{\nu=1}^{\infty} \nu P(\nu;a)\mathbf{e} = V_1(a)\mathbf{e},$$

where $V_1(a)\mathbf{e}$ is as given in formula (5.4.6). The vector β^* is therefore finite and

$$\pi^* \beta^* = c^{-1}\pi \sum_{j=0}^{c-1} \beta_j = c^{-1}\pi V_1(a)\mathbf{e} = c^{-1}\xi^* a.$$

By Theorem 3.2.1, the Markov chain P is therefore positive recurrent if and only if $\rho < 1$. •

Henceforth assuming that $\rho < 1$, all general results for stochastic matrices of $M/G/1$ type may be stated for the matrix P. The matrix G, the vector $\tilde{\mu}_1$ and other such items will be defined as in the general theory presented in Chapters 2 and 3. We shall only stress those properties that are new or specific to the present queueing model.

Since the matrices A_k, $k \geqslant 1$, are positive and A_0 has positive diagonal elements, the matrix G is positive. Unfortunately, the special structure of the matrices A_k, $k \geqslant 0$, does not impart any simplifying structure to the matrix G, so that the analysis of this model crucially involves a positive matrix of dimension cm.

It is now convenient to partition several of the vectors and matrices in an appropriate way. The row vector $\mathbf{x}$ is partitioned into vectors $\mathbf{x}_i$, $i \geqslant 0$, of dimension cm. The vector $\tilde{\mu}_1$ of formula (3.1.12) is partitioned into c m-vectors $\tilde{\mu}_1(\nu)$, $0 \leqslant \nu \leqslant c-1$. The matrix G is partitioned into c^2 blocks of order m according to

$$G = \begin{vmatrix} G(0,0) & G(0,1) & \cdots & G(0,c-1) \\ G(1,0) & G(1,1) & \cdots & G(1,c-1) \\ \cdots & \cdots & & \cdots \\ G(c-1,0) & G(c-1,1) & \cdots & G(c-1,c-1) \end{vmatrix}, \quad (5.5.59)$$

As the c rows of blocks in the matrices B_k, $k \geqslant 0$, are identical and agree with the first row of blocks in the corresponding matrices A_k, $k \geqslant 0$, it readily follows from the equality

$$G = \sum_{k=0}^{\infty} A_\nu G^k,$$

that the matrix $\sum_{k=0}^{\infty} B_k G^k$ consists of c identical rows of blocks and that each row agrees with the *first* row of the matrix G.

Let the matrix $\hat{G}$ be defined by

$$\hat{G} = \sum_{r=0}^{c-1} G(0,r). \quad (5.5.60)$$

The invariant probability vector of the positive stochastic matrix $\hat{G}$ of order m is denoted by $\hat{\mathbf{g}}$.

Theorem 5.5.5: The vector $\mathbf{X}_0$ is given by

$$\mathbf{x}_0 = [\,\hat{\mathbf{g}}\tilde{\boldsymbol{\mu}}_1(0)\,]^{-1}\,[\,\hat{\mathbf{g}}G(0,0), \hat{\mathbf{g}}G(0,1), \ldots, \hat{\mathbf{g}}G(0,c-1)\,]. \quad (5.5.61)$$

The stationary probability that an arriving customer does not have to wait is

$$\mathbf{x}_0\mathbf{e} = [\,\hat{\mathbf{g}}\tilde{\boldsymbol{\mu}}_1(0)\,]^{-1}, \quad (5.5.62)$$

and the fraction of time θ_0^* that the system is empty is given by

$$\theta_0^* = [\hat{\mathbf{g}}\tilde{\boldsymbol{\mu}}_1(0)]^{-1}\,\hat{\mathbf{g}}G(0,0)\mathbf{e}. \quad (5.5.63)$$

Proof: The particular structure of the matrices B_k, $k \geqslant 0$, entails that the matrix $\tilde{K}(z)$ (s is irrelevant here) of formula (3.2.1) consists of c identical rows of $m \times m$ blocks; each row agreeing with the first row of blocks in $\tilde{G}(z)$. It is therefore clear that the invariant probability vector $\boldsymbol{\kappa}$ of $K = \tilde{K}(z)$, is given by

$$\boldsymbol{\kappa} = [\hat{\mathbf{g}}G(0,0), \hat{\mathbf{g}}G(0,1), \ldots, \hat{\mathbf{g}}G(0,c-1)].$$

This same particular structure of $\tilde{K}(z)$ implies that the vector $\tilde{\boldsymbol{\kappa}}_1$ is of the form $[\mathbf{u},\mathbf{u}, \ldots, \mathbf{u}]$ where $\mathbf{u} = \tilde{\boldsymbol{\mu}}_1(0)$. The inner product $\boldsymbol{\kappa}\tilde{\boldsymbol{\kappa}}_1$ is therefore given by

$$\boldsymbol{\kappa}\tilde{\boldsymbol{\kappa}}_1 = \hat{\mathbf{g}}\tilde{\boldsymbol{\mu}}_1(0).$$

Formula (5.5.61) now follows from (3.2.6) and the remaining statements are clear by interpretation. •

The structure of the matrices B_k, $k \geqslant 0$, also leads to a number of obvious simplifications in the calculation of the vector $\mathbf{x}_1$. In particular, the inverse $(I - B_0)^{-1}$ is given by

$$\begin{vmatrix} I + CP(0;a) & CP(1;a) & \cdots & CP(c-1;a) \\ CP(0;a) & I + CP(1;a) & \cdots & CP(c-1;a) \\ \cdots & \cdots & & \cdots \\ CP(0;a) & CP(1;a) & \cdots & I + CP(c-1;a) \end{vmatrix},$$

where the matrix C of order m is defined by

$$C = [I - \sum_{r=0}^{c-1} P(r;a)]^{-1}.$$

We see that only the inversion of an $m \times m$ matrix is required and that storage of the inverse $(I - B_0)^{-1}$ may be avoided.

For this model, the computation of the vector $\mathbf{x}_1$ by means of (3.2.9) is not essential as only the vector $\mathbf{x}_0$ appears explicitly in the formulas for moments and probability generating functions. If it is computed, the internal accuracy check $\mathbf{x}_0 = \mathbf{x}_0 B_0 + \mathbf{x}_1 A_0$, may be performed.

Moment Formulas: As for the models in Section 4.2, there is considerable advantage in exploiting the particular structure of the blocks A_k and B_k, $k \geqslant 0$, in the computation of the moments of the queue length. To that end, we partition the stationary probability vector $\mathbf{x}$ into m−vectors $[\mathbf{y}_0, \mathbf{y}_1, \mathbf{y}_2, \cdots]$. The vector $\mathbf{x}_0$, given by Theorem 5.5.5, corresponds to the vectors $\mathbf{y}_0, \ldots, \mathbf{y}_{c-1}$.

By routine calculations, we obtain from (5.5.53) that the generating function $\mathbf{Y}(z) = \sum_{i=0}^{\infty} \mathbf{y}_i z^i$, satisfies

$$\mathbf{Y}(z)[z^c I - P^*(z;a)] = \sum_{\nu=0}^{c-1} \mathbf{y}_\nu (z^c - z^\nu) P^*(z;a). \tag{5.5.64}$$

The recursive scheme of Section 3.3 may, with minor modifications, also be used to evaluate the vectors $\mathbf{Y}^{(n)}(1-)$ and the factorial moments $L_n^* = \mathbf{Y}^{(n)}(1-)\mathbf{e}$, $n \geqslant 1$, of the queue length at an arbitrary time. We set

$$\mathbf{U}(z) = \sum_{\nu=0}^{c-1} \mathbf{y}_\nu (z^c - z^\nu) P^*(z;a), \tag{5.5.65}$$

from which we routinely obtain that

$$\mathbf{U}^{(r)}(1) = \sum_{k=1}^{r} \binom{r}{k} \mathbf{d}(k) V_{r-k}(a), \quad \text{for } r \geqslant 1, \tag{5.5.66}$$

where the matrices $V_k(a)$, $k \geqslant 0$, are as defined in (5.1.12) with $t = a$, and

$$\mathbf{d}(k) = \frac{c!}{(c-k)!} \sum_{\nu=0}^{c-1} \mathbf{y}_\nu - \sum_{\nu=k}^{c-1} \frac{\nu!}{(\nu-k)!} \mathbf{y}_\nu, \quad \text{for } 0 \leqslant k \leqslant c-1,$$

$$= 0, \quad \text{for } k \geqslant c.$$

It is also clear that $\mathbf{U}(1) = 0$, and

$$\mathbf{U}^{(r)}(1)\mathbf{e} = \sum_{k=1}^{r} \binom{r}{k} \mathbf{d}(k)\boldsymbol{\beta}_{r-k}(a),$$

where $\boldsymbol{\beta}_k(a) = V_k(a)\mathbf{e}$, $k \geqslant 1$. The matrices $V_k(a)$ and the vectors $\boldsymbol{\beta}_k(a)$ are obtained by numerical integration of the differential equations (5.4.5).

From the equation (5.5.64), it is clear that $\mathbf{Y}(1-) = \boldsymbol{\pi}$, and, after n differentiations, that

$$\mathbf{Y}^{(n)}(1)[I - P^*(1;a)] = \mathbf{U}^{(n)}(1) \tag{5.5.67}$$

$$+ \sum_{\nu=1}^{n} \binom{n}{\nu} \mathbf{Y}^{(n-\nu)}(1)\, V_\nu(a) - \sum_{\nu=1}^{\min(c,n)} \binom{n}{\nu} \frac{c!}{(c-\nu)!} \mathbf{Y}^{(n-\nu)}(1) \quad \text{for } n \geqslant 1.$$

Introducing, as in Section 3.3, the matrix

$$Z = [I - P^*(1;a) + \mathbf{e}\boldsymbol{\pi}]^{-1},$$

we obtain

$$\mathbf{Y}^{(n)}(1) = \{\, \mathbf{U}^{(n)}(1) + \sum_{\nu=1}^{n} \binom{n}{\nu} \mathbf{Y}^{(n-\nu)}(1-)\, V_\nu(a) \tag{5.5.68}$$

$$- \sum_{\nu=1}^{\min(c,n)} \binom{n}{\nu} \frac{c!}{(c-\nu)!} \mathbf{Y}^{(n-\nu)}(1)\, \}Z + L_n^* \boldsymbol{\pi}.$$

By postmultiplying by $\boldsymbol{\beta}_1(a)$ in both sides of (5.5.68) and by writing the equation (5.5.67) for $n+1$ and postmultipying by $\mathbf{e}$, we obtain upon equating the expressions for $\mathbf{Y}^{(n)}(1-)\boldsymbol{\beta}_1(a)$, that L_n^* is given by

$$L_n^* = [(n+1)(c - a\,\xi^*)]^{-1} \{\, (n+1)\theta_n + \mathbf{U}^{(n+1)}(1)\mathbf{e} \tag{5.5.69}$$

$$+ \sum_{\nu=2}^{n+1} \binom{n+1}{\nu} \mathbf{Y}^{(n+1-\nu)}(1-)\boldsymbol{\beta}_\nu(a) - \sum_{\nu=2}^{\min(c,n+1)} \binom{n+1}{\nu} \frac{c!}{(c-\nu)!} \mathbf{L}_{n-\nu+1}^* \},$$

where θ_n, $n \geqslant 1$, is defined by

$$\theta_n = \{ \mathbf{U}^{(n)}(1) + \sum_{\nu=1}^{n} \binom{n}{\nu} \mathbf{Y}^{(n-\nu)}(1-) V_\nu(a) \tag{5.5.70}$$

$$- \sum_{\nu=1}^{\min(c,n)} \binom{n}{\nu} \frac{c!}{(c-\nu)!} \mathbf{Y}^{(n-\nu)}(1) \} Z \beta_1(a).$$

Provided that moments of sufficiently high order of the realizable group size densities in the arrival process exist, the formulas (5.5.68-70) permit, in principle, the recursive computation of any number of moments of the queue length density.

If we postmultiply by $\mathbf{e}$ in the equation (5.5.67) for $n = 1$, we obtain the equality

$$\sum_{\nu=0}^{c-1} (c-\nu)\mathbf{y}_\nu \mathbf{e} + a\,\xi^* = c,$$

which, rewritten as,

$$ca^{-1}[\,1 - \sum_{\nu=0}^{c-1} \mathbf{y}_\nu \mathbf{e}\,] + a^{-1} \sum_{\nu=0}^{c-1} \nu \mathbf{y}_\nu \mathbf{e} = \xi^*, \tag{5.5.71}$$

shows that in the stationary queue, the departure rate is equal to the stationary arrival rate ξ^*.

For $n = 1$, we record the moment formulas obtained from (5.5.68-70), as well as somewhat more explicit formulas which are obtained after routine matrix calculations using (5.5.71) and

$$\pi V_1(a) = a\,\xi^* \pi + \pi R'(1)(\mathbf{e}\pi - Q^*)^{-1} [I - P^*(1;a)]. \tag{5.5.72}$$

Formula (5.5.72) is proved by the same calculations as led to (5.4.6). The second expression for L_1^* is obtained by using formula (5.4.11) for $\pi \beta_2(a)$.

We have

$$\theta_1 = [\sum_{\nu=0}^{c-1} (c-\nu)\mathbf{y}_\nu P^*(1;a) + \pi V_1(a)]\, Z \beta_1(a) - c\,\xi^* a$$

$$= \sum_{\nu=0}^{c-1} (c-\nu)\mathbf{y}_\nu (Z - I)\beta_1(a) - a\,\xi^{*2} + \pi R'(1)(\mathbf{e}\pi - Q^*)^{-1} \beta_1(a),$$

$$L_1^* = [2(c - a\,\xi^*)]^{-1} \{ 2\theta_1 + \pi\beta_2(a) - c(c-1) \tag{5.5.73}$$

$$+ 2\sum_{\nu=0}^{c-1} (c-\nu)\mathbf{y}_\nu \beta_1(a) + \sum_{\nu=0}^{c-1} [c(c-1) - \nu(\nu-1)]\, \mathbf{y}_\nu \mathbf{e} \}$$

$$= a\,\xi^* + [2(c - a\,\xi^*)]^{-1} \{\xi_2^* a + a\,\xi^*(1 + a\,\xi^*) - 2c\,\xi^*$$

$$+ 2[\, a\pi R'(1) + \sum_{\nu=0}^{c-1} (c-\nu)\mathbf{y}_\nu] (\mathbf{e}\pi - Q^*)^{-1} R'(1)\mathbf{e}$$

$$- c^2 [1 - \sum_{\nu=0}^{c-1} \mathbf{y}_\nu \mathbf{e}] - \sum_{\nu=0}^{c-1} \nu^2 \mathbf{y}_\nu \mathbf{e} \},$$

$$\mathbf{Y}'(1-) = (L_1^* - c)\pi + [\sum_{\nu=0}^{c-1} (c-\nu)\mathbf{y}_\nu P^*(1;a) + \pi V_1(a)]Z$$

$$= (L_1^* - \xi^*)\pi + \sum_{\nu=0}^{c-1} (c-\nu)\mathbf{y}_\nu (Z - I) + \pi R'(1)\, (\mathbf{e}\pi - Q^*)^{-1}.$$

In the case of Poisson arrivals, there are of course substantial simplifications, since $P^*(1;a) = Z = 1$, and $\xi^* = \lambda$, $\beta_1(a) = a\,\lambda$, $\beta_2(a) = (a\,\lambda)^2$. We then recover the well-known formula

$$L_1^* = a\,\lambda + [2(c - a\,\lambda)]^{-1} \{ a\,\lambda(1 + a\,\lambda) - c^2[1 - \sum_{\nu=0}^{c-1} y_\nu] - \sum_{\nu=0}^{c-1} \nu^2 y_\nu \}.$$

The Virtual Waiting Time Distribution: Crommelin's elegant argument, that gives the stationary distribution of the virtual waiting time for the $M/D/c$ queue, may be generalized to give the corresponding distribution for the queue with a versatile Markovian arrival process.

We first need some preliminary material. For τ with $0 \leqslant \tau < a$, let $h_{ij}(\tau)$, $i \geqslant 0$, $1 \leqslant j \leqslant m$ be the stationary probability that at time 0, the Markov process Q^* is in the state j and there are *at most* i customers in the system who will still be there at time τ. The m-vector $\mathbf{h}_i(\tau)$, $i \geqslant 0$, has the components $h_{ij}(\tau)$, $1 \leqslant j \leqslant m$.

Lemma 5.5.3: The sequences of vectors $\mathbf{h}_i(\tau)$ and $\mathbf{y}_i$, $i \geqslant 0$, are related by

$$\sum_{r=0}^{c-1} \mathbf{h}_r(\tau) P(i-r;\tau) = \sum_{r=0}^{i} \mathbf{y}_r, \quad \text{for } i \geqslant 0, \tag{5.5.74}$$

and

$$\mathbf{h}(z;\tau) = \sum_{i=0}^{\infty} \mathbf{h}_i(\tau) z^i = (1-z)^{-1}\mathbf{Y}(z)\exp[-R(z)\tau], \tag{5.5.75}$$

for $|z| < 1$, where $\mathbf{Y}(z)$ is given by (5.5.64).

Proof: The queue at time τ is made up of those customers present at time 0, who remain in the system past time τ and of those arriving during $(0,\tau]$. Since $\tau < a$, no customers who enter in $(0,\tau]$ leave prior to time τ. Since the arrival process in $(0,\tau]$ is independent of the number of those present at time 0, who survive, the equation (5.5.74) follows. Upon taking generating functions, we obtain (5.5.75).
•

For each fixed τ, the system of equations (5.5.74) is an infinite linear system with the block upper triangular coefficient matrix

$$\begin{vmatrix} P(0;\tau) & P(1;\tau) & P(2;\tau) & P(3;\tau) & \cdots \\ 0 & P(0;\tau) & P(1;\tau) & P(2;\tau) & \cdots \\ 0 & 0 & P(0;\tau) & P(1;\tau) & \cdots \\ \cdot & \cdot & \cdot & \cdot & \\ \cdot & \cdot & \cdot & \cdot & \end{vmatrix}.$$

Since $P(0;\tau) = \exp[R(0)\tau]$, for $\tau \geqslant 0$, which is a nonsingular matrix, the equations (5.5.74) uniquely determine the vectors $\mathbf{h}_i(\tau)$, $i \geqslant 0$, for each τ in $[0,a)$.

We observe that (5.5.75) is also equivalent to the differential equation

$$\mathbf{h}_\tau'(z;\tau) = -\mathbf{h}(z;\tau)R(z), \tag{5.5.76}$$

with initial condition $\mathbf{h}(z;0) = (1-z)^{-1}\mathbf{Y}(z)$. This shows that the vectors $\mathbf{h}_i(\tau)$ may also be obtained by solving an infinite system of linear differential equations on the interval $0 \leqslant \tau < a$.

For any nonnegative number t, we set $t^* = \left[\frac{t}{a}\right]$, and $\tau(t) = t - at^*$. Let $W_j(t)$ be the probability that in the stationary queue, the Markov chain Q^* is in the state j at time 0 and the virtual waiting time at time 0 does not exceed t. Customers are served in the order of arrival. The m-vector $\mathbf{W}(t)$ has components $W_j(t)$, $1 \leqslant j \leqslant m$.

Theorem 5.5.6: The vector $\mathbf{W}(t)$, $t \geqslant 0$, is given by

$$\mathbf{W}(t) = \mathbf{h}_{ct^*+c-1}[\tau(t)]. \tag{5.5.77}$$

Proof: The virtual waiting time at time 0 *exceeds* t if and only if c or more of the customers in the system at time 0 are still present at time t. In that case, there must be $c + ct^*$ or more surviving customers in the system at time $\tau(t)$, since exactly c customers depart during each of the time intervals $[\tau(t) + (\nu-1)a, \tau(t) + \nu a)$, $1 \leqslant \nu \leqslant t^*$. In order that the waiting time at time 0 does *not* exceed t, there can therefore be at most $ct^* + c - 1$ customers in the queue at time $\tau(t)$, who were already present at time 0. This immediately yields equation (5.5.77). •

We note that for any integer $t^* \geqslant 0$,

$$\mathbf{W}(t^*a) = \mathbf{h}_{ct^*+c-1}(0) = \sum_{r=0}^{ct^*+c-1} \mathbf{y}_r, \tag{5.5.78}$$

by (5.5.74) since $P(k;0) = \delta_{k0}I$, for $k \geqslant 0$. As t^* tends to infinity in (5.5.78), we find that $\mathbf{W}(\infty) = \boldsymbol{\pi}$, as is to be anticipated.

The computation that follows deals with the mean vector

$$\mathbf{b} = \int_0^\infty t d\mathbf{W}(t) = \int_0^\infty [\pi - \mathbf{W}(t)]dt \tag{5.5.79}$$

$$= \int_0^a \sum_{\nu=0}^\infty [\pi - \mathbf{h}_{c\nu+c-1}(u)]du.$$

The last equality in (5.5.79) is obtained by partitioning the domain of integration into intervals of length a and using (5.5.77). The quantity $\overline{W}_j = \pi_j^{-1} b_j$, $1 \leqslant j \leqslant m$, is clearly the conditional mean virtual waiting time, given that at time 0, the Markov process Q^* is in the state j.

The derivation of an explicit expression for the vector $\mathbf{b}$ is quite involved and requires several manipulations that are far from obvious. The quantities $\omega_k = \exp(2\pi i \frac{k}{c})$, $0 \leqslant k \leqslant c-1$, are the c-th roots of unity and the Vandermonde matrix Ω has the elements ω_k^j, for $0 \leqslant j,k \leqslant c-1$. The row vectors $\mathbf{h}^*(z)$ and $\mathbf{h}^*(j;z)$, $0 \leqslant j \leqslant c-1$, are defined by

$$\mathbf{h}^*(z) = \int_0^a \sum_{i=0}^\infty [\pi - \mathbf{h}_i(u)]z^i du, \quad \text{for } |z| < 1, \tag{5.5.80}$$

and

$$\mathbf{h}^*(j;z) = \int_0^a \sum_{\nu=0}^\infty [\pi - \mathbf{h}_{c\nu+j}(u)]z^{c\nu+j} du, \quad \text{for } |z| < 1. \tag{5.5.81}$$

The vectors $\mathbf{h}^*(\omega_j z)$ and $\mathbf{h}^*(j;z)$, $0 \leqslant j \leqslant c-1$, are the rows of the $c \times c$ matrices $\tilde{H}^*(z)$ and $H^*(z)$ respectively.

Theorem 5.5.7: The vector $\mathbf{h}^*(z)$, is given by

$$\mathbf{h}^*(z) = (1-z)^{-1}\{a\pi R(z)P^*(z;a)] + (1-z^c)\mathbf{Y}(z) \tag{5.5.82}$$

$$+ \sum_{\nu=0}^{c-1} \mathbf{y}_\nu (z^c - z^\nu)P^*(z;a)\}[P^*(z;a)R(z)]^{-1},$$

for $|z| < 1$. The vector **b** is the last row of the matrix $H^*(1-)$, which is given by

$$H^*(1-) = \Omega^{-1}\tilde{H}^*(1-). \tag{5.5.83}$$

For $1 \leqslant j \leqslant c-1$, the rows $\mathbf{h}^*(\omega_j)$ of $\tilde{H}^*(1-)$ are obtained by substituting $z = \omega_j$, into the right hand side of (5.5.82), so that

$$\mathbf{h}^*(\omega_j) = (1 - \omega_j)^{-1} a \boldsymbol{\pi} + \sum_{\nu=0}^{c-1} \mathbf{y}_\nu (1 - \omega_j^\nu) \, [R(\omega_j)]^{-1}. \tag{5.5.84}$$

The first row of $\tilde{H}^*(1-)$ is given by

$$\mathbf{h}^*(1-) = (aL_1^* - \frac{1}{2}a^2\xi^* - c)\boldsymbol{\pi} \tag{5.5.85}$$
$$+ [a\boldsymbol{\pi}R'(1) + \sum_{\nu=0}^{c-1} (c-\nu)\mathbf{y}_\nu] \, (\mathbf{e}\boldsymbol{\pi} - Q^*)^{-1},$$

provided that the second moments of all the realizable group densities in the arrival process are finite.

Proof: From (5.5.75) and (5.5.80), it is clear that for $|z| < 1$,

$$\mathbf{h}^*(z) = (1 - z)^{-1}\boldsymbol{\pi} a - (1 - z)^{-1}\mathbf{Y}(z)\int_0^a \exp[-R(z)u]\,du,$$

which is equivalent to (5.5.82), since $R(z)$ is nonsingular for $|z| < 1$, and $\mathbf{Y}(z)P^*(z;a)$ may be eliminated by using formula (5.5.64).

Provided that the limit as $z \to 1-$ exists, (5.5.79) implies that $\mathbf{b} = \mathbf{h}^*(c-1;1-)$. By elementary properties of the roots of unity, it is clear that

$$\mathbf{h}^*(\omega_j z) = \sum_{k=0}^{c-1} \omega_j^k \mathbf{h}^*(j;z), \quad \text{for } 0 \leqslant j \leqslant c-1,$$

or equivalently

$$H^*(z) = \Omega^{-1}\tilde{H}^*(z), \quad \text{for } |z| < 1.$$

The components of the vector $\mathbf{h}^*(z)$ are generating functions of sequences of *positive* numbers. They are also analytic functions inside the unit disk. Their only possible singularity on the circle $|z| = 1$, can therefore be at $z = 1$. For $1 \leqslant j \leqslant c-1$, the vectors $\mathbf{h}^*(\omega_j)$ are hence well-defined and are obtained by setting $z = \omega_j$, into the right hand side of (5.5.82). The expressions in (5.5.84) follow after elementary simplifications.

For $z = 1-$, we proceed as follows. We begin by rewriting the equation (5.5.82) as

$$(1 - z)\mathbf{h}^*(z)P^*(z;a)R(z) = a\pi R(z)P^*(z;a) + (1 - z^c)\mathbf{Y}(z)$$

$$+ \sum_{\nu=0}^{c-1} \mathbf{y}_\nu(z^c - z^\nu)P^*(z;a).$$

That equation is twice differentiated with respect to z and we set $z = 1-$. In the second equality so obtained, we also postmultiply be $\mathbf{e}$. This produces the following two equalities.

$$-\mathbf{h}^*(1-)P^*(1;a)Q^* = -c\pi + [a\pi R'(1) + \sum_{\nu=0}^{c-1} (c-\nu)\mathbf{y}_\nu]P^*(1;a),$$

and

$$-2\mathbf{h}^*(1-)P^*(1;a)R'(1)\mathbf{e} = a\pi R''(1) + 2a\pi R'(1)\beta_1(a) - 2c\mathbf{Y}'(1-)$$

$$-c(c-1) + \sum_{\nu=0}^{c-1} [c(c-1) - \nu(\nu-1)]\,\mathbf{y}_\nu\mathbf{e} + 2\sum_{\nu=0}^{c-1} (c-\nu)\mathbf{y}_\nu\beta_1(a).$$

Since $P^*(1;a) = \exp(Q^*a)$, is nonsingular, the first equation leads to

$$\mathbf{h}^*(1-) = [\mathbf{h}^*(1-)\mathbf{e} - c]\pi \tag{5.5.86}$$

$$+ [a\pi R'(1) + \sum_{\nu=0}^{c-1} (c-\nu)\mathbf{y}_\nu](\mathbf{e}\pi - Q^*)^{-1}.$$

It remains to show that

$$\mathbf{h}^*(1-)\mathbf{e} = aL_1^* - \frac{1}{2}a^2\xi^*. \tag{5.5.87}$$

To that end, we postmultiply by $-2P^*(1;a)R'(1)\mathbf{e}$ in (5.5.86) and equate the right hand side to the previously obtained expression for the quantity $-2\mathbf{h}^*(1-)P^*(1;a)R'(1)\mathbf{e}$. In the equality so obtained, we replace $\pi R''(1)\mathbf{e}$ by ξ_2^*, $\beta_1(a)$ by the expression in formula (5.4.6) with $t = a$, and $\sum_{\nu=0}^{c-1}(c-\nu)\mathbf{y}_\nu\mathbf{e}$ by $c - a\xi^*$, by virtue of formula (5.5.71).

After lengthy, but routine calculations, this leads to

$$2\xi^*\mathbf{h}^*(1-)\mathbf{e} = 2cL_1^* + c^2 - \sum_{\nu=0}^{c-1}(c^2-\nu^2)\,\mathbf{y}_\nu\mathbf{e} - a\,\xi_2^* - a\,\xi^*$$

$$-\ 2ac\,\xi^* + 2c\,\xi^* - 2[a\,\pi R'(1) + \sum_{\nu=0}^{c-1}(c-\nu)\mathbf{y}_\nu](\mathbf{e}\pi - Q^*)^{-1}R'(1)\mathbf{e},$$

and by using the second expression for L_1^* in formula (5.5.73), this may be written as

$$2\xi^*\mathbf{h}^*(1-)\mathbf{e} = 2cL_1^* - 2ac\,\xi^* - [\,2(L_1^* - a\,\xi^*)(c - a\,\xi^*) - a^2\xi^{*\,2}\,].$$

From that equality, (5.5.87) follows immediately. •

Remark *a*. Formula (5.5.87) written as

$$\mathbf{h}^*(1-)\mathbf{e} = \int_0^a (L_1^* - \xi^*u)\,du,$$

shows that $a^{-1}\mathbf{h}^*(1-)\mathbf{e}$ may be interpreted as the expected number of survivors at a point chosen at random in $(0,a)$.

Remark *b*. For $c = 1$, the vector $\mathbf{h}^*(1-)$ agrees with the vector $-\mathbf{w}'(0)$ of formula (5.5.41) for the single server queue with versatile Markovian input and *constant* service time. We then have

$$L_1^* = a\,\xi^* + [2(1 - a\,\xi^*)]^{-1}\,[\,\xi_2^* a + (a\,\xi^*)^2 - 2\xi^* + 2\mathbf{u}^* R\,'(1)\mathbf{e}\,],$$

$$W_1^* = aL_1^* - \frac{1}{2}a^2\xi^*,$$

$$-\,\mathbf{w}'(0) = (W_1^* - 1)\boldsymbol{\pi} + \mathbf{u}^*,$$

where

$$\mathbf{u}^* = [\,a\,\boldsymbol{\pi} R\,'(1) + \mathbf{y}_0\,]\,(\mathbf{e}\boldsymbol{\pi} - Q^*)^{-1}.$$

For the $M/\,D/\,1$ queue, these expressions reduce to $\mathbf{u}^* = 1$, $L_1^* = a\,\lambda\,[1 + \frac{1}{2}\lambda(1 - a\,\lambda)]$, and $W_1^* = \frac{1}{2}a^2\lambda\,[1 + \lambda(1 - a\,\lambda)^{-1}]$.

We conclude our treatment of the c –server queue with constant service times and versatile Markovian input by deriving an explicit procedure to evaluate the vector $\boldsymbol{\beta}^* = \sum_{k=1}^{\infty} kA_k\,\mathbf{e}$, which is needed in Lemma 5.5.2. If we introduce the $m \times m$ matrices

$$U_r(z) = \sum_{\nu=0}^{\infty} P(\nu c + r, a) z^{\nu}, \quad \text{for } 0 \leqslant r \leqslant c-1, \tag{5.5.88}$$

the matrix $A^*(z) = \sum_{\nu=0}^{\infty} A_{\nu} z^{\nu}$, is given by

$$\begin{vmatrix} U_0(z) & U_1(z) & U_2(z) & \cdots & U_{c-1}(z) \\ zU_{c-1}(z) & U_0(z) & U_1(z) & \cdots & U_{c-2}(z) \\ zU_{c-2}(z) & zU_{c-1}(z) & U_0(z) & \cdots & U_{c-3}(z) \\ & \cdots & & \cdots & \\ zU_1(z) & zU_2(z) & zU_3(z) & \cdots & U_0(z) \end{vmatrix}. \tag{5.5.89}$$

After routine calculations, we see that the vectors $\boldsymbol{\beta}_{\nu}$, $0 \leqslant \nu \leqslant c-1$, obtained by partitioning the vector $\boldsymbol{\beta}^*$, satisfy the equations

$$\boldsymbol{\beta}_0 = \sum_{r=0}^{c-1} U_r\,'(1)\mathbf{e}, \tag{5.5.90}$$

$$\beta_\nu = \beta_0 + \sum_{r=c-\nu}^{c-1} U_r(1)\mathbf{e}, \quad \text{for } 1 \leqslant \nu \leqslant c-1.$$

It suffices to show how the matrices $U_r(1)$ and $U_r'(1)$ may be computed. Let us set

$$F_r(z) = \sum_{\nu=0}^{\infty} P(\nu c + r, a) z^{\nu c + r} = z^r U_r(z^c), \quad \text{for } 0 \leqslant r \leqslant c-1, \quad (5.5.91)$$

then we readily obtain that

$$U_r(1) = F_r(1), \qquad (5.5.92)$$

$$U_r'(1) = c^{-1}[F_r'(1) - rF_r(1)], \quad \text{for } 0 \leqslant r \leqslant c-1,$$

so that, in turn, it suffices to develop an algorithm for the computation of the matrices $F_r(1)$ and $F_r'(1)$.

Clearly $F_r(z)$ is the matrix power series, obtained by summing the terms with indices congruent to r *mod* (c), of the matrix power series

$$P^*(z;a) = \sum_{\nu=0}^{\infty} P(\nu;a) z^\nu = \exp[R(z)a].$$

We may now write

$$P^*(\omega_k z;a) = \sum_{r=0}^{c-1} \omega_k^r F_r(z), \quad \text{for } 0 \leqslant k \leqslant c-1, \qquad (5.5.93)$$

where the quantities ω_k are the c-th roots of unity. In terms of the matrix Ω, defined earlier in this section, and by introducing the $cm \times m$ matrices $\hat{F}(z)$ and $\hat{P}(z)$, partitioned into $m \times m$ blocks as

$$\hat{F}(z) = [F_0(z), F_1(z), \ldots, F_{c-1}(z)]^T,$$

and

$$\hat{P}(z) = [P^*(z,a), P^*(\omega_1 z,a), \ldots, P^*(\omega_{c-1}z,a)]^T,$$

the equation (5.5.93) may be concisely rewritten as

$$\hat{P}(z) = (\Omega \otimes I)\,\hat{F}(z). \tag{5.5.94}$$

Since the matrix Ω is nonsingular, it follows that

$$\hat{F}(z) = (\Omega^{-1} \otimes I)\,\hat{P}(z),$$

and therefore

$$\hat{F}(1) = (\Omega^{-1} \otimes I)\,\hat{P}(1),$$

$$\hat{F}'(1) = (\Omega^{-1} \otimes I)\,\hat{P}'(1),$$

The matrices $P^*(\omega_k,a)$ and $P^{*\prime}(\omega_k,a)$ for $0 \leqslant k \leqslant c-1$, are most easily computed by numerical integration of systems of differential equations. In order to set up these equations, we need to distinguish carefully between differentiation with respect to z and to t in $P^*(z,t)$. From the equation (5.5.56) with t replacing a, we readily obtain that

$$\frac{\partial}{\partial t} P^*(\omega_k,t) = P^*(\omega_k,t)R(\omega_k),$$

as well as

$$\frac{\partial}{\partial t} M_1(\omega_k,t) = M_1(\omega_k,t)R(\omega_k) + P^*(\omega_k,t)R'(\omega_k),$$

for $0 \leqslant k \leqslant c-1$, where

$$M_1(\omega_k,t) = \left[\frac{\partial}{\partial t} P^*(\omega_k,t)\right]_{z=\omega_k} = P^{*\prime}(\omega_k,t),$$

for $0 \leqslant k \leqslant c-1$. The initial conditions are $P^*(\omega_k,0) = I$, and

$M_1(\omega_k,0) = 0$, for $0 \leqslant k \leqslant c-1$.

Remark: The actual numerical procedure for the evaluation of β^* is straightforward. We begin by inverting the matrix Ω and then we integrate the differential equations for $P^*(\omega_k,t)$ and $M_1(\omega_k,t)$ up to $t = a$. Next we evaluate the matrices $F_k(1)$ and $F_k'(1)$ for the desired value of a. Finally, we compute the matrices $U_k'(1)$ and obtain β^* by substitution into the equations (5.5.90).

NOTES, REFERENCES AND COMMENTS ON CHAPTER 5

Section 5.1

The idea of synthesizing new probability distributions (and stochastic processes) from the exponential distribution goes back to the work of Erlang. It was further developed by Kosten, see e.g. [K-096]. $PH-$distributions were first systematically described in Jensen [J-029], but the powerful matrix formalism for this class was developed in Neuts [N-026]; it is leisurely discussed in Chapter 2 of Neuts [N-042]. The historical development of this subject and the relationship of $PH-$distributions to other classes of probability distributions useful in stochastic models, such as Cox distributions and mixtures of Erlang distributions of successive orders k and common scale parameter, are also discussed there. $PH-$distributions have found use in a variety of applications. The bibliography [N-068] lists, with brief annotations, all references known to date.

An entirely parallel development for discrete $PH-$densities proceeds from the absorption times in discrete parameter Markov chains. The Bernoulli process is there the analogue of the Poisson process and various constructions, such as the discrete $PH-$renewal process and the Markov-modulated Bernoulli process are easily carried out and play a useful role in the study of discrete time queues. Such queues have increasing appeal as models for features of communication systems.

Methods which replace numerical integrations, either of ordinary integrals or of differential equations, by recursive schemes are important in computational mathematics. The examples in the chapter show that, for problems involving PH-distributions, such iterative schemes are often available. The derivation of such schemes for other integrals requires some skill in using the matrix formalism for $PH-$distributions. Related methods of great interest to computational probability are the Laguerre transform [K-036,K-037,S-078], the use of mixtures of Erlang distributions with the same parameter and increasing orders [L-054-L-056,O-019,S-051], and uniformization to solve systems of linear differential equations with constant coefficients.

The $PH-$renewal process and the Markov-modulated Poisson process had been used in earlier discussions of specific queueing models, [N-021,N-027,N-030,N-033,N-036,N-037]. Particular cases of both processes had served as input processes to queues since the earliest days of queueing theory. The introduction of the versatile Markovian point process in Neuts [N-037] was intended to show that these and many other point processes could be handled by a unified matrix formalism. As is shown by examples in the chapter, special and often simpler results can be derived by particularizing the results of the unified treatment.

The closure property for mixtures of convolutions of the form (5.1.17) is proved as Theorem 2.2.5 in Neuts [N-042]. The classical properties of the Kronecker product of matrices, used in many calculations involving PH-distributions, are discussed in Bellman [B-028], Marcus and Minc [M-011] and other texts on matrix algebra. The rapid increase in the dimensions of matrices obtained by forming Kronecker products is of no consequence to theoretical developments. In performing numerical computations, it is important to work with smaller matrices by exploiting the special structure of Kronecker products, as is suggested by the equations (5.1.29). Further examples of such calculations may be found in Neuts and Chakravarthy [N-044], Neuts and Meier [N-045] and Neuts [N-066]. Extensive theoretical use of the formalism of Kronecker products is made in Neuts [N-059], a discussion of the ***caudal characteristic curve*** for classes of queues with matrix-geometric solutions.

Theorem 5.1.5 is the natural extension to PH-renewal processes of elementary properties of counts associated with a pair of independent Poisson processes. The superposition of two PH-renewal processes is the only case of any generality in which counts related to superpositions of renewal processes yield to analytic treatment. See Neuts and Latouche [N-061]. Many special probability distributions with matrix parameters may be defined in relation to this model. Their study appears to be mathematically routine and could be undertaken if warranted by applications. See Problem 5.1.8.

Problem 5.1.1: Consider two independent, irreducible discrete time Markov chains with m and n states respectively. To avoid uninteresting issues, assume that at least one of these chains is aperiodic. Suppose that one state, say state 1, is common to both chains. Show that the point process of the epochs when both chains visit state 1, (coincidences) is a discrete PH-renewal process. Show without calculations that the mean time between coincidences is $(\pi_1 \theta_1)^{-1}$, where π_1 and θ_1 are the steady-state probabilities of state 1 in the two chains. The representation of the PH-density of the time between coincidences involves Kronecker products of matrices. Use properties of the Kronecker product to compute that density, without having to store or compute with matrices of high order. •

This problem is related to the material in Kopocińska and Kopociński [K-095] and in Neuts [N-066].

Problem 5.1.2: Consider an irreducible Markov chain with n states and infinitesimal generator Q. Let E be a proper subset of the state space containing m states. Let X be a positive random variable, independent of the Markov chain and having a distribution of phase type with representation $(\boldsymbol{\alpha},T)$. Suppose that the initial state of the Markov chain is chosen from the set E according to some probability vector $\boldsymbol{\theta}$. Let Y be the time when the *total accumulated time spent in the set* E, reaches X. Show that Y has a $PH-$distribution, construct a representation and discuss how the lower order moments of Y may be efficiently computed. •

Problem 5.1.3: For a $PH-$distribution with irreducible representation $\boldsymbol{\alpha},T$, show that the vectors $\boldsymbol{\alpha}\exp(T\mathbf{u})$ and $\exp(T\mathbf{u})\mathbf{T}^{\circ}$ are strictly positive for all $u > 0$. Prove this and deduce from this property that the matrices $P(k;t)$, $k \geqslant 1$, are positive for all $t > 0$. •

Problem 5.1.4: Let E be a proper subset of the transient states in the absorbing Markov chain describing the phase type distribution with representation $(\boldsymbol{\alpha},T)$. Show that the total time spent in the set E prior to absorption has a PH-distribution and construct a representation, Neuts and Meier [N-045]. •

One of the most important uses of PH-distributions is in the synthesis of probability distributions of the durations of more complex operations. With some experience, the necessary matrix representations can be written down directly and "by interpretation". In teaching such matters, I ask the students to visualize two or more black boxes, each describing absorbing Markov chains, which are restarted upon the occurrence of events in the other chains. The reader may wish to prove Theorem 5.1.4 by such an interpretation, which is clearly more insightful than the technical proof in [N-042]. The matrix formalism elegantly reflects the modular nature of such constructions which are analogous to those of classical linear systems theory. Several examples are discussed in [N-045] and others are stated in the following problems.

Problem 5.1.5: A task has an Erlang distribution with parameters m and μ, and there is an independent Poisson process of system failures of rate θ. If a failure occurs during one of the first k failures, $k < m$, the task is abandoned. If a failure occurs during one of the later phases, the task enters a recovery mode and the durations of the successive recovery periods are independent with common PH-distribution of representation (γ,Γ). Exhibit representations for the

PH-distributions of the total duration of the task and of the total time spent in recovery mode. •

Problem 5.1.6: The duration X of a task has an Erlang distribution with parameters m and μ and there is an independent Poisson process of system failures of rate θ. If a failure occurs during phase k, $1 \leqslant k \leqslant m$, of the Erlang distribution, the task is taken over by another processor, which requires an Erlang distributed time Y with parameters $n(k)$ and σ to complete the operation. The orders $n(k)$ are strictly decreasing in k. A system failure occurring during the second period Y terminates the task (termination mode 2). Successful completions, at time X or at time $X + Y$ are called termination modes 0 or 1 respectively.

Set up an absorbing Markov chain with $m + n(1) + 3$ states of which three are absorbing states to describe the various possible transitions during the task. Compute the probabilities of terminations in modes 0, 1 and 2 and discuss how the conditional distributions of the durations of the task, given the various termination modes may be computed. Write a computer code which, starting with the input parameters, sets up the necessary matrices and proceeds to the computation of various interesting probabilities related to this model. •

Problem 5.1.7: The stationary waiting time distribution $W(\cdot)$ (at arrivals) in an $M/\,PH/\,1$ queue has a *PH*-distribution whose representation is given in formula (5.1.21). The stationary distribution of the *time-in-system* is the convolution $W * F(\cdot)$, for which a representation involving $2m$ phases may readily be constructed. Show, by using the matrix formalism of *PH*-distributions, that this distribution also has the representation $(\boldsymbol{\alpha}, T + \rho \mathbf{T}^{\circ} \boldsymbol{\pi})$, which involves only m phases, (L. Gün, private communication). •

Problem 5.1.8: As shown in Neuts and Latouche [N-061], the superposition of two independent *PH*-renewal processes (with $m(1)$ and $m(2)$ phases), and various counting processes associated with it, may be studied in relation to a Markov renewal process with $m(1) + m(2)$ states. Let N_i, $i \geqslant 1$, be the numbers of renewals in the second process between successive renewals in the first. Express the joint probability density of N_i, $1 \leqslant i \leqslant n$ in terms of the representations of the two interrenewal distributions. Study the moments and correlations of these random variables and prepare the computer codes needed for their evaluation. Use Theorem 5.1.5 and the approach in [N-061]. •

A useful class of semi-Markovian point processes, whose superpositions can be studied in a tractable manner, is introduced in Lucantoni,

Meier-Hellstern and Neuts [L-053]. Superpositions of PH-renewal processes are a particular case. In Latouche [L-004], a different approach to Markov renewal processes with PH-distributed sojourn times is discussed.

Section 5.2

The $PH/G/1$ queue is analytically so tractable that it deserves to play a prominent model in texts on queueing theory. Moreover, many important equations for the $GI/G/1$ queue may be obtained by a continuity argument so that, even from a theoretical viewpoint, the restriction to PH-distributed interarrival times is minor. Extensive numerical experience with a model such as the $PH/PH/1$ queue greatly enhances one's understanding of its physical behavior. See Ramaswami and Latouche [R-015].

Theorem 5.2.3 and its consequences are of great interest. They suggest that a generalization of the methods of this book to nonlinear operator equations, similar to that carried out by Tweedie [T-069], could lead to a fully Markovian analysis of the $GI/G/1$ queue. If such a generalization is successful, we believe that the classical Lindley equation [L-027] and the resulting Wiener-Hopf analysis [C-047,D-003,D-004,H-044,H-061,K-063,M-069,P-044,R-053] could be obtained as consequences.

The mnemonic recurrence relations for the moments of the waiting times in the $GI/G/1$ queue were derived earlier by Marshall [M-015] by path function arguments. Clearly, all difficulties in implementing these formulas follow from the boundary behavior of the model as, in general, moments of the idle period are analytically intractable.

Problem 5.2.1: For the $PH/G/1$ queue, let $\Psi_{j,j'}(x;y,k)$ be the conditional probability that a busy period, which starts with an *amount of work* x at time $t = 0$, and with the arrival phase j, terminates no later than time y, $y \geqslant x$, with the arrival process in the phase j', and involves the service of k, $k \geqslant 0$, additional customers. Define $\Psi^*(x;s,z)$ as the transform matrix

$$\Psi^*(x;s,z) = \sum_{k=0}^{\infty} \int_x^{\infty} e^{-sy}\, d\Psi(x;y,k)\, z^k .$$

Show that

$$\Psi^*(x_1 + x_2; s, z) = \Psi^*(x_1; s, z)\Psi^*(x_2; s, z),$$

for all nonnegative x_1 and x_2.

By considering the cases where arrivals occur in $(0,x)$ and where there are none, write an integral equation for the matrix $\Psi^*(x; s, z)$. Deduce from that equation the explicit formula

$$\Psi^*(x; s, z) = \exp\{[T + T^\circ \alpha \tilde{G}(z, s) - sI]\, x\},$$

and use it to establish for the $PH/G/1$ queue, the analogue

$$\tilde{G}(z, s) = \int_0^\infty \exp\{[T + T^\circ \alpha \tilde{G}(z, s) - sI]\, x\} dH(x),$$

of Takács' equation for the $M/G/1$ queue. •

Remark: The result in Problem 5.2.1 is not a ready consequence of the classical equation for the matrix $\tilde{G}(z, s)$. This new equation has several useful implications. In particular, it may be used to prove *analytically* the factorization results for the waiting time distribution in the $GI/G/1$ queue with vacations, established by earlier by different methods. See [L-053] and [N-067]. Further extensions are given in Problem 5.5.2.

Section 5.3

Arrivals at the transitions of a finite state Markov chain were considered in Ushizawa [U-001]. The book by Grandell [G-034] is an excellent reference on doubly stochastic Poisson processes. The construction of the time-stationary version of the Markov-modulated Poisson process is an application of the corresponding construction for positive recurrent Markov renewal processes, as discussed in Pyke [P-053,P-054] and Çinlar [C-025]. The theorem establishing the stringent conditions for a doubly stochastic Poisson process to be a renewal process was proved by Kingman [K-065]. Its application to the Markov-modulated Poisson process is new. The problem of fitting a Markov-modulated Poisson process to data merits much further investigation. For the case with two rates, an adaptive algorithm was proposed by Meier-Hellstern [M-028]. General related statistical methodology is treated in Karr [K-005]. Uses of the Markov-modulated Poisson process in queues are further examined in Burman and Smith [B-100], Heffes and Lucantoni [H-036], Knessl, Matkowsky, Schuss and Tier [K-078,K-079], Naor and Yechiali [N-008], Neuts [N-021,N-036,N-037,N-042,N-067], Purdue [P-048,P-049,P-052], Ramaswami [R-006,R-013], Regterschot and de Smit [R-026], Sengupta [S-028,S-030], Van Hoorn and Seelen [V-005] and Yechiali [Y-003].

The problems in this section are based on our ongoing investigation of mathematically tractable descriptors of the physical behavior of useful point processes. Among such descriptors are the *moments of the counting process* $N(\cdot)$, discussed in the text, the *peakedness functional,* the *power spectral density* of a square wave associated with the point process and the *caudal characteristic curve,* which was introduced in Neuts [N-059]. All these descriptors are interpretable only after computer implementation of the matrix formulas which provide the natural analytic language for their expressions. In all cases, experience is essential in relating numerical results to physical behavior.

Problem 5.3.1: From the homogeneous Poisson process, defined on the entire real line, one obtains a stationary stochastic process $X(t), -\infty < t < +\infty$, called the *random telegraph wave,* by defining $X(t)$ to be +1 or -1 on alternating intervals between Poisson events. That process is a standard example in texts on spectral methods for second order processes. One may easily show that the autocovariance function $E[X(t)X(t+\tau)] = R(\tau)$, is given by $\frac{1}{4}e^{-2\lambda|\tau|}$, so that the corresponding power spectral density, which is the inverse Fourier transform

$$R^*(\omega) = \int_{-\infty}^{+\infty} e^{-i\omega\tau} R(\tau)\,d\tau,$$

of $R(\tau)$ evaluated at $\omega = 2\pi f$, is given by

$$S(f) = \frac{1}{4}\,\frac{\lambda}{\lambda^2 + \pi^2 f^2},$$

for all real values of f.

Define the random telegraph wave corresponding to a time-stationary (Q, Λ)-source and show that the corresponding formulas are given by

$$(\tau) = \frac{1}{4}\pi \exp[\,(Q - 2\Lambda)\,|\,\tau\,|\,]\mathbf{e},$$

and

$$S(f) = \pi\Lambda[\,4\pi^2 f^2 I + (Q - 2\Lambda)^2\,]^{-1}\mathbf{e}.$$

Establish the existence of the inverse in the expression for $S(f)$ and write an efficient computer code to evaluate that function for $f \geqslant 0$. Plot the ratio of the power spectral densities $S(f)$ and $S^*(f)$ of a homogeneous Poisson process of the same fundamental rate as the (Q,Λ)-source for selected examples and qualitatively interpret the graphs so obtained. •

Problem 5.3.2: A *Bernoulli thinning* with parameter p of a stationary point process is the point process obtained by saving each point of the original process with probability p or deleting it with probability $1-p$, according to independent Bernoulli trials. Show that a Bernoulli thinning of a stationary Markov-modulated Poisson process is again a Markov-modulated Poisson process with fundamental mean $E_p = (p\lambda^*)^{-1}$. Next, rescale the thinned process by the time scale transformation $t \rightarrow p^{-1}t$. Show that the power spectral density $S(f)$ of the telegraph wave obtained from the rescaled thinning is given by the formula in Problem 5.3.1 with Q replaced by $p^{-1}Q$. Which changes do you expect to see in the power spectral density as the parameter p decreases from one to zero? •

Problem 5.3.3: Consider a stationary renewal process with underlying probability distribution $F(\cdot)$, which is not of lattice type. Denote by μ_1' and by $\phi(\cdot)$ respectively the mean and the Laplace-Stieltjes transform of $F(\cdot)$. Construct the random telegraph wave for the stationary renewal process and show that its power spectral density $S(f)$ is given by

$$S(f) = \frac{1}{4\mu_1'\pi^2 f^2}\,\frac{1 - |\,\phi(2\pi i f)\,|^2}{1 + |\,\phi(2\pi i f)\,|^2 + 2\mathrm{Re}\phi(2\pi i f)},$$

for all real f. Discuss the behavior of $S(f)$ at $f = 0$. When $F(\cdot)$ is a PH-distribution with representation (α, T), show that $S(f)$ may be written as

$$S(f) = \frac{1}{4\mu_1'} \frac{2\psi_1(f)[1 - 2\pi^2 f^2 \psi_1(f)] - \psi_2^2(f)}{[1 - 2\pi^2 f^2 \psi_1(f)]^2 + \pi^2 f^2 \psi_2^2(f)},$$

where

$$\psi_1(f) = \alpha[4\pi^2 f^2 I + T^2]^{-1}\mathbf{e},$$

and

$$\psi_2(f) = \alpha[4\pi^2 f^2 I + T^2]^{-1}\mathbf{T}^\circ.$$

Establish the existence of the inverse in the preceding expressions, which are very well suited for numerical computations. •

Problem 5.3.4: The (exponential) *peakedness function* $\Psi(\mu)$ of a stationary point process is the ratio of the variance and the mean of the number of busy servers at an arbitrary time in an infinite server queue with exponential service times of rate μ and the given point process as input. Show that for a (Q, Λ)-source, the function $\Psi(\mu)$ is given by

$$\Psi(\mu) = 1 - \mu^{-1}\lambda^* + \lambda^{*\,-1}\pi\Lambda(\mu I - Q)^{-1}\Lambda\mathbf{e},$$

where $\lambda^* = \pi\Lambda\mathbf{e}$. Verify that

$$1 + \lambda^{*\,-1}\pi\Lambda(\mu I - Q)^{-1}\Lambda\mathbf{e},$$

is the expected number of events in $[0,\mu)$ in the point process where 0 is an arbitrary *arrival epoch*. We may therefore write

$$\Psi(\mu) = E_{arr}[N(\mu)] - E_{stat}[N(\mu)],$$

since $\mu^{-1}\lambda^*$ is the expected number of events in $[0,\mu)$ in the stationary (Q, Λ)–source. That equality is valid for any regular Markov renewal arrival process. Prove that assertion and use it to show that the peakedness function for the rescaled Bernoulli thinning of the Markov-modulated Poisson process, described in Problem 5.3.2, is given by the same explicit formula but with Q replaced by $p^{-1}Q$.

The "service time" distribution in the infinite service queue need not be exponential. For a general delay distribution $H(\cdot)$, the peakedness becomes a functional of H which, for some particular choices of H is numerically tractable. The peakedness functional is used in communications engineering as a measure of the burstiness of a point process. See Heffes and Holtzman [H-033]. A discussion of its general properties and further references may be found in Eckberg [E-004,E-005]. We note that the moments of $N(t)$ and the peakedness functional quantify *average local behavior* of the point process, while the square wave spectrum and the caudal characteristic curve are more sensitive to its global properties.

Section 5.4

In the following problems, we describe a class of Markov renewal processes which includes the PH-renewal process and the Markov-modulated Poisson process as special cases.

Problem 5.4.1: Consider an $m-$state irreducible infinitesimal generator Q and write Q as $C + D$, where D is a nonnegative, nonzero matrix. C is a matrix with negative diagonal elements and nonnegative off-diagonal elements. We construct a point process as follows. At time $t = 0$, an initial state i_0 is chosen according to the probability vector $\boldsymbol{\alpha}$. The sojourn time in that state has an exponential distribution with parameter $-Q_{i_0 i_0}$. At the end of that first interval, either an arrival occurs and the next state i_1 is then selected with conditional probability $(D_{i_0 i_1})(-Q_{i_0 i_0})^{-1}$, where i_1 may be equal to i_0, *or* no arrival occurs and the next state i_1, which must be different from i_0, is then selected with conditional probability $(C_{i_0 i_1})(-Q_{i_0 i_0})^{-1}$. The sojourn time in the state i_1 is independent of the past and is exponential with parameter $-Q_{i_1 i_1}$. The preceding construction is repeated at the end of that sojourn time and so on indefinitely. We keep track of the times between arrivals and of the state of the Markov chain Q at arrivals, to obtain a Markov renewal sequence.

Based on this informal description, give a formal definition of the Markov renewal sequence. As in the case of the Markov-modulated Poisson process, show that the bivariate sequence of the times between successive arrivals and of the states immediately following arrivals, is also a Markov renewal sequence and determine its transition probability matrix. Show that the $PH-$renewal process is obtained by setting $C = T$, and $D = \mathbf{T}^{\circ}\boldsymbol{\alpha}$, whereas the Markov-modulated Poisson process corresponds to the choices $C = Q - \Lambda$ and $D = \Lambda$. [L-053] •

Problem 5.4.2: Construct the time-stationary version of the process in Problem 5.4.1. State and prove the analogue of Theorem 5.3.3. •

Problem 5.4.3: For the point process in Problem 5.4.1, let $J(t)$ be the "state" at time t and $N(t)$ the number of arrivals in the interval $(0,t]$. Set $N(0) = 0$. Show that the process $\{[J(t),N(t)]\}$, $t \geqslant 0$, is a Markov process and discuss the differential equations, probability generating functions and moment formulas for that process (as was done for the cases treated in the text). •

Problem 5.4.4: Consider an alternating PH-renewal arrival process with underlying probability distributions of representations (β,S) and (γ,L). Each renewal generates one arrival. Show that the process so obtained is a particular case of the process of Problem 5.4.1 and determine the matrices C and D. •

Problem 5.4.5: Consider an arrival process with a first order Markovian dependence between the successive interarrival intervals $\{X_n, n \geqslant 1\}$, by specifying that the random variables X_n have Erlang distributions with the same scale parameters and orders I_n, which form an irreducible Markov chain on the integers $1, \ldots, m$. Define this process formally and show that it may be viewed as a particular case of the process defined in Problem 5.4.1. •

Problem 5.4.6: Consider the time-stationary version of the process in Problem 5.4.1 and derive the probability distributions of the forward and backward recurrence times at time t. Derive the probability distribution of the interarrival interval to which t belongs. •

Problem 5.4.7: Establish the generalizations of the formulas in Problem 5.3.1 for the power spectral density of a random telegraph wave constructed from the arrival process in Problem 5.4.1. Examine the cases of a $PH-$renewal and of an alternating $PH-$renewal process in detail and simplify the formulas as much as possible. •

Problem 5.4.8: Use the general result to be derived in Problem 5.4.7 to compute the ratio of the power spectral densities of the random telegraph waves of the departure process and of the (Poisson) arrival process for an $M/PH/1$ queue of finite capacity K and also for the $M/M/c$ queue in which at most $K+c$ customers can be present. Examine various cases and discuss the inferences concerning the behavior of these queues which can be drawn from the resulting graphs. •

Problem 5.4.9: Show that the product process of two independent processes of the type introduced in Problem 5.4.1, is again a process of that type and express its parameter matrices in Kronecker products. Prove that the power spectral density for the superposition of the two point processes is the *convolution* of the power spectral densities of the component processes. Write a computer code to evaluate the power spectral density of a (Q, Λ)–source, use the preceding result and a fast Fourier transform algorithm to examine how fast the power spectral density of the superposition of n independent copies of the (Q, Λ)–source approaches that of the limiting Poisson process. For each n, the time scale should be adjusted so that the fundamental rate of the superposition remains the same as for each component process. •

Problem 5.4.10: Consider the vacation model of Problem 1.1.1, but with the process of Problem 5.4.1 as arrival process. Show that the matrices A_k and B_k satisfy matrix recurrence relations similar to the scalar formulas of the simpler case with Poisson arrivals. When the service time distribution is of phase type, use the formalism of Kronecker products to compute these matrices without numerical integrations. Explore the use of uniformization in the corresponding computations for general service time distributions. See [L-052,L-053] and [R-012]. •

Problem 5.4.11: *Quasi periods of a point process:* The covariance formula (5.4.13) may be used to identify time spans between windows in which there are likely to be many events in the point process. This may be done as follows: Compute and plot the function

$$\Psi(t) = (\lambda^* t)^{-1}\sigma^2(t),$$

which tends to a constant as t tends to infinity. In interesting cases, $\Psi(t)$ will have one or more local maxima. Select, for example, the values of t corresponding to the first local maximum and to the global maximum. Next, use formula (5.4.13) with $t = t'$ and search for values of t_1 for which there is maximum correlation between the counts $N(t)$ and $N(t_1 + 2t) - N(t + t_1)$. Test your code on examples whose physical behavior is well understood. Can you suggest alternative ways of choosing the length t of the window?

Section 5.5

The theory of the single server queue with versatile Markovian input and general, independent service times was developed by Ramaswami in his Ph. D. dissertation and published in [R-006]. Our presentation essentially follows his treatment. The $PH/G/1$ queue is a particular case of special interest, as a number of final results which do not explicitly depend on the PH–formalism are (by continuity) also valid for the $GI / G / 1$ queue. See Neuts [N-060]. The continuity argument, which is not proved in detail in this book, is highly intuitive, but its formal proof is lengthy and quite technical. It consists in essence in establishing that the functionals, such as the moments of the stationary waiting time, are continuous on the set of $PH-$distributions. These form a dense set in the class of all probability distributions on $[0,\infty)$ and the unique continuous extension theorem of general topology then guarantees that results established for the $PH / G / 1$ queue, which do not depend on the $PH-$formalism, also hold for general interarrival time distributions. We refer to Barbour [B-021], Helm and Schassberger [H-038], Hordijk and Schassberger [H-072] and Whitt [W-007] for detailed discussions.

Problem 5.5.1: In the example dealing with the single server queue with a Markov-modulated arrival process (the $MMPP / G / 1$ queue), it is stated that, when the service time distribution is of phase type, the matrices A_k and B_k may be evaluated *without performing numerical integrations.* Use properties of the Kronecker product to establish matrix recurrence relations for these sequences of matrices. Show further how these recurrence relations may be used to avoid storage of these coefficient matrices, of course, at the expense of greater computational effort. An alternative is to adapt the algorithms proposed by Ramaswami and Lucantoni [L-052,R-012,R-013] to this model. •

Problem 5.5.2: State and prove the analogous results to those in Problem 5.2.1 for the queue with a Markov-modulated Poisson arrival process and a general service time distribution. •

Remark: The same analysis, but with much more notation, can extend the forms of the equations of interest to a single server queue with versatile Markovian input and general service times.

Problem 5.5.3: Consider a single server $MMPP / G / 1$ queue with a (Q, Λ) source as its arrival process. The service times of successive customers may depend on the state of the Markov process Q at the beginnings of these services, but are mutually conditionally independent, given these states. The service time distributions of all customers, starting service in the "phase" j of the process Q are identical, are denoted by $H_j(\cdot)$ and have finite non-zero means α_j, $1 \leqslant j \leqslant m$. The matrices $A_k(\cdot)$, $k \geqslant 0$, are given by

$$A_k(x) = \int_0^x d\Delta[\mathbf{H}(u)]P(k;u), \quad \text{for } x \geqslant 0.$$

The matrix $\Delta[\mathbf{H}(\cdot)]$ is an $m \times m$ diagonal matrix with the probability distributions $H_j(\cdot)$ as its diagonal elements. The matrices $P(k;u)$ are as defined in formula (5.4.1), but for the special case of a Markov-modulated Poisson process. In order to distinguish between the stationary probability vector of Q and the invariant probability vector of the matrix A, we denote the former by $\boldsymbol{\theta}$ and the latter by $\boldsymbol{\pi}$. Notice that these vectors are different in general. They agree in the particular case where the service time distributions do not depend on the state of the Markov process Q at the beginnings of services.

The queue is stable if and only if

$$\rho = \boldsymbol{\pi}\boldsymbol{\beta} < 1,$$

where the column vector $\boldsymbol{\beta}$ is given by $\boldsymbol{\beta} = \sum_{k=1}^{\infty} kA_k(\infty)\mathbf{e}$. Show that the vector $\boldsymbol{\beta}$ may be explicitly written as

$$\boldsymbol{\beta} = \lambda^* \boldsymbol{\alpha} + (I - A)(\mathbf{e}\boldsymbol{\theta} - Q)^{-1}\Lambda\mathbf{e},$$

and that the traffic intensity ρ for the *MMPP / G / 1* queue is given by

$$\rho = \lambda^*(\boldsymbol{\pi}\boldsymbol{\alpha}),$$

where $\lambda^* = \boldsymbol{\theta}\Lambda\mathbf{e}$, is the fundamental arrival rate of the input process.

Prove that, for the general version of the stable queue, the vectors $\tilde{\boldsymbol{\mu}}_1$ and $\tilde{\boldsymbol{\mu}}_1^*$ of formulas (3.1.12) and (3.1.13) may be written as

$$(1 - \rho)\tilde{\boldsymbol{\mu}}_1 = \mathbf{e} + (I - G)(\mathbf{e}\boldsymbol{\theta} - Q)^{-1}\Lambda\mathbf{e} + \lambda^*(I - G)Z\boldsymbol{\alpha},$$

$$\tilde{\boldsymbol{\mu}}_1^* = (\boldsymbol{\pi}\boldsymbol{\alpha})\tilde{\boldsymbol{\mu}}_1 + (1 - \rho)^{-1}(I - G)Z\boldsymbol{\alpha},$$

All symbols in these formulas have their usual significance. In particular, the matrix Z is the inverse of $I - A + \mathbf{e}\boldsymbol{\pi}$. Discuss the simplifications which arise when the distributions $H_j(\cdot)$ are the same. •

The model of Problem 5.5.3 was discussed by transform methods in Neuts [N-021]. The recent results stated in Problems 5.5.2 and 5.5.3, which are based on the methods of this book, are given in Neuts [N-

067]. It should be noted that the equation asked for in Problem 5.5.2 does not hold when the service times depend on the phase at the starts of services.

Problem 5.5.4: The following is a nice decision problem, which involves quite detailed computations for the *MMPP* / *G* / 1 queue. The manager of an *M* / *G* / 1 queue has to decide whether to accept a secondary input stream of customers, with the same service time distribution as the *M* / *G* / 1 queue. That input stream has a regular succession of brief, but high intensity bursts of arrivals, which may be modelled as an *MMPP* with a cyclic matrix Q of order $m+1$. The first m states have exponential sojourn times with parameter μ. The state $m+1$ has an exponential sojourn time with a larger parameter θ. Arrivals occur at a quite high rate λ', only during the sojourns in state $m+1$. (Note that this arrival process is also a *PH* – renewal process.)

The secondary input process will be accepted provided that the resulting queue remains stable and that the waiting times of customers of the original Poisson stream (of rate λ) are not too severely affected. Formalize a variety of acceptance criteria, such as the mean waiting time of such customers and various percentiles of that waiting time distribution. Study these criteria numerically as functions of the arrival rate λ.

We notice that the secondary arrival stream consists of Poisson arrivals during brief exponential intervals, separated by Erlang distributed gaps. Study the effect of the variability of these gaps by varying the order of the Erlang distributions while keeping the mean gap length the same.

Suppose that, with the secondary stream is accepted, the orginal customers will try to "beat the system" by avoiding arrivals close to the bursty arrivals of the other stream. In order to model this behavior, we imagine that the Poisson stream of rate λ will be replaced by an *MMPP* of rates λ_j, $1 \leqslant j \leqslant m+1$, operative during the various phases of the Markov chain Q. This rearrangement of the original input should, however, conserve the arrival rate λ. Propose various such rearrangements as ways of "beating the system" and assess their impact on the criteria considered earlier. Note that the problem of determining *optimal rates* λ_j, is a massive nonlinear programming problem whose objective function requires computation of the selected criteria. •

The model of Problem 5.5.4 is an example of the interplay of engineering design decisions with the analysis of stochastic models. The algorithms for the steady-state characteristics of the *MMPP* / *G* / 1 become the tools for the study of various implications of model parameters and modes of operation. By developing first an efficient algorithm for the *MMPP* / *G* / 1 model and then interacting with it in a conversational mode, the design problem can be effectively investigated. With imagination, the reader will find that variations upon this theme are abundantly available for many other models treated in this book.

The elegant arguments, which are extended here to the c −server queue, were first proposed for the $M/D/c$ queue by Crommelin [C-085,C-086]. Crommelin also demonstrated a clear awareness of the numerical difficulties associated with the analysis of queues by techniques of complex analysis. Erlang arrivals are considered in [X-001]. It is clear that a numerical implementation of our analysis for even moderate values of the number of servers c remains a monumental task. All numerical studies of the c −server queue with constant service time appear to have proceeded by very high truncations of the matrix P, followed by massive iterative computations. While we have performed numerical studies for many of the other models, discussed in this book, we have not done so for the c −server queue with constant service times. We can therefore not comment on the performance of the proposed algorithms. Some of the matrix-analytic derivations in our treatment are offered primarily for their methodological interest. Steps that are entirely elementary for the scalar case $M/D/c$ become quite demanding in the matrix case. For example, the lengthy and technical derivations in the proof of Theorem 5.5.7 extend, what is in the scalar case, a simple application of de l'Hospital's rule. Until the demands of applications warrant a numerical implementation, our treatment of this model should be viewed as somewhat of a matrix-analytic *tour de force,* whose study may benefit the reader in other contexts.

Problem 5.5.5: For the stationary c −server queue, obtain a matrix formula for the probability p_j that j service completions occur during the interval $(0,a)$, $0 \leqslant j \leqslant c$. Show that the expected number of departures during the interval $(0,a)$ is equal to $\xi^* a$. •

6

Selected Special Models

6.1. INTRODUCTION

The general structural properties of Markov renewal processes of $M/G/1$ type, discussed in Chapters 2 and 3, provide a methodological framework for a wide variety of stochastic models. In Chapters 3 and 4, we have applied these properties to several classical queues with Poisson or versatile Markovian input processes which are of recognized applicability. Variants of these and other special models, described in the notes following the preceding chapters, offer many opportunities for further elaboration and study of the theory presented so far. It should be remembered, however, that the special models of actual applications each raise their own particular questions, so that further work is usually required to express the needed quantities in terms of the constructs of the general theory.

In this final chapter, we present analyses of some selected models in greater detail and with special emphasis on results which are significant to the applications in which these arise. These are offered as examples of the types of refinements, which the user of special structural models may wish to explore beyond the results of the general theory. The reasons for the selection of these specific models are briefly stated in the following listing.

A. Discrete Models

1. **The Odoom-Lloyd-Ali Khan-Gani Dam:** For this model, the extreme sparsity of the transition probability and its special structure lead to highly explicit results. In particular, the matrix G and the vector $\mathbf{x}_0$ are given by explicit matrix formulas. This model is therefore chosen to illustrate the advantages of a thorough ***exploitation of special structure.***

2. **A Data Communications Model:** This model deals with a queue with constant service time in which the input is the superposition of a number of independent two-state discrete-time Markov chains. It is selected as an illustration of the utility, particularly in data communications, of queues with quantized time. The input

process is an example of a discrete analogue of the versatile Markovian process and the analysis makes extensive use of the formalism of Kronecker products, which arise naturally in the construction of product chains of independent Markov processes.

3. **A Poor Man's Satellite.** For some purposes of data transmission, it is possible to bounce signals off ionization layers created by the arrivals of meteorite showers to the Earth's upper atmosphere. These layers so fulfill one of the tasks of communication satellites. Although meteorite showers arrive frequently, the useful life time of the ionization layer, created by each shower, is very short. Moreover, there is a transmission delay so that, when the layer vanishes, data packets that are under way are lost and must be retransmitted. The ionization layer may be viewed as an intermittently available server processing a queue of packets waiting at the transmitting node. It is appropriate to describe that queue by a discrete time model in which the time unit corresponds to the time required to send one packet. Under plausible assumptions on the alternating periods of availability and absence of the ionization layer, we may model the queue of packets as a Markov chain of $M/G/1$ type. The service interruptions cause the matrix G for this model to have the reducible form discussed in Section 3.4. We selected this model for its imaginative modelling features and to illustrate a case in which reducibility arises naturally. Our discussion is inspired by the work of Robert, Mitrani and King on this application and follows the analysis of Chandramouli, Neuts and Ramaswami.

B. A Continuous Model

Two Servers in Series with a Finite Intermediate Buffer: Continuous time models are already amply represented by the examples in the earlier chapters. We therefore discuss only one more system, consisting of two units in series separated by a finite capacity buffer. This model is selected as it illustrates the phenomenon of *blocking*. In addition, although this model is merely an $M/SM/1$ queue with a modified boundary, significant analytic difficulties result from seemingly minor boundary modifications. In the discussion of the distributions of the queue lengths in continuous time and of the waiting times, these analytic difficulties are readily apparent. Our treatment may serve as a guide to the study of similar models.

6.2. DISCRETE MODELS

A. The Odoom-Lloyd-Ali Khan-Gani Dam.

The dam is considered at equally spaced times and the difference between these successive times is chosen as the unit of time. The amount of water removed per unit of time is constant. It is chosen as the unit in which the inflows and the content of the dam are measured. The inflow X_n during the time interval $(n-1, n)$ is added to the dam just prior to time n and, if possible, one unit is drained at time n. The inflows $\{X_n, n \geqslant 0\}$ form a Markov chain with states $0, 1, \ldots, m$ and transition probability matrix $B = \{b_{jj'}\}$, $0 \leqslant j, j' \leqslant m$. The stochastic matrix B is irreducible. As discussed in Section 2.1, the sequence of random pairs $\{(Y_n, X_n)\}$, $n \geqslant 0$, where Y_n is the content of the dam at time $n+$, is a Markov chain of $M/G/1$ type, whose transition probability matrix P has the structure

$$\begin{vmatrix} A_0+A_1 & A_2 & A_3 & A_4 & A_5 & \cdots \\ A_0 & A_1 & A_2 & A_3 & A_4 & \cdots \\ 0 & A_0 & A_1 & A_2 & A_3 & \cdots \\ 0 & 0 & A_0 & A_1 & A_2 & \cdots \\ \cdot & \cdot & \cdot & \cdot & \cdot & \end{vmatrix}.$$

In our discussion, it is advantageous to use the matrix P_1, displayed in (2.1.5), rather than the essentially equivalent Markov chain with transition probability matrix P_2. Let π be the invariant probability vector of the matrix B, and β, the column vector with components

$$\beta_j = \sum_{r=0}^{m} r b_{jr}, \quad \text{for } 0 \leqslant j \leqslant m,$$

then the Markov chain P is positive recurrent, if and only if

$$\rho = \pi\beta = \sum_{r=1}^{m} r\pi_r < 1.$$

When $\rho < 1$, the highly degenerate structure of the matrices A_k implies that the matrix G is zero except for its first column whose elements are all equal to one. In order to exploit the special structure of the matrices A_k further, we introduce some special notation. We first partition the matrix B in the form

$$B = \begin{vmatrix} b_{00} & \mathbf{d} \\ \mathbf{c} & C \end{vmatrix},$$

and we partition the vector π as $[\pi(0), \pi(1)]$. By straightforward calculations, we then obtain that

$$\pi(0) = [\,1 + \mathbf{d}(I - C)^{-1}\mathbf{e}\,]^{-1}, \tag{6.2.1}$$

and

$$\pi(1) = [\,1 + \mathbf{d}(I - C)^{-1}\mathbf{e}\,]^{-1}\,\mathbf{d}(I - C)^{-1}.$$

Of the matrix $G(z)$, only the first column is nonzero. We shall write $G_0(z)$ for its first element and write the remaining elements as the column vector $\gamma(z)$. We also introduce diagonal matrices $\Delta[G_0(z)]$, $\Delta_1[G_0(z)]$, and $\Delta_2[G_0(z)]$, whose diagonal elements are respectively given by

$$1,\ G_0(z),\ G_0^2(z),\ \ldots,\ G_0^{m-1}(z),$$

$$0,\ 1,\ 2G_0(z),\ 3G_0^2(z),\ \ldots,\ (m-1)G_0^{m-2}(z),$$

and

$$0,\ 0,\ 2,\ 2\cdot 3G_0(z),\ 3\cdot 4G_0^2(z),\ \ldots,\ (m-2)(m-1)G_0^{m-3}(z).$$

Clearly, for $z = 1$, $\Delta(1) = I$, while $\Delta_1(1) = \Delta_1$ and $\Delta_2(1) = \Delta_2$ have the diagonal elements $0, 1, \ldots, (m-1)$ and $0, 0, 1\cdot 2, \ldots, (m-2)(m-1)$.

The classical nonlinear matrix equation for $G(z)$ may then be written as

$$G_0(z) = zb_{00} + z\,\mathbf{d}\Delta[G_0(z)]\,\gamma(z), \tag{6.2.2}$$

$$\gamma(z) = z\mathbf{c} + zC\,\Delta[G_0(z)]\,\gamma(z).$$

Differentiating twice in the equations (6.2.2) and setting $z = 1$, leads after some simplifications, to the following expressions for the mean and the second factorial moments of the probability generating functions arising in the first column of $G(z)$:

$$G_0'(1) = \{\pi(0)[\,1 - \mathbf{d}(I - C)^{-1}\Delta_1\mathbf{e}\,]\}^{-1}, \tag{6.2.3}$$

$$\gamma'(1) = (I - C)^{-1}\mathbf{e} + (I - C)^{-1}C\,\Delta_1\mathbf{e}G_0'(1).$$

and

$$G_0''(1) = [\,1 - \mathbf{d}(I - C)^{-1}\Delta_1\mathbf{e}\,]^{-1}\mathbf{d}(I - C)^{-1} \tag{6.2.4}$$

$$\cdot\{\,3\Delta_1\mathbf{e}G_0'(1) + \Delta_2\mathbf{e}G_0'^{\,2}(1) + 2\gamma'(1) + \Delta_1\gamma'(1)G_0'(1)\,\},$$

$$\gamma''(1) = (I - C)^{-1}C\{\,(2I + \Delta_1G_0'(1))\gamma'(1)$$

$$+ (3\Delta_1\mathbf{e}G_0'(1) + \Delta_2\mathbf{e}G_0'^{\,2}(1)) + \Delta_1\mathbf{e}G_0''(1)\,\}.$$

Formulas (6.2.3) and (6.2.4) are clearly in a form that is easily implemented computationally. Particularly for the second moments, the corresponding formulas for general Markov chains of $M/G/1$ type are much more belabored. These formulas find an immediate application in the use of the central limit theorem to construct a normal approximation to the distribution of *the time until first emptiness,* starting from an initial state (k,j) with a large value of the dam content k. To see this, we note that the number of time units to reach the level 0 from the state (k,j) has a (discrete) probability density, whose generating function is given by the element with index j in the first column of the matrix $G^k(z)$. It is clear that the matrix $G^k(z)$ equals $G_0^{k-1}(z)\;G(z)$, so that the generating function of interest is given by

$G_0^k(z)$, for $j = 0$, and by $G_0^{k-1}(z)\gamma_j(z)$, for $1 \leqslant j \leqslant m$. With the stated formulas for the first and second moments, the application of the central limit theorem is entirely routine.

In order to obtain an explicit expression for the vector $\mathbf{x}_0$, we note that the matrix $K(z)$ is given by

$$K(z) = zA_0 + zA_1 + zA_2G(z) + \cdots zA_m G^{m-1}(z), \tag{6.2.5}$$

a matrix of which only the first two columns contain non-zero elements. As already observed in Section 2.1, this reflects the fact that the states (0,2), . . . , $(0,m)$ are *ephemeral.* The most convenient way of "ignoring" those states is to set all but the first two components of $\mathbf{x}_0$ equal to zero. Only the 2×2 matrix in the Northwest corner of the matrix $K(z)$ is relevant to the analysis. That matrix $K_1(z)$ is given by

$$\begin{vmatrix} zb_{00} + z\sum_{k=2}^{m} b_{0k} G_0^{k-2}(z)G_k(z) & zb_{01} \\ zb_{10} + z\sum_{k=2}^{m} b_{1k} G_0^{k-2}(z)G_k(z) & zb_{11} \end{vmatrix}.$$

The standard analysis to derive the vector $\mathbf{x}_0$ from the transformed semi-Markov matrix $K_1(z)$ could now be applied, but in this case, that analysis can be greatly simplified. We begin by noting that $K_1(1)$ is simply the matrix

$$\begin{vmatrix} 1 - b_{01} & b_{01} \\ 1 - b_{11} & b_{11} \end{vmatrix},$$

so that, provided that $b_{01} \neq 0$, the vector κ is given by

$$\kappa(0) = (1 - b_{11} + b_{01})^{-1}(1 - b_{11}), \qquad \kappa(1) = (1 - b_{11} + b_{01})^{-1}b_{01}.$$

When $b_{01} = 0$, a period with zero inflow can only be followed by another such period or by a period with an inflow of at least two units. A moment's reflection now shows that the state (0,1) is then also ephemeral and may be ignored. We notice, however, that the preceding formula then correctly leads to $\kappa(0) = 1$, so that we may formally proceed with an $(m+1)$–vector $\boldsymbol{\kappa}$ in which the first two components are given as above and all other components are zero.

We now know that the vector $\mathbf{x}_0 = k^* \boldsymbol{\kappa}$, where k^* is a constant, which we shall explicitly determine next by a direct argument. Evaluating the vector of generating functions $\mathbf{X}(z) = \sum_{i=0}^{\infty} \mathbf{x}_i z^i$, we obtain that

$$\mathbf{X}(z)[\, zI - A(z)\,] = (z-1)\mathbf{x}_0 A_0 = k^*(z-1)\boldsymbol{\kappa} A_0. \tag{6.2.6}$$

Setting $z = 1$, we readily obtain that $\mathbf{X}(1) = \boldsymbol{\pi}$. Differentiation, followed by setting $z = 1$, and postmultiplication by $\mathbf{e}$ yields that

$$1 - \rho = k^* \boldsymbol{\kappa} A_0 \mathbf{e},$$

which leads to

$$k^* = (1-\rho)\,\frac{1 - b_{11} + b_{01}}{b_{00}b_{01} - b_{11}b_{10} + b_{10}},$$

so that the vector $\mathbf{x}_0$ is explicitly determined.

By differentiating twice in (6.2.6) and performing the same calculations as in Section 3.3, we obtain explicit simple expressions for the vector $\mathbf{X}'(1)$ and for the global mean content L_1 of the dam. There are worthwhile simplifications due to the simple form of the right hand side in (6.2.5) and because of the relations

$$A'(1) = B\Delta_1, \qquad A''(1) = B\Delta_2.$$

We obtain that

$$\mathbf{X}'(1) = (L_1 - 1)\boldsymbol{\pi} + \mathbf{x}_0 A_0 Z + \boldsymbol{\pi}\Delta_1 Z, \tag{6.2.7}$$

where

$$L_1 = [2(1-\rho)]^{-1}[2\mathbf{x}_0 A_0 Z \boldsymbol{\beta} - 2\rho + 2\boldsymbol{\pi}\Delta_1 Z\boldsymbol{\beta} + \boldsymbol{\pi}\Delta_2\mathbf{e}]. \qquad (6.2.8)$$

The quantities $\pi_j^{-1}X_j(1)$ are the conditional mean contents of the steady-state dam at the ends of periods with inflow j. A plot of these easily computed quantities provides us with a crude, but informative picture of the effect of the variability in the input process on the average level of the dam. More informative graphs are obtained by plotting selected percentiles of the conditional distributions of the stationary dam content at these same epochs. These graphs require, as is discussed in Section 3.6, the evaluation of a sufficiently large number of the vectors $\mathbf{x}_i$ into which the invariant probability vector of P_1 is partitioned. The minimal amount of set-up computation required and the special structure of P_1 make this model a particularly attractive candidate for the implementation of the algorithms proposed in this book.

B. A Data Communications Model.

A number of discrete Markov chain models of $M/G/1$ type arise naturally in data communications engineering, notably in situations where all operations, that is, arrivals of "customers" and the completions of their processing times occur at integral multiples of a basic time unit. In the model, selected for discussion here, the server processes a single packet (customer) every unit of time, provided that the buffer (queue) is not empty. The input process consists of N independent, discrete-time sources, described by independent two-state Markov chains with irreducible transition probability matrices $B(r)$, $1 \leqslant r \leqslant N$, given by

$$B(r) = \begin{vmatrix} b_{00}(r) & b_{01}(r) \\ b_{10}(r) & b_{11}(r) \end{vmatrix}.$$

The source r transmits a packet to the buffer (queue) if and only if it visits its state 1. At a visit to the state 0, no arrival from that source occurs. During any given time unit, at most N packets can therefore arrive to the queue. The input process is fully characterized by the *product chain* of the N independent two-state sources. The product chain has 2^N states, which can conveniently be represented by the strings of zeros and ones of length N, listed in lexicographic order. The transition probability matrix A of the product chain is then given by the Kronecker product

$$A = B(1) \otimes B(2) \otimes \cdots \otimes B(N). \tag{6.2.9}$$

In order to define the Markov chain of $M/G/1$ type, which describes the number of packets in the buffer and the states of all N sources, we describe the operation of the system in detail and we define the following random variables. Let X_n be the number of packets in the system at time $n+$ and let J_n be a string of N zeros or ones, which represent the states of each of the N sources. By $\sigma(J_n)$, we denote the *number of ones in the string* J_n. At time n, one packet leaves the buffer (if possible) and thereupon all N sources are checked. Those in state 1 add a packet to the buffer; those in state 0 do not. Each source now undergoes one transition and the outcomes of the N transitions define the state J_{n+1}. We see that

$$X_{n+1} = (X_n - 1)^+ + \sigma(J_n), \quad \text{for } n \geqslant 0. \tag{6.2.10}$$

It is now readily seen that the sequence $\{(X_n, J_n)\}$ is a Markov chain of $M/G/1$ type with transition probability matrix P of the form

$$P = \begin{vmatrix} A_0 & A_1 & A_2 & A_3 & A_4 & \cdots \\ A_0 & A_1 & A_2 & A_3 & A_4 & \cdots \\ 0 & A_0 & A_1 & A_2 & A_3 & \cdots \\ 0 & 0 & A_0 & A_1 & A_2 & \cdots \\ \cdot & \cdot & \cdot & \cdot & \cdot & \end{vmatrix},$$

where the matrix A_j, $0 \leqslant j \leqslant N$, inherits from the matrix A those rows for which $\sigma(J) = j$. J stands for a generic N-tuple of zeros and ones. All other rows of these matrices and all matrices A_j for $j > N$, are zero.

It is clear that the matrix P has much special structure to be exploited. Doing so leads to a remarkable result of A. M. Viterbi, a fully explicit formula for the *expected number* of packets in the buffer at time $n+$ in the steady-state version of the queue. That formula is particularly useful for large N, when the computation of all steady-state probabilities is impractical. The derivation of that formula is quite lengthy and involves a number of clever manipulations of Kronecker products. In order to present the main arguments and calculations as transparently as possible, we first derive a number of auxiliary results.

Let B be the transition probability matrix of a typical source. Until we need to distinguish between the N sources, we suppress the index r, which identifies the source. The invariant probability vector π of B is given by

$$\pi_0 = (1 - b_{11})\,(2 - b_{00} - b_{11})^{-1}, \tag{6.2.11}$$

$$\pi_1 = (1 - b_{00})\,(2 - b_{00} - b_{11})^{-1}.$$

Following Viterbi, we set

$$b_{11} - b_{01} = b_{00} + b_{11} - 1 = \gamma. \tag{6.2.12}$$

The quantity γ is a measure of the deviation of the Markovian source from a Bernoulli source, since $\gamma = 0$, if and only if the occurrences of ones in the source follow independent Bernoulli trials. The quantity π_1 is a measure of the activity of the source, as it indicates the fraction of clock ticks at which the source contributes new packets.

The matrix $B(z)$, defined by

$$B(z) = \begin{vmatrix} b_{00} & b_{01} \\ zb_{10} & zb_{11} \end{vmatrix} = \begin{vmatrix} 1 & 0 \\ 0 & z \end{vmatrix} \begin{vmatrix} b_{00} & b_{01} \\ b_{10} & b_{11} \end{vmatrix} = \Delta(z)B,$$

is extensively used in what follows. We shall derive expressions for its Perron eigenvalue $\delta(z)$, for the corresponding normalized left and right eigenvectors $\mathbf{u}(z)$ and $\mathbf{v}(z)$, and for some derivatives at $z = 1-$ of these quantities. A general discussion of the significance of these objects is given in the Appendix.

As we are dealing with a 2×2 matrix, the following results are obtained by elementary, but somewhat lengthy calculations. The eigenvalue $\delta(z)$ satisfies

$$\delta^2(z) - (b_{00} + b_{11}z)\delta(z) + z(b_{00}b_{11} - b_{01}b_{10}) = 0, \tag{6.2.13}$$

and corresponds to the larger of the two real roots of that equation. Obviously $\delta(1) = 1$, and by routinely differentiating in (6.2.13) and setting $z = 1$, it follows that

$$\delta'(1) = \pi_1, \tag{6.2.14}$$

$$\delta''(1) = 2\pi_1(1 - \pi_1)\frac{\gamma}{1-\gamma}. \tag{6.2.15}$$

The normalized left and right eigenvectors $\mathbf{u}(z)$ and $\mathbf{v}(z)$ of $B(z)$ are given by

$$u_0(z) = (1 - b_{11})z[\delta(z) - b_{00} + z(1 - b_{11})]^{-1}, \tag{6.2.16}$$

$$u_1(z) = 1 - u_0(z),$$

and

$$v_0(z) = \frac{1 - b_{00}}{\delta(z) - b_{00}} v_1(z), \tag{6.2.17}$$

$$v_1(z) = \frac{\delta(z)(z - b_{00}) + b_{00}^2 - b_{00}z + (1 - b_{00})(1 - b_{11})z}{\delta(z)(b_{11}z - b_{00}) + b_{00}^2 - b_{00}b_{11}z + 2(1 - b_{00})(1 - b_{11})z}.$$

It is readily verified that $u_0(1) = 1 - u_1(1) = \pi_0$, and $v_0(1) = v_1(1) = 1$. We also need the first derivatives of the components of $\mathbf{v}(z)$ at $z = 1$. These are given, after lengthy but elementary calculations, as

$$v_0'(1) = -\pi_1(1 - \gamma)^{-1}, \quad v_1'(1) = (1 - \pi_1)(1 - \gamma)^{-1}. \tag{6.2.18}$$

In terms of the matrices $B(r;z)$ for the sources r, $1 \leqslant r \leqslant N$, we may write

$$A(z) = \sum_{r=0}^{N} A_r z^r = B(1;z) \otimes B(2;z) \otimes \cdots \otimes B(N;z). \tag{6.2.19}$$

This has the important consequence that for $0 < z \leqslant 1$, the Perron eigenvalue $\delta^*(z)$ of $A(z)$ is given by

$$\delta^*(z) = \prod_{r=1}^{N} \delta(r;z), \tag{6.2.20}$$

with the corresponding left and right eigenvectors $\mathbf{u}^*(z)$ and $\mathbf{v}^*(z)$, given by

$$\mathbf{u}^*(z) = \mathbf{u}(1;z) \otimes \mathbf{u}(2;z) \otimes \cdots \otimes \mathbf{u}(N;z), \tag{6.2.21}$$

$$\mathbf{v}^*(z) = \mathbf{v}(1;z) \otimes \mathbf{v}(2;z) \otimes \cdots \otimes \mathbf{v}(N;z).$$

These 2^N-dimensional vectors clearly also satisfy the normalization equations.

From Lemma 2.3.3 and Theorem 2.3.1, it readily follows that the queue is stable if and only if

$$\rho = \delta^{*\prime}(1-) = \sum_{r=1}^{N} \delta'(r;1-) = \sum_{r=1}^{N} \pi_1(r) < 1. \tag{6.2.22}$$

Direct application of the general theory in Chapters 2 and 3, would lead us to the discussion of the matrix G, which is, in general, a positive matrix of order 2^N. Quite routine calculations then lead to the standard equation

$$\mathbf{X}(z)[\,zI - A(z)\,] = (z-1)\mathbf{x}_0 A(z) = k(1-z)\mathbf{g}A(z),$$

for the generating function of the sequence of vectors $\mathbf{x}_i$. In that equation, k is a normalizing constant and $\mathbf{g}$ is the invariant probability vector of G. Since that equation contains the vector $\mathbf{g}$, explicit calculations of moments and other quantities appear to be limited to small values of the number N of sources.

However, in a clever exploitation of special structure, Viterbi noted that the first steady-state equation

$$\mathbf{x}_0 = \mathbf{x}_0 A_0 + \mathbf{x}_1 A_0,$$

implies that the vector $\mathbf{x}_0$ is given by

$$\mathbf{x}_0 = k\ (\ [b_{00}(1)\ \ b_{01}(1)] \otimes \cdots \otimes [b_{00}(N)\ \ b_{01}(N)]\ ) = k\mathbf{b}^*, \quad (6.2.23)$$

where the constant k is the stationary probability $\mathbf{x}_0\mathbf{e}$ that the queue is empty. We may therefore also write

$$\mathbf{X}(z)[\ zI - A(z)\ ] = k(z - 1)\mathbf{b}^* A(z), \quad (6.2.24)$$

and that formula will lead to the explicit expression for the mean queue length L_1.

From (6.2.24) we readily obtain that

$$\mathbf{X}(1) = \boldsymbol{\pi}^* = \boldsymbol{\pi}(1) \otimes \cdots \otimes \boldsymbol{\pi}(N),$$

as is to be anticipated. Upon differentiation in (6.2.24), we obtain that $\mathbf{X}'(1)$ satisfies

$$\mathbf{X}'(1)(I - A) = k\mathbf{b}^* A + \boldsymbol{\pi}^* A'(1) - \boldsymbol{\pi}^*, \quad (6.2.25)$$

which upon postmultiplication by $\mathbf{e}$, yields that $k = 1 - \rho$.

The argument used in Section 3.3 to obtain L_1, leads to the solution of large systems of linear equations, but these can be avoided by an argument which here (but not in general) leads to simpler expressions for L_1. Postmultiplication in (6.2.24) by the right eigenvector $\mathbf{v}^*(z)$ leads to the equation

$$[\ z - \delta^*(z)\ ]\,\mathbf{X}(z)\mathbf{v}^*(z) = (1 - \rho)(z - 1)\delta^*(z)\mathbf{b}^*\mathbf{v}^*(z), \quad (6.2.26)$$

and differentiating twice in that equation and setting $z = 1$, we obtain that

$$2(1 - \rho)L_1 = \delta^{*\,\prime\prime}(1) + 2(1 - \rho)\rho \quad (6.2.27)$$

$$+ 2(1 - \rho)[\ \mathbf{b}^*\mathbf{v}^{*\,\prime}(1) - \boldsymbol{\pi}^*\mathbf{v}^{*\,\prime}(1)\ ].$$

From (6.2.20) and by using the explicit expressions in (6.2.14) and (6.2.15), we obtain that

$$\delta^{*\,\prime\prime}(1) = 2\sum_{r=1}^{N} \pi_1(r)\,[1 - \pi_1(r)]\,\frac{\gamma(r)}{1-\gamma(r)} + \rho^2 - \sum_{r=1}^{N} \pi_1^2(r). \quad (6.2.28)$$

The quantity θ, given by

$$\theta = \mathbf{b}^*\mathbf{v}^{*\,\prime}(1) - \boldsymbol{\pi}^*\mathbf{v}^{*\,\prime}(1),$$

and needed in (6.2.27), is calculated by using the fact that all vectors involved are Kronecker products of two-dimensional vectors. In differentiating the vector $\mathbf{v}^*(z)$, we note that the usual product rule of elementary calculus extends to the differentiation of Kronecker products. Therefore

$$\boldsymbol{\pi}^*\mathbf{v}^{*\,\prime}(1) = \sum_{r=1}^{N} [\,\boldsymbol{\pi}_0(r)\mathbf{v}_0'(r;1) + \boldsymbol{\pi}_1(r)\mathbf{v}_1'(r;1)\,] = 0,$$

and similarly

$$\mathbf{b}^*\mathbf{v}^{*\,\prime}(1) = \sum_{r=1}^{N} [\,b_{00}(r)\mathbf{v}_0'(r;1) + b_{01}(r)\mathbf{v}_1'(r;1)\,]$$

$$= -\sum_{r=1}^{N} \pi_1(r)\,\frac{\gamma(r)}{1-\gamma(r)}.$$

Upon substitution in (6.2.27) and canceling common terms, we obtain the nice symmetric expression

$$L_1 = \rho \quad (6.2.29)$$

$$+ \frac{1}{1-\rho}\sum_{r=1}^{N}\sum_{k>r} \pi_1(r)\pi_1(k)\left[1 + \frac{\gamma(r)}{1-\gamma(r)} + \frac{\gamma(k)}{1-\gamma(k)}\right].$$

As noted by Viterbi, the quantity $\rho^{-1}L_1$ is the *steady-state mean delay of an arriving packet.* Formula (6.2.25) can be computationally implemented to evaluate the vector $\mathbf{X}'(1)$ and therefore also the conditional mean queue lengths given that specified sources are submitting packets. By exploiting the special structure of the matrices in that formula, we may efficiently solve the system of equations by successive substitutions and normalize the successive iterates by forcing the equality $\mathbf{X}'(1)\mathbf{e} = L_1$. This maintains the iterates in a compact set and guarantees convergence to the unique solution of (6.2.25) for which the sum of all components equals L_1. The right hand side of (6.2.25) is easily computable. We note that the vector $\mathbf{b}^* A$ is the Kronecker product of the N vectors $(b_{00}(r)\ b_{01}(r))B(r)$ and that $\boldsymbol{\pi}^* A'(1)$ is the sum of N vectors, given by the Kronecker products

$$\boldsymbol{\pi}(1) \otimes \cdots \otimes \pi_1(r)[b_{10}\ \ b_{11}] \otimes \cdots \otimes \boldsymbol{\pi}(N).$$

C. A Poor Man's Satellite.

We consider a single communications node sending packets to a receiving node via an intermittently available ionization layer, created by the arrival in the upper atmosphere of meteorite showers. The reflecting layer is "operative" or "inoperative" during alternating periods, which, in actuality, are typically of 0.5 and about 10 seconds in duration. During inoperative periods, the receiving station emits a probing signal, which, as soon as it is received by the sender, indicates that an operative period is under way. The sender starts emitting packets, one per time slot, until acknowledgement messages are no longer received. This indicates to the sending node that the layer has become inoperative and that, in the next operative period, any packets sent during the last M time slots must be retransmitted. The communication protocol between sender and receiver is symmetric, so that, in order to analyze the queue of packets at one node, we may treat it as a single server subject to service interruptions.

We shall treat the model as a queue in discrete time, choosing the time required to send one packet as the time unit. In actuality, the length of the operative periods is such that 200-500 packets are typically transmitted during a time period where the layer is present. A single time unit will be called a *slot*. The numbers of packets submitted to the sender during successive slots are assumed to be independent, identically distributed random variables with the lattice density $\{a_\nu, \nu \geqslant 0\}$ of finite mean m_1' and probability generating function $a(z)$. The operative and inoperative periods form an alternating renewal process of lattice type. The inoperative periods have a common discrete PH-density with irreducible representation (β, S), where β is a probability vector. The operative periods have a common PH-density, but in order to bring out the effect of the terminal slots of the operative period, during which packets sent will not be received, we need to take particular care in formalizing the parametric description of that density. This is done as follows.

We denote the first M terms of the density of the duration of the operative periods by $b_1, \ldots, b_M$ and we set

$$1 - \sum_{\nu=1}^{j} b_\nu = b_{j+1}',$$

for $j = 1, \cdots, M$. For the entire PH- density, we choose the representation with initial probability vector

$$b_{M+1}'\alpha,\ b_M, \cdots, b_1$$

where α is a probability vector, and the matrix which we display here for the representative value $M = 4$,

$$\begin{array}{c|ccccc|} & \text{m} & 4 & 3 & 2 & 1 \\ \hline \text{m} & T & \mathbf{T}^\circ & 0 & 0 & 0 \\ 4 & 0 & 0 & 1 & 0 & 0 \\ 3 & 0 & 0 & 0 & 1 & 0 \\ 2 & 0 & 0 & 0 & 0 & 1 \\ 1 & 0 & 0 & 0 & 0 & 0 \end{array}.$$

We see that this allows the durations of the operative times to have any lattice density of the form

$$b_1, \cdots, b_M,\ b_{M+1}'\ \alpha T^{k-M-1}\mathbf{T}^\circ, \quad \text{for } k \geqslant M+1.$$

It is clear that any PH-density on the positive integers can be cast into this form, so that there is no loss of generality in our choice of representation. Furthermore, in the problem at hand, the quantity b_{M+1}' should be positive, since otherwise all operative periods would be of length at most M and there could never be any successful transmission of packets. It is now useful to introduce notation for the means of some of the probability densities of interest. For the inoperative periods, which are denoted by the generic letter I, the mean γ_1' is given by

$$E(I) = \gamma_1' = \beta(I - S)^{-1}\mathbf{e}.$$

In relation to a generic operative period O, we introduce the mean τ_1' of the $PH-$density with representation $(\boldsymbol{\alpha}, T)$, which is given by

$$\tau_1' = \alpha(I - T)^{-1}\mathbf{e}.$$

The mean of an operative period is then given by

$$E(O) = \sum_{\nu=1}^{M} \nu b_\nu + b_{M+1}' \tau_1'.$$

For future use, we also note that

$$E(O - M)^+ = b_{M+1}'\tau_1', \text{ and } E(M - O)^+ = M(1 - b_M') - \sum_{\nu=1}^{M} \nu b_\nu.$$

Next we construct a Markov chain which describes the environment completely. It distinguishes between the operative periods where successful transmissions may occur, the final slots where any packets sent need to be retransmitted and the inoperative periods. That Markov chain, with transition probability matrix A, has $m + M + n$ states, where m and n are the dimensions of the matrices T and S respectively. Again for the representative case $M = 4$, the transition probability matrix A is given by

$$
\begin{array}{c|cccccc|}
 & m & 4 & 3 & 2 & 1 & n \\
\hline
m & T & T^\circ & 0 & 0 & 0 & 0 \\
4 & 0 & 0 & 1 & 0 & 0 & 0 \\
3 & 0 & 0 & 0 & 1 & 0 & 0 \\
2 & 0 & 0 & 0 & 0 & 1 & 0 \\
1 & 0 & 0 & 0 & 0 & 0 & \beta \\
n & b_5{}' S^\circ \alpha & b_4 S^\circ & b_3 S^\circ & b_2 S^\circ & b_1 S^\circ & S
\end{array} .
$$

A moment's reflection shows that the states of that Markov chain describe all physically meaningful states of the availability of the ionization layer for the transmission of packets. We see that successful transmission can occur only when the Markov chain is in one of its first m states. When $b_{M+1}{}'$ is positive and the representations (α, T) and (β, S) are irreducible, it is readily verified that the transition probability matrix A is irreducible. For future use, we partition the matrix A into blocks according to

$$
A = \begin{vmatrix} A(1,1) & A(1,2) \\ A(2,1) & A(2,2) \end{vmatrix},
$$

where the block $A(1,1)$ corresponds to the first m states of the Markov chain A.

As in all discrete time queues, we must carefully describe the order in which transactions are carried out to give a full accounting of the transitions in the corresponding Markov chain description of the queue. We assume that at the beginning of any slot, the environmental Markov chain A is in one of its states. Any arriving packets are added to the queue and at the end of the time slot, the environmental chain performs a transition. Depending on the availability of packets and on the state of the Markov chain, a packet is then successfully transmitted or not. The resulting content of the buffer (queue) and the state of the environmental chain are thereupon noted. It is readily seen that the resulting bivariate stochastic process is a Markov chain of $M/G/1$ type. Its transition probability matrix P is of the canonical form with the matrices A_k and B_k given by

$$
B_k = a_k A, \quad \text{for } k \geqslant 0, \tag{6.2.30}
$$

and

$$A_k = \begin{vmatrix} a_k A(1,1) & a_{k-1} A(1,2) \\ a_k A(2,1) & a_{k-1} A(2,2) \end{vmatrix},$$

where in the matrix A_0, we interpret a_{-1} as zero. It is important to note that, in A_0, all columns other than the first m are identically zero.

The matrices $A(z)$ and $B(z)$ of the general theory have the particularly simple forms

$$B(z) = a(z)A, \tag{6.2.31}$$

and

$$A(z) = a(z) \begin{vmatrix} A(1,1) & zA(1,2) \\ A(2,1) & zA(2,2) \end{vmatrix},$$

which result in major simplifications in the computations of the moments of the steady state queue length density.

We shall now obtain a general expression for the traffic intensity and the corresponding equilibrium condition of the queue. Later, after more detailed calculations, a fully explicit formula will be derived. Let the invariant probability vector π of the matrix A be partitioned in the form $\pi(1)$, $\pi(2)$, according to the same sets of states as for the matrix A. Also, let A^- be the matrix obtained from A by setting the first m columns equal to zero. It is then easily seen that

$$A'(1)\mathbf{e} = m_1'\mathbf{e} + A^-\mathbf{e}, \tag{6.2.32}$$

and

$$A''(1)\mathbf{e} = m_2'\mathbf{e} + 2m_1'A^-\mathbf{e}, \tag{6.2.33}$$

where m_2' is the second factorial moment of the probability density

$\{a_\nu\}$. It follows by elementary matrix algebra that

$$\rho = \pi A'(1)\mathbf{e} = m_1' + \pi(2)\mathbf{e}, \tag{6.2.34}$$

and

$$\rho_2 = \pi A''(1)\mathbf{e} = m_2' + 2m_1'\pi(2)\mathbf{e}. \tag{6.2.35}$$

We see that the equilibrium condition $\rho < 1$, is equivalent to

$$m_1' < 1 - \pi(2)\mathbf{e} = \pi(1)\mathbf{e}, \tag{6.2.36}$$

which states that the mean number of packets added per time slot should not exceed the expected number of packets successfully transmitted per slot.

The invariant vector π of the stochastic matrix A for the present model may be explicitly calculated. We shall not repeat the necessary elementary calculations, but only state the resulting formulas. We obtain that

$$\pi(1) = c\ b_{M+1}'\alpha(I - T)^{-1}, \tag{6.2.37}$$

and the vector $\pi(2)$, further partitioned in the form

$$\pi_1', \ldots, \pi_M', \pi'(2),$$

is given by

$$\pi'(2) = c\beta(I - S)^{-1}, \tag{6.2.38}$$

$$\pi_1' = c, \quad \pi_j' = cb_j', \quad \text{for } 2 \leqslant j \leqslant M,$$

where the constant c is given by

$$c = [\ b_{M+1}'\tau_1' + \gamma_1' + b_M'M + \sum_{\nu=1}^{M-1} \nu b_\nu\ ]^{-1} = [E(I) + E(O)]^{-1}.$$

Since $\pi(1)\mathbf{e} = c\, b_{M+1}'\tau_1'$, it follows from (6.2.36) that the queue is stable if and only if

$$m_1' < \frac{E(O - M)^+}{E(I) + E(O)}. \tag{6.2.39}$$

This form of the equilibrium condition was obtained by Robert, Mitrani and King, by an alternative approach and under somewhat different assumptions.

The matrix G: All but the first m columns of the matrix A_0 are zero. We are therefore dealing with the instance of reducibility of the matrix G, treated in Section 3.4. By examining cases of the most extreme sparcity of the matrices T and S, we have ascertained that the first m columns of the matrix G are positive, so that G is of the form

$$\begin{vmatrix} G(1) & 0 \\ G(2) & 0 \end{vmatrix},$$

where $G(1)$ and $G(2)$ are positive matrices. When $\rho < 1$, the matrix G is stochastic. We write $\mathbf{g}$ for the invariant probability vector of $G(1)$.

The analysis now proceeds along the general lines of Chapter 3. Only derivations of special, more explicit results obtained by using the features of the queueing model at hand are presented.

In the computation of the vector $\tilde{\mu}_1$, the matrix $\sum_{\nu=1}^{\infty} \nu A_\nu G^\circ$ may be explicitly evaluated. It is given by

$$m_1' G^\circ + A^- G^\circ.$$

This results in substantial simplifications in the formula for $\mathbf{x}_0$.

The vector $\mathbf{x}_0$: By routine calculations, we obtain that the matrix K is given by

$$K = A \begin{vmatrix} \sum_{\nu=0}^{\infty} a_\nu G^\nu(1) & 0 \\ G(2) \sum_{\nu=1}^{\infty} a_\nu G^{\nu-1}(1) & a_0 I \end{vmatrix} .$$

K is an irreducible stochastic matrix. Its invariant probability vector $\boldsymbol{\kappa}$, which is proportional to the vector $\mathbf{x}_0$ needs to be evaluated numerically.

Theorem 6.2.1: The vector $\mathbf{x}_0$ is explicitly given by

$$\mathbf{x}_0 = (1 - \rho) \, [\kappa(I - A^-)\mathbf{e}]^{-1} \, \boldsymbol{\kappa}. \tag{6.2.40}$$

Proof: The proof consists in an explicit evaluation of the constant $\boldsymbol{\kappa}\boldsymbol{\kappa}_1^*$, needed in the application of Theorem 3.2.1. By forming the matrix $K(z)$ and performing the now standard calculations of Section 3.2, we obtain that

$$\boldsymbol{\kappa}_1^* = \mathbf{e} + [A - I + m_1' G^\circ] \, [I - A + G^\circ - \sum_{\nu=1}^{\infty} \nu A_\nu G^\circ]^{-1} \mathbf{e},$$

and by routine matrix calculations, this leads to

$$\boldsymbol{\kappa}_1^* = (I - A^-)\mathbf{e} \, [\mathbf{g}, 0] \, [I - A + G^\circ - \sum_{\nu=1}^{\infty} \nu A_\nu G^\circ]^{-1} \mathbf{e}.$$

We now observe that

$$\boldsymbol{\pi} [I - A + G^\circ - \sum_{\nu=1}^{\infty} \nu A_\nu G^\circ] = (1 - \rho) \, [\mathbf{g}, 0],$$

which finally leads to

$$\boldsymbol{\kappa}_1^* = (1 - \rho)^{-1} \, (I - A^-)\mathbf{e}. \tag{6.2.41}$$

Formula (6.2.40) now follows by Theorem 3.2.1. •

Remark *a.* With an obvious partitioning of the vector $\mathbf{x}_0$, the constant $\kappa(I - A^-)\mathbf{e}$ in (6.2.40) may be rewritten as

$$\kappa(I - A^-)\mathbf{e} = (\mathbf{x}_0\mathbf{e})^{-1}\,[\,\mathbf{x}_0(1)A\,(1,1)\mathbf{e} + \mathbf{x}_0(2)A\,(2,1)\mathbf{e}\,].$$

That constant is therefore the conditional probability that, at an arbitrary epoch where the queue is empty, a packet *could have been successfully transmitted.* We similarly obtain the equality

$$1 - \rho = \mathbf{x}_0(1)A\,(1,1)\mathbf{e} + \mathbf{x}_0(2)A\,(2,1)\mathbf{e}, \tag{6.2.42}$$

which identifies $1 - \rho$ as the fraction of slots in the stationary queue in which a packet *could be successfully transmitted, but the buffer is empty.* This is an interesting observation for, in many complex queues of $M/G/1$ type, the quantity $1 - \rho$ does not have a ready probabilistic significance. The special forms of the matrices $A\,(1,1)$ and $A\,(2,1)$ in the model at hand allow us to write the preceding formulas in fully explicit forms. These are, however, much more involved without providing additional insight.

Remark *b.* By applying Theorem 3.3.1, or by direct calculation, we obtain that

$$\mathbf{X}(1-) = \sum_{i=0}^{\infty} \mathbf{x}_i = \boldsymbol{\pi}.$$

The special form of the matrix $A\,(z)$ results in some simplifications in the ingredients needed in the formulas (3.3.14), but there is not enough cancellation to lead to an insightful *analytically explicit formula,* even for the overall mean queue length L_1. The formulas (3.3.14) are, however, easily implemented computationally, so that various overall and conditional moments may be evaluated and interpreted.

The Sojourn Time of a Message: Suppose that a message of length K is added to the buffer at time $n = 0+$, in the stationary version of the queue. Assuming that packets are transmitted in order of arrival, the *sojourn time* of that message is the sum of the times required to send all packets in the system at time 0 and the K packets of the message. We shall now discuss the probability distribution of that random variable by two different approaches. The first approach, which is familiar from the continuous time models already discussed, consists in deriving the appropriate analogue of the Pollaczek-Khinchin equation. For discrete parameter queues, that approach results in a matrix difference equation, which is the analogue of the Volterra integral equations, obtained earlier. The derivation of such an equation, which is somewhat involved, has clear mathematical appeal and the resulting difference equation may be useful in asymptotic analysis. In the second approach, the sojourn time is identified as a first passage time in a Markov chain and its distribution is computed by straightforward recursive methods. That approach is conceptually simple, but its implementation often requires massive storage of intermediate results and substantial computation time. Its advantage is that it may be adapted to many other discrete parameter models.

Before we proceed with the derivation of the Pollaczek-Khinchin equation for this model, a number of preliminaries of some independent interest are discussed. It is clear that a packet can be successfully transmitted only when that Markov chain is in one of its first m (operative) states. Let $\Theta_1(z)$ be an $m \times m$ matrix, whose elements are the generating functions of the conditional probabilities that, at time ν, $\nu \geqslant 1$, a packet is successfully sent and that at its time of transmission, the Markov chain A is in the operative state j', given that its initial state is the operative state j. It is readily seen that

$$\Theta_1(z) = zA\,(1,1) + zA\,(1,2)\,[\,I - zA\,(2,2)\,]^{-1}\,zA\,(2,1). \qquad (6.2.43)$$

The matrix $\Theta_2(z)$ is similarly defined for the time to successful transmission starting from any one of the inoperative states. It is given by the formula

$$\Theta_2(z) = [\,I - zA\,(2,2)\,]^{-1}\,zA\,(2,1). \qquad (6.2.44)$$

For convenience, we introduce the matrix $\Theta(z)$, defined by

$$\Theta(z) = \begin{vmatrix} \Theta_1(z) & 0 \\ \Theta_2(z) & 0 \end{vmatrix}.$$

If we now consider the queue in steady-state at time 0, then the generating function of the joint probabilities of the time required to clear all work present at time 0 and of the final state of the Markov chain A is given by

$$\hat{\mathbf{X}}(z) = \sum_{i=0}^{\infty} \mathbf{x}_i \Theta^i(z) = [\, \mathbf{X}^*(z), \mathbf{x}_0(2) \,], \tag{6.2.45}$$

where the m −vector $\mathbf{X}^*(z)$ needs to be evaluated. By direct calculation, or after a moment's reflection, it is obvious that the second part of $\hat{\mathbf{X}}(z)$ must be given by the components of the vector $\mathbf{x}_0(2)$, which correspond to the inoperative states.

The $m \times m$ matrix $a^*(z)$, defined by

$$a^*(z) = \sum_{i=0}^{\infty} a_i \Theta_1^i(z), \tag{6.2.46}$$

plays a significant role in the calculations that follow. It is the transform of a semi-Markov matrix of lattice type and its elements correspond to the conditional distributions of the numbers of time slots to clear all packets added during a single time unit, given that transmission starts from an operative state.

Theorem 6.2.2: The vector $\mathbf{X}^*(z)$ is given by the equation

$$\mathbf{X}^*(z) [\, zI - a^*(z) \,] = \Psi(z), \tag{6.2.47}$$

where the vector $\Psi(z)$ is as given in (6.2.49).

Proof: From the steady-state equations for the global Markov chain, we obtain that

$$\sum_{i=0}^{\infty} \mathbf{x}_i \Theta^i(z) = \mathbf{x}_0 \sum_{i=0}^{\infty} B_i \Theta^i(z) + \sum_{r=1}^{\infty} \mathbf{x}_r \, [\sum_{k=0}^{\infty} A_k \Theta^k(z)] \, \Theta^{r-1}(z). \tag{6.2.48}$$

If we write the matrix $\sum_{k=0}^{\infty} A_k \Theta^k(z)$ in partitioned form, we see that only its first m columns are non-zero and consist of the blocks

$$\sum_{k=0}^{\infty} a_k A(1,1)\Theta_1^k(z) + \sum_{k=0}^{\infty} a_k A(1,2)\Theta_2(z)\Theta_1^k(z),$$

$$\sum_{k=0}^{\infty} a_k A(2,1)\Theta_1^k(z) + \sum_{k=0}^{\infty} a_k A(2,2)\Theta_2(z)\Theta_1^k(z).$$

However, it is clear from (6.2.44) that

$$A(1,2)\Theta_2(z) = z^{-1}\Theta_1(z) - A(1,1),$$

and this allows us to write the first block in the simplified form

$$z^{-1}\Theta_1(z)a^*(z).$$

The second block is similarly simplified to

$$z^{-1}\Theta_2(z)a^*(z),$$

so that

$$\sum_{k=0}^{\infty} A_k\Theta^k(z) = z^{-1}\Theta(z)\begin{vmatrix} a^*(z) & 0 \\ 0 & 0 \end{vmatrix}.$$

However,

$$\begin{vmatrix} a^*(z) & 0 \\ 0 & 0 \end{vmatrix}\Theta^{r-1}(z) = \begin{vmatrix} \Theta_1^{r-1}a^*(z) & 0 \\ 0 & 0 \end{vmatrix},$$

so that the second sum in the right hand side of (6.2.48) may be written as

$$z^{-1}\sum_{r=1}^{\infty} \mathbf{x}_r \begin{vmatrix} \Theta_1^r(z)a^*(z) & 0 \\ \Theta_2(z)\Theta_1^{r-1}(z)a^*(z) & 0 \end{vmatrix}.$$

Next, we turn to the first term in the right hand side of (6.2.48) and obtain by routine calculations that it may be written as

$$\mathbf{x}_0 A \begin{vmatrix} a^*(z) & 0 \\ \Theta_2(z)\sum_{r=1}^{\infty} a_r \Theta_1^{r-1}(z) & a_0 I \end{vmatrix}.$$

We now partition the vectors $\mathbf{x}_r$ as $[\mathbf{x}_r(1), \mathbf{x}_r(2)]$ and notice that

$$\mathbf{X}^*(z) = \sum_{i=0}^{\infty} \mathbf{x}_i(1)\Theta_1^i(z) + \sum_{i=1}^{\infty} \mathbf{x}_i(2)\Theta_2(z)\Theta_1^{i-1}(z).$$

From (6.2.45) and (6.2.48), it follows upon substitution that

$$[\,\mathbf{X}^*(z), \mathbf{x}_0(2)\,] = \mathbf{x}_0 A \begin{vmatrix} a^*(z) & 0 \\ \Theta_2(z)\sum_{r=1}^{\infty} a_r \Theta_1^{r-1}(z) & a_0 I \end{vmatrix}$$

$$+ z^{-1}[\sum_{r=1}^{\infty} \mathbf{x}_r(1)\Theta_1^r(z)a^*(z) + \sum_{r=1}^{\infty} \mathbf{x}_r(2)\Theta_2(z)\Theta_1^{r-1}(z)a^*(z), 0\,].$$

Upon simplifying the part of this equation corresponding to the right-most components, we obtain that

$$\mathbf{x}_0(2) = a_0\mathbf{x}_0(1)A\,(1,2) + a_0\mathbf{x}_0(2)A\,(2,2),$$

which tells us nothing new, as this is just one of the steady-state equations. From the first m components, however, we obtain upon simplification that

$$\mathbf{X}^*(z)\,[\,zI - a^*(z)\,] \tag{6.2.49}$$

$$= z\,[\,\mathbf{x}_0(1)A\,(1,1) + \mathbf{x}_0(2)A\,(2,1)]\,[\,a^*(z) - \sum_{i=1}^{\infty} a_i\,\Theta_1^{i-1}(z)\,]$$

$$+ \mathbf{x}_0(2)\Theta_2(z)\sum_{i=1}^{\infty} a_i\,\Theta_1^{i-1}(z) - a_0\mathbf{x}_0(1). \bullet$$

The transform equation (6.2.47) is of the same type as the classical equation for the steady-state probabilities of a stochastic matrix of $M/\,G/\,1$ type. Once the vector coefficients in the expansion of the function $\Psi(z)$ have been evaluated, it should be possible to compute those of $\mathbf{X}^*(z)$ by an adaptation of Ramaswami's recursive algorithm. In actual applications, the waiting and sojourn time distributions typically have very long tails, so that the computational task remains formidable. The asymptotic analysis of equations such as (6.2.47), which has not yet been undertaken, may lead to practical results of considerable interest to such computations.

From (6.2.45) and the significance of the matrix $\Theta(z)$ it is clear that the transform of the sojourn time distribution of a message of length K, arriving at an arbitrary time to the stationary queue, and of the state of the Markov chain A at the completion of its transmission, is given by

$$\hat{\mathbf{X}}(z)\Theta^K(z) = \mathbf{X}^*(z)\Theta_1^K(z) + \mathbf{x}_0(2)\Theta_2(z)\Theta_1^{K-1}(z). \tag{6.2.50}$$

From formulas (6.2.47-50), the lower order moments of the waiting and sojourn times may be routinely calculated. These are useful in providing stopping criteria in the following alternative algorithms for the sojourn time distribution.

The marginal density of the sojourn time clearly has the transform

$$\hat{\mathbf{X}}(z)\Theta^K(z)\mathbf{e} = \sum_{i=0}^{\infty} \mathbf{x}_i\,\Theta^{i+K}(z)\mathbf{e}. \tag{6.2.51}$$

We may, by successive convolution products, evaluate the sequences of vectors with generating functions

$$\Theta^j(z)\mathbf{e}, \quad \text{for } j \geqslant 1, \tag{6.2.52}$$

and accumulate the terms of the density corresponding to (6.2.51). This approach has the advantage that the sojourn time distributions for different values of K can be computed by using the same overhead operations. When the vector density in (6.2.52) has been computed for a given value of j, that density may be used in instances where it is needed in the formulas (6.2.51) (for the values of K of interest). When j is increased, careful programming enables us to write the new vector density over the old one, so that memory storage is efficiently used.

In a variant of this algorithm, we consider the $(m+M+n)-$vectors $\mathbf{r}_k(j)$, for $j \geqslant 0$ and $k \geqslant 0$, whose components are the probabilities that, starting with an unlimited supply of packets for transmission and with the medium in one of its $m+M+n$ states, the system successfully transmits *at most k packets* during the first j time slots. The vectors $\mathbf{r}_k(j)$ satisfy the recurrence relations

$$\mathbf{r}_k(j) = \begin{vmatrix} A(1,1) & 0 \\ A(2,1) & 0 \end{vmatrix} \mathbf{r}_{k-1}(j-1) + \begin{vmatrix} 0 & A(1,2) \\ 0 & A(2,2) \end{vmatrix} \mathbf{r}_k(j-1),$$

for $j \geqslant 1$ and $0 \leqslant k \leqslant j$, with $\mathbf{r}_k(j) = \mathbf{e}$, for all $k \geqslant j \geqslant 0$. If we denote by $y_K(j)$ the probability that the sojourn time of a message of length K *exceeds* j, then it is easily seen that

$$y_K(j) = \sum_{k=0}^{\infty} \mathbf{x}_k \mathbf{r}_{K+k-1}(j), \tag{6.2.53}$$

Indeed, a message of length K will spend more than j time slots in the system if and only if during the first j units of time, no more than $K+k-1$ packets are successfully transmitted, where k stands for the number of packets ahead of the message. Formula (6.2.53) is useful in computing the complementary sojourn time distribution, which provides the information needed in applications in the most direct manner. It is, of course, equivalent to the other formulas for the discrete density of the sojourn time.

6.3. A TANDEM QUEUE

The methods in this book may be applied to the analysis of several models for the interaction between two queues. Diverse examples of such interactions are described in the comments on the earlier chapters. Where the methods of this book are applicable to complex models, they can yield very detailed information, but the corresponding analysis becomes unavoidably complicated. To illustrate their application to a somewhat involved model, we shall now study a system of two queues in series with a finite intermediate waiting room. This model, under a variety of distributional assumptions, now has an extensive literature, which will not be reviewed here. We shall concentrate on the earliest version, in which each of the two units has a single server and the service times in the *second* unit are exponential. A phenomenon of interest is *blocking,* which occurs when customers accumulate in the intermediate buffer to the point that no further customers can exit from the first service unit. The various modes of boundary behavior involved in blocking and in emptiness of the two queues account for the analytic complexity of some elements of the matrices $A_k(\cdot)$ and $B_k(\cdot)$ arising in the embedded Markov renewal process of $M/G/1$ type. We shall see that our model is actually an $M/SM/1$ queue with modified boundary behavior following emptiness of the first unit. This boundary behavior results in major complications in the analysis of and in the resulting equations for the distributions of the queue lengths in continuous time and of the waiting and sojourn times. While no claims of algorithmic applicability of these equations can yet be made, the model may serve as a study example of the matrix analytic methods and as a guide to other cases of complex boundary behavior.

There is a first unit I to which customers arrive according to a Poisson process of rate λ. Service times in Unit I are independent and have a common probability distribution $H(\cdot)$ of finite mean α. Upon completion of service in Unit I, all customers proceed to Unit II via a finite waiting room. The capacity restriction on Unit II states that there can be at most K customers waiting in the intermediate buffer or receiving service in Unit II. If, upon completion of service in Unit I, a customer finds the waiting room full, Unit I *blocks* until a service in Unit II is completed. At that time, the blocked customer is allowed to enter the buffer and a new service in I may be initiated. The service times in Unit II are mutually independent, independent of the arrival times and of the service times in Unit I, and have an exponential distribution of parameter σ. We may assume that $K \geqslant 1$, as the case without an intermediate space is, even with general service times in the second unit, a simple variant of the elementary $M/G/1$ queue.

It is readily seen that, viewed at service completions in Unit I, this queueing system has an embedded Markov renewal process of $M/G/1$ type. For $i \geqslant 0$, $1 \leqslant j \leqslant K+1$, the state (i,j) signifies that there are i customers in Unit I. The second index j indicates the number of *II − customers* in the system, that is, *customers who have completed service in Unit I, but not yet in Unit II*. When after a service completion in I, the second index is $K+1$, the customer in Unit I is blocked until a departure occurs from Unit II. Following a service completion in I, there is always at least one II −customer present, so that the index j runs from 1 to $K+1$. At an arbitrary time point, it is possible that no II −customers are in the system.

In our discussion of this model, we initially follow the 1968 treatment by Neuts, which can now be streamlined by using developments in the general theory of models of $M/G/1$ type. The systematic use of matrix analytic methods further enables us to treat the distributions of the queue lengths and waiting times at an arbitrary time in a tractable manner. These derivations are new.

The main initial effort goes into the definition of the matrices $A_k(\cdot)$ and $B_k(\cdot)$, which are all of order $K+1$. The matrices $A_k(\cdot)$ are of lower Hessenberg form as during a service in Unit I, at most one II-customer can be created, but several such customers may depart. Specifically, we have for $k \geqslant 0$, that

$$[A_k]_{rv}(x) = \int_0^x e^{-(\lambda+\sigma)u} \frac{(\lambda u)^k}{k!} \frac{(\sigma u)^{r-v+1}}{(r-v+1)!} \, dH(u), \tag{6.3.1}$$

for $1 < v \leqslant r+1$, $r \leqslant K$,

$$[A_k]_{r1}(x) = \sum_{\nu=r}^{\infty} \int_0^x e^{-(\lambda+\sigma)u} \frac{(\lambda u)^k}{k!} \frac{(\sigma u)^{\nu}}{\nu!} \, dH(u), \tag{6.3.2}$$

for $r \leqslant K$,

$$[A_k]_{K+1,v}(x) = \int_0^x \int_0^u \sigma e^{-\sigma y} \, dy \; e^{-\lambda u} \frac{(\lambda u)^k}{k!} \tag{6.3.3}$$

$$\cdot \, e^{-\sigma(u-y)} \frac{[\sigma(u-y)]^{K-v+1}}{(K-v+1)!} \, dH(u-y),$$

for $1 < v \leqslant K+1$,

$$[A_k]_{K+1,1}(x) = \sum_{\nu=K}^{\infty} \int_0^x \int_0^u \sigma e^{-\sigma y}\, dy\ e^{-\lambda u} \frac{(\lambda u)^k}{k!} \tag{6.3.4}$$

$$\cdot\ e^{-\sigma(u-y)} \frac{[\sigma(u-y)]^\nu}{\nu!}\ dH(u-y).$$

The other elements of $A_k(\cdot)$ are zero.

It is readily seen that the matrices $A_k(\cdot)$ are of the form

$$A_k(x) = \int_0^x e^{-\lambda u} \frac{(\lambda u)^k}{k!}\ dF(u), \tag{6.3.5}$$

where $F(\cdot)$ is a semi-Markov matrix. The tandem queue is, in fact, an $M/SM/1$ queue with a modified boundary.

The expressions for the matrices $B_k(\cdot)$ are more involved, as we need to distinguish between cases where a departure from Unit *II* precedes the arrival of a new customer to Unit *I* and where the order of these occurrences is reversed. We obtain

$$[B_k]_{rv}(x) = \int_0^x \int_0^u e^{-(\lambda+\sigma)u} \frac{[\lambda(u-y)]^k}{k!} \frac{(\sigma u)^{r-v+1}}{(r-v+1)!} \lambda dy\ \ dH(u-y), \tag{6.3.6}$$

for $1 < v \leqslant r+1$, $r \leqslant K$,

$$[B_k]_{r1}(x) = \sum_{\nu=r}^{\infty} \int_0^x \int_0^u e^{-(\lambda+\sigma)u} \frac{[\lambda(u-y)]^k}{k!} \frac{(\sigma u)^\nu}{\nu!} \lambda dy\ \ dH(u-y), \tag{6.3.7}$$

for $r \leqslant K$,

$$[B_k]_{K+1,v}(x) = \int_0^x \int_0^y \int_0^{u_1} e^{-\lambda u_1} \lambda du_1 e^{-\sigma u} \sigma du\ \ e^{-\sigma(y-u)} \tag{6.3.8}$$

$$\cdot\ \frac{[\sigma(y-u)]^{K-v+1}}{(K-v+1)!} e^{-\lambda(y-u_1)} \frac{[\lambda(y-u_1)]^j}{j!}\ \ dH(y-u_1)$$

$$+ \int_0^x \int_0^y \int_0^u e^{-\lambda u_1} \lambda du_1 e^{-\sigma u} \sigma du \; e^{-\sigma(y-u)} \frac{[\sigma(y-u)]^{K-v+1}}{(K-v+1)!}$$

$$\cdot \, e^{-\lambda(y-u_1)} \frac{[\lambda(y-u_1)]^j}{j!} \; dH(y-u),$$

for $1 < v \leqslant K+1$. The first term corresponds to the case where a customer leaves Unit *II* before an arrival occurs to Unit *I*; the second term corresponds to the case where an arrival to Unit *I* occurs first. Finally, in exactly the same manner, we obtain

$$[B_k]_{K+1,1}(x) = \sum_{\nu=K}^{\infty} \int_0^x \int_0^y \int_0^{u_1} e^{-\lambda u_1} \lambda du_1 e^{-\sigma u} \sigma du \; e^{-\sigma(y-u)} \tag{6.3.9}$$

$$\cdot \, \frac{[\sigma(y-u)]^{\nu}}{\nu!} e^{-\lambda(y-u_1)} \frac{[\lambda(y-u_1)]^j}{j!} \; dH(y-u_1)$$

$$+ \sum_{\nu=K}^{\infty} \int_0^x \int_0^y \int_0^u e^{-\lambda u_1} \lambda du_1 e^{-\sigma u} \sigma du \; e^{-\sigma(y-u)}$$

$$\cdot \, \frac{[\sigma(y-u)]^{\nu}}{\nu!} e^{-\lambda(y-u_1)} \frac{[\lambda(y-u_1])^j}{j!} \; dH(y-u).$$

Laplace-Stieltjes transforms and probability generating functions of these expressions can be obtained by routine calculations. We shall present only the results that are essential to subsequent derivations. The vector $\boldsymbol{\beta}$ of formula (2.3.5) is given by

$$\lambda[\alpha, \cdots, \alpha, \alpha + \sigma^{-1}],$$

so that the traffic intensity ρ is given by

$$\rho = \lambda[\alpha + \sigma^{-1}\pi_{K+1}], \tag{6.3.10}$$

where π_{K+1} is the last component of the invariant probability vector $\boldsymbol{\pi}$ of the matrix A. The quantities π_{K+1} may be recursively computed by application of the following theorem.

Theorem 6.3.1: The quantity π_{K+1} is given by

$$\pi_{K+1} = \left[\sum_{j=0}^{K} c_j \right]^{-1}, \tag{6.3.11}$$

where the terms c_j are the coefficients in the power series expansion of the function

$$c(z) = \frac{(1-z)h(\sigma - \sigma z)}{h(\sigma - \sigma z) - z}. \tag{6.3.12}$$

Proof: If we set

$$\psi_\nu = \int_0^\infty e^{-\sigma u} \frac{(\sigma u)^\nu}{\nu!} dH(u), \quad \text{and} \quad \Psi_\nu = \sum_{j=0}^{\nu} \psi_j, \tag{6.3.13}$$

for $\nu \geqslant 0$, then the matrix A is given by

$$\begin{vmatrix} 1-\Psi_0 & \psi_0 & 0 & 0 & \cdots & 0 & 0 \\ 1-\Psi_1 & \psi_1 & \psi_0 & 0 & \cdots & 0 & 0 \\ 1-\Psi_2 & \psi_2 & \psi_1 & \psi_0 & \cdots & 0 & 0 \\ \cdot & \cdot & \cdot & \cdot & & \cdot & \cdot \\ \cdot & \cdot & \cdot & \cdot & & \cdot & \cdot \\ 1-\Psi_{K-1} & \psi_{K-1} & \psi_{K-2} & \psi_{K-3} & \cdots & \psi_1 & \psi_0 \\ 1-\Psi_{K-1} & \psi_{K-1} & \psi_{K-2} & \psi_{K-3} & \cdots & \psi_1 & \psi_0 \end{vmatrix}.$$

The equation $\pi = \pi A$ leads to

$$\pi_i = \sum_{\nu=0}^{K-i+1} \psi_\nu \pi_{i+\nu-1} + \psi_{K-i+1} \pi_{K+1},$$

for $1 \leqslant i \leqslant K+1$. Upon setting $\pi_i = \pi_{K+1} c_{K-i+1}$, we see that the constants c_i may be recursively computed by

$$c_0 = 1, \quad \text{and} \quad \psi_0 c_{i+1} = c_j - \sum_{\nu=0}^{i} \psi_\nu c_{i-\nu+1} - \psi_i,$$

for $0 \leqslant i \leqslant K-1$, and from this the equations (6.3.11) and (6.3.12) are routinely obtained. •

We note that (6.3.12) may be rewritten as a stable recurrence relation involving only additions and multiplications of positive quantities as was shown in Chapter 1 for the stationary queue length for the $M/G/1$ queue. In a few special cases, explicit expressions for the c_i, and therefore also for π_{K+1}, are available. For example, when the service time distribution $H(\cdot)$ is exponential, we have that

$$h(s) = [1 + \alpha s]^{-1},$$

which leads to

$$\pi_{K+1} = \frac{1 - \alpha\sigma}{1 - (\alpha\sigma)^{K+1}}, \quad \text{for} \quad \alpha\sigma \neq 1,$$

and $\pi_{K+1} = (K+1)^{-1}$ for $\alpha\sigma = 1$. It is clear that for small values of K, the limited capacity of the intermediate buffer severely reduces the critical arrival rate of the system.

Introducing the probability generating functions

$$\mathbf{X}(z) = \sum_{i=0}^{\infty} \mathbf{x}_i z^i, \qquad \mathbf{A}(z) = \sum_{i=0}^{\infty} \mathbf{A}_i z^i, \qquad \mathbf{B}(z) = \sum_{i=0}^{\infty} \mathbf{B}_i z^i,$$

we obtain the standard equation

$$\mathbf{X}(z)[zI - A(z)] = z\mathbf{x}_0 B(z) - \mathbf{x}_0 A(z), \tag{6.3.14}$$

from which expressions for the vector $\mathbf{X} = \mathbf{X}(1-)$ and for the moments of the stationary queue length density are routinely obtained. The components of the vector $\mathbf{X}$ give the stationary probability density of the number of II-customers at service completions in Unit I. In particular, X_{K+1} is the fraction of customers who experience blocking. The following theorem relates that quantity to the probabilities of emptiness.

Theorem 6.3.2: The probability X_{K+1} is related to the vector $\mathbf{x}_0$ by the equality

$$1 - \mathbf{x}_0\mathbf{e} = \lambda\alpha + \frac{\lambda}{\sigma}(X_{K+1} - x_{0,K+1}) + \frac{\lambda}{\sigma}\frac{\lambda}{\lambda + \sigma}x_{0,K+1}. \tag{6.3.15}$$

This relation is equivalent to the statement that the fundamental rate E^{-1} of the embedded Markov renewal process $Q(\cdot)$ is equal to the arrival rate λ.

Proof: By differentiation in (6.3.14), followed by setting $z = 1$ and by postmultiplication by $\mathbf{e}$. By evaluation of the fundamental mean E of the embedded Markov renewal process, we obtain an expression which, after routine algebra, shows that $E = \lambda^{-1}$. •

We note that the last two terms in (6.3.15) are positive, so that the probability $\mathbf{x}_0\mathbf{e}$ that the queue in Unit I is *empty* following a service completion, is always smaller that the corresponding value $1 - \lambda\alpha$ for the $M/G/1$ queue with the same parameters as Unit I.

Queue Lengths at an Arbitrary Time: We now turn to the discussion of the queue lengths at an arbitrary time point. In the interest of brevity, only the most salient intermediate results will be presented. The initial step proceeds in the standard manner by relating the probabilities of the possible queue lengths to the renewal functions of the embedded Markov renewal process. This is followed by an application of the key renewal theorem. The details of that argument can be omitted. In order to write the equations so obtained in a transparent form, it is useful to define a number of matrices and to establish some preliminary results. In displaying the structure of matrices, we shall for convenience use the representative case $K = 3$.

The matrix $F(\cdot)$ in (6.3.5) is of the form

$$\begin{vmatrix} H(x) - \Psi_0(x) & \psi_0(x) & 0 & 0 \\ H(x) - \Psi_1(x) & \psi_1(x) & \psi_0(x) & 0 \\ H(x) - \Psi_2(x) & \psi_2(x) & \psi_1(x) & \psi_0(x) \\ E_1 * [H(x) - \Psi_2(x)] & E_1 * \psi_2(x) & E_1 * \psi_1(x) & E_1 * \psi_0(x) \end{vmatrix},$$

where E_1 denotes the exponential distribution with parameter σ and

$$\psi_\nu(x) = \int_0^x e^{-\sigma u} \frac{(\sigma u)^\nu}{\nu!} dH(u), \quad \text{and} \quad \Psi_\nu(x) = \sum_{j=0}^{\nu} \psi_j(x), \tag{6.3.16}$$

so that $A(z,s) = f(s+\lambda-\lambda z)$, where $f(s)$ is the Laplace-Stieltjes transform of the semi-Markov matrix $F(\cdot)$.

Letting x tend to infinity in formulas (6.3.6-9) and after routine calculations, we see that the non-zero elements of the matrices B_i are given by the equations

$$[B_i]_{rv} = \sum_{j=0}^{r-v+1}\left(\frac{\lambda}{\lambda+\sigma}\right)\left(\frac{\sigma}{\lambda+\sigma}\right)^j \int_0^\infty e^{-(\lambda+\sigma)t}\frac{(\sigma t)^{r-v-j+1}}{(r-v-j+1)!}\frac{(\lambda t)^i}{i!}dH(t)$$

$$= \sum_{\nu=1}^{r}\left(\frac{\lambda}{\lambda+\sigma}\right)\left(\frac{\sigma}{\lambda+\sigma}\right)^{r-\nu}[A_i]_{\nu v}, \tag{6.3.17}$$

for $1 \leqslant r \leqslant K$, $1 < v \leqslant r+1$, and

$$[B_i]_{r1} = \sum_{\nu=r}^{\infty}\sum_{j=0}^{\nu}\left(\frac{\lambda}{\lambda+\sigma}\right)\left(\frac{\sigma}{\lambda+\sigma}\right)^j \int_0^\infty e^{-(\lambda+\sigma)t}\frac{(\sigma t)^{\nu-j}}{(\nu-j)!}\frac{(\lambda t)^i}{i!}dH(t)$$

$$= \sum_{\nu=1}^{r}\left(\frac{\lambda}{\lambda+\sigma}\right)\left(\frac{\sigma}{\lambda+\sigma}\right)^{r-\nu}[A_i]_{\nu 1} + \left(\frac{\sigma}{\lambda+\sigma}\right)^r \int_0^\infty e^{-\lambda t}\frac{(\lambda t)^i}{i!}dH(t),$$

for $1 \leqslant r \leqslant K$, and

$$[B_i]_{K+1,v} = \frac{\sigma}{\lambda+\sigma}[B_i]_{Kv} + \left(\frac{\lambda}{\lambda+\sigma}\right)\sum_{k=0}^{i}\left(\frac{\sigma}{\lambda+\sigma}\right)\left(\frac{\lambda}{\lambda+\sigma}\right)^k [A_{i-k}]_{Kv},$$

for $1 \leqslant v \leqslant K+1$.

Let us define the matrix C_{K+1} by

$$\left(\frac{\lambda}{\lambda+\sigma}\right)\begin{vmatrix} 1 & 0 & 0 & 0 \\ \left(\frac{\sigma}{\lambda+\sigma}\right) & 1 & 0 & 0 \\ \left(\frac{\sigma}{\lambda+\sigma}\right)^2 & \left(\frac{\sigma}{\lambda+\sigma}\right) & 1 & 0 \\ \left(\frac{\sigma}{\lambda+\sigma}\right)^3 & \left(\frac{\sigma}{\lambda+\sigma}\right)^2 & \left(\frac{\sigma}{\lambda+\sigma}\right) & 0 \end{vmatrix}.$$

The matrix N_{K+1} has the first column with elements

$$\left(\frac{\sigma}{\lambda+\sigma}\right), \left(\frac{\sigma}{\lambda+\sigma}\right)^2, \ldots, \left(\frac{\sigma}{\lambda+\sigma}\right)^{K+1},$$

and all its other columns are zero; the matrix J_{K+1} is zero, except for $[J_{K+1}]_{K+1,K} = 1$. In terms of the matrices C_{K+1}, N_{K+1}, and J_{K+1}, the equations (6.3.17) may be conveniently be rewritten as

$$B_i = C_{K+1}A_i + N_{K+1}\int_0^\infty e^{-\lambda t}\frac{(\lambda t)^i}{i!}dH(t) \tag{6.3.18}$$

$$+ \sum_{k=0}^{i}\left(\frac{\sigma}{\lambda+\sigma}\right)\left(\frac{\lambda}{\lambda+\sigma}\right)^{k+1} J_{K+1}A_{i-k}, \quad \text{for } i \geqslant 0.$$

The three terms in the right hand side of formula (6.3.18) have the following significance. The first term corresponds to the case where the system is unblocked at the first arrival to Unit I, but some II-customers remain in the system. If the system became blocked at the last departure, this means that there must be at least one departure of a II-customer before that first arrival. The second term is contributed by the case where all II-customers have cleared the system prior to the first arrival in Unit I, while the third term corresponds to the alternative where the system is blocked and one or more customers arrive to Unit I before the *blocked* customer in Unit I can enter the buffer.

From (6.3.18), we readily obtain the formula

$$\sum_{i=0}^{\infty} B_i[f(s)]^i \tag{6.3.19}$$

$$= C_{K+1}\sum_{i=0}^{\infty} A_i[f(s)]^i + N_{K+1}\int_0^\infty \exp[-\lambda t[I - f(s)]]dH(t)$$

$$+ \left(\frac{\sigma}{\lambda+\sigma}\right)\left(\frac{\lambda}{\lambda+\sigma}\right)J_{K+1}[\sum_{i=0}^{\infty} A_i[f(s)]^i][I - \frac{\lambda}{\lambda+\sigma}f(s)]^{-1},$$

which plays an important role in the derivation of equations for the waiting time distributions.

The terms in the queue length density at an arbitrary time which correspond to emptiness in Unit I or to a non-boundary transition of the embedded Markov renewal process are easily derived by calculations similar to those performed in Section 4.1. for the $M/SM/1$ queue. The emphasis of the present discussion is on the complexity induced by the boundary transitions. Application of the key renewal theorem yields that the probabilities y_{0j} that there are j II-customers in the system and none in Unit I are given by

$$y_{0j} = \sum_{\nu=j}^{K+1} x_{0\nu} \left(\frac{\lambda}{\lambda+\sigma}\right)\left(\frac{\sigma}{\lambda+\sigma}\right)^{\nu-j}, \quad \text{for} \quad 1 \leqslant j \leqslant K+1, \qquad (6.3.20)$$

$$y_{00} = \sum_{\nu=1}^{K+1} x_{0\nu} \left(\frac{\sigma}{\lambda+\sigma}\right)^{\nu}.$$

By adding these quantities, we obtain the equality

$$\sum_{j=0}^{K+1} y_{0,j} = \mathbf{x}_0\mathbf{e},$$

a relation between the emptiness probabilities, whose simplicity is unusual for a case with involved boundary states.

The case where the arbitrary time falls during a non-boundary transition is treated exactly as in the study of the $M/SM/1$ queue in Chapter 4. To the probability y_{ij}, that there are i customers who have not yet completed service in Unit I and j II-customers in the system, this case contributes a term obtained by the standard argument. For $1 \leqslant j \leqslant K$, it is given by

$$\sum_{k=1}^{i} \sum_{\nu=j}^{K} \lambda x_{k\nu} \int_0^{\infty} e^{-(\lambda+\sigma)\tau} \frac{(\lambda\tau)^{i-k}}{(i-k)!} \frac{(\sigma\tau)^{\nu-j}}{(\nu-j)!} [1 - H(\tau)] d\tau$$

$$+ \sum_{k=1}^{i} \lambda x_{k,K+1} \int_0^{\infty} \int_0^{\tau} e^{-\sigma u} \sigma du \; e^{-\lambda\tau} \frac{(\lambda\tau)^{i-k}}{(i-k)!}$$

$$\cdot\, e^{-\sigma(\tau-u)} \frac{[\sigma(\tau-u)]^{K-j}}{(K-j)!}[1 - H(\tau-u)]d\tau .$$

When $j = 0$, the corresponding expression is

$$\sum_{k=1}^{i} \sum_{\nu=1}^{K} \lambda x_{k\nu} \int_0^\infty e^{-(\lambda+\sigma)\tau} \frac{(\lambda\tau)^{i-k}}{(i-k)!} \sum_{r=\nu}^{\infty} \frac{(\sigma\tau)^r}{r!}[1 - H(\tau)]d\tau$$

$$+ \sum_{k=1}^{i} \lambda x_{k,K+1} \int_0^\infty \int_0^\tau e^{-\sigma u}\sigma du \; e^{-\lambda\tau} \frac{(\lambda\tau)^{i-k}}{(i-k)!}$$

$$\cdot \sum_{r=K}^{\infty} e^{-\sigma(\tau-u)} \frac{[\sigma(\tau-u)]^r}{r!}[1 - H(\tau-u)]d\tau ,$$

while for $j = K+1$, we obtain the simpler expression

$$\sum_{k=1}^{i} \lambda x_{k,K+1} \int_0^\infty e^{-(\lambda+\sigma)\tau} \frac{(\lambda\tau)^{i-k}}{(i-k)!} d\tau .$$

If we introduce the auxiliary transform functions

$$\theta_\nu(s) = \int_0^\infty e^{-(s+\sigma)u} \frac{(\sigma u)^\nu}{\nu!}[1 - H(u)]du , \qquad (6.3.21)$$

and

$$\Theta_\nu(x) = \sum_{r=\nu}^{\infty} \theta_r(s), \quad \text{for } 0 \leqslant \nu \leqslant K ,$$

the generating functions (on i) of the preceding expressions may be concisely related to the probability generating functions $X_\nu(z)$, which satisfy (6.3.14). For $1 \leqslant j \leqslant K$, we obtain after some calculations the expression

$$\lambda \sum_{\nu=j}^{K} [X_\nu(z) - x_{0\nu}]\, \theta_{\nu-j}(\lambda - \lambda z) \qquad (6.3.22)$$

$$+ \lambda[X_{K+1}(z) - x_{0,K+1}] \frac{\sigma}{\lambda+\sigma-\lambda z} \theta_{K-j}(\lambda - \lambda z),$$

while the corresponding expressions for $j = 0$ and $j = K+1$ are respectively given by

$$\lambda \sum_{\nu=1}^{K} [X_\nu(z) - x_{0\nu}] \Theta_\nu(\lambda - \lambda z)$$

$$+ \lambda \frac{\sigma}{\lambda+\sigma-\lambda z} [X_{K+1}(z) - x_{0,K+1}] \Theta_K(\lambda - \lambda z),$$

and

$$\lambda[X_{K+1}(z) - x_{0,K+1}] \frac{1}{\lambda+\sigma-\lambda z}.$$

When a boundary transition is in progress at the arbitrary time, we need a more laborious consideration of cases. For $1 \leqslant j \leqslant K$, the following alternatives arise:

a. The system is not blocked when Unit I becomes empty and an arrival occurs in Unit I before the departure of a II-customer.

b. The system is blocked when Unit I becomes empty, but by the time of the first arrival to that unit, some II-customers have left.

c. The system is blocked when Unit I becomes empty and one or more arrivals to Unit I precede the departure of the first II-customer. When the system unblocks, the first customer to arrive starts service in Unit I.

We leave it to the initiative of the reader to write down the terms contributed by each of these alternatives to the probability y_{ij} and to verify that they have respectively the following three generating functions:

$$\lambda z \sum_{\nu=j}^{K} x_{0\nu} \sum_{r=0}^{\nu-j} \left(\frac{\lambda}{\lambda+\sigma}\right) \left(\frac{\sigma}{\lambda+\sigma}\right)^{\nu-j-r} \theta_r(\lambda - \lambda z),$$

$$\lambda z x_{0,K+1} \left(\frac{\sigma}{\lambda+\sigma}\right) \sum_{r=0}^{K-j} \left(\frac{\lambda}{\lambda+\sigma}\right) \left(\frac{\sigma}{\lambda+\sigma}\right)^{K-j-r} \theta_r(\lambda - \lambda z), \quad (6.3.23)$$

$$\lambda z x_{0,K+1} \left(\frac{\lambda}{\lambda+\sigma}\right)\left(\frac{\sigma}{\lambda+\sigma-\lambda z}\right) \theta_{K-j}(\lambda - \lambda z).$$

For $j = 0$, we have the three alternatives of the preceding case, except that by the time point considered, all II-customers have cleared the system. There is, however, a fourth possibility. All II-customers may depart the system before the first arrival to Unit I. That alternative contributes terms with generating function

$$\lambda z \sum_{\nu=1}^{K+1} x_{0\nu} \left(\frac{\sigma}{\lambda+\sigma}\right)^{\nu} \frac{1 - h(\lambda-\lambda z)}{1 - z},$$

in addition to the three terms

$$\lambda z \sum_{\nu=1}^{K} x_{0\nu} \sum_{k=\nu}^{\infty} \sum_{r=0}^{k} \left(\frac{\lambda}{\lambda+\sigma}\right)\left(\frac{\sigma}{\lambda+\sigma}\right)^{k-r} \theta_r(\lambda - \lambda z),$$

$$\lambda z x_{0,K+1} \left(\frac{\sigma}{\lambda+\sigma}\right) \sum_{k=K}^{\infty} \sum_{r=0}^{k} \left(\frac{\lambda}{\lambda+\sigma}\right)\left(\frac{\sigma}{\lambda+\sigma}\right)^{k-r} \theta_r(\lambda - \lambda z),$$

$$\lambda z x_{0,K+1} \left(\frac{\lambda}{\lambda+\sigma}\right)\left(\frac{\sigma}{\lambda+\sigma-\lambda z}\right) \Theta_K(\lambda - \lambda z).$$

For $j = K+1$, there cannot be a service in course at the time point considered. The system was blocked when Unit I became empty and has not yet become unblocked. There will be i customers waiting in Unit I if and only if i arrivals have occurred since the last service completion in I. The probability generating function of the term contributed to $y_{i,K+1}$ is given by

$$\lambda x_{0,K+1} \left[\frac{1}{\lambda+\sigma-\lambda z} - \frac{1}{\lambda+\sigma}\right].$$

The power of appropriate matrix notation is again demonstrated by the fact that the results of the preceding derivations can be summarized in a single matrix formula. To that end, we define the

$(K+1) \times (K+1)$ matrix $\Theta^*(s)$ by

$$\Theta^*(s) = \begin{vmatrix} \Theta_1(s) & \theta_0(s) & 0 & 0 \\ \Theta_2(s) & \theta_1(s) & \theta_0(s) & 0 \\ \Theta_3(s) & \theta_2(s) & \theta_1(s) & \theta_0(s) \\ \frac{\sigma}{s+\sigma}\Theta_3(s) & \frac{\sigma}{s+\sigma}\theta_2(s) & \frac{\sigma}{s+\sigma}\theta_1(s) & \frac{\sigma}{s+\sigma}\theta_0(s) \end{vmatrix},$$

and we define the vector $\mathbf{Y}(z)$ with components

$$Y_j(z) = \sum_{i=0}^{\infty} y_{ij} z^i, \quad \text{for} \quad 0 \leqslant j \leqslant K.$$

The relationship between the vectors $\mathbf{Y}(z)$ and $\mathbf{X}(z)$, or equivalently the relation between the stationary density of the queue lengths at an arbitrary time and after service completions in Unit I, may then be summarized by the formula

$$\mathbf{Y}(z) = \mathbf{Y}(0) + \lambda[\mathbf{X}(z) - \mathbf{x}_0]\Theta^*(\lambda-\lambda z) + \lambda z\, \mathbf{x}_0 C_{K+1}\Theta^*(\lambda-\lambda z) \quad (6.3.24)$$

$$+ \lambda\left(\frac{\lambda}{\lambda+\sigma}\right)\mathbf{x}_0 J_{K+1}\Theta^*(\lambda-\lambda z) + \lambda z\, \mathbf{x}_0 N_{K+1}\frac{1 - h(\lambda-\lambda z)}{1 - z},$$

where $\mathbf{Y}(0)$ is given in (6.3.20). The probability generating function $Y_{K+1}(z)$ is simply related to $X_{K+1}(z)$ by

$$Y_{K+1}(z) = \frac{\lambda}{\lambda+\sigma-\lambda z} X_{K+1}(z). \quad (6.3.25)$$

By setting $z = 1$ in this formula, we obtain the equality

$$\sigma Y_{K+1}(1) = \lambda X_{K+1}(1),$$

which states that, in steady state, *the rate at which the system unblocks is equal to the arrival rate of customers who experience blocking.*

By setting $z = 1$ in (6.3.24) and postmultiplying by $\mathbf{e}$ we obtain, after some calculations, an expression which after appealing to (6.3.15) shows that

$$\sum_{j=0}^{K+1} Y_j(1) = 1,$$

and this provides a partial accuracy check on the rather formidable calculations which led to (6.3.24) and (6.3.25).

The Virtual Waiting Time: As for the bulk service queue, treated in Section 4.2, we introduce appropriate stationary waiting time distributions in terms of which others, such as the sojourn time in Unit I and the total sojourn time in the system, may be routinely expressed. To that end, we consider a (virtual) customer arriving at an arbitrary time and define $W_j(x)$, for $0 \leqslant j \leqslant K+1$, $x \geqslant 0$, as the probability that this customer has to wait no longer that x before coming *to the head of the line* in Unit I and that when he does, there are j II-customers in the system.

It is readily seen that the probabilities of no waiting satisfy

$$W_j(0+) = y_{0j}, \quad \text{for} \quad 0 \leqslant j \leqslant K+1, \tag{6.3.26}$$

and are therefore explicitly given by formula (6.3.20). Next, we define the transform matrix

$$\Xi(s) = \sum_{i=0}^{\infty} A_i [f(s)]^i = \int_0^{\infty} dF(u) \exp\{-\lambda u [I - f(s)]\}, \tag{6.3.27}$$

which unfortunately remains in the final Pollaczek-Khinchin equation for this model. $\Xi(s)$ is the transform of a semi-Markov matrix which corresponds to the total transition time of the Markov renewal process $F(\cdot)$, which "arrives" to the queue during a typical transition of that "service" process.

The same calculations which led to formula (4.1.16) result here in the equality

$$\sum_{i=0}^{\infty} \mathbf{x}_i [f(s)]^i = \mathbf{x}_0 \sum_{i=0}^{\infty} B_i [f(s)]^i + \sum_{i=0}^{\infty} \mathbf{x}_i \Xi(s) [f(s)]^{i-1}, \tag{6.3.28}$$

and we recall that $\sum_{i=0}^{\infty} B_i [f(s)]^i$ is given by (6.3.19). For $1 \leq j \leq K+1$, we denote by $w_j(s)$ the Laplace-Stieltjes transform of $W_j(\cdot)$, and by $\mathbf{w}(s)$ the vector with components $w_j(s)$. We notice that $W_0(x) = W_0(0+) = w_0(s)$, since the only case where the arriving customer can come to the head of the line in Unit I with no II-customers in the system, is when he arrives to a completely empty system.

Theorem 6.3.3: The transform vector $\mathbf{w}(s)$ satisfies

$$\mathbf{w}(s)[sI - \lambda I + \lambda f(s)] \tag{6.3.29}$$

$$= \lambda \mathbf{x}_0 + \mathbf{W}(0+)[sI - \lambda I + \lambda f(s)] - \lambda \mathbf{x}_0 C_{K+1} f(s) - \lambda \mathbf{x}_0 N_{K+1} h(s)$$

$$+ \left(\frac{\lambda}{\lambda+\sigma}\right)\left(\frac{\sigma}{\lambda+\sigma}\right) \lambda \mathbf{x}_0 J_{K+1} [f(s) - \Xi(s)] [I - \frac{\lambda}{\lambda+\sigma} f(s)]^{-1}$$

$$- \left(\frac{\lambda}{\lambda+\sigma}\right)\left(\frac{\sigma}{s+\sigma}\right) \lambda \mathbf{x}_0 J_{K+1} f(s).$$

Proof: As in the case of the queue length density, the complexities of this proof are all induced by the boundary transitions. As for the $M/SM/1$ queue, the principal alternative where the virtual customer arrives during a regular transition not involving the boundary, contributes a term

$$\sum_{i=0}^{\infty} \sum_{k=1}^{i} \lambda \mathbf{x}_k \int_0^{\infty} d\tau \int_0^{\tau} e^{-\lambda\tau} \frac{(\lambda\tau)^{i-k}}{(i-k)!} dF(\tau+v) F^{(i-1)}(x-v).$$

By following the same steps as in the proof of Theorem 4.1.2 and by use of formula (6.3.28), the Laplace-Stieltjes transform of this term is found to be

$$\lambda[\mathbf{x}_0 - \mathbf{x}_0 \sum_{i=0}^{\infty} B_i [f(s)]^i] [sI - \lambda I + \lambda f(s)]^{-1}. \tag{6.3.30}$$

The contributions to the probability mass functions $W_j(\cdot)$ of the boundary transitions involve four cases depending on the conditions that prevail when the virtual customer arrives. These are:

a. When the virtual customer arrives, the first service of a busy period is in course; the system was not blocked at the beginning of that service and at the completion of that first service, there are two or more II-customers in the system.

b. When the virtual customer arrives, the first service of a busy period is in course; the system was not blocked at the beginning of that service and at the completion of that first service, there is only one II-customer in the system.

c. When the virtual customer arrives, the first service of a busy period is in course and at the beginning of that service there were no II-customers in the system.

d. At the arrival of the virtual customer, Unit I which was empty but blocked at the last service completion, is still blocked and at least one customer has arrived to it before the virtual customer.

The contributions to $W_j(\cdot)$ from the first two alternatives are respectively

$$\sum_{i=0}^{\infty}\sum_{\nu=1}^{K+1}\sum_{\nu_1=1}^{\min(\nu,K)}\sum_{\nu_2=1}^{\nu_1+1}\lambda x_{0\nu}\int_0^{\infty}d\tau\int_0^{x}\int_0^{\tau}e^{-\lambda u_1}\lambda du_1$$

$$\cdot\, e^{-\sigma u_1}\frac{(\sigma u_1)^{\nu-\nu_1}}{(\nu-\nu_1)!}\, e^{-\lambda(\tau-u_1)}\frac{[\lambda(\tau-u_1)]^i}{i!}$$

$$\cdot\, e^{-\sigma(\tau+v-u_1)}\frac{[\sigma(\tau+v-u_1)]^{\nu_1-\nu_2+1}}{(\nu_1-\nu_2+1)!}dH(\tau+v-u_1)[F^{(i)}(x-v)]_{\nu_2 j},$$

and

$$\sum_{i=0}^{\infty}\sum_{\nu=1}^{K+1}\sum_{\nu_1=1}^{\min(\nu,K)}\sum_{r=\nu_1}^{\infty}\lambda x_{0,\nu}\int_0^{\infty}d\tau\int_0^{x}\int_0^{\tau}e^{-\lambda u_1}\lambda du_1$$

$$\cdot\, e^{-\sigma u_1}\frac{(\sigma u_1)^{\nu-\nu_1}}{(\nu-\nu_1)!}\, e^{-\lambda(\tau-u)}\frac{[\lambda(\tau-u)]^i}{i!}$$

$$\cdot\, e^{-\sigma(\tau+v-u_1)}\frac{[\sigma(\tau+v-u_1)]^r}{r!}dH(\tau+v-u_1)[F^{(i)}(x-v)]_{1j}\,.$$

The Laplace-Stieltjes transform of the sum of these first two terms may be written in matrix form as

$$\lambda \mathbf{x}_0 C_{K+1}\,[\Xi(s) - f(s)]\,[sI - \lambda I + \lambda f(s)]^{-1}. \tag{6.3.31}$$

The contribution to $W_j(\cdot)$ from the third alternative is given by

$$\sum_{i=0}^{\infty}\sum_{\nu=1}^{K+1}\sum_{r=\nu}^{\infty} \lambda x_{0\nu} \int_0^{\infty} d\tau \int_0^{x}\int_0^{\tau} e^{-\lambda u_1}\lambda du_1 e^{-\sigma u_1}\frac{(\sigma u_1)^r}{r!}$$

$$\cdot\, e^{-\lambda(\tau-u_1)}\frac{[\lambda(\tau-u_1)]^i}{i!}dH(\tau+v-u_1)[F^{(i)}(x-v)]_{1j}\,,$$

and its matrix transform is obtained as

$$\lambda \mathbf{x}_0 N_{K+1}\,\{\int_0^{\infty} \exp[-\lambda\tau[I - f(s)]]dH(\tau) - h(s)I\} \tag{6.3.32}$$

$$\cdot\,[sI - \lambda I + \lambda f(s)]^{-1}.$$

Finally, the contribution of the fourth alternative is given by

$$\sum_{i=1}^{\infty} \lambda x_{0,K+1} \int_0^{\infty} d\tau \int_0^{x} e^{-\sigma(\tau+v)}\sigma dv\; e^{-\lambda\tau}\frac{(\lambda\tau)^i}{i!}\,[F^{(i)}(x-v)]_{Kj}\,,$$

with the matrix Laplace-Stieltjes transform

$$\lambda\,\frac{\sigma}{s+\sigma}\,\mathbf{x}_0 J_{K+1}\,\{\,[(\lambda+\sigma)I - \lambda f(s)]^{-1} - \frac{1}{\lambda+\sigma}I\,\}. \tag{6.3.33}$$

By adding the expressions in formulas (6.3.26) and (6.3.30-33) and replacing $\sum_{i=0}^{\infty} B_i [f(s)]^i$ by the expression in (6.3.19), we obtain (6.3.29), which is the transform version of a linear Volterra integral equation similar to that obtained for the $M/SM/1$ queue but with a much more involved inhomogeneous term. •

The stationary distributions of various sojourn times of customers in the system, in Unit I, in the intermediate buffer and in Unit II may be related to the mass-functions $W_j(\cdot)$ by elementary probability arguments. To do so is left to the initiative of the reader.

NOTES, REFERENCES AND COMMENTS ON CHAPTER 6

Section 6.2

For an algorithmic viewpoint, stochastic models with discrete time offer many advantages. Their formal theory is, in most cases, entirely similar to that of the familiar continuous time versions. However, set-up computations are usually much simpler as numerical integrations are avoided. The discrete versions of many classical integral equations, such as the Pollaczek-Khinchin equation for the waiting time distribution or the nonlinear Volterra equations for the fundamental period, are matrix difference equations, which can be efficiently solved by recursive or iterative schemes. Mostly for historical reasons, the analysis and implementation of discrete models has not received much attention. In recent years, however, the performance evaluation of devices that operate at the ticks of an internal clock, have kindled interest in such models.

Ponstein [P-026] used a discrete analogue of the $GI/G/1$ queue in a study of the stationary waiting time distribution. Related results may be found in Kimura [K-056], Konheim [K-086] and Servi [S-033]. The advantages of discrete time models are discussed in Dafermos and Neuts [D-001], in conjunction with an analysis of the discrete $M/G/1$ queue. In a series of early papers in algorithmic probability, Neuts and co-authors Klimko and Heimann [H-037,K-071,N-023], illustrate iterative techniques for the solution of the discrete version of a simple queueing model. Some algorithmic aspects of the Galton-Watson process and a related queueing problem are treated in Neuts [N-040]. The text book by Hunter [H-082,H-083] emphasizes discrete-time models and addresses some of their algorithmic aspects. Further references on discrete queues, many of which may be further investigated by the methods in this book, include Alfa [A-012], Anderson, Foschini and Gopinath [A-023], Balmer [B-020], Beightler and Crisp [B-024], Beusch [B-038], Bhat [B-039], Bithell [B-060], Gergely and Török [G-024], Gopinath and Morrison [G-032], Kleinecke [K-069], Meisling [M-030], Minh [M-043,M-045-46], Minh and Blunden [M-044], Morrison [M-063], Neuts [N-023] and Takács [T-026].

Discrete dam models are discussed in Ali Khan and Gani [A-017], Ali Khan [A-018,A-019], Balagopal [B-017,B-019], Herbert [H-041], Kennedy [K-049], Neuts [N-032,N-042], Odoom and Lloyd [O-007], Pakes [P-011], Pakes and Phatarfod [P-013], Phatarfod and Mardia [P-021], Phatarfod [P-022,P-023], and Pyke Tin and Phatarfod [P-056,P-

057]. Other references discussing related models for dams are Brill [B-090], Collings [C-057], Gani [G-005-G-008], Harrison [H-019], Hsu and Bosch [H-073], Lloyd [L-028-29], Moran [M-055-M-057], Pegg and Phatarfod [P-016], Prabhu [P-041,P-045], and Takács [T-022].

The data communications model is treated in Viterbi [V-010]. By using the methods of this book, Viterbi's formula for the mean queue length can be shown to hold when, for each source, the runs of ones and zeros form an alternating renewal process in which runs of zeros have a geometric distribution and those of ones, an arbitrary discrete $PH-$distribution. The definition of the constants $\gamma(r)$ needs to be modified. It depends on the $PH-$density only through the first two moments of that distribution. That generalization was established by the author after the completion of this book.

The satellite communications model was introduced into the queueing literature by Robert, Mitrani and King [R-034] in an analysis by complex variable methods. Our treatment follows Chandramouli, Neuts and Ramaswami [C-007]. In that article, the effect of possible errors in the transmission of packets is also considered. With realistic values for the bit error rate, that feature is shown to have very little effect on the behavior of the queue. Mathematically, it adds only a minor complication. We have therefore chosen not to include it in our discussion.

A difficulty of discrete time models is the extra care needed in giving a careful account of coincident events. At each clock tick, we need to consider carefully whether we shall first add new arrivals to the queue and then account for departures or vice versa. In many applications, the arrival process nearly always involves group arrivals, as is illustrated in the models treated in the chapter. However, once a precise accounting has been made of all transactions in the model, the structure that leads to efficient numerical computations is often immediately clear.

In many discrete time models, there is no conceptual distinction between queue length and waiting time. Individual customers, who are counted in the queue length, bring integral numbers of work units. The waiting time is then simply the number of work units which need to be processed before a given customer's units will be processed. We may therefore think of the waiting time as the length of a queue of work units, rather than of customers. Relationships between characteristics of the queue length and the waiting times, which require some proof for general models, are often matters of trivial accounting for discrete queues. The distributions of waiting and sojourn times are often long-tailed, yet, in most cases the tail probabilities of these distributions are of primary importance. A promising area for research is the derivation of approximations and asymptotic expressions for such distributions. See e.g. Pakes [P-012].

Problem 6.2.1: For the data communications model of Viterbi, the vector $\mathbf{X}(1)$ is explicitly known and the vector $\mathbf{X}'(1)$ may be computed by solving the system of linear equations (6.2.25). Carry out the numerical solution of the system (6.2.25), using the special structure of the coefficient matrices so that larger values of the number N of sources can be handled. Plot the conditional mean queue lengths given that j, $0 \leq j \leq N$, sources are transmitting. Keeping ρ constant, modify the parameters of the arrival process to obtain varying degrees of imbalance in the input process, that is, start with N identical sources and next compare the computed results for various models in which some sources have much higher input rates than others. •

Problem 6.2.2: Consider the Perron eigenvalue $\delta(z)$ of the matrix $B(z)$ which arises in the discussion of the data communication model. That eigenvalue is the larger of the two solutions of the quadratic equation (6.2.13). Develop an algorithm to compute the coefficients in the MacLaurin series of the analytic function $\delta(z)$. Find numerical examples showing that $\delta(z)$ is not in general a probability generating function. •

Problem 6.2.3: Consider the Viterbi model for which the parameters of all N sources are identical. In that case, it suffices to keep account only of the number of sources which are in their state 1, so that the input process can be described in terms of an $(N+1)$−state Markov chain. Precisely define the Markov chain of $M/G/1$ type for that version of the model. Notice that the structure of its transition probability matrix has many similarities with that of the discrete dam model of Section 6.2. All matters related to non-boundary states are straightforward, but the matrices B_k for the boundary behavior are of high order and their special structure needs to be exploited in efficient algorithms. Develop an efficient code to perform detailed computations for larger values of N. Use the explicit formula (6.2.29) as an accuracy check. •

Problem 6.2.4: For the satellite communications model, verify the explicit form of the invariant probability vector $\boldsymbol{\pi}$, given in the text. •

Problem 6.2.5: The satellite communications model is a special case of a discrete time queue with service breakdowns. Let the irreducible stochastic matrix A which describes the availability of the server be partitioned as

$$A = \begin{vmatrix} A(1,1) & A(1,2) \\ A(2,1) & A(2,2) \end{vmatrix},$$

where the first set 1 of states correspond to an available server and the set 2 to states where the server is inoperative. Develop the theory to handle this general case. All services last one unit of time. •

Problem 6.2.6: The arrival process to a queue is a two-state Markov modulated Bernoulli process, that is, a process in which an arrival occurs with probability p_1 when the environment chain is in the state 1 and with probability p_2, when that chain is in its state 2. The environment is described by a two-state Markov chain with irreducible transition probability matrix A. Develop the general theory for the case where the service times have a discrete density of phase type. Write an code to implement the algorithms to compute many quantities of interest for this model. •

Problem 6.2.7: In a special case of the general model in Problem 6.2.5, the operative and inoperative periods form an alternating renewal process of phase type, with underlying distributions with representations $(\boldsymbol{\alpha}, T)$ and $(\boldsymbol{\gamma}, \Gamma)$. The matrix A is then of the form

$$A = \begin{vmatrix} T & \mathbf{T}^{\circ}\boldsymbol{\gamma} \\ \boldsymbol{\Gamma}^{\circ}\boldsymbol{\alpha} & \Gamma \end{vmatrix}.$$

Verify that the model treated in the chapter is an instance of that special case and identify the vector $\boldsymbol{\gamma}$ and the matrix Γ. Discuss in detail the special forms of the equations for $G(1)$ and $G(2)$ for this case and see to what extent these may be used to expedite numerical computations. •

Problem 6.2.8: For the model in Problem 6.2.7, derive matrix-analytic expressions for the first two moments of the stationary waiting and sojourn time densities (of messages of length K). •

Problem 6.2.9: Under the assumptions stated in Problem 6.2.7, derive the equations for the sojourn time distribution of a message of length K, similar to those discussed in the text. Perform a critical analysis (with computer implementations) of the three alternative approaches we have discussed. In the coding of each algorithm, strive for the greatest possible efficiency. Generate numerical examples and give qualitative interpretations in the manner of [C-007].

Brown and Simonian [B-092] consider the following discrete time queue, which arises in communications engineering, in models for delays in processing multiplexed traffic streams.

Problem 6.2.10: Every d time slots, a principal source sends a packet to an infinite capacity queue. d is a positive integer > 1, which in typical applications is 20 or a value of comparable magnitude. The server can process one packet per time slot. The input to the queue from all other sources is modeled as a sequence of independent integer valued random variables with common density $\{c_\nu\}$. c_ν is the probability that, during a time slot, ν packets arrive from the other sources. Packets (customers) are processed in order of arrival and, to agree on the accounting, we imagine that (when they occur), the packets from the principal source are added to the queue ahead of any others that may arrive in the same time slot. Formulate the queueing model as a Markov chain of $M/G/1$ type and discuss its mathematical analysis. Obtain expressions for important quantities related to the delay of packets coming from the principal source. •

The model in Problem 6.2.10 is one of a wide variety of models related to merged packets streams. Such problems abound in research stimulated by developments in I(ntegrated) S(ervices) D(igital) N(etworks). The analysis in Brown and Simonian proceeds by methods of complex variables. A practically important question on the model is the following: What is the probability that a packet from the main source, arriving to the stationary queue at time 0, will be the first of k *consecutive* packets from the principal source to be delayed by N time slots or more? The answer to that question appears to be analytically quite intractable, so that appropriate algorithmic procedures or fully justified approximate answers would be of interest.

Section 6.3

The tandem queue without an intermediate waiting space was shown to be a variant of the $M/G/1$ queue by Avi-Itzhak and Yadin [A-031]. Its time dependent analysis is presented in Prabhu [P-042]. The model with K waiting spaces and exponential servers in both units, was first considered in Yadin's Ph.D. thesis. General discussions of the equilibrium condition may be found in Hildebrand [H-053-54], while the embedded Markov renewal process was discussed by transform methods in Neuts [N-016, N-020] and in Suzuki [S-080-S-082]. Konheim and Reiser [K-085] examined that analysis for the case of exponential servers, which was further shown to be of matrix-geometric type in Latouche and Neuts [L-002]. That paper also deals with several other blocking and unblocking rules. See also Neuts [N-042]. Approximations are treated in Boxma and Konheim [B-081]. Discussions of related or other tandem queues may be found in Balagopal [B-019], Boxma [B-078,B-079], Disney [D-043], Disney and Chandramohan [D-047], Gün [G-049], Hsu [H-075], Ishikawa [I-006], Iyama [I-009,I-010], Makino [M-002], Morrison [M-063] Nair [N-003], Newell [N-070,N-071], Nicolas and Latouche [N-072] Nishida and Hiramatsu [N-074,N-076], Nishida, Ueshima and Yoneyama [N-075], Pack [P-001], Pearce [P-015], Shalmon and Kaplan [S-034], Tumura and Ishikawa [T-068], and Wong, Giffin and Disney [W-020].

The complexity of the expressions for the boundary matrices $B_k(x)$ is due to the fact that the transitions involved in the first arrival and the first unblocking prior to the start of the busy period may occur in different orders. By separating the idle period and the first service, we can obtain somewhat more transparent expressions.

Problem 6.3.1: Starting from an empty queue in Unit I, consider the idle period (which may start with a blocked customer) and the subsequent service in Unit I. By a probabilistic argument, show that the transform matrix

$$\tilde{B}(z;s) = \sum_{k=0}^{\infty} \int_0^{\infty} e^{-sx}\, dB_k(x)\, z^k,$$

may be written as

$$\tilde{B}(z;s) = D(z;s)E(z;s),$$

where $D(z;s)$ is a matrix of order $K+1$ which describes the duration, the number of arrivals to Unit I and the numbers of II−customers at the beginning and the end of the idle period. Verify that only one element of $D(z;s)$ depends on z. The transform matrix $E(z;s)$ of order $K+1$ and has a special structure. •

Problem 6.3.2: Derive the equilibrium condition for the tandem system in which there are K servers of rate σ in Unit II, so that blocking occurs only if all K exponential servers are busy.

Problem 6.3.3: Derive explicit expressions for the elements of matrices A_k and B_k for the case of the tandem queue where the service time distribution $H(\cdot)$ is of phase type. Adapt the result in Problem 5.1.7 to derive a simple explicit matrix formula for the quantity π_{K+1} which arises in the expression for the traffic intensity ρ. •

References

[A-001] Abolnikov, L. M. Investigation of a class of discrete Markov processes. *Engin. Cybern.*, 15, 51-63, 1977.

[A-002] Abolnikov, L. M. and Dzhalalov, E. A. Feedback queueing systems: duality principle and optimization. *Automation and remote control*, 39, 11-20, 1978.

[A-003] Abolnikov, L. M. and Postan, M. Ya. On duality relationships in queueing systems with group appearance, group servicing, and feedback. *Engin. Cybern.*, 18, 64-71, 1980.

[A-004] Ackroyd, M. H. Iterative computation of the $M/G/1$ queue length distribution via the discrete Fourier transform. *IEEE Trans. Comm.*, COM-28, 1929-32, 1980.

[A-005] Ackroyd, M. H. $M/G/1$ queue analysis via the inverse Poisson transform. *IEE Proc., 129 - Part E*, 119-22, 1982.

[A-006] Ackroyd, M. H. and Kanyangarara, R. Skinner's method for computing bounds on distributions and the numerical solution of continuous-time queueing problems. *IEEE Trans. Comm.*, COM-30, 1746-49, 1982.

[A-007] Ackroyd, M. H. $M/M/1$ transient state occupancy probabilities via the discrete Fourier transform. *IEEE Trans. Comm.*, COM-30, 557-9, 1982.

[A-008] Ackroyd, M. H. Stationary and cyclostationary finite buffer behaviour computation via Levinson's method. *AT&T Bell Labs Tech. J.*, 63, 2159-70, 1984.

[A-009] Adiri, I. Cyclic queues with bulk arrivals. *J. ACM*, 20, 416-28, 1973.

[A-010] Albin, S. L. On Poisson approximations for superposition arrival processes in queues. *Mgmt. Sci.*, 28, 126-37, 1982.

[A-011] Aleksandrov, A. M. A queueing system with repeated orders. *Engin. Cybern.*, 12, 1-4, 1974.

[A-012] Alfa, A. S. Time-inhomogeneous bulk server queue in discrete time: A transportation type problem. *Opns. Res.*, 30, 650-8, 1982.

[A-013] Ali, H. Two results in the theory of queues. *J. Appl. Prob.*, 7, 219-26, 1970.

[A-014] Ali, O. M. E. and Neuts, M. F. A service system with two stages of waiting and feedback of customers. *J. Appl. Prob.*, 21, 404-13, 1984.

[A-015] Ali, O. M. E. and Neuts, M. F. A message queue with secondary jobs: A case of reducibility in matrix-geometric solutions. *Belgian J. O.R., Stat. and C. S.*, 26, 11-26, 1986.

[A-016] Ali, O. M. E. and Neuts, M. F. A queue with service times dependent on their order within the busy periods. *Stoch. Mod.*, 2, 67-96, 1986.

[A-017] Ali Khan, M. S. and Gani, J. Infinite dams with inputs forming a Markov chain. *J. Appl. Prob.*, 5, 72-84, 1968.

[A-018] Ali Khan, M. S. Finite dams with inputs forming a Markov chain. *J. Appl. Prob.*, 7, 291-303, 1970.

[A-019] Ali Khan, M. S. Infinite dams with discrete additive inputs. *J. Appl. Prob.*, 14, 170-80, 1977.

[A-020] Allen, A. O. *Probability, Statistics and Queueing Theory, with Computer Science Applications.* New York: Academic Press, 1978.

[A-021] Allsop, R. E. An analysis of delays to vehicle platoons at traffic signals. *4-th. Internat. Sym. on the Theory of Road Traffic, Karlsruhe, Germany*, 98-104, 1969.

[A-022] Anderson, M. Q. Optimal admission pricing policies for $M / E_k / 1$ queues. *Nav. Res. Logist. Quart.*, 27, 57-64, 1980.

[A-023] Anderson, R. R.; Foschini, G. J. and Gopinath, B. A queueing model for a hybrid data multiplexer. *Bell Syst. Tech. J.*, 58, 279-300, 1979.

[A-024] Asmussen, S. Equilibrium properties of the $M / G / 1$ queue. *Z. Wahrsch. verw. Gebiete*, 58, 267-81, 1981.

[A-025] Asmussen, S. Time-dependent approximations in some queueing systems with imbedded Markov chains related to random walks. *Preprint No. 6, Inst. Math. Statist., Univ. Copenhagen, Copenhagen, Denmark, May 1981.*

[A-026] Asmussen, S. Conditioned limit theorems relating a random walk to its associate, with applications to risk reserve processes and the $GI / G / 1$ queue. *Adv. Appl. Prob.*, 14, 143-170, 1982.

[A-027] Asmussen, S. Risk theory in a Markovian environment. *Scand. Actuarial J.*, (forthcoming).

[A-028] Asmussen, S. The heavy traffic limit of a class of Markovian queueing models. *Oper. Res. Letters, 6, 301-306, 1987.*

[A-029] Asmussen, S. *Applied Probability and Queues.* New York: John Wiley and Sons, 1987.

[A-030] Assaf, D.; Langberg, N. A.; Savits, T. H. and Shaked, M. Multivariate phase-type distributions. *Opns. Res.*, 32, 688-702, 1984.

[A-031] Avi-Itzhak, B. and Yadin, M. A sequence of two queues with no intermediate queue. *Mgmt. Sci.*, 11, 553-64, 1965.

[A-032] Avi-Itzhak, B.; Maxwell, W. L. and Miller, L. W. Queuing with alternating priorities. *Opns. Res.*, 13, 306-18, 1965.

[A-033] Avi-Itzhak, B. and Heyman, D. P. Approximate queueing models for multi-programming computer systems. *Opns. Res.*, 21, 1212-30, 1973.

[B-001] Baba, Y. Algorithmic methods for $PH / PH / 1$ queues with batch arrivals or batch services. *Report No. 104, Research Reports on Information Sciences, Tokyo Institute of Technology, Tokyo, Japan, November 1981.*

[B-002] Baba, Y. An algorithmic solution to the $M^X / PH / c$ queue. *Report No. 112, Research Reports on Information Sciences, Tokyo Institute of Technology, Tokyo, Japan, February 1982.*

[B-003] Baba, Y. The $M^X / G / 1$ queue with finite waiting room. *J. Oper. Res. Soc. Japan*, 27, 260-72, 1984.

[B-004] Baccelli, F.; Boyer, P. and Hebuterne, G. Single-server queues with impatient customers. *Adv. Appl. Prob.*, 16, 887-905, 1984.

[B-005] Baccelli, F. and Trivedi, K. S. A single server queue in a hard-real-time environment. *OR Letters*, 4, 161-8, 1985.

[B-006] Bagchi, T. P. and Templeton, J. G. C. *Numerical Methods in Markov Chains and Bulk Queues.* New York: Springer-Verlag, Lecture Notes in Economics and Mathematical Systems, 72, 1972.

[B-007] Bagchi, T. P. and Templeton, J. G. C. Finite waiting space bulk queueing systems. *J. of Eng. Math.*, 7, 313-7, 1973.

[B-008] Bagchi, T. P. and Templeton, J. G. C. A note on the $M^X / G^Y / 1,K$ bulk queueing system. *J. Appl. Prob.*, 10, 901-6, 1973.

[B-009] Bagchi, T. P. and Templeton, J. G. C. Some finite waiting space bulk queueing systems. In *"Proceedings of the Conference on Mathematical Methodology in the Theory of Queues, Kalamazoo, Mich."*, New York: Springer-Verlag, 133-7, 1974.

[B-010] Bahary, E. and Kolesar, P. Multilevel bulk service queues. *Opns. Res.*, 20, 406-20, 1972.

[B-011] Bailey, N. T. J. On queueing processes with bulk service. *J. Roy. Statist. Soc. Ser. B*, 16, 80-87, 1954.

[B-012] Bailey, N. T. J. A continuous time treatment of a simple queue using generating functions *J. Roy. Statist. Soc. Ser. B,* 16, 288-91, 1954.

[B-013] Balachandran, K. R. Parametric priority rules: An approach to optimization in priority queues. *Opns. Res.,* 18, 526-40, 1970.

[B-014] Balachandran, K. R. Purchasing priorities in queues. *Mgmt. Sci.,* 18, 319-26, 1972.

[B-015] Balachandran, K. R. Control policies for a single server system. *Mgmt. Sci.,* 19, 1013-18, 1973.

[B-016] Balachandran, K. R., and Tijms, H. On the $D-$policy for the $M/G/1$ queue. *Mgmt. Sci.,* 21, 1073-76, 1975.

[B-017] Balagopal, K. Dryness of discrete dams: Comments on a paper by Tin and Phatarfod. *J. Appl. Prob.,* 15, 858-64, 1978.

[B-018] Balagopal, K. Some limit theorems for the general semi-Markov storage model. *J. Appl. Prob.,* 16, 607-17, 1979.

[B-019] Balagopal, K. On queues in discrete regenerative environments, with application to the second of two queues in series. *Adv. Appl. Prob.,* 11, 851-69, 1979.

[B-020] Balmer, D. W. A single server queue in discrete time with customers served in random order. *J. Appl. Prob.,* 9, 862-67, 1972.

[B-021] Barbour, A. D. Networks of queues and the method of stages. *Adv. Appl. Prob.,* 8, 584-91, 1976.

[B-022] Bartoszewicz, J and Rolski, T. Queueing systems with a reserve service channel. *Zast. Matemat.,* 11, 439-48, 1970.

[B-023] Bear, D. An approximation for the delay distribution of the $M/G/N$ queue and an exact analysis for the $M/E_2/N$ queue with random service. *Proc. Ninth Internat. Teletraffic Conf., Torremolinos, Spain,* Paper No. 411, 1979.

[B-024] Beightler, C. S. and Crisp, R. M. Jr. A discrete-time queuing analysis of conveyor-serviced production stations. *Opns. Res.,* 16, 986-1001, 1968.

[B-025] Bell, C. E. Characterization and computation of optimal policies for operating an $M/G/1$ queuing system with removable server. *Opns. Res.,* 19, 208-18, 1971.

[B-026] Bell, C. E. Efficient operation of optional-priority queuing systems. *Opns. Res.,* 21, 777-86, 1973.

[B-027] Bell, C. E. Optimal operation of an $M/G/1$ priority queue with removable server. *Opns. Res.,* 21, 1281-90, 1973.

[B-028] Bellman, R. *Introduction to Matrix Analysis.* New York: McGraw Hill Book Co., 1960.

[B-029] Benes, V. E. On queues with Poisson arrivals. *Ann. Math. Statist.,* 28, 670-77, 1957.

[B-030] Benes, V. E. Combinatory methods and stochastic Kolmogorov equations in the theory of queues with one server. *Trans. Amer. Math. Soc.,* 94, 282-94, 1960.

[B-031] Benes, V. E. General stochastic processes in traffic systems with one server. *Bell Syst. Tech. J.,* 39, 127-60, 1960.

[B-032] Benes, V. E. *General Stochastic Processes in the Theory of Queues.* Reading MA: Addison-Wesley, 1963.

[B-033] Bergmann, R. and Stoyan, D. On exponential bounds for the waiting-time distribution function in $GI/G/1$. *J. Appl. Prob.,* 13, 411-7, 1976.

[B-034] Bergmann, R. Qualitative properties and bounds for the serial covariances of waiting times in single-server queues. *Opns. Res.,* 27, 1168-79, 1979.

[B-035] Bergmann, R., Daley, D. J.; Rolski, T. and Stoyan, D. Bounds for cumulants of waiting-times in $GI/G/1$ queues. *Math. Operationsforsch. u. Statist.,* 10, 257-63, 1979.

[B-036] Berman, M. and Westcott, M. On queueing systems with renewal departure processes. *Adv. Appl. Prob.,* 15, 657-73, 1983.

[B-037] Berman, O., Larson, R. C. and Chiu, S. S. Optimal server location on a network operating as an $M/G/1$ queue. *Opns. Res.,* 33, 746-71, 1985.

[B-038] Beusch, J. U. A general model of a single-channel queue: discrete and continuous cases. *Opns. Res.,* 15, 1131-44, 1967.

[B-039] Bhat, U. N. On single-server bulk-queueing processes with binomial input. *Opns. Res.,* 12, 527-33, 1964.

[B-040] Bhat, U. N. On the busy period of a single server bulk queue with a modified service mechanism. *Calcutta Statist. Assoc. Bull.,* 13, 163-71, 1964.

[B-041] Bhat, U. N. Imbedded Markov chain analysis of single server bulk queues. *J. Austral. Math. Soc.,* 4, 244-63, 1964.

[B-042] Bhat, U. N. Customer overflow in queues with finite waiting space. *Austral. J. Statist.,* 7, 15-19, 1965.

[B-043] Bhat, U. N. On a stochastic process occurring in queueing systems. *J. Appl. Prob.,* 2, 467-69, 1965.

[B-044] Bhat, U. N. *A Study of the Queueing Systems GI / M / 1 and M / G / 1*. Lecture Notes in Opns. Res. and Math. Econ., No. 2. New York: Springer-Verlag, 1968.

[B-045] Bhat, U. N. Queueing systems with first-order dependence. *Opsearch,* 6, 1-24, 1969.

[B-046] Bhat, U. N. Sixty years of queueing theory. *Mgmt. Sci.,* 15, 280-94, 1969.

[B-047] Bhat, U. N. A controlled transportation queueing process. *Mgmt. Sci.,* 16, 446-52, 1970.

[B-048] Bhat, U. N. Some queueing systems with mixed disciplines. *SIAM J. Appl. Math.,* 18, 415-32, 1970.

[B-049] Bhat, U. N. and Nance, R. E. Busy period analysis of a time-sharing system modeled as a semi-Markov process. *J. ACM,* 18, 221-38, 1971.

[B-050] Bhat, U. N. *Elements of Applied Stochastic Processes.* New York: John Wiley and Sons, 1972.

[B-051] Bhat, U. N. Two measures for describing queue behavior. *Opns. Res.,* 20, 357-72, 1972.

[B-052] Bhat, U. N. and Subba Rao, S. A statistical technique for the control of traffic intensity in the queueing systems *M / G /* 1 and *GI / M /* 1. *Opns. Res.,* 20, 955-66, 1972.

[B-053] Bhat, U. N. Some problems in finite queues. In *"Proceedings of the Conference on Mathematical Methodology in the Theory of Queues, Kalamazoo, Mich.",* New York: Springer-Verlag, 139-56, 1974.

[B-054] Bhat, U. N.; Wheeler, A. C. and Fischer, M. J. On the existence of limiting distributions in stochastic systems with secondary inputs. *Opsearch,* 11, 81-89, 1974.

[B-055] Bhat, U. N. Imbedded Markov chains in queueing systems *M / G /* 1 and *GI / M /* 1 with limited waiting room. *Metrika,* 22, 153-60, 1975.

[B-056] Bhat, U. N. and Fischer, M. J. Multichannel queueing systems with heterogeneous classes of arrivals. *Nav. Res. Logist. Quart.,* 23, 271-82, 1976.

[B-057] Bhat, U. N. Theory of queues: Problems, developments and trends. In *"The Handbook of Operations Research",* Chapt. III-2, S. E. Elmaghhraby and J. J. Moder, eds., New York: Van Nostrand - Reinbold Co., 1978.

[B-058] Bhat, U. N.; Shalaby, M. and Fischer, M. J. Approximation techniques in the solution of queueing problems. *Nav. Res. Logist.*

Quart., 26, 311-26, 1979.

[B-059] Bhat, U. N. and Nance, R. E. An evaluation of *CPU* efficiency under dynamic quantum allocation. *J. ACM,* 26, 761-78, 1979.

[B-060] Bithell, J. F. A class of discrete-time models for the study of hospital admission systems. *Opns. Res.*, 17, 48-69, 1969.

[B-061] Blanc, J. P. C. Asymptotic analysis of a queueing system with a two-dimensional state space. *J. Appl. Prob.*, 21, 870-86, 1984.

[B-062] Bloemena, A. R. On a queueing process with a certain type of bulk service. *Bull. Inst. Statist.*, 37, 219-27, 1960.

[B-063] Blomqvist, N. The covariance function of the $M/G/1$ queueing system. *Skandinavisk Aktuarietidskrift,* 50, 157-74, 1967.

[B-064] Blomqvist, N. Estimation of waiting-time parameters in the $GI/G/1$ queuing system - Part I: General results. *Skandinavisk Aktuarietidskrift,* 51, 178-97, 1968.

[B-065] Blomqvist, N. Estimation of waiting-time parameters in the $GI/G/1$ queuing system - Part II: Heavy traffic approximation. *Skandinavisk Aktuarietidskrift,* 52, 125-36, 1969.

[B-066] Blomqvist, N. Serial correlation in a simple dam process. *Opns. Res.*, 21, 966-73, 1973.

[B-067] Bocharov, P. and Naumov V. Matrix-geometric stationary distribution for the $PH/PH/1/r$ queue. *Elektron. Inf. Verarb. Kybern.*, EIK 22, 179-186, 1986.

[B-068] Bocharov, P. and Litvin, V. G. Ways to analyze and design service systems with phase type distributions (in Russian, with English summary). *Automation and Remote Control,* 5, 5-23, 1986.

[B-069] Boel, R. Martingale methods for the semi-Markov analysis of queues with blocking. *Stochastics,* 5, 115-33, 1981.

[B-070] Borel, E. Sur l'emploi du theorème de Bernoulli pour faciliter le calcul d'une infinité de coéfficients. Application au problème de l'attente à un guichet. *Comp. Rend. Acad. Sci. Paris,* 214, 452-56, 1942.

[B-071] Borovkov, A. A. *Stochastic Processes in Queueing Theory.* New York: Springer-Verlag, 1976.

[B-072] Borthakur, A. and Medhi, J. A queueing system with arrivals and services in batches of variable size. *Cah. Centre d'études Rech. Opér.*, 16, 117-26, 1974.

[B-073] Borthakur, A. On the busy period of a bulk queueing system with a general bulk service rule. *Opsearch,* 12, 40-46, 1975.

[B-074] Boudreau, P. E.; Griffin, J. S. and Kac, M. An elementary queueing problem. *Amer. Math. Monthly,* 69, 713-24, 1962.

[B-075] Boxma, O. J. The single-server queue with random service output. *J. Appl. Prob.,* 12, 763-78, 1975.

[B-076] Boxma, O. J. On the longest service time in a busy period of the *M / G /* 1 queue. *Stoch. Proc. Appl.,* 8, 93-100, 1978.

[B-077] Boxma, O. J. Two queues in series with non-overlapping service times. *Proc. Ninth Internat. Teletraffic Conf., Torremolinos, Spain,* Paper No. 412, 1979.

[B-078] Boxma, O. J. On a tandem queueing model with identical service times at both counters, I. *Adv. Appl. Prob.,* 11, 603-15, 1979.

[B-079] Boxma, O. J. On a tandem queueing model with identical service times at both counters, II. *Adv. Appl. Prob.,* 11, 644-659, 1979.

[B-080] Boxma, O. J. The longest service time in a busy period. *Zeitschrift f. Oper. Res.,* 24, 235-42, 1980.

[B-081] Boxma, O. J., and Konheim, A. G. Approximate analysis of exponential queueing systems with blocking. *Acta Informatica,* 15, 19-66, 1981.

[B-082] Boxma, O. J. The cyclic queue with one general and one exponential server. *Adv. Appl. Prob.,* 15, 857-73, 1983.

[B-083] Boxma, O. J. Joint distribution of sojourn time and queue length in the *M / G /* 1 queue with (in)finite capacity. *Euro. J. Oper. Res.,* 16, 246-56, 1984.

[B-084] Brandwajn, A. An iterative solution of two-dimensional birth and death processes. *Opns. Res.,* 27, 595-605, 1979.

[B-085] Brandwajn, A. A finite difference equations approach to a priority queue. *Opns. Res.,* 30, 74-81, 1982.

[B-086] Branford, A. On a property of finite-state birth and death processes. *J. Appl. Prob.,* 23, 859-66, 1986.

[B-087] Breny, H. Quelques propriétés des files d'attente ou les clients arrivent en grappes. *Mem. Soc. Roy. Sc. de Liége,* VI, Fac.4, 7-65, 1961.

[B-088] Breny, H. Sur un point de la théorie des files d'attente. *Ann. Soc. Sc. de Bruxelles,* 76, 5-12, 1962.

[B-089] Brill, P. H. and Posner, M. J. M. Level crossings in point processes applied to queues: Single-server case. *Opns. Res.,* 25, 662-74, 1977.

[B-090] Brill, P. H. An embedded level crossing technique for dams and queues. *J. Appl. Prob.,* 16, 174-86, 1979.

[B-091] Brockwell, P. J. A storage model in which the net growth-rate is a Markov chain. *J. Appl. Prob.,* 9, 129-39, 1972.

[B-092] Brown, P. and Simonian, A. Perturbation of a periodic flow in a synchronous server. In *"Performance '87",* Proc. 12-th IFIP Internat. Symp., Brussels, 1987, Courtois, P-J. and Latouche, G. eds., Amsterdam: North Holland Publ., 89-112, 1988.

[B-093] Brown, Th. $M/G/1$ round robin discipline. *Computing,* 22, 225-41, 1979.

[B-094] Brown, Th. Determination of the conditional response for quantum allocation algorithms. *J. ACM,* 29, 448-60, 1982.

[B-095] Burke, P. J. Equilibrium delay distribution for one channel with constant holding time, Poisson input and random service. *Bell Syst. Tech. J.,* 38, 1021-31, 1959.

[B-096] Burke, P. J. Random service, finite-source delay distribution for one server with constant holding time. *Opns. Res.,* 14, 695-8, 1966.

[B-097] Burke, P. J. The overflow distribution for constant holding time. *Bell Syst. Tech. J.,* 50, 3195-210, 1971.

[B-098] Burke, P. J. Delays in single-server queues with batch input. *Opns. Res.,* 23, 830-4, 1975.

[B-099] Burman, D. Y. and Smith, D. R. Asymptotic analysis of a queueing model with bursty traffic. *Bell Syst. Tech. Journ.,* 62, 1433-53, 1983.

[B-100] Burman, D. Y. and Smith, D. R. An asymptotic analysis of a queueing system with Markov-modulated arrivals. *Opns. Res.,* 34, 105-19, 1986.

[B-101] Bux, W. Single server queues with general interarrival and phase-type service time distributions - Computational algorithms. *Proc. Ninth Internat. Teletraffic Conf., Torremolinos, Spain,* Paper No. 413, 1979.

[C-001] Calo, S. B. and Schwartz, S. C. An upper bound for the equilibrium mean wait in a stationary $GI/G/1$ queue. *Math. of O.R.,* 2, 240-3, 1977.

[C-002] Cao, J-H. and Cheng, K. Analysis of $M/G/1$ queueing system with repairable service station. (in Chinese) *Acta Math. Appl. Sinica,* 5, 113-27, 1982.

[C-003] Cao, J-H. Analysis of a model of servicing machines with a repairable service equipment (in Chinese). *Chinese J. of Oper. Res.,* 2,

53-5, 1983.

[C-004] Carroll, J. L.; van de Liefvoort, A. and Lipsky, L. Solutions of $M/G/1//N$-type loops with extensions to $M/G/1$ and $GI/M/1$ queues. *Opns. Res.,* 30, 490-514, 1982.

[C-005] Carter, G. M., and Cooper, R. B. Queues with service in random order. *Opns. Res.,* 20, 389-405, 1972.

[C-006] Chakravarthy, S. and Neuts, M. F. Algorithms for the design of finite capacity service systems. *Nav. Res. Logist.,* (forthcoming).

[C-007] Chandramouli, Y.; Neuts, M. F. and Ramaswami, V. A queueing model for meteor burst packet communication systems. *IEEE Trans. Comm., 1989,* (forthcoming).

[C-008] Chang, W. Queues with feedback for time-shared computer systems. *Opns. Res.,* 16, 613-27, 1968.

[C-009] Charlot, F. and Pujolle, G. Recurrence in single server queues with impatient customers. *Ann. Inst. Henri Poincaré,* 14, 399-410, 1978.

[C-010] Chatterjee, U. and Mukkerjee, S. P. Two bulk queueing models with vacation periods. *Cah. Centre d'études Rech. Opér.,* 29, 85-93, 1987.

[C-011] Chaudhry, M. L. Some queueing problems with phase-type service. *Opns. Res.,* 13, 466-77, 1965.

[C-012] Chaudhry, M. L. Limited space queueing with arrivals correlated. *J. Roy. Nav. Sci. Serv.,* 21, 272-8, 1966.

[C-013] Chaudhry, M. L. The theory of bulk-arrival bulk-service queues. *Opsearch,* 9, 103-21, 1972.

[C-014] Chaudhry, M. L. and Templeton, J. G. C. Bounds for the least positive root of a characteristic equation in the theory of queues. *INFOR,* 11, 177-9, 1973.

[C-015] Chaudhry, M. L. Marriage between the supplementary variable technique and the imbedded Markov chain technique-I. *Trans. Eight Prague Conf. on Info. Th., Statist. Dec. Funct., Random Proc.,* Vol. A, 133-41, 1978.

[C-016] Chaudhry, M. L. The queueing system $M^X/G/1$ and its ramifications. *Nav. Res. Logist. Quart.,* 26, 667-74, 1979.

[C-017] Chaudhry, M. L. and Templeton, J. G. C. The queuing system $M/G^B/1$ and its ramifications. *Euro. J. Oper. Res.,* 6, 56-60, 1981.

[C-018] Chaudhry, M. L. and Templeton, J. G. C. *A First Course in Bulk Queues.* New York: John Wiles & Sons, 1983.

[C-019] Cheong, C. K. and Teugels, J. L. On a semi-Markov generalization of the random walk. *Stoch. Proc. Appl.*, 1, 53-66, 1973.

[C-020] Choo, Q. H. and Conolly, B. W. New results in the theory of repeated orders queueing systems. *J. Appl. Prob.*, 16, 631-40, 1979.

[C-021] Çinlar, E. Time dependence of queues with semi-Markovian service. *J. Appl. Prob.*, 4, 356-64, 1967.

[C-022] Çinlar, E. Queues with semi-Markovian arrivals. *J. Appl. Prob.*, 4, 365-79, 1967.

[C-023] Çinlar, E., and Disney, R. L. Streams of overflows from a finite queue. *Opns. Res.*, 15, 131-34, 1967.

[C-024] Çinlar, E. Decomposition of a semi-Markov process under a state-dependent rule. *SIAM J. Appl. Math.*, 15, 252-63, 1967.

[C-025] Çinlar, E. Markov renewal theory. *Adv. Appl. Prob.*, 1, 123-87, 1969.

[C-026] Çinlar, E. *Introduction to Stochastic Processes.* Englewood Cliffs, NJ: Prentice-Hall, 1975.

[C-027] Clarke, A. B., and Disney, R. L. *Probability and Random Processes for Engineers and Scientists.* New York: John Wiley and Sons, 1970.

[C-028] Coffman, E. G. Jr. and Hofri, M. A class of *FIFO* queues arising in computer systems. *Opns. Res.*, 26, 864-80, 1978.

[C-029] Cohen, J. W. Applications of derived Markov chains in queueing theory. *Appl. Sci. Res.*, B 10, 269-303, 1962.

[C-030] Cohen, J. W. A note on the delay problem in crossing a traffic stream. *Statistica Neerlandica*, 17, 3-11, 1963.

[C-031] Cohen, J. W. On two integral equations of queueing theory. *J. Appl. Prob.*, 4, 343-55, 1967.

[C-032] Cohen, J. W. The distribution of the maximum number of customers present simultaneously during a busy period for the queueing systems $M/G/1$ and $GI/M/1$. *J. Appl. Prob.*, 4, 162-79, 1967.

[C-033] Cohen, J. W. Extreme value distributions for the $M/G/1$ and $G/M/1$ queueing systems. *Ann. Inst. H. Poincaré*, B4, 83-98, 1968.

[C-034] Cohen. J. W. and Greenberg, I. Distribution of crossings of level K in a busy cycle of the $M/G/1$ queue. *Ann. Inst. H. Poincaré*, 4, 75-81, 1968.

[C-035] Cohen, J. W. Single server queue with uniformly bounded virtual waiting time. *J. Appl. Prob.*, 5, 93-122, 1968.

[C-036] Cohen, J. W. Single server queues with restricted accessibility. *J. Engin. Math.,* 3, 265-84, 1969.

[C-037] Cohen, J. W. *The Single Server Queue.* London: North Holland Publishing Co., 1969.

[C-038] Cohen. J. W. On the busy periods for the $M/G/1$ queue with finite and with infinite waiting room. *J. Appl. Prob.,* 8, 821-7, 1971.

[C-039] Cohen, J. W. On the tail of the stationary waiting time distribution and limit theorems for the $M/G/1$ queue. *Ann. Inst. H. Poincaré,* 8, 255-63, 1972.

[C-040] Cohen, J. W. The suprema of the actual and virtual waiting times during a busy cycle of the $K_m/K_n/1$ queueing system. *Adv. Appl. Prob.,* 4, 339-56, 1972.

[C-041] Cohen, J. W. Asymptotic relations in queueing theory. *Stoch. Proc. and Appl.,* 1, 107-24, 1973.

[C-042] Cohen, J. W. Some results on regular variation for distributions in queueing and fluctuation theory. *J. Appl. Prob.,* 10, 343-53, 1973.

[C-043] Cohen, J. W. Some aspects of queueing theory. *Statistica Neerlandica,* 28, 55-67, 1974.

[C-044] Cohen, J. W. *On Regenerative Processes in Queueing Theory.* New York: Springer-Verlag, Lecture Notes in Economics and Mathematical Systems, 121, 1976.

[C-045] Cohen, J. W. On a single-server queue with group arrivals. *J. Appl. Prob.,* 13, 619-22, 1976.

[C-046] Cohen, J. W. On the optimal switching level for an $M/G/1$ queueing system. *Stoch. Proc. Appl.,* 4, 297-316, 1976.

[C-047] Cohen, J. W. The Wiener-Hopf technique in applied probability. In *"Perspectives in Probability and Statistics",* papers in honour of M. S. Bartlett, J. Gani, ed., London: Academic Press, 145-56, 1976.

[C-048] Cohen, J. W. On up- and downcrossings. *J. Appl. Prob.,* 14, 405-10, 1977.

[C-049] Cohen, J. W. and Rubinovitch, M. On level crossings and cycles in dam processes. *Math. of O.R.,* 2, 297-310, 1977.

[C-050] Cohen, J. W. Properties of the process of level crossings during a busy cycle of the $M/G/1$ queueing system. *Math. of O.R.,* 3, 133-44, 1978.

[C-051] Cohen, J. W. On the maximal content of a dam and logarithmically concave renewal functions. *Stoch. Proc. Appl.,* 6, 291-304, 1978.

[C-052] Cohen, J. W. and Boxma, O. J. The $M/G/1$ queue with alternating service formulated as a Riemann-Hilbert problem. In *"Performance '81"*, F. J. Kylstra, ed., North-Holland Publ., 181-99, 1981.

[C-053] Cohen, J. W. and Hooghiemstra, G. Brownian excursion, the $M/M/1$ queue and their occupation times. *Math. of O.R.,* 6, 608-29, 1981.

[C-054] Cohen, J. W. On the $M/G/2$ queueing model. *Stoch. Proc. Appl.,* 12, 231-48, 1982.

[C-055] Colard, J. P. and Latouche, G. Algorithmic analysis of Markovian model for a system with batch and interactive jobs. *Opsearch,* 17, 12-32, 1980.

[C-056] Coleman, R. D. Use of a gate to reduce the variance of delays in queues with random service. *Bell Syst. Tech. J.,* 52, 1403-22, 1973.

[C-057] Collings, P. S. Dams with random outputs. *J. Appl. Prob.,* 11, 858-61, 1974.

[C-058] Conolly, B. W. On the busy period in relation to the single server queueing process with Erlangian input and general service time. *Giornale dell'Istituto Italiano degli Attuari,* 26, 217-32, 1963.

[C-059] Conolly, B. W. On randomized random walks. *SIAM Rev.,* 13, 81-99, 1971.

[C-060] Conolly, B. W. *Lecture Notes on Queueing Systems.* New York: Halstead, 1975.

[C-061] Conway, R.W.; Maxwell, W. L., and Miller, L. W. *Theory of Scheduling.* Reading MA: Addison-Wesley, 1967.

[C-062] Cooper, R. B., and Murray, G. Queues served in cyclic order. *Bell Syst. Tech. J.,* 48, 675-89, 1969.

[C-063] Cooper, R. B. Queues served in cyclic order: Waiting times. *Bell Syst. Tech. J.,* 49, 399-413, 1970.

[C-064] Cooper, R. B. Derivation from queueing theory of an identity related to Abel's generalization of the binomial theorem, which is useful in graph theory. *Mgmt. Sci.,* 19, 582-4, 1973.

[C-065] Cooper, R. B., and Tilt, B. On the relationship between the distribution of maximal queue length in the $M/G/1$ queue and the mean busy period in the $M/G/1/n$ queue. *J. Appl. Prob.,* 13, 195-199, 1976.

[C-066] Cooper, R. B. *Introduction to Queueing Theory.* 2nd ed., New York: North Holland Publishing Co., 1981.

[C-067] Cosmetatos, G. P. Approximate explicit formulae for the average queueing time in the processes $(M/D/r)$ and $(D/M/r)$. *INFOR,* 13, 328-32, 1975.

[C-068] Cosmetatos, G. P. Some approximate equilibrium results for the multiserver queue $(M/G/r)$. *Oper. Res. Quart.,* 27, 615-20, 1976.

[C-069] Courtois, P. J. and Georges, J. On a single-server finite queuing model with state-dependent arrival and service processes. *Opns. Res.,* 19, 424-35, 1971.

[C-070] Courtois, P. J. The $M/G/1$ finite capacity queue with delays. *IEEE Trans. on Comm.,* COM-28, 165-71, 1980.

[C-071] Cowan, R. An analysis of the fixed-cycle traffic-light problem. *J. Appl. Prob.,* 18, 672-83, 1981.

[C-072] Cox, D. R. A use of complex probabilities in the theory of stochastic processes. *Proc. Camb. Phil. Soc.,* 51, 313-19, 1955.

[C-073] Cox, D. R. The analysis of non-Markovian stochastic processes by the inclusion of supplementary variables. *Proc. Camb. Phil. Soc.,* 51, 433-41, 1955.

[C-074] Cox, D. R., and Smith, W. L. *Queues.* London: Methuen's Monographs on Applied Probability and Statistics, 1961.

[C-075] Cox, D. R. *Renewal Theory.* London: Methuen's Monographs on Applied Probability and Statistics, 1962.

[C-076] Crabill, T. B. Sufficient conditions for positive recurrence and recurrence of specially structured Markov chains. *Opns. Res.,* 16, 858-67, 1968.

[C-077] Craven, B. D. Some results for the bulk service queue. *Austral. J. Statist.,* 5, 49-56, 1963.

[C-078] Craven, B. D. Asymptotic transient behaviour of the bulk service queue. *J. Austral. Math. Soc.,* 3, 503-12, 1963.

[C-079] Craven, B. D. Serial dependence of a Markov process. *J. Austral. Math. Soc.,* 5, 299-314, 1965.

[C-080] Craven. B. D. Asymptotic correlation in a queue. *J. Appl. Prob.,* 6, 573-83, 1969.

[C-081] Craven, B. D. The spectral density of a Markov process. *J. Appl. Prob.,* 10, 520-7, 1973.

[C-082] Craven, B. D. and Shanbhag, D. N. The number of customers in a busy period. *Res. Rept. 140/BDC & DNS 1, Manchester-Sheffield School of Prob. and Statist., August 1973.*

[C-083] Craven, B. D. Asymptotic rate of a Markov process. *Adv. Appl. Prob.,* 6, 732-46, 1974.

[C-084] Cromie, M. V.; Chaudhry, M. L. and Grassmann, W. K. Further results for the queueing system $M^X / M / c$. *Opns. Res.,* 30, 755-63, 1979.

[C-085] Crommelin, C. D. Delay probability formulae when the holding times are constant. *Post Office Electrical Engineer's Journal,* 25, 41-50, 1932.

[C-086] Crommelin, C. D. Delay probability formulae. *Post Office Electrical Engineer's Journal,* 26, 266-74, 1934.

[C-087] Curry, G. L. and Feldman, R. M. An $M / M / 1$ queue with a general bulk service rule. *Nav. Res. Logist.,* 32, 595-603, 1985.

[D-001] Dafermos, S. C. and Neuts, M. F. A single server in discrete time. *Cah. Centre d'études Rech. Opér.,* 13, 23-40, 1971.

[D-002] Daganzo, C. F. Traffic delay at unsignalized intersections: Clarification of some issues. *Trans. Sci.,* 11, 180-9, 1977.

[D-003] Dagsvik, J. The general bulk queue as a matrix factorisation problem of the Wiener-Hopf type. Part I. *Adv. Appl. Prob.,* 7, 636-46, 1975.

[D-004] Dagsvik, J. The general bulk queue as a matrix factorisation problem of the Wiener-Hopf type. Part II. *Adv. Appl. Prob.,* 7, 647-55, 1975.

[D-005] Daley, D. J. Single-server queueing systems with uniformly limited queueing time. *J. Austral. Math. Soc.,* 4, 489-505, 1964.

[D-006] Daley, D. J. General customer impatience in the queue $GI / G / 1$. *J. Appl. Prob.,* 2, 186-205, 1965.

[D-007] Daley, D. J. The correlation structure of the output process of some single server queueing systems. *Ann. Math. Statist.,* 39, 1007-19, 1968.

[D-008] Daley, D. J. and Moran, P. A. P. Two-sided inequalities for waiting and queue size distributions in $GI / G / 1$. *Teor. Veroyatnost i Prim.,* 13, 338-41, 1968.

[D-009] Daley, D. J. The serial correlation coefficients of waiting times in a stationary single server queue. *J. Austral. Math. Soc.,* 8, 683-99, 1968.

[D-010] Daley, D. J. Total waiting time in a busy period of a stable single server queue. *J. Appl. Prob.,* 6, 550-64, 1969.

[D-011] Daley, D. J. and Jacobs, D. R. The total waiting time in a busy period of a stable single-server queue, II. *J. Appl. Prob.,* 6, 565-72, 1969.

[D-012] Daley, D. J. Characterizing pure loss $GI / G / 1$ queues with renewal output. *Proc. Camb. Phil. Soc.,* 75, 103-7, 1974.

[D-013] Daley, D. J. Notes on queueing output processes. In *"Proceedings of the Conference on Mathematical Methodology in the Theory of Queues, Kalamazoo, Mich.",* New York: Springer-Verlag, 351-8, 1974.

[D-014] Daley, D. J. Further second-order properties of certain single-server queueing systems. *Stoch. Proc. Appl.,* 3, 185-91, 1975.

[D-015] Daley, D. J. and Shanbhag, D. N. Interdependent inter-departure times in $M / G / 1 / N$ queues. *J. Roy. Statist. Soc. Ser. B,* 37, 259-63, 1975.

[D-016] Daley, D. J. Queueing output processes. *Adv. Appl. Prob.,* 8, 395-415, 1976.

[D-017] Daley, D. J. Inequalities for moments of tails of random variables, with a queueing application. *Z. Wahrsch. verw. Gebiete,* 41, 139-43, 1977.

[D-018] Daley, D. J. and Rolski, T. Some comparability results for waiting times in single- and multiple-server queues. *J. Appl. Prob.,* 21, 887-900, 1984.

[D-019] Daley, D. J. A lower bound for mean characteristics in $E_k / G / 1$ and $GI / E_k / 1$ queues. *Optimization,* 17, 117-24, 1986.

[D-020] Daniels, H. E. Mixtures of geometric distributions. *J. Roy. Statist. Soc. Ser. B,* 23, 409-13, 1961.

[D-021] Danielyan, E. A. and Dimitrov, B. N. On the $M / G / 1$ waiting line with two types of breakdown. (in Russian) *Mathematica Balkanica,* 2, 21-37, 1972.

[D-022] Darroch, J. N. On the traffic-light queue. *Ann. Math. Statist.,* 35, 380-88, 1964.

[D-023] Darroch, J. N.; Newell, G. F. and Morris, R. W. J. Queues for a vehicle-actuated traffic light. *Oper. Res.,* 12, 882-95, 1964.

[D-024] d'Avignon, G. R. and Disney, R. L. Queues with instantaneous feedback. *Mgmt. Sci.,* 24, 168-80, 1977.

[D-025] Davis, R. H. Waiting-time distribution of a multi-server, priority queuing system. *Opns. Res.*, 14, 133-6, 1966.

[D-026] Dehon, M. and Latouche, G. A geometric interpretation of the relations between the exponential and generalized Erlang distributions. *Adv. Appl. Prob.*, 14, 885-97, 1982.

[D-027] Delbrouck, L. E. N. A multiserver queue with enforced idle times. *Opns. Res.*, 17, 506-18, 1969.

[D-028] Delbrouck, L. E. N. A feedback queueing system with batch arrivals, bulk service and queue-dependent service time. *J. ACM*, 17, 314-23, 1970.

[D-029] Delbrouck, L. E. N. The law of averages as a computing tool. *Nav. Res. Logist. Quart.*, 19, 149-58, 1972.

[D-030] Delbrouck, L. E. N. Convexity properties, moment inequalities and asymptotic exponential approximations for delay distributions in $GI / G / 1$ systems. *Stoch. Proc. Appl.*, 3, 193-208, 1975.

[D-031] Delbrouck, L. E. N. Approximations for certain congestion functions in single server queueing systems. *Proc. Eight Internat. Teletraffic Conf., Melbourne, Australia,* Paper No. 233, 1976.

[D-032] Delbrouck, L. E. N. Numerical approximations for compound geometric distributions with applications in queueing theory. *J. Appl. Prob.*, 15, 202-8, 1978.

[D-033] De Meyer, A. and Teugels, J. L. On the asymptotic behaviour of the distributions of the busy period and service time in $M / G / 1$. *J. Appl. Prob.*, 17, 802-13, 1980.

[D-034] Descloux, A. On overflow processes of trunk groups with Poisson inputs and exponential service times. *Bell Syst. Tech. J.*, 42, 383-97, 1963.

[D-035] de Smit, J. H. A. The transient behaviour of the queue at a fixed cycle traffic light. *Transp. Res.*, 4, 1-14, 1971.

[D-036] de Smit, J. H. A. The single server semi-Markov queue. *Stoch. Proc. Appl.*, 22, 37-50, 1986.

[D-037] Dewess, M. On applications of dominated variation to the queueing theory. *Math. Operationsforsch. u. Statist.*, 9, 601-9, 1978.

[D-038] Dick, R. S. Some theorems on a single-server queue with balking. *Opns. Res.*, 18, 1193-1206, 1970.

[D-039] Dimitrov, B. N. Asymptotic expansions of characteristics for queueing systems of the $M / G / 1$ type. (in Russian with an English summary) *Bull. Inst. de Math., Acad. Bulgare des Sciences*, 15, 237-63, 1973.

[D-040] Dimitrov, B. N. and Karapenev, Khr. K. The number of served calls in a queueing system with finite source and nonreliable server. (in Bulgarian, with English and Russian summaries) *University Annual, Applied Mathematics,* 10, 139-50, 1974.

[D-041] Dimitrov, B. N. Some characteristics of a system with a limited number of sources, unreliable server and absolute priority. (in Russian) *Serdika,* 1, 158-66, 1975.

[D-042] Dimitrov, B. N. and Petrova, A. Single server queueing system with Poisson input and exponentially bounded waiting time in the queue. *Proc. Fifth Conf. on Probability Theory, Brasov, Romania, 1974,* Bereanu, Iosifescu and Popescu (eds.), Editura Academiei Republicii Socialiste Romania, Bucharest, 367-73, 1977.

[D-043] Disney, R. L. A matrix solution for the two server queue with overflow. *Mgmt. Sci.,* 19, 254-65, 1972.

[D-044] Disney, R.L.; Farrell, R. L., and De Morais, P. R. A characterization of $M/G/1$ queues with renewal departure processes. *Mgmt. Sci.,* 19, 1222-28, 1973.

[D-045] Disney, R. L. and Cherry, W. P. Some topics in queueing network theory. In *"Proceedings of the Conference on Mathematical Methodology in the Theory of Queues, Kalamazoo, Mich."*, New York: Springer-Verlag, 23-44, 1974.

[D-046] Disney, R. L.; McNickle, D. C., and Simon, B. The $M/G/1$ queue with instantaneous Bernoulli feedback. *Nav. Res. Logist. Quart.,* 27, 635-44, 1980.

[D-047] Disney, R. L. and Chandramohan, J. A correction note on "Two finite $M/M/1$ queues in tandem: A matrix solution for the steady state". *Opsearch,* 17, 181-3, 1980.

[D-048] Disney, R. L. A note on sojourn times in $M/G/1$ queues with instantaneous, Bernoulli feedback. *Nav. Res. Logist. Quart.,* 28, 679-84, 1981.

[D-049] Disney, R. L. and Ott, T. J. (eds.) *Applied Probability - Computer Science: The Interface. (2 Vols.)* Proc. of the Conference at Boca Raton, Florida, January 1981, Boston: Birkhäuser, 1982.

[D-050] Disney, R. L.; König, D. and Schmidt, V. Stationary queue-length and waiting-time distributions in single-server feedback queues. *Adv. Appl. Prob.,* 16, 437-46, 1984.

[D-051] Doshi, B. T. Optimal control of the service rate in an $M/G/1$ queueing system. *Adv. Appl. Prob.,* 10, 682-701, 1978.

[D-052] Doshi, B. T. and Lipper, E. H. Comparisons of service disciplines in a queueing system with delay dependent customer behaviour. In *"Applied Probability - Computer Science: The Interface"*, Proc. of the Conference at Boca Raton, Florida, January 1981, R. L. Disney and T. J. Ott (eds), Boston: Birkhäuser, Vol II, 269-304, 1982.

[D-053] Doshi, B. T. An $M/G/1$ queue with a hybrid discipline. *Bell Syst. Tech. J.*, 62, 1251-71, 1983.

[D-054] Doshi, B. T. A note on stochastic decomposition in a $GI/G/1$ queue with vacations or set-up times. *J. Appl. Prob.*, 22, 419-28, 1985.

[D-055] Doshi, B. T. Queueing systems with vacations: A survey. *Queueing Systems*, 1, 29-66, 1986.

[D-056] Downton, F. Waiting times in bulk service queues. *J. Roy. Statist. Soc. Ser. B*, 17, 256-61, 1955.

[D-057] Downton, F. On limiting distributions arising in bulk service queues. *J. Roy. Statist. Soc. Ser. B*, 18, 265-74, 1956.

[D-058] Dunne, M. C. Traffic delay at a signalized intersection with binomial arrivals. *Trans. Sci.*, 1, 24-31, 1967.

[D-059] Durr, L. A single-server priority queuing system with general holding times, Poisson input, and reverse-order-of-arrival queuing discipline. *Opns. Res.*, 17, 351-8, 1969.

[E-001] Easton, G. D. and Chaudhry, M. L. The queueing system $E_k/M(a,b)/1$ and its numerical analysis. *Comput. and Oper. Res.*, 9, 197-205, 1982.

[E-002] Eckberg, A. E. Sharp bounds on Laplace-Stieltjes transforms, with applications to various queueing problems. *Math. of O.R.*, 2, 135-42, 1977.

[E-003] Eckberg, A. E. The single server queue with periodic arrival process and deterministic service times. *IEEE Trans. Comm.*, COM-27, 556-62, 1979.

[E-004] Eckberg, A. E. Generalized peakedness of teletraffic processes. *Proc. Tenth Internat. Teletraffic Conf., Montreal, Canada,* Paper No. 4.4B-3, 1983.

[E-005] Eckberg, A. E. Approximations for bursty (and smoothed) arrival queueing delays based on generalized peakedness. *Proc. Eleventh Internat. Teletraffic Conf., Kyoto, Japan,* Paper No. 3.1A-4, 1985.

[E-006] Egenolf, F. J. Queue discipline *NIFO* for a tree structured generalized $M/G/1$ - queueing system and its application (*NIFO* =

Nearest in first out). *Proc. Eight Internat. Teletraffic Conf., Melbourne, Australia,* Paper No. 234, 1976.

[E-007] Eisenberg, M. Two queues with changeover times. *Opns. Res.,* 19, 386-401, 1971.

[E-008] Eisenberg, M. Queues with periodic service and changeover times. *Opns. Res.,* 20, 440-51, 1972.

[E-009] Eisenberg, M. Two queues with alternating service. *SIAM J. Appl. Math.,* 36, 287-303, 1979.

[E-010] El-Affendi and Kouvatsos, D. D. A maximum entropy analysis of the $M/G/1$ and $GI/M/1$ queueing systems at equilibrium. *Acta Informatica,* 19, 339-55, 1983.

[E-011] Elsner, L. Verfahren zur Berechnung des Spektralradius nichtnegativer, irreduzibler Matrizen I. *Computing,* 8, 32-39, 1971.

[E-012] Elsner, L. Verfahren zur Berechnung des Spektralradius nichtnegativer, irreduzibler Matrizen II. *Computing,* 9, 69-73, 1972.

[E-013] Enns, E. G. The trivariate distribution of the maximum queue length, the number of customers served and the duration of the busy period for the $M/G/1$ queueing system. *J. Appl. Prob.,* 6, 154-61, 1969.

[E-014] Enns, E. G. Some waiting-time distributions for queues with multiple feedback and priorities. *Opns. Res.,* 17, 519-25, 1969.

[E-015] Enns, E. G. The distribution of a number of random variables associated with the absorbing Markov chain. *Austral. J. Statist.,* 14, 79-83, 1972.

[E-016] Enns, E. G. The busy period and the probability that a particular queue is in service in a single server queueing system with N queues, arbitrary priorities, and a general probabilistic inter-queue transition matrix. *Proc. Seventh Internat. Teletraffic Conf., Stockholm, Sweden,* Paper No. 438, 1973.

[E-017] Erlander, S. The remaining busy period for a single server queue with Poisson input. *Opns. Res.,* 13, 734-46, 1965.

[E-018] Evans, D. H.; Herman, R. and Weiss, G. H. The highway merging and queuing problem. *Opns. Res.,* 12, 832-37, 1964.

[E-019] Ewens, W. J. and Finch, P. D. A generalized single-server queue with Erlang input. *Biometrika,* 49, 242-5, 1962.

[F-001] Fabens, A. J. The solution of queueing and inventory models by semi-Markov processes. *J. Roy. Statist. Soc. Ser. B,* 23, 113-27, 1961.

[F-002] Fabens, A. J. and Perera, A. G. A. D. A correction to "The solution of queueing and inventory model by semi-Markov processes". *J. Roy. Statist. Soc. Ser. B,* 25, 455-56, 1963.

[F-003] Fainberg, M. A. and Stoyan, D. A further remark on a paper of K. T. Marshall on bounds for $GI/G/1$. *Math. Operationsforsch. u. Statist.,* 9, 145-50, 1976.

[F-004] Fakinos, D. The expected remaining service time in a single server queue. *Opns. Res.,* 30, 1014-8, 1982.

[F-005] Falin, G. I. A single-line system with secondary orders. *Engin. Cybern.,* 17, 76-83, 1979.

[F-006] Falin, G. I. Quasi-input in the $M/G/1$ queue. *Adv. Appl. Prob.,* 16, 695-6, 1984.

[F-007] Federgruen, A. and Tijms, H. C. Computation of the stationary distribution of the queue size in an $M/G/1$ queueing system with variable service rate. *J. Appl. Prob.,* 17, 515-22, 1980.

[F-008] Federgruen, A. and Green, L. An $M/G/1$ queue in which the number of servers required is random. *J. Appl. Prob.,* 21, 583-602, 1984.

[F-009] Federgruen, A. and Green, L. Queueing systems with service interruptions. *Opns. Res.,* 34, 752-68, 1986.

[F-010] Feller, W. *An Introduction to Probability Theory and its Applications. Vol. 1.* 3d. ed. New York: John Wiley and Sons, 1968.

[F-011] Feller, W. *An Introduction to Probability Theory and its Applications. Vol. 2.* 2d. ed. New York: John Wiley and Sons, 1971.

[F-012] Finch, P. D. The effect of the size of the waiting room on a simple queue. *J. Roy. Statist. Soc. Ser. B,* 20, 182-6, 1957.

[F-013] Finch, P. D. A probability limit theorem with application to a generalisation of queueing theory. *Acta Math. Acad. Sci. Hung.,* 10, 317-25, 1959.

[F-014] Finch, P. D. The output process of the queueing system $M/G/1$. *J. Roy. Statist. Soc. Ser. B,* 21, 375-80, 1959.

[F-015] Finch, P. D. On the transient behaviour of a simple queue. *J. Roy. Statist. Soc. Ser. B,* 22, 277-84, 1960.

[F-016] Finch, P. D. On the busy period in the queueing system $GI/G/1$. *J. Austral. Math. Soc.,* 2, 217-28, 1961.

[F-017] Finch, P. D. A note on the queueing system $GI/E_k/1$. *J. Appl. Prob.,* 6, 708-10, 1969.

[F-018] Fischer, M. J. The waiting time in the $E_k / M / 1$ queueing system. *Opns. Res.*, 22, 898-902, 1974.

[F-019] Fischer, M. J. An approximation to queueing systems with interruptions. *Mgmt. Sci.*, 24, 338-44, 1977.

[F-020] Fischer, M. J. Analysis and design of loop systems via a diffusion approximation. *Opns. Res.*, 25, 269-78, 1977.

[F-021] Fischer, M. J. A queueing analysis of an integrated telecommunications system with priorities. *INFOR*, 15, 277-88, 1977.

[F-022] Fischer, M. J. Two queueing systems sharing the same finite waiting room. *Nav. Res. Logist. Quart.*, 25, 667-79, 1978.

[F-023] Fischer, M. J. A queueing model for a technical control facility. *AIIE Trans.*, 10, 76-80, 1978.

[F-024] Fischer, M. J. Data performance in a system where data packets are transmitted during voice silent periods - Single channel case. *IEEE Trans. Comm.*, COM-27, 1371-75, 1979.

[F-025] Fleming, P. J. An approximate analysis of sojourn times in the $M / G / 1$ queue with round-robin discipline. *AT&T Bell Labs. Tech. Journ.*, 63, 1521-35, 1984.

[F-026] Foley, R. D. and Disney, R. L. Queues with delayed feedback. *Adv. Appl. Prob.*, 15, 162-82, 1983.

[F-027] Fond, S. and Ross, S. M. A heterogeneous arrival and service queueing loss model. *Nav. Res. Logist. Quart.*, 25, 483-8, 1978.

[F-028] Fontana, B. and Diaz Berzosa, C. Stationary queue length distributions in an $M / G / 1$ queue with two non-preemptive priorities and general feedback. In " *Performance of Computer-Communication Systems*", H. Rudin and W. Bux (eds), Amsterdam: North-Holland, 333-347, 1984.

[F-029] Fontana, B. and Diaz Berzosa, C. $M / G / 1$ queue with N-priorities and feedback: Joint queue length distributions and response time distribution for any particular sequence. In " *Teletraffic Issues in an Advanced Information Society*", *ITC*-11, M. Akiyama (ed), Amsterdam: North-Holland, 452-58, 1985.

[F-030] Foster, F. G. On stochastic matrices associated with certain queueing processes. *Ann. Math. Statist.*, 24, 355-60, 1953.

[F-031] Foster, F. G. Queues with batch arrivals I. *Acta Math. Acad. Sci. Hungar.*, 12, 1-10, 1961.

[F-032] Foster, F. G., and Perera, A. G. A. D. Queues with batch arrivals II. *Acta Math. Acad. Sci. Hungar.*, 16, 275-87, 1965.

[F-033] Franken, P.; König, D.; Arndt, U., and Schmidt, V. *Queues and Point Processes.* Berlin: Akademie-Verlag, 1981.

[F-034] Fredericks, A. A. A class of approximations for the waiting time distribution in a $GI/G/1$ queueing system. *Bell Syst. Tech. J.,* 61, 295-325, 1982.

[F-035] Fryer, M. C. and Winsten, C. B. An algorithm to compute the equilibrium distributions of a one-dimensional bounded random walk. *Opns. Res.,* 34, 449-54, 1986.

[F-036] Fuhrmann, S. W. A note on the $M/G/1$ queue with server vacations. *Opns. Res.,* 32, 1368-73, 1984.

[F-037] Fuhrmann, S. W. and Cooper, R. B. Stochastic decompositions in the $M/G/1$ queue with generalized vacations. *Opns. Res.,* 33, 1117-29, 1985.

[F-038] Fuhrmann, S. W. and Cooper, R. B. Application of decomposition principle in $M/G/1$ vacation model to two continuum cyclic queueing models - Especially token-ring *LAN*s. *A T & T Tech. Journ.,* 64, 1091-99, 1985.

[F-039] Fujiki, M. and Murao, Y. Queueing model with regular service interruptions. *Proc. Eight Internat. Teletraffic Conf., Melbourne, Australia,* Paper No. 232, 1976.

[F-040] Fujisawa, T. and Sugizaki, Y. Steady-state solution of a single server queue with Poisson arrivals and multiple transactions. *Math. Operationsforsch. u. Statist.,* 12, 233-41, 1981.

[G-001] Gallisch, E. On monotone optimal policies in a queueing model of $M/G/1$ type with controllable service time distribution. *Adv. Appl. Prob.,* 11, 870-87, 1979.

[G-002] Gani, J. Problems in the probability theory of storage systems. *J. Roy. Statist. Soc. Ser. B,* 19, 181-206, 1957.

[G-003] Gani, J, and Prabhu, N. U. Stationary distributions of the negative exponential type for the infinite dam. *J. Roy. Statist. Soc. Ser. B,* 19, 342-51, 1957.

[G-004] Gani, J. and Prabhu, N. U. The time-dependent solution for a storage model with Poisson input. *J. Math. and Mech.,* 8, 653-64, 1959.

[G-005] Gani, J. First emptiness of two dams in parallel. *Ann. Math. Statist.,* 32, 219-29, 1961.

[G-006] Gani, J. The time-dependent solution for a dam with ordered Poisson inputs. In *"Studies in Applied Probability and Management Science",* Arrow, Karlin, Scarf, eds., Stanford: Stanford University

Press, 101-09, 1962.

[G-007] Gani, J. A note on the first emptiness of dams with Markovian inputs. *J. Math. Anal. Appl.*, 26, 270-74, 1969.

[G-008] Gani, J. Recent advances in storage and flooding theory. *Adv. Appl. Prob.*, 1, 90-110, 1969.

[G-009] Gantmacher, F. R. *The Theory of Matrices*. New York: Chelsea, 1959.

[G-010] Gaver, D. P. The influence of servicing times in queuing processes. *Opns. Res.*, 2, 139-49, 1954.

[G-011] Gaver, D. P. Imbedded Markov chain analysis of a waiting-line process in continuous time. *Ann. Math. Statist.*, 30, 698-720, 1959.

[G-012] Gaver, D. P. Jr. and Miller, R. G. Jr. Limiting distributions for some storage problems. In *"Studies in Applied Probability and Management Science"*, Arrow, Karlin, Scarf, eds., Stanford: Stanford University Press, 110-26, 1962.

[G-013] Gaver, D. P. A waiting line with interrupted service including priorities. *J. Roy. Statist. Soc. Ser. B*, 24, 73-91, 1962.

[G-014] Gaver, D. P. A comparison of queue disciplines when service orientation times occur. *Nav. Res. Logist. Quart.*, 10, 219-35, 1963.

[G-015] Gaver, D. P. An absorption probability problem. *J. Math. Anal. Appl.*, 9, 384-93, 1964.

[G-016] Gaver, D. P.; Jacobs, P. A. and Latouche, G. Finite birth-and-death models in randomly changing environments. *Adv. Appl. Prob.*, 16, 715-31, 1984.

[G-017] Gaver, D. P. and Jacobs, P. A. On inference and transient response for $M/G/1$ models. In *"Teletraffic Analysis and Computer Performance Evaluation"*, O. J. Boxma, J. W. Cohen, H. C. Tijms, eds., Amsterdam: Elseviers Science, North Holland Publ., 163-70, 1986.

[G-018] Gaver, D. P. and Jacobs, P. A. Nonparametric estimation of the probability of a long delay in the $M/G/1$ queue. *Tech. Rept. NPS55-86-022, Naval Postgraduate School, Monterey, Calif., November 1986.*

[G-019] Gavish, B. and Schweitzer, P. J. The Markovian queue with bounded waiting time. *Mgmt. Sci.*, 23, 1349-57, 1977.

[G-020] Gebhardt, D. Die Ermittlung von Kenngröszen für das Wartesystem $M/G/1$ mit beschränktem Warteraum. *Z. f. Oper. Res.*, 17, 207-16, 1973.

[G-021] Gelenbe, E. and Mitrani, I. *Analysis and Synthesis of Computer Systems.* New York: Academic Press, 1980.

[G-022] Gelenbe, E. and Iasnogorodski, R. A queue with server of walking type (autonomous service). *Ann. Inst. H. Poincaré,* 16, 63-73, 1980.

[G-023] Georganas, N. D. Buffer behavior with Poisson arrivals and bulk geometric service. *IEEE Trans. Comm.*, COM-24, 938-40, 1976.

[G-024] Gergely, T. and Török, T. L. On the busy period of discrete-time queues. *J. Appl. Prob.*, 11, 853-7, 1974.

[G-025] Giffin, W. C. *Queueing: Basic Theory and Applications.* Columbus, Ohio : Grid Publishing, 1978.

[G-026] Gilchrist, R. The inter-arrival times of accepted customers in an $M/G/1$ queue with finite capacity. *J. Appl. Prob.*, 13, 633-8, 1976.

[G-027] Gillent, F. and Latouche, G. Semi-explicit solutions for $M/PH/1$-like queueing systems. *Euro. J. Oper. Res.*, 13, 151-60, 1983.

[G-028] Gnedenko, B. V., and Kovalenko, I. N. *Introduction to Queueing Theory.* Jerusalem: Israel Program for Scientific Translations, 1968.

[G-029] Gnedenko, B. V. and König, D. *Handbuch der Bedienungstheorie I: Grundlagen und Methoden.* Berlin: Akademie Verlag, 1982.

[G-030] Goel, L. R. Limited space heterogeneous queueing problem with an additional special feeding source. *Math. Operationsforsch. u. Statist.*, 8, 119-25, 1977.

[G-031] Gopinath, B. and Morrison, J. A. Single server queues with correlated inputs. In *"Computer Performance"*, Proc. Int. Symp. Comp. Perf. Model., Meas., and Eval., Yorktown Heights, New York, K. M. Chandy and M. Reiser, eds., New York: North Holland Publ., 263-77, 1977.

[G-032] Gopinath, B. and Morrison, J. A. Discrete-time single server queues with correlated inputs. *Bell Syst. Tech. J.*, 56, 1743-68, 1977.

[G-033] Goyal, T. L. and Harris, C. M. Maximum-likelihood estimates for queues with state-dependent service. *Sankhya, Ser. A,* 34, 65-80, 1972.

[G-034] Grandell, J. *Doubly Stochastic Processes.* Berlin: Springer-Verlag, 1976.

[G-035] Grassmann, W. K. The steady state behaviour of the $M / E_k / 1$ queue, with state dependent arrival rates. *INFOR,* 12, 163-73, 1974.

[G-036] Grassmann, W. K. The $GI / PH / 1$ queue: A method to find the transition matrix. *INFOR,* 20, 144-56, 1982.

[G-037] Grassmann, W. K. and Chaudhry, M. L. A new method to solve steady state equations. *Nav. Res. Logist. Quart.,* 29, 461-73, 1982.

[G-038] Grassmann, W. K., Taksar, M. I. and Heyman, D. P. Regenerative analysis and steady-state distributions for Markov chains. *Opns. Res.,* 33, 1107-16, 1985.

[G-039] Green, L. A queueing system with general use and limited use servers. *Opns. Res.,* 33, 168-82, 1985.

[G-040] Greenberg, I. Some duality results in the theory of queues. *J. Appl. Prob.,* 6, 99-121, 1969.

[G-041] Greenberg, I. Distribution-free analysis of the $M / G / 1$ and $G / M / 1$ queues. *Opns. Res.,* 21, 629-35, 1973.

[G-042] Greenberg, I. An approximation for the waiting time distribution in single server queues. *Nav. Res. Logist. Quart.,* 27, 223-30, 1980.

[G-043] Griffiths, J. D. A mathematical model of a nonsignalized pedestrian crossing. *Transp. Res.,* 15, 222-32, 1981.

[G-044] Grinstein, J. and Rubinovitch, M. Queues with random service output: The case of Poisson arrivals. *J. Appl. Prob.,* 11, 771-84, 1974.

[G-045] Gross, D., and Harris, C. M. On one-for-one-ordering inventory policies with state-dependent leadtimes. *Opns. Res.,* 19, 735-60, 1971.

[G-046] Gross, D.; Harris, C. M. and Lechner, J. A. Stochastic inventory models with bulk demand and state-dependent leadtimes. *J. Appl. Prob.,* 8, 521-34, 1971.

[G-047] Gross, D., and Harris, C. M. *Fundamentals of Queueing Theory.* New York: John Wiley and Sons, 1974.

[G-048] Gross, D. and Miller, D. R. The randomization technique as a modeling tool and solution procedure for transient Markov processes. *Opns. Res.,* 32, 343-61, 1984.

[G-049] Gün, L. Tandem queueing systems subject to blocking with phase type servers: Analytic solutions and approximations. *Tech. Res. Rept. TR-87-02, Systems Research Center, University of Maryland, 1987.*

[H-001] Haight, F. A. *Mathematical Theories of Traffic Flow.* New York: Academic Press, 1963.

[H-002] Haight, F. A. The discrete busy period distribution for various single server queues. *Zast. Matemat.,* 8, 37-46, 1965.

[H-003] Haight, F. A. *Applied Probability.* New York: Plenum Publishing Company, 1981.

[H-004] Halfin, S. Steady-state distribution for the buffer content of an $M/G/1$ queue with varying service rate. *SIAM J. Appl. Math.,* 23, 356-63, 1972.

[H-005] Halfin, S. and Segal, M. A priority queueing model for a mixture of two types of customers. *SIAM J. Appl. Math.,* 23, 369-79, 1972.

[H-006] Hajek, B. Birth-and-death processes on the integers with phases and general boundaries. *J. Appl. Prob.,* 19, 488-99, 1982.

[H-007] Hannibalsson, I. and Disney, R. L. An $M/M/1$ queue with delayed feedback. *Nav. Res. Logist. Quart.,* 24, 281-91, 1977.

[H-008] Hanschke, T. The $M/G/1/1$ queue with repeated attempts and different types of feedback effects. *OR Spektrum,* 7, 209-15, 1985.

[H-009] Harris, C. M. Queues with stochastic service rates. *Nav. Res. Logist. Quart.,* 14, 219-30, 1967.

[H-010] Harris, C. M. A queueing system with multiple service time distributions. *Nav. Res. Logist. Quart.,* 14, 231-9, 1967.

[H-011] Harris, C. M. Queues with state-dependent stochastic service rates. *Opns. Res.,* 15, 117-30, 1967.

[H-012] Harris, C. M. The Pareto distribution as a queue service discipline. *Opns. Res.,* 16, 307-13, 1968.

[H-013] Harris, C. M. Correction to "Queues with state-dependent stochastic service rates". *Opns. Res.,* 16, 885-6, 1968.

[H-014] Harris, C. M. Some results for bulk-arrival queues with state-dependent service times. *Mgmt. Sci.,* 16, 313-26, 1970.

[H-015] Harris, C. M. On queues with state-dependent Erlang service. *Nav. Res. Logist. Quart.,* 18, 103-10, 1971.

[H-016] Harris, C. M. and Marlin, P. G. A note on feedback queues with bulk service. *J. ACM,* 19, 727-33, 1972.

[H-017] Harris, C. M. Some new results in the statistical analysis of queues. In *"Proceedings of the Conference on Mathematical Methodology in the Theory of Queues, Kalamazoo, Mich."*, New York: Springer-Verlag, 157-83, 1974.

[H-018] Harrison, J. M. and Lemoine, A. J. On the virtual and actual waiting time distributions of a *GI / G / 1* queue. *J. Appl. Prob.*, 13, 833-36, 1976.

[H-019] Harrison, J. M. Some stochastic bounds for dams and queues. *Math. of O.R.*, 2, 54-63, 1977.

[H-020] Harrison, J. M. and Lemoine, A. J. Limit theorems for periodic queues. *J. Appl. Prob.*, 14, 566-76, 1977.

[H-021] Hashida, O. On the busy period in the queueing system with finite capacity. *J. Oper. Res. Soc. Japan*, 15, 115-37, 1972.

[H-022] Hasofer, A. M. On the single-server queue with non-homogeneous Poisson input and general service time. *J. Appl. Prob.*, 1, 369-84, 1964.

[H-023] Hasofer, A. M. Some perturbation results for the single server queue with Poisson input. *J. Appl. Prob.*, 2, 462-6, 1965.

[H-024] Hawkes, A. G. Time-dependent solution of a priority queue with bulk arrivals. *Opns. Res.*, 13, 586-95, 1965.

[H-025] Hawkes, A. G. Queueing for gaps in traffic. *Biometrika*, 52, 79-85, 1965.

[H-026] Hawkes, A. G. Delay at traffic intersections. *J. Roy. Statist. Soc. Ser. B*, 28, 202-12, 1966.

[H-027] Hawkes, A. G. Gap acceptance in road traffic. *J. Appl. Prob.*, 5, 84-92, 1968.

[H-028] Heathcote, C. R. On the queueing process *M / G / 1*. *Ann. Math. Statist.*, 32, 770-3, 1961.

[H-029] Heathcote, C. R. Divergent single server queues. *Proc. Symp. on Congestion Theory*, W. L. Smith and W. E. Wilkinson, eds., Chapel Hill: Univ. of N. Carolina Monographs, 108-36, 1965.

[H-030] Heathcote, C. R. On the maximum of the queue *GI / M / 1*. *J. Appl. Prob.*, 2, 206-14, 1965.

[H-031] Heathcote, C. R. and Winer, P. An approximation for the moments of waiting times. *Opns. Res.*, 17, 175-86, 1969.

[H-032] Heffes, H. Analysis of first-come first-served queueing systems with peaked inputs. *Bell Syst. Tech. J.*, 52, 1215-28, 1973.

[H-033] Heffes, H. and Holtzman, J. M. Peakedness of traffic carried by a finite trunk group with renewal input. *Bell Syst. Tech. J.*, 52, 1617-42, 1973.

[H-034] Heffes, H. On the output of a *GI / M / N* queuing system with interrupted Poisson input. *Opns. Res.*, 24, 530-42, 1976.

[H-035] Heffes, H. A class of data traffic processes - Covariance function characterization and related queueing results. *Bell Syst. Tech. J.*, 59, 897-929, 1980.

[H-036] Heffes, H. and Lucantoni, D. M. A Markov modulated characterization of packetized voice and data traffic and related statistical multiplexer performance. *IEEE J. on Selected Areas in Communication, Special Issue on Network Performance Evaluation*, SAC-4, 6, 856-868, 1986.

[H-037] Heimann, D. and Neuts, M. F. The single server in discrete time - Numerical analysis IV. *Nav. Res. Logist. Quart.*, 20, 753-66, 1973.

[H-038] Helm, W. E. and Schassberger, R. Insensitive generalized semi-Markov schemes with point process input. *Math. of O.R.*, 7, 129-38, 1982.

[H-039] Henderson, J. and Finch, P. D. A note on the queueing system $E_k / G / 1$. *J. Appl. Prob.*, 7, 473-75, 1970.

[H-040] Henderson, W. Alternative approaches to the analysis of the $M / G / 1$ and $G / M / 1$ queues. *J. Oper. Res. Soc. Japan*, 15, 92-101, 1972.

[H-041] Herbert, H. G. An infinite discrete dam with dependent inputs. *J. Appl. Prob.*, 9, 404-13, 1972.

[H-042] Herzog, U. Solution of queuing problems by a recursive technique. *IBM J. Res. and Developmt.*, 19, 295-300, 1975.

[H-043] Herzog, U. Optimal scheduling strategies for real-time computers. *IBM J. Res. and Developmt.*, 19, 494-505, 1975.

[H-044] Heyde, C. C. On the stationary waiting time distribution in the $GI / G / 1$ queue. *J. Appl. Prob.*, 1, 175-78, 1964.

[H-045] Heyde, C. C. On the growth of the maximum queue length in a stable queue. *Opns. Res.*, 19, 447-52, 1971.

[H-046] Heyman, D. P. Optimal operating policies for $M / G / 1$ queueing systems. *Opns. Res.*, 16, 362-82, 1968.

[H-047] Heyman, D. P. and Marshall, K. T. Bounds on the optimal operating policy for a class of single-server queues. *Opns. Res.*, 16, 1138-46, 1968.

[H-048] Heyman, D. P. A priority queueing system with server interference. *SIAM J. Appl. Math.*, 17, 74-82, 1969.

[H-049] Heyman, D. P. An approximation for the busy period of the $M / G / 1$ queue using a diffusion approximation. *J. Appl. Prob.*, 11, 159-69, 1974.

[H-050] Heyman, D. P. The T-policy for the $M / G / 1$ queue. *Mgmt. Sci.*, 23, 775-78, 1977.

[H-051] Heyman, D. P., and Sobel, M. J. *Stochastic Models in Operations Research, Vols. I and II.* New York: McGraw-Hill, 1981.

[H-052] Heyman, D. P. On Ross's conjectures about queues with non-stationary Poisson arrivals. *J. Appl. Prob.*, 19, 245-9, 1982.

[H-053] Hildebrand, D. K. Stability of finite queue, tandem service systems. *J. Appl. Prob.*, 4, 571-83, 1967.

[H-054] Hildebrand, D. K. On the capacity of tandem server, finite queue, service systems. *Opns. Res.*, 16, 72-82, 1968.

[H-055] Hillier, F. S., and Boling, R. W. Finite queues in series with exponential or Erlang service times: A numerical approach. *Opns. Res.*, 15, 286-303, 1971.

[H-056] Hillier, F. S., and Lieberman, G. J. *Introduction to Operations Research.* 3rd ed. San Francisco: Holden-Day, 1980.

[H-057] Hillier, F. S., and Yu, O. S. *Queueing Tables and Graphs* New York: North Holland Publishing Co., 1981.

[H-058] Hofri, M. Disk scheduling: *FCFS* vs. *SSTF* revisited. *Comm. ACM*, 23, 645-53, 1980.

[H-059] Hofri, M. Analysis of interleaved storage via a constant-service queueing system with Markov-chain-driven input. *J. ACM*, 31, 628-48, 1984.

[H-060] Hokstad, P. A supplementary variable technique applied to the $M / G / 1$ queue. *Scand. J. Statist.*, 2, 95-98, 1975.

[H-061] Hokstad, P. The use of Wiener-Hopf decomposition in the study of waiting time and busy period distribution for the $G / G / 1$ queue. *Scand. J. Statist.*, 3, 79-85, 1976.

[H-062] Hokstad, P. Asymptotic behaviour of the $E_k / G / 1$ queue with finite waiting room. *J. Appl. Prob.*, 14, 358-66, 1977.

[H-063] Hokstad, P. A $M / G / 1$ priority queue. *INFOR*, 16, 158-70, 1978.

[H-064] Hokstad, P. Approximations for the $M / G / m$ queue. *Opns. Res.*, 26, 510-23, 1978.

[H-065] Hokstad, P. A single server queue with constant service time and restricted accessibility. *Mgmt. Sci.*, 25, 205-8, 1979.

[H-066] Hokstad, P. On the relationship of the transient behaviour of a general queueing model to its idle and busy period distributions. *Math. Operationsforsch. u. Statist.*, 10, 421-9, 1979.

[H-067] Hokstad, P. On the steady-state solution of the $M/G/2$ queue. *Adv. Appl. Prob.*, 11, 240-55, 1979.

[H-068] Hooke, J. A. On some limit theorems for the $GI/G/1$ queue. *J. Appl. Prob.*, 7, 634-40, 1970.

[H-069] Hooke, J. A. A priority queue with low-priority arrivals general. *Opns. Res.*, 20, 373-80, 1972.

[H-070] Hooke, J. A. Some heavy-traffic limit theorems for a priority queue with general arrivals. *Opns. Res.*, 20, 381-8, 1972.

[H-071] Hordijk, A. and Tijms, H. C. A simple proof of the equivalence of the limiting distributions of the continuous-time and the embedded process of the queue size in the $M/G/1$ queue. *Statistica Neerlandica*, 30, 97-100, 1976.

[H-072] Hordijk, A. and Schassberger, R. Weak convergence for generalized semi-Markov processes. *Stoch. Proc. Appl.*, 11, 271-91, 1982.

[H-073] Hsu, G. H. and Bosch, K. Finite dams with double level of release. *Zeitschrift f. Oper. Res.*, 27, 83-106, 1983.

[H-074] Hsu, J. Buffer behavior with Poisson arrival and geometric output processes. *IEEE Trans. Comm. Tech.*, COM-22, 1940-41, 1971.

[H-075] Hsu, J. Behavior of tandem buffers with geometric input and Markovian output. *IEEE Trans. Comm. Tech.*, COM-24, 358-61, 1976.

[H-076] Hunt, G. C. Sequential arrays of waiting lines. *Opns. Res.*, 4, 674-83, 1956.

[H-077] Hunter, J. J. On the moments of Markov renewal processes. *Adv. Appl. Prob.*, 1, 188-210, 1969.

[H-078] Hunter, J. J. Two queues in parallel with exponential type semi-Markovian inputs. *Opsearch*, 14, 29-37, 1977.

[H-079] Hunter, J. J. Generalized inverses and their application to applied probability problems. *Lin. Algebra and Appl.*, 45, 157-98, 1982.

[H-080] Hunter, J. J. Filtering of Markov renewal queues, I: Feedback queues. *Adv. Appl. Prob.*, 15, 349-75, 1983.

[H-081] Hunter, J. J. Filtering of Markov renewal queues, II: Birth-death queues. *Adv. Appl. Prob.*, 15, 376-91, 1983.

[H-082] Hunter, J. J. *Mathematical Techniques of Applied Probability, Vol. I: Discrete Time Models - Basic theory.* New York: Academic Press, 1983.

[H-083] Hunter, J. J. *Mathematical Techniques of Applied Probability, Vol. II: Discrete Time Models - Techniques and Applications.* New York: Academic Press, 1983.

[H-084] Hunter, J. J. Filtering of Markov renewal queues, III: Semi-Markov processes embedded in feedback queues. *Adv. Appl. Prob.*, 16, 422-36, 1984.

[H-085] Hunter, J. J. Filtering of Markov renewal queues, IV: Flow processes in feedback queues. *Adv. Appl. Prob.*, 17, 386-407, 1985.

[H-086] Hunter, J. J. The non-renewal nature of the quasi-input process in the $M/G/1/\infty$ queue. *J. Appl. Prob.*, 23, 803-11, 1986.

[I-001] Iglehart, D. L. Limit theorems for queues with traffic intensity one. *Ann. Math. Statist.*, 36, 1437-49, 1965.

[I-002] Iglehart, D. L. Functional limit theorems for the $GI/G/1$ queue in light traffic. *Adv. Appl. Prob.*, 3, 269-81, 1971.

[I-003] Iglehart, D. L. Extreme values in the $GI/G/1$ queue. *Ann. Math. Statist.*, 43, 627-35, 1972.

[I-004] Iglehart, D. L. Weak convergence in queueing theory. *Adv. Appl. Prob.*, 5, 570-94, 1973.

[I-005] Ignall, E. and Kolesar, P. Some stationary properties of the dispatch system $M/G/1$. *INFOR*, 10, 292-98, 1972.

[I-006] Ishikawa, A. On tandem queueing system having a general service times. *TRU Math.*, 9, 85-9, 1973.

[I-007] Iversen, V. B. Decomposition of an $M/D/rk$ queue with *FIFO* into k $E_k/D/r$ queues with *FIFO*. *Oper. Res. Letters*, 2, 20-1,1983.

[I-008] Iyama, T. The behavior of some design factors in a parallel production line. *J. Oper. Res. Soc. Japan*, 21, 226-43, 1978.

[I-009] Iyama, T. The efficiency of two-stage series lines. *J. Oper. Res. Soc. Japan*, 22, 321-37, 1979.

[I-010] Iyama, T. The efficiency of two-stage line. *J. Oper. Res. Soc. Japan*, 24, 237-56, 1981.

[J-001] Jackson, R. R. P. Queueing system with phase type service. *Opns. Res. Quart.*, 5, 109-20, 1954.

[J-002] Jackson, R. R. P. Queueing processes with phase-type service. *J. Roy. Statist. Soc. Ser. B*, 18, 129-32, 1956.

[J-003] Jackson, R. R. P., and Nickols, D. J. Some equilibrium results for the queueing process $E_k/M/1$. *J. Roy. Statist. Soc. Ser. B*, 18, 275-79, 1956.

[J-004] Jagerman, D. L. Approximate mean waiting time times in transient $GI/G/1$ queues. *Bell Syst. Tech. J.*, 61, 2003-22, 1982.

[J-005] Jagerman, D. L. Waiting time convexity in the $M/G/1$ queue. *AT&T Tech. J.*, 64, 33-41, 1985.

[J-006] Jagerman, D. L. Laplace transform inequalities with application to queueing. *AT&T Tech. J.*, 64, 1755-63, 1985.

[J-007] Jagerman, D. L. Certain Volterra integral equations in queueing. *Stoch. Models*, 1, 239-56, 1985.

[J-008] Jaiswal, N. K. Time-dependent solution of the bulk-service queueing problem. *Opns. Res.*, 8, 773-81, 1960.

[J-009] Jaiswal, N. K. A bulk-service queuing problem with variable capacity. *J. Roy. Statist. Soc. Ser. B*, 23, 143-8, 1961.

[J-010] Jaiswal, N. K. Distribution of busy periods for the bulk-service queuing problem. *Defense Sci. J.*, 12, 309-16, 1962.

[J-011] Jaiswal, N. K. Preemptive resume priority queue with Erlangian inputs. *Indian J. of Math.*, 4, 53-70, 1962.

[J-012] Jaiswal, N. K. *Priority Queues*. New York: Academic Press, 1968.

[J-013] Jankiewicz, M. Explicit formulas for transition intensities in the queueing system $E_2/E_2/n$. *Zast. Matemat.*, 13, 187-98, 1972.

[J-014] Jankiewicz, M. Limit distribution of the number of items in the queueing system E2/E2/n. *Zast. Matemat.*, 13, 547-51, 1973.

[J-015] Jankiewicz, M. Cyclic system with preemptive priority. *Zast. Matemat.*, 14, 165-76, 1974.

[J-016] Jankiewicz, M. and Kopociński, B. Steady-state distributions of piecewise Markov processes. *Zast. Matemat.*, 15, 25-32, 1976.

[J-017] Jankiewicz, M. and Rolski, T. Piecewise Markov processes on a general state space. *Zast. Matemat.*, 15, 421-36, 1977.

[J-018] Jankiewicz, M. Extended piecewise Markov processes in continuous time. *Zast. Matemat.*, 16, 175-95, 1978.

[J-019] Jankiewicz, M. Extended piecewise Markov processes in discrete time. *Zast. Matemat.*, 16, 197-205, 1978.

[J-020] Janssen, J. The semi-Markov model in risk theory. *Advances in Operations Research, North-Holland, 613-21, 1977.*

[J-021] Janssen, J. Some prediction problems in semi-Markov queueing models and related topics. *Oper. Res. Verfahren*, 29, 702-12, 1978.

[J-022] Janssen, J. Some explicit results for semi-Markov models in risk theory and in queuing theory. *Oper. Res. Verfahren*, 33, 217-31, 1979.

[J-023] Janssen, J. Some transient results on the $M/SM/1$ special semi-Markov model in risk and queueing theories. *Astin Bull.*, 11, 41-

51, 1980.

[J-024] Janssen, J. On the interaction between risk and queueing theories. *Blätter,* 15, 383-95, 1982.

[J-025] Janssen, J. Stationary semi-Markov models in risk and queueing theories. *Scand. Actuarial J., 199-210, 1982.*

[J-026] Janssen, J. Modèles de risque semi-Markoviens. *Cah. Centre d'études Rech. Opér.,* 24, 261-80, 1982.

[J-027] Jenkins, J. H. Stationary joint distributions arising in the analysis of the $M/G/1$ queue by the method of the imbedded Markov chain. *J. Appl. Prob.,* 3, 512-20, 1966.

[J-028] Jenkins, J. H. On the correlation structure of the departure process of the $M/E_k/1$ queue. *J. Roy. Statist. Soc. Ser. B,* 28, 336-44, 1966.

[J-029] Jensen, A. *A Distribution Model applicable to Economics.* Copenhagen: Munksgaard, 1954.

[J-030] Jo, K. Y. and Stidham, S. Jr. Optimal service-rate control of $M/G/1$ queueing systems using phase methods. *Adv. Appl. Prob.,* 15, 616-37, 1983.

[J-031] Jung, M. M. Busy period distribution of a single server queueing system, to which traffic is offered via a clock-pulse operated gate. *Proc. Ninth Internat. Teletraffic Conf., Torremolinos, Spain,* Paper No. 513, 1979.

[J-032] Jung, M. M. and de Boer, J. Waiting time distribution for call-processing tasks in an *SPC* telephone system, offered to the processor via a clock-pulse operated gate. *Philips Telecommunication Review,* 39, 90-101, 1981.

[K-001] Kambo, N. S. and Chaudhry, M. L. Distribution of busy period for the bulk-service queueing system $E_k^{a,b}/M/1$. *Comput. and Oper. Res.,* 11, 267-74, 1984.

[K-002] Karlin, S.; Miller, R. G., and Prabhu, N. U. Note on a moving single server problem. *Ann. Math. Statist.,* 30, 243-46, 1959.

[K-003] Karlin, S. and Fabens, A. J. Generalized renewal functions and stationary inventory models. *J. Math. Anal. and Appl.,* 5, 461-87, 1962.

[K-004] Karlin, S., and Taylor, H. M. *A First Course in Stochastic Processes.* 2d. ed. New York: Academic Press, 1975.

[K-005] Karr, A. F. *Point Processes and their Statistical Inference.* New York: Marcel Dekker, 1986.

[K-006] Kaspi, H. and Perry, D. Inventory systems for perishable commodities with renewal input and Poisson output. *Adv. Appl. Prob.*, 16, 402-21, 1984.

[K-007] Kawamura, T. Single server queue with Erlangian input and holding time. *Yokohama Math. J.*, 12, 39-61, 1964.

[K-008] Keilson, J. and Kooharian, A. On time dependent queueing processes. *Ann. Math. Statist.*, 31, 104-12, 1960.

[K-009] Keilson, J. The homogeneous random walk on the half-line and the Hilbert problem. *Bull. Inst. Int. Stat., 33rd session, Paris, 1-14, 1961.*

[K-010] Keilson, J. The use of Green's functions in the study of bounded random walks with applications to queuing theory. *J. Math. and Phys.*, 41, 42-52, 1962.

[K-011] Keilson, J. The general bulk queue as a Hilbert problem. *J. Roy. Statist. Soc. Ser. B*, 24, 344-58, 1962.

[K-012] Keilson, J. Non-stationary Markov walks on the lattice. *J. Math. and Phys.*, 41, 205-11, 1962.

[K-013] Keilson, J. Queues subject to service interruption. *Ann. Math. Statist.*, 33, 1314-22, 1962.

[K-014] Keilson, J. The first passage time density for homogeneous skip-free walks on the continuum. *Ann. Math. Statist.*, 34, 1003-11, 1963.

[K-015] Keilson, J. A gambler's ruin type problem in queuing theory. *Opns. Res.*, 11, 570-76, 1963.

[K-016] Keilson, J. On the asymptotic behaviour of queues. *J. Roy. Statist. Soc. Ser. B*, 25, 464-76, 1963.

[K-017] Keilson, J. An alternative to Wiener-Hopf methods for the study of bounded processes. *J. Appl. Prob.*, 1, 85-120, 1964.

[K-018] Keilson, J. On the ruin problem for the generalized random walk. *Opns. Res.*, 12, 504-06, 1964.

[K-019] Keilson, J. Some comments on single server queuing methods and some new results. *Proc. Camb. Phil. Soc.*, 60, 237-51, 1964.

[K-020] Keilson, J. and Wishart, D. M. G. A central limit theorem for processes defined on a finite Markov chain. *Proc. Camb. Phil. Soc.*, 60, 547-67, 1964.

[K-021] Keilson, J. The role of Greens's functions in congestion theory. *Proc. Symp. on Congestion Theory*, W. L. Smith and W. E. Wilkinson, eds., Chapel Hill: Univ. of N. Carolina Monographs, 43-71, 1965.

[K-022] Keilson, J. and Wishart, D. M. G. Boundary problems for additive processes defined on a finite Markov chain. *Proc. Camb. Phil. Soc.*, 61, 173-90, 1965.

[K-023] Keilson, J. *Green's Function Methods in Probability Theory.* Griffin's Statistical Monographs and Courses, no. 17. London: Charles Griffin & Co., 1965.

[K-024] Keilson, J. The ergodic queue length distribution for queueing systems with finite capacity. *J. Roy. Statist. Soc. Ser. B,* 28, 190-201, 1966.

[K-025] Keilson, J. A technique for discussing the passage time distribution for stable systems. *J. Roy. Statist. Soc. Ser. B,* 28, 477-86, 1966.

[K-026] Keilson, J. and Wishart, D. M. G. Addenda to processes defined on a finite Markov chain. *Proc. Camb. Phil. Soc.*, 63, 187-93, 1967.

[K-027] Keilson, J.; Cozzolino, J. and Young, H. A service system with unfilled requests repeated. *Opns. Res.*, 16, 1126-37, 1968.

[K-028] Keilson, J. A note on the waiting-time distribution for the $M/G/1$ queue with last-come-first-served discipline. *Opns. Res.*, 16,1230-31, 1968.

[K-029] Keilson, J. A queue model for interrupted communication. *Opsearch,* 6, 59-67, 1969.

[K-030] Keilson, J. An intermittent channel with finite storage. *Opsearch,* 6, 109-117, 1969.

[K-031] Keilson, J. On the matrix renewal function for Markov renewal processes. *Ann. Math. Statist.*, 40, 1901-907, 1969.

[K-032] Keilson, J. and Subba Rao, S. A process with chain dependent growth rate. *J. Appl. Prob.*, 7, 699-711, 1970.

[K-033] Keilson, J. and Subba Rao, S. A process with chain dependent growth rate. Part II: The ruin and ergodic probabilities. *Adv. Appl. Prob.*, 3, 315-38, 1971.

[K-034] Keilson, J. Exponential spectra as a tool for the study of server-systems with several classes of customers. *J. Appl. Prob.*, 15, 162-70, 1978.

[K-035] Keilson, J. *Markov Chain Models - Rarity and Exponentiality.* New York: Springer-Verlag, 1979.

[K-036] Keilson, J., and Nunn, W. R. Laguerre transformation as a tool for the numerical solution of integral equations of convolution type. *Appl. Math. and Comp.*, 5, 313-59, 1979.

[K-037] Keilson, J. and Sumita, U. Waiting time distribution response to traffic surges via the Laguerre transform. In *"Applied Probability - Computer Science: The Interface"*, Proc. of the Conference at Boca Raton, Florida, January 1981, R. L. Disney and T. J. Ott, eds, Boston: Birkhäuser, Vol II, 109-33, 1982.

[K-038] Keilson, J. and Sumita, U. The depletion time for $M/G/1$ systems and a related limit theorem. *Adv. Appl. Prob.,* 15, 420-43, 1983.

[K-039] Keilson, J. and Sumita, U. Evaluation of the total time in system in a preemptive/resume priority queue via a modified Lindley process. *Adv. Appl. Prob.,* 15, 840-56, 1983.

[K-040] Keilson, J. and Servi, L. D. Dynamics if the $M/G/1$ vacation model. *Opns. Res.,* 35, 575-82, 1987.

[K-041] Kella, O. and Yechiali, U. Waiting times in the non-preemptive priority $M/M/c$ queue. *Stoch. Models,* 1, 257-62, 1985.

[K-042] Kemeny, J., and Snell, J. L. *Finite Markov Chains.* Princeton, N.J.: Van Nostrand Publishing Co., 1960.

[K-043] Kemeny, J. G. Slowly spreading chains of the first kind. *J. of Math. Anal. and Appl.,* 15, 295-310, 1966.

[K-044] Kendall, D. G. Some problems in the theory of queues. *J. Roy. Statist. Soc. Ser. B,* 13, 151-85, 1951.

[K-045] Kendall, D. G. Stochastic processes occurring in the theory of queues and their analysis by the method of imbedded Markov chains. *Ann. Math. Statist.,* 24, 338-54, 1953.

[K-046] Kendall, D. G. Some problems in the theory of dams. *J. Roy. Statist. Soc. Ser. B,* 19, 207-12, 1957.

[K-047] Kendall, D. G. Some recent work and further problems in the theory of queues. *Theory of Prob. and Its Appl.,* 9, 1-15, 1964.

[K-048] Kennedy, D. P. Limiting diffusions for the conditioned $M/G/1$ queue. *J. Appl. Prob.,* 11, 355-62, 1974.

[K-049] Kennedy, D. P. On the time to first overflow in dams with inputs forming a Markov chain. *J. Appl. Prob.,* 15, 171-8, 1978.

[K-050] Kesten, H. and Runnenburg, J. T. Priority in waiting line problems I and II. *Koninkl. Ned. Akad. Wetensch. Proc. Ser. A,* 60, 312-36, 1957.

[K-051] Khinchine, A. Ya. *Mathematical Methods in the Theory of Queueing.* 2nd ed. New York: Hafner, 1969.

[K-052] Kimura, T.; Ohno, K. and Mine, H. Diffusion approximation for $GI/G/1$ queueing systems with finite capacity: I - The first overflow time. *J. Oper. Res. Soc. Japan,* 22, 41-68, 1979.

[K-053] Kimura, T.; Ohno, K. and Mine, H. Diffusion approximation for $GI/G/1$ queueing systems with finite capacity: II - The stationary behaviour. *J. Oper. Res. Soc. Japan,* 22, 301-20, 1979.

[K-054] Kimura, T.; Ohno, K. and Mine, H. Approximate analysis of optimal operating policies for a $GI/G/1$ queueing system. *Mem. Faculty Engin. Kyoto Univ.,* 42, 377-90, 1980.

[K-055] Kimura, T. Optimal control of an $M/G/1$ queuing system with removable server via diffusion approximation. *Euro. J. Oper. Res.,* 8, 390-98, 1981.

[K-056] Kimura, T. An analytical interpretation of the discrete approximation for the $GI/G/1$ queue. *Optimization,* 16, 285-96, 1985.

[K-057] King, R. A. The covariance structure of the departure process from $M/G/1$ queues with finite waiting lines. *J. Roy. Statist. Soc. Ser. B,* 33, 401-6, 1971.

[K-058] Kingman, J. F. C. A convexity property of positive matrices. *Quart. J. Math.,* 12, 283-4, 1961.

[K-059] Kingman, J. F. C. The single server queue in heavy traffic. *Proc. Camb. Phil. Soc.,* 57, 902-4, 1961.

[K-060] Kingman, J. F. C. On queues in heavy traffic. *J. Roy. Statist. Soc. Ser. B,* 24, 383-92, 1962.

[K-061] Kingman, J. F. C. On queues in which customers are served in random order. *Proc. Camb. Phil. Soc.,* 58, 79-90, 1962.

[K-062] Kingman, J. F. C. Some inequalities for the $GI/G/1$ queue. *Biometrika,* 49, 315-24, 1962.

[K-063] Kingman, J. F. C. The use of Spitzer's identity in the investigation of the busy period and other quantities in the $GI/G/1$ queue. *J. Austral. Math. Soc.,* 2, 345-56, 1962.

[K-064] Kingman, J. F. C. A martingale inequality in the theory of queues. *Proc. Camb. Phil. Soc.,* 59, 359-61, 1964.

[K-065] Kingman, J. F. C. On doubly stochastic Poisson processes. *Proc. Camb. Phil. Soc.,* 60, 923-30, 1964.

[K-066] Kingman, J. F. C. Approximations for queues in heavy traffic. *Proc. of the NATO Conf. on Queueing Theory, Lisbon, Portugal,* 73-78, 1966.

[K-067] Kingman, J. F. C. On the algebra of queues. *J. Appl. Prob.,* 3, 285-326, 1966.

[K-068] Kingman, J. F. C. Inequalities in the theory of queues. *J. Roy. Statist. Soc. Ser. B,* 32, 102-10, 1970.

[K-069] Kleinecke, D. C. Discrete time queues at a periodic traffic light. *Opns. Res.,* 12, 809-14, 1964.

[K-070] Kleinrock, L. *Queueing Systems. Vol. 1, Theory. Vol. 2, Computer Applications.* New York: John Wiley and Sons, 1975.

[K-071] Klimko, E. and Neuts, M. F. The single server in discrete time - Numerical analysis II. *Nav. Res. Logist. Quart., 20, 305-19, 1973.*

[K-072] Klimov, G. P. *Bedienungsprozesse.* Basel: Birkhäuser Verlag, 1979.

[K-073] Knepley, J. E. and Fischer, M. J. A numerical solution for some computational problems occurring in queueing theory. In *"Algorithmic Methods in Probability"*, TIMS Studies in the Management Sciences, no. 7. London: North Holland Publishing Co., 271-85, 1977.

[K-074] Knessl, C.; Matkowsky, B. J.; Schuss, Z. and Tier, C. Asymptotic analysis of a state-dependent $M/G/1$ queueing system. *SIAM J. Appl. Math., 483-505, 1986.*

[K-075] Knessl, C.; Matkowsky, B. J.; Schuss, Z. and Tier, C. A finite capacity single-server queue with customer loss. *Stoch. Mod.,* 2, 97-121, 1986.

[K-076] Knessl, C.; Matkowsky, B. J.; Schuss, Z. and Tier, C. System crash in a finite capacity $M/G/1$ queue. *Stoch. Mod.,* 2, 171-201, 1986.

[K-077] Knessl, C.; Matkowsky, B. J.; Schuss, Z. and Tier, C. Distribution of the maximum buffer content during a busy period for state-dependent $M/G/1$ queues. *Stoch. Mod.,* 3, 191-226, 1987.

[K-078] Knessl, C.; Matkowsky, B. J.; Schuss, Z. and Tier, C. A Markov modulated $M/G/1$ queue I: Stationary distribution. *Queueing Systems,* 1, 355-74, 1987.

[K-079] Knessl, C.; Matkowsky, B. J.; Schuss, Z. and Tier, C. A Markov modulated $M/G/1$ queue II: Busy period and time for buffer overflow. *Queueing Systems,* 1, 375-99, 1987.

[K-080] Kobayashi, H. and Konheim, A. G. Queueing models for computer communications analysis. *IEEE Trans. Comm.,* COM-25, 2-28, 1977.

[K-081] Konheim, A. G. The stationary waiting time distribution for a single-server queue. *J. Soc. Indust. Appl. Math.,* 13, 966-76, 1965.

[K-082] Konheim, A. G. and Meister, B. Service in a loop system. *J. ACM,* 19, 92-108, 1972.

[K-083] Konheim, A. G. Service epochs in a loop system. *Proc. Symp. Computer-Communications Networks and Teletraffic, 125-43, 1972.*

[K-084] Konheim, A. G. Distributions of queue lengths and waiting times in a loop with two-way traffic. *J. Comp. Syst. Sci.,* 7, 506-21, 1973.

[K-085] Konheim, A. G., and Reiser, M. A queueing model with finite waiting room and blocking. *J. ACM,* 23, 328-41, 1976.

[K-086] Konheim, A. G. An elementary solution to the queuing system $G/G/1$. *SIAM J. of Computing,* 4, 540-45, 1976.

[K-087] Konheim, A. G. and Reiser, M. Finite capacity queuing systems with applications in computer modeling. *SIAM J. Computing,* 7, 210-29, 1978.

[K-088] König, D., and Stoyan, D. *Methoden der Bedienungstheorie.* Braunschweig: Vieweg, 1976.

[K-089] König, D.; Rykow W. W. and Schmidt, V. On a single server-queue with multiple transaction. *Optimization,* 16, 125-36, 1985.

[K-090] Kopocińska, I. Imbedded Markov chains in two examples of queueing models. *Zast. Matemat.,* 8, 29-36, 1965.

[K-091] Kopocińska, I. On a $M/G/1$ queueing model with feedback. *Zast. Matemat.,* 9, 161-71, 1965.

[K-092] Kopocińska, I. and Kopociński, B. Queueing systems with feedback. *Bull. Acad. Polon. Sciences, Ser. Math., Astr., et Phys.,* 19, 397-401, 1971.

[K-093] Kopocińska, I. and Kopociński, B. Queueing systems with feedback. *Zast. Matemat.,* 12, 373-84, 1972.

[K-094] Kopocińska, I. Piecewise Markov processes in discrete time and their certain extensions. *Zast. Matemat.,* 16, 23-38, 1977.

[K-095] Kopocińska, I. and Kopociński, B. On coincidences in renewal streams. *Zast. Matemat.,* 19, 169-80, 1987.

[K-096] Kosten, L. and ten Broeke, A. M. On the accuracy of measurements of waiting times in the single server system with arbitrary distribution of holding times. *Proc. Fifth Internat. Teletraffic Conf., New York,* 435-41, 1967.

[K-097] Kotiah, T. C. T.; Thompson, J. W.; and Waugh, W. A. O'N. Use of Erlangian distributions for single server queueing systems. *J. Appl. Prob.,* 6, 584-93, 1969.

[K-098] Kotiah, T. C. T., and Slater, N. B. On two-server Poisson queues with two types of customers. *Opns. Res.*, 21, 597-603, 1973.

[K-099] Kovalenko, I. N. Some queueing problems with restrictions. *Th. of Prob. and its Appl.*, 6, 204-8, 1961.

[K-100] Krakowski, M. Conservation methods in queuing theory. *R.A.I.R.O.*, 7, 63-84, 1973.

[K-101] Krakowski, M. Arrival and departure processes in queues - Pollaczek-Khintchine formulas for bulk arrivals and bounded systems. *R.A.I.R.O.*, 8, 45-56, 1974.

[K-102] Kramer, W. and Lagenbach-Belz, M. Approximate formulae for the delay in the queueing system $GI / G / 1$. *Proc. Eight Internat. Teletraffic Conf., Melbourne, Australia,* Paper No. 235, 1976.

[K-103] Kuczura, A. Queues with mixed renewal and Poisson inputs. *Bell Syst. Tech. J.*, 51, 1305-26, 1972.

[K-104] Kuczura, A. The accuracy of call-congestion measurements for loss systems with renewal input. *Bell Syst. Tech. J.*, 51, 2197-2208, 1972.

[K-105] Kuczura, A. Batch input to a multiserver queue with constant service times. *Bell Syst. Tech. J.*, 52, 83-99, 1973.

[K-106] Kuczura, A. The interrupted Poisson process as an overflow process. *Bell Syst. Tech. J.*, 52, 437-48, 1973.

[K-107] Kuczura, A. Loss systems with mixed renewal and Poisson inputs. *Opns. Res.*, 21, 787-95, 1973.

[K-108] Kuczura, A. Piecewise Markov processes. *SIAM J. Appl. Math.*, 24, 169-81, 1973.

[K-109] Kuczura, A. and Bajaj, D. A method of moments for the analysis of a switched communication network's performance. *IEEE Trans. Comm.*, COM-25, 185-93, 1977.

[K-110] Kühn, P. J. Multiqueue systems with nonexhaustive cyclic service. *Bell Syst. Tech. J.*, 58, 671-98, 1979.

[K-111] Kulkarni, V. G. Letter to the Editor. *J. Appl. Prob.*, 19, 901-4, 1982.

[K-112] Kyprianou, E. K. On the quasi-stationary distribution of the virtual waiting time in queues with Poisson arrivals. *J. Appl. Prob.*, 8, 494-507, 1971.

[L-001] Lambotte, J. P. Processus semi-markoviens et files d'attente. *Cah. Centre d'études Rech. Opér.*, 10, 21-31, 1968.

[L-002] Latouche, G., and Neuts, M. F. Efficient algorithmic solutions to exponential tandem queues with blocking. *SIAM J. Algebraic and Discrete Meth.*, 1, 93-106, 1980.

[L-003] Latouche, G. Queues with paired customers. *J. Appl. Prob.*, 18, 684-96, 1981.

[L-004] Latouche, G. A phase-type semi-Markov point process. *SIAM J. Algebraic and Discrete Meth.*, 3, 77-90, 1982.

[L-005] Latouche, G.; Jacobs, P. A. and Gaver, D. P. Finite Markov chain models skip-free in one direction. *Nav. Res. Logist. Quart.*, 31, 571-88, 1984.

[L-006] Latouche, G. An exponential semi-Markov process with applications to queueing theory. *Stoch. Models*, 1, 137-170, 1985.

[L-007] Latouche, G. A note on two matrices occurring in the solution of quasi-birth-and-death processes. *Stoch. Models*, 3, 251-7, 1987.

[L-008] Lavenberg, S. S. The steady-state queueing time distribution for the $M / G / 1$ finite capacity queue. *Mgmt. Sci.*, 21, 501-6, 1975.

[L-009] Lee, T. T. $M / G / 1 / N$ queue with vacation time and exhaustive service. *Opns. Res.*, 32, 774-84, 1984.

[L-010] LeGall, P. *Les Systèmes avec ou sans Attente et les Processus Stochastiques.* Paris: Dunod, 1962.

[L-011] Lehoczky, J. P. A note on the first emptiness time of an infinite reservoir with inputs forming a Markov chain. *J. Appl. Prob.*, 8, 276-84, 1971.

[L-012] Lehoczky, J. P. Traffic intersection control and zero-switch queues under conditions of Markov chain dependent input. *J. Appl. Prob.*, 9, 382-95, 1972.

[L-013] Lehoczky, J. P. First-emptiness problems in applied probability under first-order dependent input and general output conditions. *J. Appl. Prob.*, 10, 330-42, 1973.

[L-015] Lehoczky, J. P. and Gaver, D. P. Performance evaluation of voice/data queueing systems. In *"Applied Probability - Computer Science: The Interface"*, Proc. of the Conference at Boca Raton, Florida, January 1981, R. L. Disney and T. J. Ott, eds, Boston: Birkhäuser, Vol I, 329-46, 1982.

[L-016] Lemoine, A. J. Some limiting results for a queueing model with feedback. *Sankhya, Ser. A*, 36, 293-304, 1974.

[L-017] Lemoine, A. J. Limit theorems for generalized single server queues. *Adv. Appl. Prob.*, 6, 159-74, 1974.

[L-018] Lemoine, A. J. On two stationary distributions for the stable *GI / G / 1* queue. *J. Appl. Prob.,* 11, 849-52, 1974.

[L-019] Lemoine, A. J. A queueing system with heterogeneous servers and autonomous traffic control. *Opns. Res.,* 23, 681-86, 1975.

[L-020] Lemoine, A. J. Limit theorems for generalized single server queues: The exceptional system. *SIAM J. Appl. Math.,* 28, 596-606, 1975.

[L-021] Lemoine, A. J. On random walks and stable *GI / G / 1* queues. *Math. of O.R.,* 1, 159-64, 1976.

[L-022] Lemoine, A. J. Some limiting results for a queueing model with feedback II. *Sankhya,* Ser. A, 39, 82-93, 1977.

[L-023] Lemoine, A. J. On queues with periodic Poisson input. *J. Appl. Prob.,* 18, 889-900, 1981.

[L-024] Levy, H. and Kleinrock L. A queue with starter and a queue with vacations: Delay analysis by decomposition. *Opns. Res.,* 34, 426-36, 1986.

[L-025] Levy, Y., and Yechiali, U. Utilization of idle time in an *M / G / 1* queueing system. *Mgmt. Sci.,* 22, 202-11, 1975.

[L-026] Levy, Y. A class of scheduling policies for real-time processors with switching system applications *Proc. Eleventh Internat. Teletraffic Conf., Kyoto, Japan,* 4.3B, 1-7, 1985.

[L-027] Lindley, D. V. The theory of queues with a single server. *Proc. Camb. Phil. Soc.,* 48, 277-89, 1952.

[L-028] Lloyd, E. H. Reservoirs with serially correlated inflows. *Technometrics,* 5, 85-93, 1963.

[L-029] Lloyd, E. H. A note on the time-dependent and the stationary behaviour of a semi-infinite reservoir subject to a combination of Markovian inflows. *J. Appl. Prob.,* 8, 708-15, 1971.

[L-030] Loris-Teghem, J. Condition nécessaire d'ergodisme pour un processus stochastique lié à un système d'attente à arrivées et services en groupes d'effectif aléatoire. *Acad. Roy. Belgique, Bull. Classe des Sciences, 5e. Serie,* 52, 382-9, 1966.

[L-031] Loris-Teghem, J. Systèmes d'attente à plusieurs guichets at à arrivées et services en groupes d'effectif aléatoire. *Cah. Centre d'études Rech. Opér.,* 8, 98-111, 1966.

[L-032] Loris-Teghem, J. Un systeme d'attente à arrivées et services en groupes d'effectif aléatoire. *Cah. Centre d'études Rech. Opér.,* 8, 179-91, 1966.

[L-033] Loris-Teghem, J. Etude transitoire de systèmes d'attente où le serveur peut être occupé en l'absence de clients. *Actes du Sixième Congres de l'A.F.I.R.O, Nancy,* 1, 221-30, 1967.

[L-034] Loris-Teghem, J. Une application d'un theorème de G. Baxter à la détermination de la distribution des temps d'attente dans un modèle *GI / G /* 1 generalisé. *Cah. Centre d'études Rech. Opér.,* 10, 79-83, 1968.

[L-035] Loris-Teghem, J. On the waiting time distribution in a generalized *GI / G /* 1 queueing system. *J. Appl. Prob.,* 8, 241-51, 1971.

[L-036] Loris-Teghem, J. Un traitement algébrique du modèle d'attente *GI / M /* 2. *Cah. Centre d'études Rech. Opér.,* 13, 57-62, 1971.

[L-037] Loris-Teghem, J. On the waiting time distribution in a generalized queueing system with uniformly bounded sojourn times. *J. Appl. Prob.,* 9, 642-9, 1972.

[L-038] Loris-Teghem, J. On the waiting time distribution in some generalized queueing systems. *Bull. Soc. Math. Belgique,* 25, 11-24, 1973.

[L-039] Loris-Teghem, J. An algebraic approach to the waiting time process in GI/M/S. *J. Appl. Prob.,* 10, 181-91, 1973.

[L-040] Loris-Teghem, J. Hysteretic control of an *M / G /* 1 queueing system with two service time distributions and removable server. In *"Point Processes and Queuing problems: Proceedings of a Colloquium, Keszthely, Hungary, 1978",* London: North Holland Publ. Co., 291-305, 1981.

[L-041] Loris-Teghem, J. and Manya, N. Analysis of a queuing system with group arrivals and state dependent service times, related to a stochastic continuous-review (s,S) inventory model. *Euro. J. Oper. Res.,* 11, 82-92, 1982.

[L-042] Loris-Teghem, J. Imbedded and non-imbedded stationary distributions in a finite capacity queueing system with removable server. *Cah. Centre d'études Rech. Opér.,* 26, 87-94, 1984.

[L-043] Loris-Teghem, J. Analysis of single server queueing systems with vacation periods. *Belg. J. Oper. Res., Stat and C.S.,* 25, 47-54, 1985.

[L-044] Loris-Teghem, J. Vacation policies in an *M / G /* 1 type queueing system with vacation periods. *Queueing Systems,* 3, 41-52, 1988.

[L-045] Loulou, R. An explicit upper bound for the mean busy period in a *GI / G /* 1 queue. *J. Appl. Prob.,* 15, 452-5, 1978.

[L-046] Love, R. F. Steady-state solution of the queueing system Ew/M/s with batch service. *Opns. Res.,* 18, 160-71, 1970.

[L-047] Loynes, R. M. Stationary waiting-time distributions for single-server queues. *Ann. Math. Statist.,* 33, 1323-39, 1962.

[L-048] Loynes, R. M. A continuous-time treatment of certain queues and infinite dams. *J. Austral. Math. Soc.,* 2, 484-98, 1962.

[L-049] Lucantoni, D. M., and Neuts, M. F. Numerical methods for a class of Markov chains arising in queueing theory. *Tech. Rept. no. 78/10. Appl. Math. Inst., Univ. of Delaware, Newark, 1978.*

[L-050] Lucantoni, D. M. A $GI / M / C$ queue with a different service rate for customers who need not wait - An algorithmic solution. *Cah. Centre d'études Rech. Opér.,* 24, 5-20, 1982.

[L-051] Lucantoni, D. M. *An Algorithmic Analysis of a Communication Model with Retransmission of Flawed Messages.* Pitman Research Notes in Mathematics No. 81, 160 pages, 1983.

[L-052] Lucantoni, D. M. and Ramaswami, V. Efficient algorithms for solving the non-linear matrix equations arising in phase type queues. *Stoch. Models,* 1, 29-52, 1985.

[L-053] Lucantoni, D. M.; Meier-Hellstern, K. S. and Neuts, M. F. A single server queue with server vacations and a class of non-renewal arrival processes. *Adv. Appl. Prob.,* (forthcoming).

[L-054] Luchak, G. The solution of single channel queueing equations characterized by a time-dependent Poisson distributed arrival rate and a general class of holding times. *Opns. Res.,* 4, 711-32, 1956.

[L-055] Luchak, G. The distribution of the time required to reduce to some preassigned level a single-channel queue characterized by a time-dependent Poisson arrival rate and a general class of holding times. *Opns. Res.,* 5, 205-9, 1957.

[L-056] Luchak, G. The continuous-time solution of the equations of the single-channel queue with a general class of service-time distribution by the method of generating functions. *J. Roy. Statist. Soc. Ser. B,* 20, 176-81, 1958.

[M-001] Machihara, F. and Keilson, J. Hyperexponential waiting time structure in hyperexponential $H_K / H_L / 1$ systems. *J. Oper. Res. Soc. Japan,* 28, 242-51, 1985.

[M-002] Makino, T. On the mean passage time concerning some queuing problems of the tandem type. *J. Oper. Res. Soc. Japan,* 7, 17-47, 1964.

[M-003] Makino, T. On a study of output distribution. *J. Oper. Res. Soc. Japan,* 8, 109-33, 1966.

[M-004] Makis, V. A note on optimal control limit for a batch service queueing system: Average cost case. *Opsearch,* 21, 113-6, 1984.

[M-005] Makis, V. On the steady-state distributions in a controlled queueing system with bounded waiting time. *Probl. Control and Information Th.,* 14, 261-8, 1985.

[M-006] Mandelbaum, A. and Yechiali, U. The conditional residual service time in the $M/G/1$ queue. *Tech. Rept., Dept. of Statistics, Tel Aviv University, 1980.*

[M-007] Mandelbaum, A. and Yechiali, U. Optimal entering rules for a customer with wait option at an $M/G/1$ queue. *Mgnt. Sci.,* 29, 174-87, 1983.

[M-008] Marchal, W. G. and Harris, C. M. A modified Erlang approach to approximating $GI/G/1$ queues. *J. Appl. Prob.,* 13, 118-26, 1976.

[M-009] Marchal, W. G. An approximate formula for waiting time in single server queues. *AIIE Trans.,* 8, 473-4, 1976.

[M-010] Marchal, W. G. Some simpler bounds on the mean queuing time. *Opns. Res.,* 26, 1083-8, 1978.

[M-011] Marcus, M. and Minc, H. *A Survey of Matrix Theory and Matrix Inequalities.* Boston: Allyn and Bacon, 1964.

[M-012] Marlin, P. G. On the ergodic theory of Markov chains. *Opns. Res.,* 21, 617-22, 1973.

[M-013] Marshall, K. T. Some inequalities in queuing. *Opns. Res.,* 16, 651-65, 1968.

[M-014] Marshall, K. T. Bounds for some generalizations of the $GI/G/1$ queue. *Opns. Res.,* 16, 841-48, 1968.

[M-015] Marshall, K. T. Some relationships between the distributions of waiting time, idle time and interoutput time in the $GI/G/1$ queue. *SIAM J. Appl. Math.,* 16, 324-7, 1968.

[M-016] Marshall, K. T. and Wolff, R. W. Customer average and time average queue lengths and waiting times. *J. Appl. Prob.,* 8, 535-42, 1971.

[M-017] Massey, W. A. and Morrison, J. A. Calculation of steady-state probabilities for content of buffer with correlated inputs. *Bell Syst. Tech. J.,* 57, 3097-117, 1978.

[M-018] Matthews, D. E. Probabilistic priorities in the $M/G/1$ queue. *Nav. Res. Logist. Quart.*, 24, 457-62, 1977.

[M-019] Matthews, D. E. A simple method for reducing queueing times in $M/G/1$. *Opns. Res.*, 27, 318-23, 1979.

[M-020] Mazumdar, S. On priority queues in heavy traffic. *J. Roy. Statist. Soc. Ser. B*, 111-14, 1970.

[M-021] McNeil, D. R. A solution to the fixed-cycle traffic light problem for compound Poisson arrivals. *J. Appl. Prob.*, 5, 624-35, 1968.

[M-022] McNickle, D. C. The number of departures from a semi-Markov queue. *J. Appl. Prob.*, 11, 825-8, 1974.

[M-023] Medhi, J. and Borthakur, A. On a two server Markovian queue with a general bulk services rule. *Cah. Centre d'études Rech. Opér.*, 14, 151-58, 1972.

[M-024] Medhi, J. Waiting time distribution in a Poisson queue with a general bulk service rule. *Mgmt. Sci.*, 21, 777-82, 1975.

[M-025] Medhi, J. Further results on the waiting time distribution of a Poisson queue with a general bulk service rule. *Cah. Centre d'études Rech. Opér.*, 21, 183-91, 1979.

[M-026] Medhi, J. *Stochastic Processes.* New Delhi: Wiley Eastern Ltd., 1982.

[M-027] Medhi, J. *Recent Developments in Bulk Queueing Models.* New Delhi: Wiley Eastern Ltd., 1984.

[M-028] Meier-Hellstern, K. S. A fitting algorithm for Markov-modulated Poisson processes having two arrival rates. *Euro. J. Oper. Res.*, 29, 370-7, 1987.

[M-029] Meilijson, I. and Yechiali, U. On optimal right-of-way policies at a single-server station when insertion of idle times is permitted. *Stoch. Proc. Appl.*, 6, 25-32, 1977.

[M-030] Meisling, T. Discrete-time queueing theory. *Opns. Res.*, 6, 96-105, 1958.

[M-031] Mercer, A. A queue with random arrivals and scheduled bulk departures. *J. Roy. Statist. Soc. Ser. B*, 30, 185-9, 1968.

[M-032] Mevert, P. A priority system with setup times. *Opns. Res.*, 16, 602-12, 1968.

[M-033] Michel, J. A. and Coffman Jr., E. G. Synthesis of a feedback queueing discipline for computer operation. *J. ACM*, 21, 3 29-39, 1974.

[M-034] Miller, D. R. Computation of steady-state probabilities for $M/M/1$ priority queues. *Opns. Res.*, 29, 945-58, 1981.

[M-035] Miller, D. R. Steady-state algorithmic analysis of $M/M/c$ two-priority queues with heterogeneous rates. In *"Applied Probability - Computer Science: The Interface"*, Proc. of the Conference at Boca Raton, Florida, January 1981, R. L. Disney and T. J. Ott, eds, Boston: Birkhäuser, Vol II, 207-25, 1982.

[M-036] Miller, H. D. A convexity property in the theory of random variables defined on a Markov chain. *Ann. Math. Statist.*, 32, 1260-70, 1961.

[M-037] Miller, L. W. A note on the busy period of an $M/G/1$ finite queue. *Opns. Res.*, 23, 1179-82, 1975.

[M-038] Miller, R. G. Priority queues. *Ann. Math. Statist.*, 31, 86-103, 1960.

[M-039] Mine, H. and Ohno, K. An optimal rejection time for an $M/G/1$ queuing system. *Opns. Res.*, 19, 194-207, 1971.

[M-040] Mine, H. and Ohno, K. Traffic light queues as a generalization to queueing theory. *J. Appl. Prob.*, 8, 480-93, 1971.

[M-041] Mine, H. and Ohno, K. Traffic light queues with dependent arrivals as a generalization of queueing theory. *J. Appl. Prob.*, 9, 630-41, 1972.

[M-042] Mine, H.; Ohno, K. and Koizumi, T. A single-server queue with service times depending on the order of services. *Opns. Res.*, 24, 188-90, 1976.

[M-043] Minh, D. L. A discrete time, single server queue from a finite population. *Mgmt. Sci.*, 23, 756-67, 1977.

[M-044] Minh, D. L. and Blunden, W. R. Time-inhomogeneous signalized intersection as a discrete infinite dam. *Proc. Seventh Internat. Symp. Transp. and Traffic Theory,* T. Sasaki, and T. Yamaoka, eds., Kyoto, Japan, 271-80, 1977.

[M-045] Minh, D. L. The discrete-time single-server queue with time-inhomogeneous compound Poisson input and general service time distribution. *J. Appl. Prob.*, 15, 590-601, 1978.

[M-046] Minh, D. L. A discrete-time single server queue with set-up time. *Stoch. Proc. Appl.*, 8, 181-97, 1978.

[M-047] Minh, D. L. The actual waiting time of each customer in a $GI/G/1$ queue. *J. Appl. Prob.*, 16, 910-6, 1979.

[M-048] Minh, D. L. Analysis of the exceptional queueing system by the use of regenerative processes and analytical methods. *Math. of O.R.*, 5, 147-59, 1980.

[M-049] Minh, D. L. The $GI / G / 1$ queue with uniformly limited virtual waiting times; The finite dam. *Adv. Appl. Prob.*, 12, 501-16, 1980.

[M-050] Minh, D. L. The single-server queue with uniformly limited actual waiting times. *Math. Operationsforsch. u. Statist.*, 12, 607-21, 1981.

[M-051] Minh, D. L. and Sorli, R. M. Simulating the $GI / G / 1$ queue in heavy traffic. *Opns. Res.*, 31, 966-71, 1983.

[M-052] Mitchell, W. Optimal service-rate selection in an $M / G / 1$ queue. *SIAM J. Appl. Math.*, 24, 19-35, 1973.

[M-053] Mitrany, I. L. and Avi-Itzhak, B. A many-server queue with service interruptions. *Opns. Res.*, 16, 628-38, 1968.

[M-054] Moore, S. C., and Bhat, U. N. On a computational approach to some value functions in an $M / E_z / 1$ queue. *A. I. I. E. Trans.*, 7, 73-76, 1975.

[M-055] Moran, P. A. P. A probability theory of dams and storage systems: Modifications of the release rule. *Austral. J. Appl. Sci.*, 6, 117-30, 1955.

[M-056] Moran, P. A. P. A probability theory of a dam with a continuous release. *Quart. J. Math.* (2) 7, 130-37, 1956.

[M-057] Moran, P. A. P. *The Theory of Storage.* London: Methuen's Monographs on Applied Probability and Statistics, 1959.

[M-058] Mori, M. Some bounds for queues. *J. Oper. Res. Soc. Japan,* 18, 152-81, 1975.

[M-059] Mori, M. Transient behaviour of the mean waiting time and its exact forms in $M / M / 1$ and $M / D / 1$. *J. Oper. Res. Soc. Japan,* 19, 14-31, 1976.

[M-060] Morimura, H. On the number of served customers in a busy period. *J. Oper. Res. Soc. Japan,* 4, 67-75, 1961-2.

[M-061] Morimura, H. On the relation between the distributions of the queue size and the waiting time. *Kodai Math. Sem. Rep.,* 14, 6-19, 1962.

[M-062] Morris, R. J. T. An algorithmic technique for a class of queueing models with packet switching applications. *IEEE Internat. Conf. on Communications, June 1981,* Paper 41.2.

[M-063] Morrison, J. A. Two discrete-time queues in tandem. *IEEE Trans. Comm.,* COM-27, 563-573, 1979.

[M-064] Morrison, J. A. Analysis of some overflow problems with queueing. *Bell Syst. Tech. J.,* 59, 1427-62, 1980.

[M-065] Morrison, J. A. Some traffic overflow problems with a large secondary queue. *Bell Syst. Tech. J.*, 59, 1463-82, 1980.

[M-066] Morrison, J. A. An overflow system in which queuing takes precedence. *Bell Syst. Tech. J.*, 60, 1-12, 1981.

[M-067] Morse, P. M. Stochastic processes of waiting lines. *Opns. Res.*, 3, 255-61, 1955.

[M-068] Morse, P. *Queues, Inventories and Maintenance.* New York: John Wiley and Sons, 1958.

[M-069] Müller, I. and Siegel, G. About two homogeneous Wiener-Hopf integral equations and their interpretation for delayed random walks and queueing models *GI / G /* 1 with "warming-up". *Elektron. Inf. Verarb. Kybern.*, EIK 10, 606-26, 1974.

[M-070] Murari, K. and Parkash, K. A queue with correlated interarrival time distributions. *Cah. Centre d'études Rech. Opér.*, 23, 73-80, 1981.

[M-071] Murdoch, J. *Queueing theory: Worked Examples and Problems.* London: The Macmillan Free Press, 1978.

[N-001] Nair, S. S. and Neuts, M. F. A priority rule based on the ranking of the service times for the *M / G /* 1 queue. *Opns. Res.*, 17, 466-77, 1969.

[N-002] Nair, S. S. Semi-Markov analysis of two queues in series attended by a single server. *Bull. Soc. Math. Belgique,* 22, 355-67, 1970.

[N-003] Nair, S. S. A single server tandem queue. *J. Appl. Prob.*, 8, 95-109, 1971.

[N-004] Nair, S. S. and Neuts, M. F. An exact comparison of the waiting times under three priority rules. *Opns. Res.*, 19, 414-23, 1971.

[N-005] Nair, S. S. and Neuts, M. F. Distribution of the occupation time and virtual waiting time of a general class of bulk queues. *Sankhya,* Ser. A, 34, 17-22, 1972.

[N-006] Nair, S. S. Alternating priority queues with non-zero switch rule. *Comput. and Oper. Res.*, 3, 337-46, 1976.

[N-007] Nance, R. E.; Bhat, U. N. and Claybrook, B. G. Busy period analysis of a time-sharing system: Transform inversion. *J. ACM,* 19, 543-63, 1972.

[N-008] Naor, P., and Yechiali, U. Queueing problems with heterogeneous arrivals and service. *Opns. Res.*, 19, 722-34, 1971.

[N-009] Neal, S. R. and Kuczura, A. A theory of traffic-measurement errors for loss systems with renewal input. *Bell Syst. Tech. J.*, 52, 967-90, 1973.

[N-010] Neuts, M. F. The distribution of the maximum queue length of a Poisson queue during a busy period. *Opns. Res.*, 12, 281-85, 1964.

[N-011] Neuts, M. F. The single server queue with Poisson input and semi-Markov service times. *J. Appl. Prob.*, 3, 202-30, 1966.

[N-012] Neuts, M. F. The busy period of a queue with batch service. *Opns. Res.*, 5, 815-19, 1966.

[N-013] Neuts, M. F. Semi-Markov analysis of a bulk queue. *Bull. Soc. Math. Belgique*, 27, 28-42, 1966.

[N-014] Neuts, M. F. A general class of bulk queues with Poisson input. *Ann. Math. Statist.*, 38, 759-70, 1967.

[N-015] Neuts, M. F. Two Markov chains arising from examples of queues with state-dependent service times. *Sankhya, Ser. A*, 29, 259-64, 1967.

[N-016] Neuts, M. F. Two queues in series with a finite intermediate waiting room. *J. Appl. Prob.*, 5, 123-42, 1968.

[N-017] Neuts, M. F. The joint distribution of the virtual waiting time and the residual busy period for the $M/G/1$ queue. *J. Appl. Prob.*, 5, 224-29, 1968.

[N-018] Neuts, M. F. and Yadin, M. The transient behavior of the queue with alternating priorities with special reference to the waiting times. *Bull. Soc. Math. Belgique*, 20, 343-76, 1968.

[N-019] Neuts, M. F. The queue with Poisson input and general service times treated as a branching process. *Duke Math. J.*, 36, 215-32, 1969.

[N-020] Neuts, M. F. Two queues in series, treated in terms of a Markov renewal branching process. *Adv. Appl. Prob.*, 2, 110-49, 1970.

[N-021] Neuts, M. F. A queue subject to extraneous phase changes. *Adv. Appl. Prob.*, 3, 78-119, 1971.

[N-022] Neuts, M. F. and Purdue, P. Multivariate semi-Markov matrices. *J. Austral. Math. Soc.*, 13, 107-13, 1971.

[N-023] Neuts, M. F. The single server in discrete time - Numerical analysis I. *Nav. Res. Logist. Quart.*, 20, 297-304, 1973.

[N-024] Neuts, M. F. and Klimko, E. The single server in discrete time - Numerical analysis III. *Nav. Res. Logist. Quart.*, 20, 557-67, 1973.

[N-025] Neuts, M. F. The Markov renewal branching process. In *"Proceedings of the Conference on Mathematical Methodology in the*

Theory of Queues, Kalamazoo, Mich.", New York: Springer-Verlag, 1-21, 1974.

[N-026] Neuts, M. F. Probability distributions of phase type. In *"Liber Amicorum Prof. Emeritus H. Florin"*, Dept. of Math., University of Louvain, Belgium, 173-206, 1975.

[N-027] Neuts, M. F. Computational uses of the method of phases in the theory of queues. *Computers and Math. with Appl.*, 1, 151-66, 1975.

[N-028] Neuts, M. F. Moment formulas for the Markov renewal branching process. *Adv. Appl. Prob.*, 8, 690-711, 1976.

[N-029] Neuts, M. F. Some explicit formulas for the steady-state behavior of the queue with semi-Markovian service times. *Adv. Appl. Prob.*, 9, 141-57, 1977.

[N-030] Neuts, M. F. Algorithms for the waiting time distributions under various queue disciplines in the $M/G/1$ queue with service time distributions of phase type. In *"Algorithmic Methods in Probability"*, TIMS Studies in the Management Sciences, no. 7. London: North Holland Publishing Co., 177-97, 1977.

[N-031] Neuts, M. F. The $M/G/1$ queue with several types of customers and change-over times. *Adv. Appl. Prob.*, 9, 604-44, 1977.

[N-032] Neuts, M. F. Queues solvable without Rouché's theorem. *Opns. Res.*, 27, 767-81, 1978.

[N-033] Neuts, M. F. Renewal processes of phase type. *Nav. Res. Logist. Quart.*, 25, 445-54, 1978.

[N-034] Neuts, M. F. The second moments of the absorption times in the Markov renewal branching process. *J. Appl. Prob.*, 15, 707-14, 1978.

[N-035] Neuts, M. F. The $M/M/1$ queue with randomly varying arrival and service rates. *Opsearch*, 15, 139-57, 1978.

[N-036] Neuts, M. F. Further results on the $M/M/1$ queue with randomly varying rates. *Opsearch*, 15, 158-68, 1978.

[N-037] Neuts, M. F. A versatile Markovian point process. *J. Appl. Prob.*, 16, 764-79, 1979.

[N-038] Neuts, M. F. Some algorithms for queues with group arrivals or group services. *Proc. Tenth Annual Pittsburgh Conference. Modeling and Simulation*, Vol. 10, pt. 2. Pittsburgh, Pa.: Instrument Society of America, 311-14, 1979.

[N-039] Neuts, M. F., and Lucantoni, D. M. A Markovian queue with N servers subject to breakdowns and repairs. *Mgmt. Sci.*, 25, 849-61,

1979.

[N-040] Neuts, M. F. Computational problems related to the Galton-Watson process. In *"Computational Probability"*, Proceedings of the 1975 Brown Actuarial Research Conference on Computational Probability, P. M. Kahn, ed., New York: Academic Press, 11-37, 1980.

[N-041] Neuts, M. F. The probabilistic significance of the rate matrix in matrix-geometric invariant vectors. *J. Appl. Prob.*, 17, 291-6, 1980.

[N-042] Neuts, M. F. *Matrix-Geometric Solutions in Stochastic Models: An Algorithmic Approach.* Baltimore: The Johns Hopkins University Press, 1981.

[N-043] Neuts, M. F. Stationary waiting time distributions in the $GI/PH/1$ queue. *J. Appl. Prob.*, 18, 901-12, 1981.

[N-044] Neuts, M. F. and Chakravarthy, S. The single server queue with platooned arrivals and service of phase type. *Euro. J. Oper. Res.*, 8, 379-89, 1981.

[N-045] Neuts, M. F. and Meier, K. S. On the use pf phase type distributions in reliability modelling of systems with two components. *OR Spektrum*, 2, 227-34, 1981.

[N-046] Neuts, M. F. and Takahashi, Y. Asymptotic behavior of the stationary distributions in the $GI/PH/c$ queue with heterogeneous servers. *Z. Wahrsch. verw. Gebiete*, 57, 441-52, 1981.

[N-047] Neuts, M. F. Explicit steady-state solutions to some elementary queueing models. *Opns. Res.*, 30, 480-89, 1982.

[N-048] Neuts, M. F. and Kumar, S. Algorithmic solution of some queues with overflows. *Mgmt. Sci.*, 28, 925-35, 1982.

[N-049] Neuts, M. F. and Nadarajan, R. A multi-server queue with thresholds for the acceptance of customers into service. *Opns. Res.*, 30, 948-60, 1982.

[N-050] Neuts, M. F. The c-server queue with constant service times and a versatile Markovian arrival process. In *"Applied probability - Computer science: The Interface"*, Proc. of the Conference at Boca Raton, Florida, January 1981, R. L. Disney and T. J. Ott, eds, Boston: Birkhäuser, Vol I, 31-70, 1982.

[N-051] Neuts, M. F. On the coefficient of variation of mixtures of probability distributions. *Comm. Statist.*, B 11, 649-657, 1982.

[N-052] Neuts, M. F. The abscissa of convergence of the Laplace-Stieltjes transform of a PH-distribution. *Commun. Statist.-Simula. Computa.*, 13, 367-373, 1984.

[N-053] Neuts, M. F. The gambler's ruin problem with Markov-dependent trials: An example in computational probability. *The Math. Scientist,* 9, 25-36, 1984.

[N-054] Neuts, M. F. Problems of reducibility in structured Markov chains of $M/G/1$ type and related queueing models in communications engineering. In *"Mathematical Computer Performance and Reliability",* G. Iazeolla, P. J. Courtois and A. Hordijk, eds., Amsterdam: North-Holland, 139-51, 1984.

[N-055] Neuts, M. F. Matrix-analytic methods in queuing theory. *Euro. J. of Oper. Res.,* 15, 2-12, 1984.

[N-056] Neuts, M. F. and Ramalhoto, M. F. A service model in which the server is required to search for customers. *J. Appl. Prob.,* 21, 157-66, 1984.

[N-057] Neuts, M. F. A queueing model for a storage buffer in which the arrival rate is controlled by a switch with random delay. *Performance Evaluation,* 5, 243-56, 1985.

[N-058] Neuts, M. F. The $M/G/1$ queue with a limited number of admissions or a limited admission period during each service time. *Stoch. Mod.,* 1, 361-391, 1985.

[N-059] Neuts, M. F. The caudal characteristic curve of queues. *Adv. Appl. Prob.,* 18, 221-54, 1986.

[N-060] Neuts, M. F. A new informative embedded Markov renewal process for the $PH/G/1$ queue. *Adv. Appl. Prob.,* 18, 535-557, 1986.

[N-061] Neuts, M. F. and Latouche, G. The superposition of two PH-renewal processes. In *"Semi-Markov Models: Theory and Applications",* J. Janssen, ed., London: Plenum Publishers, 131-77, 1986.

[N-062] Neuts, M. F. Generalizations of the Pollaczek-Khinchin integral equation in the theory of queues. *Adv. Appl. Prob.,* 18, 952-90, 1986.

[N-063] Neuts, M. F. Transform-free equations for the stationary waiting time distributions in the queue with Poisson arrivals and bulk services. In *"Statistical and Computational Issues in Probability Modeling",* S. L. Albin and C. M. Harris, eds, Basel: J. C. Baltzer AG, Publishers, Annals of Oper. Res., 8, 3-26, 1987.

[N-064] Neuts, M. F. Profile curves for the $M/G/1$ queue with group arrivals. *Stoch. Mod.,* 4, 277-298, 1988.

[N-065] Neuts, M. F. and Chandramouli, Y. Statistical group testing with queueing involved. *Queueing Systems,* 2, 19-39, 1987.

[N-066] Neuts, M. F. An algorithm for the distribution of the time between coincidences of two independent PH-renewal processes. *Zast. Matemat.*, (forthcoming).

[N-067] Neuts, M. F. The fundamental period of the queue with Markov-modulated arrivals. In *A Volume in Honor of Professor Samuel Karlin, 1989,* (forthcoming).

[N-068] Neuts, M. F. Phase-type distributions: A bibliography. *1989,* (forthcoming).

[N-069] Newell, G. F. Queues for a fixed-cycle traffic light. *Ann. Math. Statist.*, 31, 589-97, 1960.

[N-070] Newell, G. F. Approximate behavior of tandem queues. *Research rept. Inst. of Transp. and Traffic Engin., University of California, Berkeley, chaps. 1-4, 1975.*

[N-071] Newell, G. F. Approximate behavior of tandem queues. *Research rept. Inst. of Transp. and Traffic Engin., University of California, Berkeley, chaps. 5-8, 1977.*

[N-072] Nicolas, B. and Latouche, G. Blocking in tandem queues. *Belg. J. Oper. Res., Stat and C.S.*, 25, 29-38, 1985.

[N-073] Nilsson, A. An extended busy period analysis of the $M/G/1$ and $G/M/m$ queueing systems. *Ericsson Technics,* 32, 173-94, 1974.

[N-074] Nishida, T. and Hiramatsu, T. Commutative tandem queue with finite waiting room. *J. Oper. Res. Soc. Japan,* 20, 194-202, 1977.

[N-075] Nishida, T.; Ueshima, Y. and Yoneyama, K. Two-stage tandem queue with two types of customers. *Technology Reports of the Osaka University,* 28, 343-9, 1978.

[N-076] Nishida, T. and Hiramatsu, T. Commutative tandem queue with correlated two servers. *Math. Japonica,* 24, 73-80, 1979.

[N-077] Niu, S. C. and Cooper, R. B. Duality and other results for the $M/G/1$ and $GI/M/1$ queues via a new ballot theorem. *Math. of O.R.*, (forthcoming).

[N-078] Nollan, V. *Semi-Markovsche Prozesse.* Deutsche Taschenbücher Nr. 31, Thun-Frankfurt am Main: Verlag Harri Deutsch, 1980.

[O-001] O'Donovan, T. M. The queue $M/G/1$ with semipreemptive-priority queuing discipline. *Opns. Res.*, 20, 434-39, 1972.

[O-002] O'Donovan, T. M. Distribution of attained and residual service in general queuing systems. *Opns. Res.*, 22, 570-5, 1974.

[O-003] O'Donovan, T. M. Direct solutions of $M/G/1$ processor-sharing models. *Opns. Res.*, 22, 1232-35, 1974.

[O-004] O'Donovan, T. M. The queue $M/G/1$ when jobs are scheduled within generations. *Opns. Res.*, 23, 821-24, 1975.

[O-005] O'Donovan, T. M. Conditional response times in $M/M/1$ processor-sharing models. *Opns. Res.*, 24, 382-5, 1976.

[O-006] O'Donovan, T. M. Direct solutions of $M/G/1$ priority queues. *R.A.I.R.O.*, 10, 107-11, 1976.

[O-007] Odoom, S., and Lloyd, E. H. A note on the equilibrium distribution of levels in a semi-infinite reservoir subject to Markovian inputs and unit withdrawals. *J. Appl. Prob.*, 2, 215-22, 1965.

[O-008] Ohno, K. and Mine, H. Traffic light queues with departure headways depending upon positions. *J. Oper. Res. Soc. Japan,* 17, 145-69, 1974.

[O-009] Ohno, K. Computational algorithm for a fixed cycle traffic signal and new approximate expressions for average delay. *Transp. Sci.*, 12, 29-47, 1978.

[O-010] Oliver, R. M. Tables of the waiting time distribution for the constant service queue $(M/D/1)$. *Int. J. Comp. Math.*, 2, 35-56, 1968.

[O-011] Ortega, J. M., and Rheinboldt, W. C. *Iterative solution of nonlinear equations in several variables.* New York: Academic Press, 1970.

[O-012] Osone, T. and Fujisawa, T. On the queueing system $M/G^{[m]}/1$ with the maximum batch size m. *Rep. Univ. Electr-Comm.*, 30-2, (Sci. & Tech. Sect.), 225-31, 1980.

[O-013] Osone, T. and Fujisawa, T. An application of an imbedded level crossing technique to reserve for contingencies. *Rep. Univ. Electr-Comm.*, 31-2, (Sci. & Tech. Sect.), 225-36, 1981.

[O-014] Ott, T. J. The covariance function of the virtual waiting-time in an $M/G/1$ queue. *Adv. Appl. Prob.*, 9, 158-68, 1977.

[O-015] Ott, T. J. The stable $M/G/1$ queue in heavy traffic and its covariance function. *Adv. Appl. Prob.*, 9, 169-86, 1977.

[O-016] Ott, T. J. Some more results for the stable $M/G/1$ queue in heavy traffic. *J. Appl. Prob.*, 16, 187-97, 1979.

[O-017] Ott, T. J. On the $M/G/1$ queue with additional inputs. *J. Appl. Prob.*, 21, 129-42, 1984.

[O-018] Ott, T. J. The sojourn-time distribution in the $M/G/1$ queue with processor sharing. *J. Appl. Prob.*, 21, 360-78, 1984.

[O-019] Ott, T. J. On the stationary waiting-time distribution in the $GI/G/1$ queue, I: Transform methods and almost-phase-type distributions. *Adv. Appl. Prob.*, 19, 240-65, 1987.

[O-020] Ott, T. J. The single-server queue with independent GI/G and M/G input streams. *Adv. Appl. Prob.*, 19, 266-86, 1987.

[P-001] Pack, C. D. The optimum design of a tandem computer buffer in a system for remote data collection. *IEEE Trans. Comm.*, COM-22, 1501-6, 1974.

[P-002] Pack, C. D. The output of an $M/D/1$ queue. *Opns. Res.*, 23, 750-60, 1975.

[P-003] Pack, C. D. A comparison of the output process of an $M/D/1$ queue as measured from an arbitrary departure epoch to that measured from an arbitrary instant in time. *SIAM J. Appl. Math., 28, 367-75, 1975.*

[P-004] Pack, C. D. The output of multiserver queuing systems. *Opns. Res., 26, 492-505, 1978.*

[P-005] Pagurek, B. and Woodside, C. M. A recursive algorithm for computing serial correlations of times in an $M/G/1$ queue. In *"Algorithmic Methods in Probability"*, TIMS Studies in the Management Sciences, No. 7. London: North Holland Publishing Co., 161-75, 1977.

[P-006] Pagurek, B. and Woodside, C. M. The sum of serial correlations of waiting and system times in $GI/G/1$ queues. *Opns. Res.*, 27, 755-66, 1979.

[P-007] Paige, C.C.; Styan, G. P. H.; and Wachter, P. G. Computation of the stationary distribution of a Markov chain. *J. Statist. Comput. Simul.*, 4, 173-86, 1975.

[P-008] Pakes, A. G. A note on the queueing systems $GI/D/1$ and $D/G/1$. *J. Appl. Prob.*, 7, 465-8, 1970.

[P-009] Pakes, A. G. The correlation coefficients of the queue lengths of some stationary single server queues. *J. Austral. Math. Soc.*, 12, 35-46, 1971.

[P-010] Pakes, A. G. On the busy period of the modified $GI/G/1$ queue. *J. Appl. Prob.*, 10, 192-7, 1973.

[P-011] Pakes, A. G. On dams with Markovian inputs. *J. Appl. Prob.*, 10, 317-29, 1973.

[P-012] Pakes, A. G. On the tails of waiting-time distributions. *J. Appl. Prob.*, 12, 555-64, 1975.

[P-013] Pakes, A. G., and Phatarfod, R. M. The limiting distribution for the infinitely deep dam with a Markovian input. *Stoch. Proc. Appl.*, 8, 199-209, 1978.

[P-014] Pearce, C. A queueing system with non-recurrent input and batch servicing. *J. Appl. Prob.*, 2, 442-8, 1965.

[P-015] Pearce, C. The series queue $M/G/1 \cdots / M/1$ with finite waiting room in the first stage. *Studia Sc. Math. Hungar.*, 6, 41-8, 1971.

[P-016] Pegg, P. A. and Phatarfod, R. M. Dams with additive inputs revisited. *J. Appl. Prob.*, 14, 367-74, 1977.

[P-017] Perros, H. G. On the $M/C_k/1$ queue. *Performance Evaluation*, 3, 83-93, 1983.

[P-018] Perry, D. An inventory system for perishable commodities with random lifetime. *Adv. Appl. Prob.*, 17, 234-6, 1985.

[P-019] Pestalozzi, G. A queue with Markov-dependent service times. *J. Appl. Prob.*, 5, 461-6, 1968.

[P-020] Peters, P. E. Delays for a *LIFO* queue with constant service time. *Opns. Res.*, 16, 1147-51, 1968.

[P-021] Phatarfod, R. M. and Mardia, K. V. Some results for dams with Markovian inputs. *J. Appl. Prob.*, 10, 166-80, 1973.

[P-022] Phatarfod, R. M. The bottomless dam. *J. Hydrology*, 40, 337-63, 1979.

[P-023] Phatarfod, R. M. A note on the infinitely deep dam with a Markovian input. *J. Appl. Prob.*, 16, 917-22, 1979.

[P-024] Pollaczek, F. Über eine Aufgabe der Wahrscheinlichkeitstheorie. *Math. Z.*, 32, 64-100; 729-50, 1930.

[P-025] Pollaczek, F. Über das Warteproblem. *Math. Z.*, 38, 492-537, 1934.

[P-026] Ponstein, J. Theory and numerical solution of a discrete queueing problem. *Statistica Neerlandica*, 20, 139-52, 1974.

[P-027] Posner, M. Single-server queues with service time dependent on waiting time. *Opns. Res.*, 21, 610-6, 1973.

[P-028] Powell, B. A. and Avi-Itzhak, B. Queuing systems with enforced idle time. *Opns. Res.*, 15, 1145-56, 1967.

[P-029] Powell, W. B. and Humblet, P. Queue length and waiting time transforms for bulk arrival, bulk service queues with a general control strategy. *Res. Rept. No. EES-81-7, School of Engineering and Applied Science, Princeton University, September 1981.*

[P-030] Powell, W. B. Numerical methods in the study of waiting times for bulk arrival, bulk service queues with Poisson and non-Poisson arrivals. *Res. Rept. No. EES-82-5, School of Engineering and Applied Science, Princeton University, May 1982.*

[P-031] Powell, W. B. Analysis of vehicle holding and cancellation strategies in bulk arrival, bulk service queues. *Trans. Sci.,* 19, 352-77, 1985.

[P-032] Powell, W. B. Iterative algorithms for bulk arrival, bulk service queues with Poisson and non-Poisson arrivals. *Trans. Sci.,* 20, 65-80, 1985.

[P-033] Powell, W. B. Approximate, closed form moment formulas for bulk arrival, bulk service queues. *Trans. Sci.,* 20, 13-23, 1986.

[P-034] Powell, W. B. and Humblet, P. The bulk service queue with a general control strategy: Theoretical analysis and a new computational procedure. *Opns. Res.,* 34, 267-75, 1986.

[P-035] Prabhu, N. U. Some results for the queue with Poisson arrivals. *J. Roy. Statist. Soc. Ser. B,* 22, 104-7, 1960.

[P-036] Prabhu, N. U. Application of storage theory to queues with Poisson arrivals. *Ann. Math. Statist.,* 31, 475-82, 1960.

[P-037] Prabhu, N. U. and Bhat, U. N. Further results for the queue with Poisson arrivals. *Opns. Res.,* 11, 380-6, 1963.

[P-038] Prabhu, N. U. and Bhat, U. N. Some first passage problems and their application to queues. *Sankhya, Ser. A,* 25, 281-92, 1963.

[P-039] Prabhu, N. U. Time-dependent results in storage theory. *J. Appl. Prob.,* 1, 1-46, 1964.

[P-040] Prabhu, N. U. *Queues and Inventories: A Study of their Basic Stochastic Processes.* New York: John Wiley and Sons, 1965.

[P-041] Prabhu, N. U. Unified results and methods for queues and dams. *Proc. Symp. on Congestion Theory,* W. L. Smith and W. E. Wilkinson, eds., Chapel Hill: Univ. of N. Carolina Monographs, 317-36, 1965.

[P-042] Prabhu, N. U. Transient behavior of a tandem queue. *Mgmt. Sci.,* 13, 631-39, 1966.

[P-043] Prabhu, N. U. Some comments on Sven Erlander's paper: The remaining busy period for a single server queue with Poisson input. *Opns. Res.,* 15, 357-8, 1967.

[P-044] Prabhu, N. U. Wiener-Hopf techniques in queueing theory. In *"Proceedings of the Conference on Mathematical Methodology in the Theory of Queues, Kalamazoo, Mich."*, New York: Springer-Verlag,

81-90, 1974.

[P-045] Prabhu, N. U. *Stochastic Storage Processes: Queues, Insurance Risk and Dams.* New York: Springer-Verlag, 1980.

[P-046] Proudfoot, A. D. and Lampard, D. G. A random walk problem with correlation. *J. Appl. Prob.*, 9, 436-40, 1972.

[P-047] Purdue, P. Non-linear matrix integral equations of Volterra type in queueing theory. *J. Appl. Prob.*, 10, 644-51, 1973.

[P-048] Purdue, P. The single server queue in a Markovian environment. In *"Proceedings of the Conference on Mathematical Methodology in the Theory of Queues, Kalamazoo, Mich."*, New York: Springer-Verlag, 359-65, 1974.

[P-049] Purdue, P. The $M/M/1$ queue in a Markovian environment. *Opns. Res.*, 22, 562-69, 1974.

[P-050] Purdue, P. A queue with Poisson input and semi-Markov service times: Busy period analysis. *J. Appl. Prob.*, 12, 353-57, 1975.

[P-051] Purdue, P. Applications of branching processes in the theory of queues. *Proc. EURO II*, M. Roubens, ed., Amsterdam: North-Holland, 375-81, 1977.

[P-052] Purdue, P. The single server queue in a random environment. *Oper. Res. Verfahren*, 33, 363-72, 1979.

[P-053] Pyke, R. Markov renewal processes: Definitions and preliminary properties. *Ann. Math. Statist.*, 32, 1231-42, 1961.

[P-054] Pyke, R. Markov renewal processes with finitely many states. *Ann. Math. Statist.*, 32, 1243-59, 1961.

[P-055] Pyke, R. and Schaufele, R. A. Limit theorems for Markov renewal processes. *Ann. Math. Statist.*, 35, 1746-64, 1964.

[P-056] Pyke Tin and Phatarfod, R. M. On infinite dams with inputs forming a stationary process. *J. Appl. Prob.*, 11, 553-61, 1974.

[P-057] Pyke Tin and Phatarfod, R. M. Some exact results for dams with Markovian inputs. *J. Appl. Prob.*, 13, 329-37, 1976.

[R-001] Rade, L. A model for interaction of a Poisson process and a renewal process and its relation with queueing theory. *J. Appl. Prob.*, 9, 451-6, 1972.

[R-002] Rajeswari, A. R. Nonpreemptive priority queue with binomial input. *Opns. Res.*, 16, 416-21, 1968.

[R-003] Raju, S. N. and Bhat, U. N. Recursive relations in the computation of the equilibrium results of finite queues. In *"Algorithmic Methods in Probability"*, TIMS Studies in the Management Sciences, no. 7. London: North Holland Publishing Co., 247-70, 1977.

[R-004] Ramanarayanan, R. Markovian queueing systems with two waiting rooms. *Cah. Centre d'études Rech. Opér.*, 24, 61-70, 1982.

[R-005] Ramaswami, V., and Lucantoni, D. M. On the merits of an approximation to the busy period of the *GI* / *G* / 1 queue. *Mgmt. Sci.*, 25, 285-89, 1979.

[R-006] Ramaswami, V. The *N* / *G* / 1 queue and its detailed analysis. *Adv. Appl. Prob.*, 12, 222-61, 1980.

[R-007] Ramaswami, V. and Neuts, M. F. A duality theorem for phase type queues. *Ann. Prob.*, 5, 974-85, 1980.

[R-008] Ramaswami, V. Algorithms for a continuous-review (s, S) inventory system. *J. Appl. Prob.*, 18, 461-72, 1981.

[R-009] Ramaswami, V. Queues with correlated interarrival times - Comments on a paper by Murari and Parkash. *Cah. Centre d'études Rech. Opér.*, 24, 21-5, 1982.

[R-010] Ramaswami, V. and Lucantoni, D. M. Algorithmic analysis of a dynamic priority queue. In *"Applied probability - Computer science: The Interface"*, Proc. of the Conference at Boca Raton, Florida, January 1981, R. L. Disney and T. J. Ott, eds, Boston: Birkhäuser, Vol II, 157-206, 1982.

[R-011] Ramaswami, V. The busy period of queues which have a matrix-geometric steady state probability vector. *Opsearch,* 19, 238-61, 1982.

[R-012] Ramaswami, V. and Lucantoni, D. M. Stationary waiting time distributions in queues with phase type service and in quasi-birth-and-death processes. *Stoch. Models,* 1, 125-36, 1985.

[R-013] Ramaswami, V. Independent Markov processes in parallel. *Stoch. Models,* 1, 419-32, 1985.

[R-014] Ramaswami, V. and Latouche, G. A general class of Markov processes with explicit matrix-geometric solutions. *OR Spektrum,* 8, 209-18, 1986.

[R-015] Ramaswami, V. and Latouche, G. An experimental evaluation of the matrix-geometric method for the *PH* / *PH* / 1 queue. *Stoch. Mod.,* 5, 1989, (forthcoming).

[R-016] Ramaswami, V. A stable recursion for the steady state vector in Markov chains of *M* / *G* / 1 type. *Stoch. Models,* 4, 183-188, 1988.

[R-017] Ramaswami, V. Nonlinear matrix equations in applied probability - Solution techniques and open problems. *SIAM Review,* (forthcoming).

[R-018] Rao, B. M. S. and Posner, M. J. M. On the output process of an $M / M / 1$ queue with randomly varying system parameters. *OR Letters,* 3, 191-7, 1984.

[R-019] Rao, B. M. S. and Posner, M. J. M. Algorithmic and approximation analyses of the split and match queue. *Stoch. Models,* 1, 433-56, 1985.

[R-020] Rao, B. M. S. and Posner, M. J. M. On the departure process of an $M / PH / 1$ queue. *SCIMA,* 14, 77-84, 1985.

[R-021] Rao, B. M. S. and Posner, M. J. M. Parallel exponential queues with dependent service rates. *Comp. and O.R.,* 13, 681-92, 1986.

[R-022] Rao, B. M. S. and Posner, M. J. M. Algorithmic and approximation analyses of the shorter queue model. *Nav. Res. Logist.,* 34, 381-98, 1987.

[R-023] Rath, J. H. and Sheng, D. Approximations for overflows from queues with a finite waiting room. *Opns. Res.,* 27, 1208-16, 1979.

[R-024] Ravichandran, N. and Gravey, A. Some comments on the simple queue. *Adv. Appl. Prob.,* 16, 933-5, 1984.

[R-025] Rege, K. M. and Sengupta, B. A priority-based admission scheme for a multiclass queueing system. *AT&T Tech. J.,* 64, 1731-53, 1985.

[R-026] Regterschot, G. J. K. and de Smit, J. H. A. The queue $M / G / 1$ with Markov modulated arrivals and services. *Math. of O.R.,* 11, 465-83, 1986.

[R-027] Reich, E. Departure processes. *Proc. Symp. on Congestion Theory,* W. L. Smith and W. E. Wilkinson, eds., Chapel Hill: Univ. of N. Carolina Monographs, 439-57, 1965.

[R-028] Reiser, M. and Kobayashi, H. Accuracy of the diffusion approximation for some queueing systems. *IBM J. Res. Develop.,* 18, 110-24, 1974.

[R-029] Reiser, M., and Konheim, A. G. Finite capacity queueing systems with applications in computer modeling. *SIAM J. on Computing,* 7, 210-29, 1978.

[R-030] Restrepo, R. A. A queue with simultaneous arrivals and Erlang service time distribution. *Opns. Res.,* 13, 375-81, 1965.

[R-031] Riordan, J. Delays for last-come-first-served service and the busy period. *Bell Syst. Tech. J.,* 40, 785-93, 1961.

[R-032] Riordan, J. *Stochastic Service Systems.* New York: John Wiley and Sons, 1962.

[R-033] Rizzuto, G. T. and Boullion, T. L. Busy period in batch queues. *Commun. Statist. - Theor. Meth.*, 12, 2655-62, 1983.

[R-034] Robert, P.; Mitrani, I. and King, P. J. B. Analysis of a meteor scatter communication protocol. In *"Performance '87"*, Proc. 12-th IFIP Internat. Symp., Brussels, 1987, Courtois, P-J. and Latouche, G. eds., Amsterdam: North Holland Publ., 469-79, 1988.

[R-035] Roes, P. B. M. The finite dam. *J. Appl. Prob.*, 7, 316-26, 1970.

[R-036] Roes, P. B. M. The finite dam II. *J. Appl. Prob.*, 7, 599-616, 1970.

[R-037] Rolski, T. and Stoyan, D. On the comparison of waiting times in $GI/G/1$ queues. *Opns. Res.*, 24, 197-200, 1976.

[R-038] Rose, M. The $(S-1,S)$ inventory model with arbitrary backordered demand and constant delivery times. *Opns. Res.*, 20, 1020-32, 1972.

[R-039] Rosenlund, S. I. On the length and number of served customers of the busy period of a generalised $M/G/1$ queue with finite waiting room. *Adv. Appl. Prob.*, 5, 379-89, 1973.

[R-040] Rosenlund, S. I. An $M/G/1$ model with finite waiting room in which a customer remains during part of a service. *J. Appl. Prob.*, 10, 778-85, 1973.

[R-041] Rosenlund, S. I. Busy period of a finite queue with phase type service. *J. Appl. Prob.*, 12, 201-4, 1975.

[R-042] Rosenlund, S. I. On the $M/M/m$ queue with finite waiting room. *Bull. Soc. Roy. Sci. Liege*, 44, 42-55, 1975.

[R-043] Rosenlund, S. I. Busy periods in time-dependent $M/G/1$ queues. *Adv. Appl. Prob.*, 8, 195-208, 1976.

[R-044] Rosenlund, S. I. Recurrent emptiness in a finite queue with increasing arrival rate. *J. Appl. Prob.*, 13, 423-6, 1976.

[R-045] Rosenlund, S. I. Busy periods and other aspects of $GI/M/m$ and $M/G/1$ queues. *University of Goteborg, Dept. of Mathematics, Goteborg, Sweden, 1983.*

[R-046] Ross, S. M. *Applied Probability Models with Optimization Applications.* San Francisco: Holden-Day, 1970.

[R-047] Ross, S. M. *Introduction to Probability Models.* New York: Academic Press, 1972.

[R-048] Ross, S. M. Bounds on the delay distribution in $GI/G/1$ queues. *J. Appl. Prob.*, 11, 417-21, 1974.

[R-049] Ross, S. M. Average delay in queues with non-stationary Poisson arrivals. *J. Appl. Prob.*, 15, 602-9, 1978.

[R-050] Rossberg, H. J. Eine neue Methode zur Behandlung der Integralgleichung von Lindley und ihrer Verallgemeinung durch Finch. *Elektron. Inf. Verarb. Kybern.*, EIK 3, 215-38, 1967.

[R-051] Rossberg, H. J. Erhältungssätze fur vollmonotone Verteilungsdichten beim Lindleyschen Warteschlangenmodell. *Math. Nachr.*, 34, 79-94, 1967.

[R-052] Rossberg, H. J. Optimale Eigenschaften einiger Wartesysteme bei regelmäszigen Eingang bzw. konstanten Bedienungszeiten. *ZAMM*, 48, 395-403, 1968.

[R-053] Rossberg, H. J. Ausbedeutung einer bekannten Wiener-Hopf-Faktorisierung beim Wartemodell $G / G / 1$ und einer mit ihm zusammenhängenden Irrfahrt. *Math. Operationsforsch. u. Statist.*, 2, 129-46, 1971.

[R-054] Rossberg, H. J. and Siegel, G. Die Bedeutung von Kingmans Integralgleichungen bei der Approximation der stationären Wartezeitverteilung im Modell $GI / G / 1$ mit und ohne Verzögerung beim Beginn einer Beschäftigungsperiode. *Math. Operationsforsch. u. Statist.*, 5, 687-99, 1974.

[R-055] Rossberg, H. J. and Siegel, G. The $GI / G / 1$ model with warming-up time. *Zast. Matemat.*, 14, 17-26, 1974.

[R-056] Rossberg, H. J. and Siegel, G. On Kingman's integral inequalities for approximations of waiting time distribution in the queuing model $GI / G / 1$ with and without warming-up time. *Zast. Matemat.*, 14, 27-30, 1974.

[R-057] Rossberg, H. J. and Siegel, G. A new method for the study of the stationary waiting time distribution for a $GI / G / 1$ queue with warm-up (in Russian with Polish summary). *Zast. Matemat.*, 14, 537-47, 1975.

[R-058] Rubinovitch, M. The output of a buffered data communication system. *Stoch. Proc. Appl.*, 1, 375-82, 1973.

[R-059] Rudemo, M. Point processes generated by transitions of Markov chains. *Adv. Appl. Prob.*, 5, 262-86, 1973.

[R-060] Ruiz-Pala, E.; Avila-Beloso, C., and Hines, W. W. *Waiting-line models: An Introduction to their Theory and Application.* New York: Rheinhold, 1970.

[R-061] Runnenburg, J. Th. Probabilistic interpretation of some formulae in queueing theory. *Bull. Inst. Internat. Statist.*, 37, 405-14, 1960.

[R-062] Runnenburg, J. Th. On the use of the method of collective marks in queueing theory. *Proc. Symp. on Congestion Theory,* W. L. Smith and W. E. Wilkinson, eds., Chapel Hill: Univ. of N. Carolina Monographs, 399-438, 1965.

[R-063] Ryba, T. Remark on a service system with mixed input streams. *Opns. Res.,* 20, 452-4, 1972.

[S-001] Saaty, T. L. *Elements of Queueing Theory with Applications.* New York: McGraw-Hill Book Co., 1961.

[S-002] Sahin, I. and Bhat, U. N. A stochastic system with scheduled secondary inputs. *Opns. Res.,* 19, 436-46, 1971.

[S-003] Sahin, I. Equilibrium behavior of a stochastic system with secondary input. *J. Appl. Prob.,* 8, 252-60, 1971.

[S-004] Sahin, I. On the single server queue with preemptive service interruptions. *J. Appl. Prob.,* 8, 835-7, 1971.

[S-005] Sakasegawa, H. Numerical tables of the queueing system $E_k / E_2 / s$. *Inst. Stat. Math. Sci., Comp. Sc.* Monograph No. 10, Tokyo, Japan, 1978.

[S-006] Sakate, M.; Noguchi, S. and Oizumi, J. An analysis of the $M/G/1$ queue under round-robin scheduling. *Opns. Res.,* 19, 371-85, 1971.

[S-007] Sandhu, D. and Posner, M.J.M. A priority $M/G/1$ queue with application to voice/data communication. *Euro. J. of Oper. Res.,* (forthcoming).

[S-008] Schäl, M. Markov renewal processes with auxiliary paths. *Ann. Math. Statist.,* 41, 1604-23, 1970.

[S-009] Schäl, M. The analysis of queues with state-dependent parameters by Markov renewal processes. *Adv. Appl. Prob.,* 3, 155-75, 1971.

[S-010] Schassberger, R. *Warteschlangen.* New York: Springer-Verlag, 1973.

[S-011] Schassberger, R. A broad analysis of single server priority queues with two independent input streams, one of them Poisson. *Adv. Appl. Prob.,* 6, 666-88, 1974.

[S-012] Schassberger, R. An aggregation principle for computing invariant probability vectors for semi-Markovian models. In *"Mathematical Computer Performance and Reliability",* G. Iazeolla, P. J. Courtois and A. Hordijk, eds., Amsterdam: North-Holland, 259-275, 1984.

[S-013] Schassberger, R. Residence time in the $M/G/1$ processor-sharing queue. *Adv. Appl. Prob.,* 16, 202-13, 1984.

[S-014] Schellhaas, H. On the T -policy for an $M/G/1$ queue. *Oper. Res. Verfahren,* 29, 750-63, 1978.

[S-015] Schellhaas, H. Semi-regenerative processes with unbounded rewards. *Math. of O.R.,* 4, 70-8, 1979.

[S-016] Schellhaas, H. Computation of the state probabilities in $M/G/1$ queues with state dependent input and state dependent service. *OR Spektrum,* 5, 223-8, 1983.

[S-017] Schellhaas, H. Computation of the state probabilities in a class of semi-regenerative queueing models. In *"Semi-Markov Models: Theory and Applications",* J. Janssen, ed., London: Plenum Publishers, 111-30, 1986.

[S-018] Schlee-Kössler, W. A single server queue with two types of customers and alternating service in pieces of l_1 and l_2. In *"Point Processes and Queuing problems: Proceedings of a Colloquium, Keszthely, Hungary, 1978",* 343-57, 1981, London: North Holland Publ. Co.

[S-019] Scholl, M. and Kleinrock, L. On the $M/G/1$ queue with rest periods and certain service-independent queueing disciplines. *Opns. Res.,* 31, 705-19, 1983.

[S-020] Schrage, L. E. and Miller, L. W. The queue $M/G/1$ with the shortest remaining processing time discipline. *Opns. Res.,* 14, 670-84, 1966.

[S-021] Schrage, L. The queue $M/G/1$ with feedback to lower priority queues. *Mgmt. Sci.,* 13, 466-74, 1967.

[S-022] Schrage, L. A proof of the optimality of the shortest remaing processing time discipline. *Opns. Res.,* 16, 687-9, 1968.

[S-023] Scott, M. Queueing with control on the arrival of certain type of customers. *Canad. Opns. Res. J.,* 8, 75-86, 1970.

[S-024] Scott, M. Queuing with two servers working in shifts. *Opns. Res.,* 20, 421-4, 1972.

[S-025] Seal, H. L. *Stochastic Theory of a Risk Business.* New York: John Wiley and Sons, 1969.

[S-026] Seal, H. L. *Survival Probabilities: The Goal of Risk Theory.* New York: John Wiley and Sons, 1978.

Seelen, L. P. An algorithm for $Ph/Ph/c$ queues. *Euro. J. of Oper. Res.,* 23, 118-27, 1986.

[S-027] Sengupta, B. The spatial requirement of an $M/G/1$ queue, or: How to design for buffer space. In *"Modelling and Performance Evaluation Methodology "* F. Baccelli and G. Fayolle, eds., Heidelberg:

Springer-Verlag, 547-64, 1984.

[S-028] Sengupta, B. Sojourn time distributions for the $M/M/1$ queue in a Markovian environment. *Euro. J. of Oper. Res.*, 32, 140-9, 1987.

[S-029] Sengupta, B. A perturbation method for solving some queues with processor sharing discipline. *J. Appl. Prob.*, (forthcoming).

[S-030] Sengupta, B. A queue with service interruptions in an alternating random environment. *Manuscript, AT&T Bell Laboratories, 1987.*

[S-031] Sengupta, B. Markov processes whose steady state distribution is matrix-exponential with an application to the $GI/PH/1$ queue. *Adv. Appl. Prob.*, 21, 1989 (forthcoming).

[S-032] Serfozo, R. F. Extreme values of queue lengths in $M/G/1$ and $GI/M/1$ systems. *Math. of O.R.*, 13, 1988, (forthcoming).

[S-033] Servi, L. D. $D/G/1$ queues with vacations. *Opns. Res.*, 34, 619-29, 1986.

[S-034] Shalmon, M. and Kaplan, M. A. A tandem network of queues with deterministic service and intermediate arrivals. *Opns. Res.*, 32, 753-73, 1984.

[S-035] Shanbhag, D. N. Some remarks concerning the departure process of a queue with Poisson arrivals and no balking. *Opns. Res.*, 15, 972-5, 1967.

[S-036] Shanthikumar, J. G. Effects of opportune maintenance in a single machine job shop. *Engineer, Quart. J. Inst. Of Engineers, Sri Lanka,* 7, 22-25, 1979.

[S-037] Shanthikumar, J. G. On a single-server queue with state dependent service. *Nav. Res. Logist. Quart.*, 26, 305-9, 1979.

[S-038] Shanthikumar, J. G. Some analyses on the control of queues using level crossings of regenerative processes. *J. Appl. Prob.*, 17, 814-21, 1980.

[S-039] Shanthikumar, J. G. Analysis of the control of queues with shortest processing time service discipline. *J. Oper. Res. Soc. Japan,* 23, 341-52, 1980.

[S-040] Shanthikumar, J. G. and Buzacott, J. A. On the approximations to the single server queue. *Int. J. Prod. Res.*, 18, 761-73, 1980.

[S-041] Shanthikumar, J. G. $M/G/1$ queues with scheduling within generations and removable server. *Opns. Res.*, 29, 1010-18, 1981.

[S-042] Shanthikumar, J. G. Optimal control of an $M/G/1$ priority queue via $N-$control. *Amer. J. Math. and Mgmt. Sci.*, 1, 191-212, 1981.

[S-043] Shanthikumar, J. G. and Sargent, R. G. An algorithmic solution for a queueing model of a computer system with interactive and batch jobs. *Performance Evaluation,* 1, 344-57, 1981.

[S-044] Shanthikumar, J. G. On the buffer behavior with Poisson arrivals, priority service, and random server interruptions. *IEEE Trans. Computers,* C-30, 781-6, 1981.

[S-045] Shanthikumar, J. G. On reducing time spent in $M/G/1$ systems. *Euro. J. Oper. Res.,* 9, 286-94, 1982.

[S-046] Shanthikumar, J. G. A recursive algorithm to generate joint probability distribution of arrivals from exponential sources during a random time interval. *Information Processing Letters,* 14, 214-7, 1982.

[S-047] Shanthikumar, J. G. Analysis of a single server queue with time- and operation-dependent server failures. *Adv. in Mgmt. Studies,* 1, 339-59, 1982.

[S-048] Shanthikumar, J. G. and Jeya Chandra, M. Application of level crossing analysis to discrete state processes in queueing systems. *Nav. Res. Logist. Quart.,* 29, 593-608, 1982.

[S-049] Shanthikumar, J. G. Comparison of dispatch policies for a single server queueing model with limited operational control. *Int. J. Prod. Res.,* 22, 389-403, 1984.

[S-050] Shanthikumar, J. G. Analyses of priority queues with server control. *Opsearch,* 21, 183-92, 1984.

[S-051] Shanthikumar, J. G. Bilateral phase type distributions. *Nav. Res. Logist. Quart.,* 32, 119-36, 1985

[S-052] Shanthikumar, J. G. and Sumita, U. On the busy period distributions of $M/G/1/K$ queues with state dependent arrivals and $FCFS/LCFS-P$ service disciplines. *J. Appl. Prob.,* 22, 912-19, 1985.

[S-053] Shanthikumar, J. G. On stochastic decomposition in $M/G/1$ type queues with generalized server vacations. *Opns. Res.,* 36, 566-69, 1988.

[S-054] Sharma, S. D. On a continuous/discrete time queueing system with arrivals in batches of variable size and correlated departures. *J. Appl. Prob.,* 12, 115-29, 1974.

[S-055] Shaw, L. Busy period control of queues based on waiting times at arrivals. *Opns. Res.,* 24, 543-63, 1976.

[S-056] Shogan, A. W. A single server queue with arrival rate dependent on server breakdowns. *Nav. Res. Logist. Quart.,* 26, 487-97, 1979.

[S-057] Shore, J. E. Information theoretic approximations for $M/G/1$ and $G/G/1$ queuing systems. *Acta Informatica,* 17, 43-61, 1982.

[S-058] Siegel, C. Results on a transient queue. *Comm. Pure. Appl. Math.*, 21, 371-84, 1968.

[S-059] Siegel, G. The stationary waiting time and other variables in single-server queues with specialities at the beginning of a busy period. *Zast. Matemat.*, 13, 465-79, 1973.

[S-060] Siegel, G. Abschätzungen für die Wartezeitverteilungen und ihre Momente beim Wartemodell $GI/G/1$ mit und ohne "Erwärmung". *ZAMM*, 54, 609-19, 1974.

[S-061] Siegel, G. Verallgemeinerungen einer Irrfahrt im R_1 und deren Bedeutung für Bedienungssysteme mit verzögertem Bedienungsbeginn. *Math. Operationsforsch. u. Statist.*, 5, 465-86, 1974.

[S-062] Sim, S. H. and Templeton, J. G. C. Steady state results for the $M/M(a,b)/c$ batch-service system. *Euro. J. Oper. Res.*, 21, 260-7, 1985.

[S-063] Simon, B. and Disney, R. L. Markov renewal processes: Some conditions for equivalence. *NZOR*, 12, 19-29, 1984.

[S-064] Singh, V. P. Finite waiting space bulk service system. *J. Engin. Math.*, 5, 241-8, 1971.

[S-065] Siskind, V. The fixed cycle traffic light problem: a note on a paper by McNeil. *J. Appl. Prob.*, 7, 245-8, 1970.

[S-066] Sivazlian, B. D. Approximate optimal solution for a D-policy in an $M/G/1$ queuing system. *A.I.I.E. Trans.*, 11, 341-43, 1979.

[S-067] Skinner, C. E. A priority queuing system with server-walking time. *Opns. Res.*, 15, 278-85, 1967.

[S-068] Smith, D. R. A new proof of the optimality of the shortest remaining processing time discipline. *Opns. Res.*, 26, 197-9, 1978.

[S-069] Smith, W. L. Distribution of queueing times. *Proc. Camb. Phil. Soc.*, 49, 449-61, 1953.

[S-070] Snyder, P. M. and Stewart, W. J. Explicit and iterative numerical approaches to solving queueing models. *Opns. Res.*, 33, 183-202, 1985.

[S-071] Sobel, M. J. Optimal average-cost policy for a queue with start-up and shut-down costs. *Opns. Res.*, 17, 145-162, 1969.

[S-072] Soriano, A. On the problem of batch arrivals and its application to a scheduling system. *Opns. Res.*, 14, 398-408, 1966.

[S-073] Srinivasan, S. K.; Subramanian, R. and Vasudevan, R. Correlation functions in queueing theory. *J. Appl. Prob.*, 9, 604-16, 1972.

[S-074] Stanford, R. E. Reneging phenomena in single channel queues. *Math. of O.R.*, 4, 162-78, 1979.

[S-075] Stidham, S., Jr. Regenerative processes in the theory of queues, with applications to the alternating-priority queue. *Adv. Appl. Prob.*, 4, 542-77, 1972.

[S-076] Subba Rao, S. Queuing models with balking, reneging and interruptions. *Opns. Res.*, 13, 596-608, 1965.

[S-077] Subba Rao, S. and Jaiswal, N. K. On a class of queuing problems and discrete transforms. *Opns. Res.*, 17, 1062-76, 1969.

[S-078] Sumita, U. The matrix Laguerre transform. *Appl. Math. and Comp.*, 15, 1-28, 1984.

[S-079] Suzuki, T. On a queueing process with service depending on queue-length. *Commentariorum Mathematicorum Univ. S. Pauli,* 10, 1-12, 1961.

[S-080] Suzuki, T. Two queues in series. *J. Oper. Res. Soc. Japan,* 5, 149-55, 1963.

[S-081] Suzuki, T. On a tandem queue with blocking. *J. Oper. Res. Soc. Japan,* 6, 137-57, 1964.

[S-082] Suzuki, T. Ergodicity of a tandem queue with blocking. *J. Oper. Res. Soc. Japan,* 7, 68-75, 1964.

[S-083] Sykes, J. S. Simplified analysis of an alternating-priority queuing model with setup times. *Opns. Res.*, 18, 1182-92, 1970.

[S-084] Syski, R. *Introduction to Congestion Theory in Telephone Engineering.* London: Oliver and Boyd, 1960.

[T-001] Takács, L. Investigation of waiting time problems by reduction to Markov processes. *Acta Math. Acad. Sci. Hungar.*, 6, 101-29, 1955.

[T-002] Takács, L. The probability law of the busy period for two types of queuing processes. *Opns. Res.*, 9, 402-7, 1961.

[T-003] Takács, L. Transient behavior of single-server queueing processes with Erlang input. *Trans. Amer. Math. Soc.*, 100, 1-28, 1961.

[T-004] Takács, L. The distribution of the virtual waiting time for a single-server queue with Poisson input and general service times. *Opns. Res.*, 11, 261-64, 1961.

[T-005] Takács, L. The transient behavior of a single server queueing process with a Poisson input. *Proc. Fourth Berkeley Symp. on Math. Stat. and Prob.*, Berkeley: Univ. of California Press, 2, 535-67, 1961.

[T-006] Takács, L. *Introduction to the Theory of Queues.* New York: Oxford University Press, 1962.

[T-007] Takács, L. A generalization of the ballot problem and its application in the theory of queues. *J. Amer. Statist. Assoc.*, 57, 327-37, 1962.

[T-008] Takács, L. The time dependence of a single-server queue with Poisson input and general service times. *Ann. Math. Statist.*, 33, 1340-48, 1962.

[T-009] Takács, L. A combinatorial method in the theory of queues. *J. Soc. Indust. Appl. Math.*, 10, 691-94, 1962.

[T-010] Takács, L. The stochastic law of the busy period for a single server queue with Poisson input. *J. Math. Anal. and Appl.*, 6, 33-42, 1963.

[T-011] Takács, L. The limiting distribution of the virtual waiting time and the queue size for a single-server queue with recurrent input and general service times. *Sankhya,* Ser. A, 25, 91-100, 1963.

[T-012] Takács, L. Delay distributions for one line with Poisson input, general holding times and various orders of service. *Bell Syst. Tech. J.*, 42, 487-503, 1963.

[T-013] Takács, L. A single-server queue with feedback. *Bell Syst. Tech. J.*, 42, 505-19, 1963.

[T-014] Takács, L. Priority queues. *Opns. Res.*, 12, 63-74, 1964.

[T-015] Takács, L. Occupation time problems in the theory of queues. *Opns. Res.*, 12, 753-67, 1964.

[T-016] Takács, L. A combinatorial method in the theory of Markov chains. *J. Math. Anal. Appl.*, 9, 153-61, 1964.

[T-017] Takács, L. Combinatorial methods in the theory of queues. *Rev. Int. Statist. Inst.*, 32, 207-19, 1964.

[T-018] Takács, L. Combinatorial methods in the theory of dams. *J. Appl. Prob.*, 1, 69-76, 1964.

[T-019] Takács, L. *Combinatorial Methods in the Theory of Stochastic Processes.* New York: John Wiley and Sons, 1967.

[T-020] Takács, L. The distribution of the content of finite dams. *J. Appl. Prob.*, 4, 151-61, 1967.

[T-021] Takács, L. Two queues attended by a single server. *Opns. Res.*, 16, 639-50, 1968.

[T-022] Takács, L. On dams of finite capacity. *J. Austral. Math. Soc.*, 8, 161-70, 1968.

[T-023] Takács, L. On inverse queuing processes. *Zast. Matemat.*, 10, 213-24, 1969.

[T-024] Takács, L. A fundamental identity in the theory of queues. *Ann. Inst. Stat. Math.*, 22, 339-48, 1970.

[T-025] Takács, L. The distribution of the occupation time for single-server queues. *Opns. Res.*, 19, 1494-1501, 1971.

[T-026] Takács, L. Discrete queues with one server. *J. Appl. Prob.*, 8, 691-707, 1971.

[T-027] Takács, L. Occupation time problems in the theory of queues. In *"Proceedings of the Conference on Mathematical Methodology in the Theory of Queues, Kalamazoo, Mich."*, New York: Springer-Verlag, 91-131, 1974.

[T-028] Takács, L. A single-server queue with limited virtual waiting time. *J. Appl. Prob.*, 11, 612-17, 1974.

[T-029] Takács, L. Combinatorial and analytic methods in the theory of queues. *Adv. Appl. Prob.*, 7, 607-35, 1975.

[T-030] Takács, L. On a formula of Harald Cramér. *Scand. Actuarial J.*, 65-72, 1975.

[T-031] Takács, L. A storage process with semi-Markov input. *Adv. Appl. Prob.*, 7, 830-44, 1975.

[T-032] Takács, L. On fluctuation problems in the theory of queues. *Adv. Appl. Prob.*, 8, 548-83, 1976.

[T-033] Takács, L. A Banach space of matrix functions and its application in the theory of queues. *Sankhya*, Ser. A, 38, 201-11, 1976.

[T-034] Takács, L. On the busy periods of single-server queues with Poisson input and general service times. *Opns. Res.*, 24, 564-71, 1976.

[T-035] Takács, L. A queuing model with feedback. *R.A.I.R.O.*, 11, 345-54, 1977.

[T-036] Takahashi, Y. A lumping method for numerical calculations of stationary distributions of Markov chains. *Research rept., Information Sciences no. B18, Tokyo Institute of Technology, Tokyo, Japan, 1975.*

[T-037] Takahashi, Y., and Takami, Y. A numerical method for the steady-state probabilities of a $GI/G/c$ queueing system in a general class. *J. Oper. Res. Soc. Japan*, 19, 147-57, 1976.

[T-038] Takahashi, Y. Asymptotic exponentiality of the tail of the waiting time distribution in a $PH/PH/c$ queue. *Adv. Appl. Prob.*, 13, 619-30, 1981.

[T-039] Talman, A. J. J. A simple proof of the optimality of the best $N-$policy in the $M/G/1$ queueing problem with removable server. *Statistica Neerlandica*, 32, 143-50, 1979.

[T-040] Tambouratzis, D. G. The modified $M/G/1$ queue. *Bull. Soc. Math. Grece,* 14, 176-99, 1973.

[T-041] Tambouratzis, D. G. A generalization of the $M/G/1$ queue. *Bull. Greek Math. Soc.,* 20, 106-20, 1979.

[T-042] Tan, H. H. Another martingale bound on the waiting-time distribution in $GI/G/1$ queues. *J. Appl. Prob.,* 16, 454-7, 1979.

[T-043] Tanner, J. C. The delay to pedestrians crossing a road. *Biometrika,* 38, 383-92, 1953.

[T-044] Tanner, J. C. A problem of interference between two queues. *Biometrika,* 40, 58-69, 1953.

[T-045] Tanner, J. C. A theoretical analysis of delays at an uncontrolled intersection. *Biometrika,* 49, 163-70, 1962.

[T-046] Tcha, D. W. and Pliska, S. R. Optimal control of single-server queuing networks and multi-class $M/G/1$ queues with feedback. *Opns. Res.,* 25, 248-58, 1977.

[T-047] Teghem, J.; Loris-Teghem, J. and Lambotte, J. P. *Modèles d'Attente $M/G/1$ et $GI/M/1$ à Arrivées et Services en Groupes.* Lecture Notes in Opns. Res. and Math. Econ., No. 8. New York: Springer-Verlag, 1969.

[T-048] Teghem Jr, J. Properties of $(0,K)$ policy in an $M/G/1$ queue and optimal joining rules in an $/M/M/1$ queue with removable server. *Proc. 7 th. IFORS Int. Conf. on Oper. Res.,* 229-59, 1975.

[T-049] Teghem Jr, J. Optimal control of queues: removable servers. *Belg. J. Oper. Res., Stat and C.S.,* 25, 99-128, 1985.

[T-050] Teghem Jr, J. Optimal control of a removable server in an $M/G/1$ queue with finite capacity. *Euro. J. Oper. Res.,* 31, 358-67, 1987.

[T-051] Teugels. J. L. Exponential ergodicity in Markov renewal processes. *J. Appl. Prob.,* 5, 387-400, 1968.

[T-052] Teugels, J. L. and Neuts, M. F. Exponential ergodicity of the $M/G/1$ queue. *SIAM J. Appl. Math.,* 17, 921-29, 1969.

[T-053] Teugels, J. L. A bibliography on semi-Markov processes. *J. Comp. Appl. Math.,* 2, 125-44, 1976.

[T-054] Teugels, J. L. A second bibliography on semi-Markov processes. In *"Semi-Markov Models: Theory and Applications",* J. Janssen, ed., London: Plenum Publishers, 507-84, 1986.

[T-055] Thiagarajan, T. R. and Harris, C. M. Statistical tests for exponential service from $M/G/1$ waiting-time data. *Nav. Res. Logist. Quart.,* 26, 511-20, 1973.

[T-056] Tijms, H. C. Optimal control of the workload in an $M/G/1$ queueing system with removable server. *Math. Operationsforsch. u. Statist.*, 7, 933-43, 1976.

[T-057] Tijms, H. C. On a switch-over policy for controlling the workload in a queueing system with two constant service rates and fixed switch-over costs. *Zeitschrift f. Oper. Res.*, 21, 19-23, 1977.

[T-058] Tijms, H. C. and van der Duyn Schouten, F. A. Inventory control with two switch-over levels for a class of $M/G/1$ queueing systems with variable arrival and service rate. *Stoch. Proc. Appl.*, 6, 213-22, 1978.

[T-059] Tijms, H. C. and Van Hoorn, M. H. Algorithms for the state probabilities and waiting times in single server queueing systems with random and quasirandom input and phase-type service times. *OR Spektrum*, 2, 145-52, 1981.

[T-060] Tijms, H. C.; Van Hoorn, M. H. and Federgruen, A. Approximations for the steady state probabilities in the $M/G/c$ queue. *Adv. Appl. Prob.*, 13, 186-206, 1981.

[T-061] Tijms, H. C. and Van Hoorn, M. H. Computational methods for single server and multiserver queues with Markovian input and general service times. In *"Applied probability - Computer science: The Interface"*, Proc. of the Conference at Boca Raton, Florida, January 1981, R. L. Disney and T. J. Ott, eds, Boston: Birkhäuser, Vol II, 71-102, 1982.

[T-062] Tin, P. A queueing system with Markov-dependent arrivals. *J. Appl. Prob.*, 22, 688-96, 1985.

[T-063] Tomko, J. Semi Markov analysis of an $E_k/G/1$ queue with finite waiting room. In *"Point Processes and Queuing problems: Proceedings of a Colloquium, Keszthely, Hungary, 1978"*, 381-89, 1981, London: North Holland Publ. Co.

[T-064] Towsley, D. and Wolf, J. K. On the statistical analysis of queue lengths and waiting times for statistical multiplexers with *ARQ* retransmission schemes. *IEEE Trans. Comm.*, COM-27, 693-702, 1979.

[T-065] Truslove, A. L. Queue length for the $E_k/G/1$ queue with finite waiting room. *Adv. Appl. Prob.*, 7, 215-26, 1975.

[T-066] Truslove, A. L. The busy period of the $E_k/G/1$ queue with finite waiting room. *Adv. Appl. Prob.*, 7, 416-30, 1975.

[T-067] Tu, H. Y. and Kumin, H. A convexity result for a class of $GI/G/1$ queueing systems. *Opns. Res.*, 31, 948-50, 1983.

[T-068] Tumura, Y and Ishikawa, A. Numerical calculation of the tandem queueing system. *TRU Math.*, 14, 57-70, 1978.

[T-069] Tweedie, R. L. Operator-geometric stationary distributions for Markov chains with application to queueing models. *Adv. Appl. Prob.*, 14, 368-91, 1982.

[U-001] Ushijima, T. A queueing system with Markov arrivals. *J. Oper. Res. Soc. Japan*, 15, 167-93, 1972.

[V-001] van den Berg, J. L.; Boxma, O. J. and Groenendijk, W. P. Sojourn times in the $M/G/1$ queue with deterministic feedback. *Stoch. Mod.*, 5, 1989, (forthcoming).

[V-002] van der Duyn Scouten, F. A. An $M/G/1$ queueing model with vacation times. *Zeitschrift f. Oper. Res.*, 22, 95-105, 1978.

[V-003] Vanderperre, E. J. On the busy period for the $M/G/1$ queue with finite waiting room. *R.A.I.R.O.*, 8, 141-44, 1974.

[V-004] Vanderperre, E. J. A bivariate distribution for the system $M/G/1$ with finite waiting room. *Cah. Centre d'études Rech. Opér.*, 17, 91-4, 1975.

[V-005] Van Hoorn, M. H. Algorithms for the state probabilities in a general class of single server queueing systems with group arrivals. *Mgmt. Sci.*, 27, 1178-87, 1981.

[V-006] Van Hoorn, M. H. and Tijms, H. C. Approximations for the waiting time distribution of the M/G/c queue. *Performance Evaluation*, 2, 22-8, 1982.

[V-007] Van Hoorn, M. H. and Seelen, L. P. The $SPP/G/1$ queue: A single server queue with a switched Poisson process as input process. *OR Spektrum*, 5, 207-18, 1984.

[V-008] Varga, R. S. *Matrix Iterative Analysis.* Englewood Cliffs, New Jersey: Prentice-Hall, 1962.

[V-009] Varma, R. S. and Jain, J. L. Queueing models. *J. of Math. Sci.*, 2, 49-74, 1967.

[V-010] Viterbi, A. M. Approximate analysis of time-synchronous packet networks. *IEEE J. on Selected Areas in Communication, Special Issue on Network Performance Evaluation*, SAC-4, 6, 879-890, 1986.

[V-011] Vlach, T. L. and Disney, R. L. The departure process from the $GI/G/1$ queue. *J. Appl. Prob.*, 6, 704-7, 1969.

[W-001] Wallstrom, B. A queueing system with time-outs and random departures. *Proc. Eight Internat. Teletraffic Conf., Melbourne, Australia*, Paper No. 231, 1976.

[W-002] Weber, R. R. A note on waiting times in single server queues. *Opns. Res.,* 31, 950-1, 1983.

[W-003] Weiss, G. H. A survey of some recent research in road traffic. *Proc. Symp. on Congestion Theory,* W. L. Smith and W. E. Wilkinson, eds., Chapel Hill: Univ. of N. Carolina Monographs, 253-88, 1965.

[W-004] Welch, P. D. On a generalized $M/G/1$ queueing process in which the first customer of each busy period receives exceptional service. *Opns. Res.,* 12, 736-52, 1964.

[W-005] Welch, P. D. On pre-emptive resume priority queues. *Ann. Math. Statist.,* 35, 600-12, 1967.

[W-006] White, J. A.; Schmidt, J. W., and Bennett, G. K. *Analysis of Queueing Systems.* New York: Academic Press, 1975.

[W-007] Whitt, W. Continuity of generalized semi-Markov processes. *Math. of O.R.,* 5, 494-501, 1980.

[W-008] Whitt, W. Comparing counting processes and queues. *Adv. Appl. Prob.,* 13, 207-20, 1981.

[W-009] Whitt, W. Approximating a point process by a renewal process: The view through a queue, an indirect approach. *Mgmt. Sci.,* 27, 619-36, 1981.

[W-010] Whitt, W. Approximating a point process by a renewal process, I: Two basic methods. *Opns. Res.,* 30, 125-47, 1982.

[W-011] Wignall, T. K. and Enns, E. G. The joint stationary multivariate queue length distribution in a single server queueing system with N queues, arbitrary priorities, and a general probabilistic inter-queue transition matrix. *Mgmt. Sci.,* 19, 778-82, 1973.

[W-012] Wignall, T. K. Priority queuing systems with and without feedback. *Opns. Res.,* 21, 764-76, 1973.

[W-013] Wijngaard, J. A direct numerical method for a class of queueing problems. *Mgmt. Sci.,* 24, 1441-47, 1978.

[W-014] Wikarski, D. An algorithm for the solution of linear equation systems with block structure. *Elektron. Inf. Verarb. Kybern.,* EIK 16, 10-2, 1980.

[W-015] Wikarski, D. Decomposition of stationary continuous-time Markov chains - A new approach. *Elektron. Inf. Verarb. Kybern.,* EIK 20, 355-73, 1984.

[W-016] Wishart, D. M. G. Queuing systems in which the discipline is "last-come, first-served". *Opns. Res.,* 8, 591-99, 1960.

[W-017] Wolf, D. Approximation of the invariant probability measure of an infinite stochastic matrix. *Adv. Appl. Prob.*, 12, 710-26, 1980.

[W-018] Wolf, D. Approximating the stationary waiting time distribution function of $GI / G / 1$-queues with arithmetic interarrival time and service time distribution function. *OR Spektrum*, 4, 135-48, 1982.

[W-019] Wolff, R. W. Poisson arrivals see time averages. *Opns. Res.*, 30, 223-31, 1982.

[W-020] Wong, B.; Giffin, W. and Disney, R. L. Two finite $/ M / M / 1$ queues in tandem: A matrix solution for the steady state. *Opsearch*, 14, 1-18, 1977.

[W-021] Woodside, C. M. and Pagurek, B. An algorithm for computing serial correlations of times in $GI / G / 1$ queues with rational arrival processes. *Mgmt. Sci.*, 25, 54-63, 1979.

[X-001] Xerocostas, D. A. and Demertzes, C. Steady state solution of the $E_k / D / r$ queueing model. *OR Spektrum*, 2, 47-51, 1982.

[Y-001] Yadin, M. and Naor, P. Queuing systems with a removable server. *Oper. Res. Quart.*, 14, 393-405, 1963.

[Y-002] Yadin, M., and Syski, R. Randomization of intensities in a Markov chain. *Adv. Appl. Prob.*, 11, 397-421, 1979.

[Y-003] Yechiali, U. A queueing-type birth-and-death process defined on a continuous-time Markov chain. *Opns. Res.*, 21, 604-9, 1973.

[Y-004] Yechiali, U. A new derivation of the Khintchine-Pollaczek formula. *Proc. 7 th. IFORS Int. Conf. on Oper. Res.*, 261-64, 1976.

[Y-005] Yeo, G. F. Single server queues with modified service mechanisms. *J. Austral. Math. Soc.*, 2, 499-507, 1962.

[Y-006] Yeo, G. F. Preemptive priority queues. *J. Austral. Math. Soc.*, 3, 491-502, 1963.

[Y-007] Yeo, G. F. and Weesakul, B. Delays to road traffic at an intersection. *J. Appl. Prob.*, 1, 297-310, 1964.

[Y-008] Young, J. P. Administrative control of multiple-channel queuing systems with parallel input streams. *Opns. Res.*, 14, 145-56, 1966.

Appendix: Properties of Semi-Markov Matrices

1. THE MAXIMAL EIGENVALUE OF THE TRANSFORM MATRIX

In this appendix, we consider a number of general properties of semi-Markov matrices, which are of importance to the matrix-analytic methodology, discussed in this book. Let $A(\cdot)$ be an $m \times m$ irreducible semi-Markov matrix. Its elements are probability mass-functions on $[0,\infty)$ and the matrix $A = A(\infty)$, is an irreducible stochastic matrix, whose invariant probability vector is denoted by $\boldsymbol{\pi}$.

For every nonnegative s, the matrix

$$\tilde{A}(s) = \int_0^\infty e^{-sx}\, dA(x), \tag{A.1.1}$$

is an irreducible, nonnegative matrix. It has a uniquely defined Perron-Frobenius eigenvalue $\delta(s)$, which is an analytic function of s for $s > 0$, since it is a simple solution to a polynomial equation with analytic coefficients. Clearly $\delta(0+) = 1$. Since $\delta(s)$ is a simple eigenvalue, we may choose corresponding left and right eigenvectors $\mathbf{u}(s)$ and $\mathbf{v}(s)$, whose components are analytic functions of s for $s > 0$. We may further choose the vectors $\mathbf{u}(s)$ and $\mathbf{v}(s)$ so that the normalizing conditions

$$\mathbf{u}(s)\mathbf{v}(s) = 1, \tag{A.1.2}$$

$$\mathbf{u}(s)\mathbf{e} = 1, \tag{A.1.3}$$

$$\mathbf{u}(0+) = \boldsymbol{\pi}, \tag{A.1.4}$$

$$\mathbf{v}(0+) = \mathbf{e}, \tag{A.1.5}$$

are satisfied. It is easily seen that these conditions determine the vectors $\mathbf{u}(s)$ and $\mathbf{v}(s)$ uniquely. By virtue of the definitions, the following equations hold:

$$[\tilde{A}(s) - \delta(s)I]\mathbf{v}(s) = 0, \tag{A.1.6}$$

$$\mathbf{u}(s)[\tilde{A}(s) - \delta(s)I] = 0. \tag{A.1.7}$$

In this book, the monotonicity properties of $\delta(s)$ and the derivatives of $\delta(s)$ and of the vectors $\mathbf{u}(s)$ and $\mathbf{v}(s)$ play an important role. As they are also of interest to other considerations, they merit a unified and systematic discussion. We begin by proving the following important result, due to J. F. C. Kingman.

Theorem A.1.1: The function $\log \delta(s)$ is convex and decreasing on $[0,\infty)$.

Proof: Since each of the elements of the matrix $\tilde{A}(s)$ are either identically zero or positive, decreasing on $[0,\infty)$, a classical monotonicity property of the Perron-Frobenius eigenvalue implies that $\delta(s)$ and therefore its logarithm are decreasing on $[0,\infty)$.

a. The logarithm of the positive elements of $\tilde{A}(s)$ *is convex.* This is immediate, since by Hölder's inequality

$$\int_0^\infty \exp\{-[ps_1 + (1-p)s_2]x\}dA_{ij}(x)$$
$$\leqslant \left[\int_0^\infty \exp[-s_1 x]dA_{ij}(x)\right]^p \left[\int_0^\infty \exp[-s_2 x]dA_{ij}(x)\right]^{1-p},$$

for $0 \leqslant p \leqslant 1$.

Let now C be the class consisting of the function which is identically zero and of the positive functions on $[0,\infty)$, whose logarithm is convex. We have already shown that the elements of $\tilde{A}(\cdot)$ belong to the class C.

b. The class C *is closed under addition.* This again follows from Hölder's inquality, since if $f \in C$ and $g \in C$ and $h = f + g$, then for all p with $0 \leqslant p \leqslant 1$,

$$h[ps_1 + (1-p)s_2] = f[ps_1 + (1-p)s_2] + g[ps_1 + (1-p)s_2]$$

$$\leqslant f^p(s_1)f^{1-p}(s_2) + g^p(s_1)g^{1-p}(s_2)$$

$$\leqslant [f(s_1) + g(s_1)]^p [f(s_2) + g(s_2)]^{1-p} = h^p(s_1)h^{1-p}(s_2).$$

c. The class C is closed under multiplication and all positive powers of elements of C belong to C. This is verified by noting that positive linear combinations of convex functions are convex.

d. The limit superior of sequences of functions in the class C belongs to C. Let $f_n \in C$, for $n \geqslant 0$, and let $f = \limsup_n f_n$, then for all p with $0 \leqslant p \leqslant 1$,

$$f[ps_1 + (1-p)s_2] \leqslant \limsup_n \{ f_n^p(s_1) f_n^{1-p}(s_2) \}$$

$$\leqslant \limsup_n f_n^p(s_1) \limsup_n f_n^{1-p}(s_2) = f^p(s_1) f^{1-p}(s_2).$$

e. The elements of $\tilde{A}^n(s)$ belong to C. This is obvious, since they are sums of products of elements of C.

f. The function $[\text{trace } \tilde{A}^n(s)]^{1/n}$ belong to C, for all $n \geqslant 0$. This is immediate by *e* and *c*.

Since $\delta(s) = \limsup_n [\text{trace } \tilde{A}^n(s)]^{1/n}$, the stated conclusion follows. •

2. THE DERIVATIVES OF $\delta(s)$, $\mathbf{u}(s)$ and $\mathbf{v}(s)$.

As is shown by the following example, the function $\delta(s)$ is, in general, not the Laplace-Stieltjes transform of a probability distribution on $[0,\infty)$. Yet in many calculations and asymptotic results, the derivatives of $\delta(s)$ at $s = 0+$, play a role analogous to that of the corresponding derivatives of Laplace-Stieltjes transforms.

Example: Let the semi-Markov matrix $Q(\cdot)$ be defined by

$$Q(x) = \begin{vmatrix} 0 & F_1(x) \\ F_2(x) & 0 \end{vmatrix},$$

where $F_1(\cdot)$ and $F_2(\cdot)$ are probability distributions on $[0,\infty)$ with Laplace-Stieltjes transforms $f_1(s)$ and $f_2(s)$ and chosen so that the product $f_1(s)f_2(s)$ is not the square of the Laplace-Stieltjes transform of a probability distribution. Clearly $\delta(s) = [f_1(s)f_2(s)]^{\frac{1}{2}}$, it is then not a Laplace-Stieltjes transform of a probability distribution. The choice of such a pair of probability distributions is easy; we may set for example,

$$f_1(s) = e^{-s}, \quad f_2(s) = \frac{1}{2}[1 + e^{-s}].$$

We denote by $\delta^{[n]}(0+)$, $\mathbf{u}^{[n]}(0+)$ and $\mathbf{v}^{[n]}(0+)$ the n-th derivatives at $s = 0+$ of $\delta(s)$, $\mathbf{u}(s)$ and $\mathbf{v}(s)$ (provided they exist). $\tilde{A}^{[n]}(s)$, is the matrix of the n-th derivatives of the elements of the matrix $\tilde{A}(s)$. The matrix $\tilde{A}^{[n]}(0+)$ is finite if and only if the moment matrix $\int_0^\infty x^n \, dA(x)$ is finite. The vectors $\boldsymbol{\beta}_n^*$ are defined by

$$\boldsymbol{\beta}_n^* = (-1)^n \tilde{A}^{[n]}(0+)\mathbf{e}. \tag{A.2.1}$$

The matrix Z is the inverse of the matrix $I - A + \mathbf{e}\boldsymbol{\pi}$.

Theorem A.2.1: The quantities $\delta^{[n]}(0+)$, $\mathbf{u}^{[n]}(0+)$, $\mathbf{v}^{[n]}(0+)$ are finite for each $n \geq 0$, for which the matrix $\tilde{A}^{[n]}(0+)$ is finite. The triples $\{\delta^{[n]}(0+), \mathbf{u}^{[n]}(0+), \mathbf{v}^{[n]}(0+)\}$ are recursively given by

$$\delta^{[0]}(0+) = 1, \qquad \mathbf{u}^{[0]}(0+) = \boldsymbol{\pi}, \qquad \mathbf{v}^{[0]}(0+) = \mathbf{e}, \tag{A.2.2}$$

$$\delta^{[1]}(0+) = \delta'(0+) = \boldsymbol{\pi}\tilde{A}'(0+)\mathbf{e} = -\boldsymbol{\pi}\boldsymbol{\beta}_1^*,$$

$$\mathbf{u}^{[1]}(0+) = \boldsymbol{\pi}[\tilde{A}'(0+) - \delta'(0+)I]Z = \boldsymbol{\pi}\tilde{A}'(0+)Z + (\boldsymbol{\pi}\boldsymbol{\beta}_1^*)\boldsymbol{\pi},$$

$$\mathbf{v}^{[1]}(0+) = Z[\tilde{A}\,'(0+) - \delta\,'(0+)I]\mathbf{e} = (\boldsymbol{\pi}\boldsymbol{\beta}_1^*)\mathbf{e} - Z\boldsymbol{\beta}_1^*,$$

and for $n \geqslant 2$,

$$\delta^{[n]}(0+) = \sum_{\nu=1}^{n}\binom{n}{\nu}\boldsymbol{\pi}\tilde{A}^{[\nu]}(0+)\mathbf{v}^{[n-\nu]}(0+) - \sum_{\nu=1}^{n-1}\binom{n}{\nu}\boldsymbol{\pi}\mathbf{v}^{[n-\nu]}(0+)\delta^{[\nu]}(0+),$$

$$\mathbf{u}^{[n]}(0+) = \sum_{\nu=0}^{n-1}\binom{n}{\nu}\mathbf{u}^{[\nu]}(0+)[\tilde{A}^{[n-\nu]}(0+) - \delta^{[n-\nu]}(0+)I]Z,$$

$$\mathbf{v}^{[n]}(0+) = Z\sum_{\nu=1}^{n}\binom{n}{\nu}[\tilde{A}^{[\nu]}(0+) - \delta^{[\nu]}(0+)I]\mathbf{v}^{[n-\nu]}(0+)$$
$$- \sum_{\nu=1}^{n-1}\binom{n}{\nu}\mathbf{u}^{[\nu]}(0+)\mathbf{v}^{[n-\nu]}(0+)]\mathbf{e}.$$

Proof: The stated expressions for $n = 0$ are obvious. By differentiating n times in (A.1.6), we get that for $s > 0$,

$$\sum_{\nu=0}^{n}\binom{n}{\nu}[\tilde{A}^{[\nu]}(s) - \delta^{[\nu]}(s)I]\mathbf{v}^{[n-\nu]}(s) = 0. \tag{A.2.3}$$

Premultiplying by $\mathbf{u}(s)$ and letting $s \to 0+$, we obtain

$$\delta^{[n]}(0+) = \sum_{\nu=1}^{n}\binom{n}{\nu}\boldsymbol{\pi}\tilde{A}^{[\nu]}(0+)\mathbf{v}^{[n-\nu]}(0+) - \sum_{\nu=1}^{n-1}\binom{n}{\nu}\boldsymbol{\pi}\mathbf{v}^{[n-\nu]}(0+)\delta^{[\nu]}(0+), \tag{A.2.4}$$

which is finite, provided that $\tilde{A}^{[n]}(0+)$ is finite.

If we let s tend to 0+ is (A.2.3), we obtain the singular system of linear equations

$$(I - A)\mathbf{v}^{[n]}(0+) = \sum_{\nu=1}^{n}\binom{n}{\nu}[\tilde{A}^{[\nu]}(0+) - \delta^{[\nu]}(0+)I]\mathbf{v}^{[n-\nu]}(0+). \tag{A.2.5}$$

Adding $\mathbf{e}\boldsymbol{\pi}\mathbf{v}^{[n]}(0+)$, to both sides of that equation and recalling that the matrix $I - A + \mathbf{e}\boldsymbol{\pi}$ is nonsingular, it follows that

$$\mathbf{v}^{[n]}(0+) = Z\sum_{\nu=1}^{n}\binom{n}{\nu}[\tilde{A}^{[\nu]}(0+) - \delta^{[\nu]}(0+)I]\mathbf{v}^{[n-\nu]}(0+) + [\boldsymbol{\pi}\mathbf{v}^{[n]}(0+)]\mathbf{e}, \tag{A.2.6}$$

since $Z\mathbf{e} = \mathbf{e}$. In order to obtain the inner product $\boldsymbol{\pi}\mathbf{v}^{[n]}(0+)$ in terms of earlier terms in the recurrence, we differentiate n times in (A.1.2) and let s tend to 0+ to obtain

$$\boldsymbol{\pi}\mathbf{v}^{[n]}(0+) = -\sum_{\nu=1}^{n}\binom{n}{\nu}\mathbf{u}^{[\nu]}(0+)\mathbf{v}^{[n-\nu]}(0+). \tag{A.2.7}$$

We note that $\boldsymbol{\pi}\mathbf{v}'(0+) = 0$. Clearly, substitution of (A.2.7) into (A.2.6) yields the stated recurrence for $\mathbf{v}^{[n]}(0+)$. Similar computations applied to (A.1.7) yield the recursion formula for $\mathbf{u}^{[n]}(0+)$. In that case, equation (A.1.3) implies that $\mathbf{v}^{[n]}(0+)\mathbf{e} = 0$, for $n \geqslant 1$. •

Corollary A.2.1: The quantity $\delta''(0+)$ is explicitly given by

$$\delta''(0+) = \boldsymbol{\pi}\boldsymbol{\beta}_2^* - 2(\boldsymbol{\pi}\boldsymbol{\beta}_1^*)^2 - 2\boldsymbol{\pi}\tilde{A}'(0+)Z\boldsymbol{\beta}_1^*, \tag{A.2.8}$$

where $\boldsymbol{\beta}_2^* = \tilde{A}''(0+)\mathbf{e}$.

Proof: By direct computation. •

Remark a: Analytic expressions for the lower values of n may be worked out, but they soon become extremely complicated. The recursion formulas (A.2.2) are in a convenient form for numerical computation, but for values of n larger than five, cautious programming is needed to guard against loss of significance.

Remark b: It readily follows from the convexity of $\log \delta(s)$ that $\sigma^{*2} = \delta''(0+) - \delta'^2(0+) \geqslant 0$. The following theorem clarifies the significance of σ^{*2}, as an *asymptotic variance,* and also shows that, except in highly degenerate cases, that quantity is positive.

Theorem A.2.2: The variance V_n of the probability distribution with the Laplace-Stieltjes transform $\boldsymbol{\pi}\tilde{A}^{*n}(s)\mathbf{e}$ satisfies

$$V_n = n\sigma^{*2} + o(n), \quad \text{for } n \to \infty. \tag{A.2.9}$$

Proof: The first derivative of $\pi \tilde{A}^{*n}(s)\mathbf{e}$ is given by

$$\pi \sum_{r=0}^{n-1} \tilde{A}^{*r}(s)\tilde{A}'(s)\tilde{A}^{*n-r-1}(s)\mathbf{e},$$

which for $s \to 0+$, tends to $-n\pi\beta_1^*$. A second differentiation, followed by letting s tend to $0+$, yields the second moment in the form

$$n\pi\beta_2^*\mathbf{e} + 2\sum_{r=1}^{n-1}\sum_{k=0}^{r-1} \pi A^{*\prime}(0+)A^{r-k-1}A^{*\prime}(0+)\mathbf{e}.$$

We now note that

$$\sum_{r=1}^{n-1}\sum_{k=0}^{r-1} A^{\nu}(I - A + \mathbf{e}\pi) = \sum_{r=1}^{n-1} [I - A^{r} + r\mathbf{e}\pi]$$

$$= (n-1)I + \frac{1}{2}(n-1)(n-2)\mathbf{e}\pi - A(I - A^{n-1})Z,$$

which, after routine simplifications, leads to

$$\frac{1}{n}V_n \to \pi\beta_2^* + 2\pi A^{*\prime}(0+)ZA^{*\prime}(0+)\mathbf{e} - 3(\pi\beta_1^*)^2,$$

which proves the stated result. •

In many applications, we need the lattice analogues of the recursion formulas in Theorem A.2.1. The semi-Markov matrix $A(\cdot)$ is then a sequence of substochastic matrices $\{A_k, k \geqslant 0\}$, whose sum A is an irreducible stochastic matrix. The matrix

$$A^*(z) = \sum_{k=0}^{\infty} A_k z^k,$$

is then irreducible and nonnegative for $0 < z \leqslant 1$. Its Perron-Frobenius eigenvalue is denoted by $\delta(z)$ and we define corresponding left and right eigenvectors $\mathbf{u}(z)$ and $\mathbf{v}(z)$, whose components are positive, analytic functions of z for $0 < z < 1$. The normalizing equations

$\mathbf{u}(z)\mathbf{v}(z) = \mathbf{u}(z)\mathbf{e} = 1$, $\mathbf{u}(1-) = \boldsymbol{\pi}$, $\mathbf{v}(1-) = \mathbf{e}$, hold. The vector $\boldsymbol{\beta}$ is defined by $\boldsymbol{\beta} = \sum_{k=1}^{\infty} kA_k\mathbf{e}$, and the matrices $A^{*[n]}$ by

$$A^{*[n]} = \sum_{k=n}^{\infty} \frac{k!}{(k-n)!} A_k, \quad \text{for } n \geqslant 1,$$

and $A^{*[0]} = A$. The following theorem is proved in the same manner as Theorem A.2.1.

Theorem A.2.3: The quantities $\delta^{[n]}(1-)$, $\mathbf{u}^{[n]}(1-)$, $\mathbf{v}^{[n]}(1-)$ are finite for each $n \geqslant 0$, for which the matrix $A^{*[n]}(1-)$ is finite. The triples $\{\delta^{[n]}(1-), \mathbf{u}^{[n]}(1-), \mathbf{v}^{[n]}(1-)\}$ are recursively given by

$$\delta^{[0]}(1-) = 1, \qquad \mathbf{u}^{[0]}(1-) = \boldsymbol{\pi}, \qquad \mathbf{v}^{[0]}(1-) = \mathbf{e}, \tag{A.2.10}$$

$$\delta^{[1]}(1-) = \delta'(1-) = \boldsymbol{\pi}\boldsymbol{\beta},$$

$$\mathbf{u}^{[1]}(1-) = \boldsymbol{\pi} A^{*[1]}(1-)Z - (\boldsymbol{\pi}\boldsymbol{\beta})\boldsymbol{\pi},$$

$$\mathbf{v}^{[1]}(1-) = Z\boldsymbol{\beta} - (\boldsymbol{\pi}\boldsymbol{\beta})\mathbf{e},$$

and for $n \geqslant 2$,

$$\delta^{[n]}(1-) = \sum_{\nu=1}^{n} \binom{n}{\nu} \boldsymbol{\pi} A^{*[\nu]}\mathbf{v}^{[n-\nu]}(1-) - \sum_{\nu=1}^{n-1} \binom{n}{\nu} \boldsymbol{\pi}\mathbf{v}^{[n-\nu]}(1-)\delta^{[\nu]}(1-),$$

$$\mathbf{u}^{[n]}(1-) = \sum_{\nu=0}^{n-1} \binom{n}{\nu} \mathbf{u}^{[\nu]}(1-)[A^{*[n-\nu]} - \delta^{[n-\nu]}(1-)I]Z,$$

$$\mathbf{v}^{[n]}(1-) = Z\sum_{\nu=1}^{n} \binom{n}{\nu} [A^{*[\nu]} - \delta^{[\nu]}(1-)I]\mathbf{v}^{[n-\nu]}(1-)$$
$$- [\sum_{\nu=1}^{n-1} \binom{n}{\nu} \mathbf{u}^{[\nu]}(1-)\mathbf{v}^{[n-\nu]}(1-)]\mathbf{e}.$$

Corollary A.2.2: The quantity $\delta''(0+)$ is explicitly given by

$$\delta''(0+) = \pi\beta_2 - 2(\pi\beta)^2 + 2\pi A^{*[1]}Z\beta, \qquad \text{(A.2.11)}$$

where $\beta_2 = A^{*[2]}\mathbf{e}$.

Remark: Theorem A.2.2 provides the expression for the asymptotic variance, needed in the analogue of the simple central limit theorem for semi-Markov matrices. If, for simplicity, we assume that the matrix A is *aperiodic,* then the classical asymptotic formula

$$\tilde{A}^n(s) = \delta^n(s)\mathbf{v}(s)\mathbf{u}(s) + o[\delta^n(s)], \quad \text{for } n \to \infty,$$

and a verbatim repetition of the proof of the elementary central limit theorem implies that

$$\lim_{n \to \infty} A^{(n)}(c_n + d_n x) = \Phi(x)\mathbf{e}\pi,$$

where $\Phi(\cdot)$ is the standard normal distribution and the normalizing constants are

$$c_n = n\pi\beta_1^*, \qquad d_n = [n\sigma^{*2}]^{1/2}.$$

NOTES, REFERENCES AND COMMENTS ON THE APPENDIX

Section A.1

The convexity of the logarithm of $\delta(s)$ is proved in Kingman [K-058]. Its extension to the multivariate case is given in Neuts and Purdue [N-022]. An interesting, though apparently very difficult, question is whether there is an appropriate space of generalized functions in which a meaning can be given to the Laplace-Stieltjes inverse of $\delta(s)$.

Apart from its theoretical use in the proofs of the lemmas leading to Theorem 2.3.1, the properties of $\delta(s)$ led to major simplifications in the derivation of Viterbi's formula for the mean queue length in the data communications model in Section 6.1.

Section A.2

The recurrence formulas in Theorems A.2.1 and A.2.3 were first presented in Neuts [N-029]. As was pointed out in Chapter 3, their primary use in [N-029] was to serve in the derivation of moment formulas for the $M/SM/1$ and related models. If the existence of moments of a sufficiently high order for the input distributions can be ascertained, the recurrence relations in Chapter 3 are more efficient in numerical computations. The central limit theorem was proved by Keilson and Wishart [K-020]. It can be fruitfully applied to many models treated in this book. An example may be found in [N-053].

Index